# ELEMENTS OF THERMODYNAMICS
# AND HEAT TRANSFER

ELEMENTS OF

# THERMODYNAMICS
## AND HEAT TRANSFER

**EDWARD F. OBERT**
PROFESSOR OF MECHANICAL ENGINEERING
UNIVERSITY OF WISCONSIN

**ROBERT L. YOUNG**
PROFESSOR OF MECHANICAL ENGINEERING
UNIVERSITY OF TENNESSEE

SECOND EDITION

McGRAW-HILL BOOK COMPANY, INC.
New York    Toronto    London
1962

ELEMENTS OF THERMODYNAMICS AND HEAT TRANSFER

47592

*The streets are filled with human toys*
*wound up for three-score years*

# PREFACE

This revision has the same objective as the earlier edition—to present a simple yet rigorous approach to the fundamentals of thermodynamics and heat transfer aimed directly at engineering applications.

It is the experience of the authors that most students do not understand thermodynamics because of the confusion arising from the number of concepts and equations that may be presented in the earlier chapters, and also because of the jargon of poorly defined terms (in particular, the "path" function and the "inexact" differential). In this text, a clearly defined path is established: Definitions are emphasized first to form the foundation of the subject, and upon this foundation arise the First Law, the principle of reversibility, and then the Second and Third Laws. Upon this framework, the secondary phases are based: the properties of real fluids and gases, the concept of the ideal gas, and the applications of thermodynamics. However, considerable latitude in approach has been allowed the instructor by the choice and arrangement of topics in each chapter so that he can follow his own desires in the presentation.

Sufficient material has been included to fulfill the needs of the undergraduate curriculum in thermodynamics. To survey the subject in one semester it is suggested that Chapters 1 through 12 be covered, supplemented with assignments from other chapters to illustrate the applications. For example, after the study of Chapter 9, the Carnot and Rankine cycle analysis of Chapter 15 might well be included.

This book can be viewed as a condensation of the senior author's text "Concepts of Thermodynamics,"† amplified by applications and a rather extensive survey of heat transfer. However, many of the derivations have been made different from those in the parent text to emphasize anew the concept of the function—the theme of the present series of texts. A third text of intermediate level, "Thermodynamics," is in preparation and will be a revision of the older book of the same title published in 1948. It can be remarked that the sequence of topics in the older texts, along with the then innovations of an introductory

† McGraw-Hill Book Company, Inc., New York, 1960.

vii

chapter on dimensions and units and an emphasis on non-steady-flow processes, is followed again in the newer series.

Let us say that, because the humanities have been increasingly neglected in the engineering curricula, an attempt to interest the student in this area has been made by adding material to the textbook from the scrapbook of the senior author. Experience has shown that students will read (and remember) these unassigned excerpts more thoroughly than they will the assigned text material. Too, each quotation has some relationship to the chapter material.

The allied subject of heat transfer is presented in the last two chapters of the text. Chapter 19 emphasizes the fundamentals of heat transfer, while several advanced topics are presented in Chapter 20. Throughout, examples and problems have been chosen to demonstrate the application of heat transfer to realistic engineering problems.

In preparing these two chapters, it has been recognized that the time available in the undergraduate thermodynamics sequence for heat transfer may vary greatly from one school to another. If six to eight class periods are available, assignment of Articles 19-1 through 19-17 (omitting, if desired, Articles 3 and 6) will provide a proper understanding of the modes of heat transfer and the means for performing simple heat-transfer circulations. Ten to fifteen class periods should suffice for a thorough coverage of Chapter 19 and at least one of the advanced topics in Chapter 20. Material approximately equivalent to that given in the typical undergraduate course in heat transfer can be obtained by a thorough study of both chapters. If both are included, it is recommended that the topics in Chapter 20 be inserted during the study of Chapter 19, as suggested in the footnotes.

It is believed that proper acknowledgment has been made throughout the text for the work of others, but the authors would be most grateful if any oversight is brought to their attention. Much of the material, as well as the methods of presentation, originated in the older editions of 1948 and 1949, and here the help of Irwin T. Wetzel and William H. Roberts was invaluable. For the present edition, Richard A. Gaggioli, I. Carl Romer, and Samuel S. Lestz reviewed the material on thermodynamics and made many helpful suggestions. Comments by Clark H. Lewis and Charles C. Oehring were of great assistance in the preparation of the chapters on heat transfer. Each of the authors owes much to his friends and, individually, to his very closest relation.

*Edward F. Obert*
*Robert L. Young*

# CONTENTS

work—steady-flow system. 6. Process equations. 7. Indicated work. 8. Other work problems.

cept. 5. Convection heat transfer with change of phase. 6. An electrical analogy for radiation. 7. Radiation instrumentation.

# TERMINOLOGY

1. *Capital* letters designate, in general, extensive quantities.
2. *Lower-case* letters designate, in general, specific quantities.
3. *Reference* or *atmospheric* or *stagnation* values have subscript zero.
4. *Standard-state* values have superscript degree.
5. The first derivative with respect to time is signified by a *dot* over the symbol.
6. The basic dimension of energy is $FL$.
7. Equations involving a change in chemical composition are translated as soon as feasible to show extensive values (capital letters) for the properties to warn of the chemical reaction.

| Symbol | Concept | Dimension | Introductory or explanatory article |
|---|---|---|---|
| $a$, $A$, $b$, $B$, $c$, $C$..... | Constants | | |
| $a$............... | Acceleration | $L/\tau^2$ | 1-3 |
| | Specific Helmholtz free energy | $FL/M$ | 8-1 |
| | Velocity of sound | $L/\tau$ | A-4 |
| atm.............. | Atmospheres (pressure) | $F/L^2$ | 1-8 |
| $A$............... | Area | $L^2$ | 1-1 |
| | Helmholtz free energy | $FL$ | 8-1, 8-5 |
| $\mathfrak{a}$............... | Availability of heat | $FL$ or $FL/M$ | 7-8 |
| $\mathfrak{A}$............... | Availability function | $FL$ or $FL/M$ | 12-2, 12-3 |
| AF............... | Air-fuel ratio | None | 16-1 |
| Bi............... | Biot number | None | 20-3 |
| Btu.............. | British thermal unit (IT) | $FL$ | 2-4, B-1 |
| $c$............... | Heat capacity per unit mass | $FL/MT$ | 2-11, 3-5 |
| | Velocity of light | $L/\tau$ | 2-5, 19-9 |
| $c_m$............... | Humid heat capacity (per $lb_m$ dry air) | $FL/MT$ | 14-3 |
| $c_n$............... | Polytropic "heat" capacity | $FL/MT$ | 11-4 |
| $c_p$............... | Heat capacity, constant pressure | $FL/MT$ | 2-11 |
| $c_v$............... | Heat capacity, constant volume | $FL/MT$ | 2-11 |
| cal.............. | Calorie (defined) | $FL$ | B-1 |
| cfm.............. | Cubic feet per minute | $L^3/\tau$ | |
| CP.............. | Coefficient of performance | None | 4-4, 18-1, 18-8 |

| Symbol | Concept | Dimension | Introductory or explanatory article |
|---|---|---|---|
| C................. | Celsius (centigrade) scale | $T$ | 3-4 |
| C................. | Capacity | .......... | 17-4, 18-3 |
|  | Coefficient of discharge | None | 13-8 |
| $C_V$............. | Velocity coefficient | None | 13-8 |
| D................. | Characteristic length | $L$ | 19-7 |
|  | Diameter | $L$ |  |
|  | Displacement | .......... | 17-4 |
|  | Hydraulic diameter | $L$ | 12-5 |
| $D_e$............. | Hydraulic diameter | $L$ | 19-7 |
| e................. | Napierian log base |  |  |
| e................. | Energy per unit confined mass | $FL/M$ | 4-1, 5-4 |
| $e_f$............. | Energy per unit flowing mass | $FL/M$ | 5-3 |
| emf............. | Electromotive force | $FL/q$ | A-1, 6-8 |
| E................. | Energy of confined mass | $FL$ | 2-1, 4-1, 5-4 |
|  | Modulus of elasticity | $F/L^2$ | 6-8 |
| $E_f$............. | Energy of flowing mass | $FL$ | 5-3 |
| $\varepsilon$................. | Electromotive force | $FL/q$ | A-1, 6-8 |
|  | Effectiveness | None | 12-2 |
| f................. | Friction factor | None | 12-5 |
| F................. | Fahrenheit temperature scale | $T$ | 3-4 |
| FA................. | Fuel-air ratio | None | 16-1 |
| F................. | Force | $F$ | 1-2, 1-3 |
|  | Force function | $F$ | 13-19 |
|  | Geometric factor | None | 19-17 |
| $\bar{F}$................. | Geometric-reflection factor | None | 19-17 |
| F................. | Faraday | $q$ | A-1, 6-8 |
| $\mathfrak{F}$................. | Friction | $FL$ or $FL/M$ | 4-3b, 7-7, 12-4 |
|  | Geometric-reflection-emissivity factor | None | 19-1 |
| Fo................. | Fourier number | None | 20-3 |
| $FE$............. | Flow energy | $FL/M$ | 5-3 |
| g................. | Gram | $M$ or $F$ | 1-2 |
| g................. | Gravitational acceleration | $L/\tau^2$ | 1-2 |
|  | Specific Gibbs free energy | $FL/M$ | 8-1 |
| $g_0$............. | Standard gravity acceleration | $L/\tau^2$ | 1-2, 1-5 |
| $g_c$............. | Dimensional constant | $ML/F\tau^2$ | 1-3, 1-7 |
| G................. | Gibbs free energy | $FL$ | 8-1, 8-5 |
|  | Irradiation | $F/\tau L$ | 20-6 |
|  | Mass velocity | $M/L^2\tau$ | 12-5, 19-8 |
| Gr................. | Grashof number | None | 19-7 |
| h................. | Planck's constant | $FL\tau$ | 19-9 |
|  | Specific enthalpy | $FL/M$ | 2-11 |
|  | Surface heat-transfer coefficient | $F/\tau LT$ | 19-1 |
| $h_c$............. | Surface convection coefficient | $F/\tau LT$ | 19-1 |
| $h_r$............. | Surface radiation coefficient | $F/\tau LT$ | 19-1 |

| Symbol | Concept | Dimension | Introductory or explanatory article |
|--------|---------|-----------|-------------------------------------|
| $H$ . . . . . . . . . . . . . . | Enthalpy | $FL$ | 2-11 |
|  | Enthalpy per unit mass dry air | $FL/M$ | 14-2 |
| $H$ . . . . . . . . . . . . . . | Head | $L$ | 13-10 |
| ihp . . . . . . . . . . . . . . | Indicated horsepower | $FL/\tau$ | 6-7 |
| int . . . . . . . . . . . . . . | International units | . . . . . . . . . . | 1-2 |
| $I$ . . . . . . . . . . . . . . | Electric current $(dq/d\tau)$ | $q/\tau$ | A-1 |
|  | Intensity | $F/\tau L$ | 19-12 |
| IT . . . . . . . . . . . . . . | International Steam Table Conference units | . . . . . . . . . . | 2-4 |
| $\mathscr{I}$ . . . . . . . . . . . . . . | Unavailable energy function | $FL$ or $FL/M$ | 7-9, 12-2 |
| $J$ . . . . . . . . . . . . . . | Joule's unit conversion factor | None | 2-4 |
|  | Radiosity | $F/\tau L$ | 20-6 |
| $k$ . . . . . . . . . . . . . . | $c_p/c_v$ | None | 2-11 |
|  | Boltzmann's constant $= R_0/N_0$ | $FL/T$ |  |
|  | Thermal conductivity | $F/T\tau$ | 19-1, 19-2, 19-3 |
| $K$ . . . . . . . . . . . . . . | Kelvin temperature scale | $T$ | 3-4 |
| $K_p$ . . . . . . . . . . . . . . | Equilibrium constant | . . . . . . . . . . | 16-7 |
| $KE$ . . . . . . . . . . . . . . | Kinetic energy | $FL$ or $FL/M$ | 2-2, 5-1, 5-3 |
| $\mathbf{Kn}$ . . . . . . . . . . . . . . | Knudsen number | None | 19-6 |
| $lb_f$ . . . . . . . . . . . . . . | Pound force | $F$ | 1-2 |
| $lb_m$ . . . . . . . . . . . . . . | Pound mass | $M$ | 1-2 |
| ln . . . . . . . . . . . . . . | Logarithm to the base e | None |  |
| $L$ . . . . . . . . . . . . . . | Coordinate of an open system | $L$ | 2-7, 5-1 |
|  | Length | $L$ | 1-1, 1-2 |
| $m$ . . . . . . . . . . . . . . | Mass of an object | $M$ | 1-2, 1-3, 2-5 |
|  | Mass within a system | $M$ | 4-1, 5-2, 5-6 |
| $m_f$ . . . . . . . . . . . . . . | Mass in flow | $M$ | 5-2, 5-5, 5-6 |
| $m_0$ . . . . . . . . . . . . . . | Rest mass of an object | $M$ | 2-5 |
| mep . . . . . . . . . . . . . . | Mean effective pressure | $F/L^2$ | 6-7, 17-4 |
| $\mathbf{M}$ . . . . . . . . . . . . . . | Mach number | None | A-5, 13-1, 19-8 |
| $M$ . . . . . . . . . . . . . . | Molecular weight | None | 1-2, 11-9, 16-1 |
|  | Mass | $M$ | 1-3 |
| $n$ . . . . . . . . . . . . . . | Number of moles | None | 3-5 |
|  | Polytropic exponent | None | 3-5 |
|  | Number of equivalents | None | 6-8 |
|  | Speed, rpm | $1/\tau$ |  |
| $N_0$ . . . . . . . . . . . . . . | Avogadro's number | $1/M$ | A-1 |
| $\mathbf{Nu}$ . . . . . . . . . . . . . . | Nusselt number | None | 19-7 |
| $p$ . . . . . . . . . . . . . . | Pressure | $F/L^2$ | 1-8 |
| $p_0$ . . . . . . . . . . . . . . | Atmospheric pressure | $F/L^2$ | 1-8, 6-1 |
|  | Stagnation pressure | $F/L^2$ | 13-1, 13-14 |
| psi . . . . . . . . . . . . . . | Pound force per square inch | $F/L^2$ | 1-8 |
| $PE$ . . . . . . . . . . . . . . | Potential energy | $FL$ or $FL/M$ | 2-2, 5-3 |
| $\mathbf{Pr}$ . . . . . . . . . . . . . . | Prandtl number | None | 19-7 |
| $q$ . . . . . . . . . . . . . . | Heat transfer per unit mass | $FL/M$ | 4-2, 5-7 |
|  | Electric charge | $q$ | A-1, 6-8 |

| Symbol | Concept | Dimension | Introductory or explanatory article |
|---|---|---|---|
| $Q$............... | Heat, in general | | |
| | Heat, system | $FL$ | 2-13, 4-2, 5-5 |
| $r$............... | Coordinate direction | $L$ | 19-4 |
| | Recovery factor | None | 13-13, 19-8 |
| $r_p$............... | Pressure ratio | None | 11-7, 17-1 |
| $r_v$............... | Specific volume ratio | None | 11-7, 17-1 |
| $r_W$............... | Work ratio | None | 15-1 |
| R............... | Rankine temperature scale | $T$ | 3-4 |
| $R$............... | Electrical resistance | $FL\tau/q^2$ | A-1 |
| | Specific or universal gas constant | $FL/MT$ | 3-5 |
| $R_t$............... | Thermal resistance | $T/FL$ | 19-5 |
| $R_0$............... | Universal gas constant (used to emphasize mole unit of mass) | $FL/MT$ | 3-4, B-8 |
| **Re**............... | Reynolds number | None | A-2, A-3, 19-7 |
| $s$............... | Specific entropy | $FL/MT$ | 7-4 |
| $s_0$............... | Stagnation entropy | $FL/MT$ | 13-2 |
| $S$............... | Entropy | $FL/T$ | 7-3 |
| **St**............... | Stanton number | None | 19-7 |
| $t$............... | Temperature, Celsius or Fahrenheit scale | $T$ | 3-4 |
| | Temperature on any scale | $T$ | 2-1, 3-2 |
| | Temperature, bulk, mixing cup | $T$ | 19-7 |
| $T$............... | Absolute temperature (Kelvin or Rankine scale) | $T$ | 3-4, 7-2 |
| $T_0$............... | Atmospheric temperature | $T$ | 6-1, 7-8, 12-2 |
| | Stagnation temperature | $T$ | 13-1, 13-13 |
| $u$............... | Specific internal energy | $FL/M$ | 2-3 |
| $U$............... | Internal energy | $FL$ | 2-3, 2-5 |
| | Over-all heat-transfer coefficient | $F/\tau LT$ | 19-20 |
| $v$............... | Specific volume | $L^3/M$ | 1-8 |
| $V$............... | Velocity | $L/\tau$ | 2-2, 2-5, 5-1 |
| | Volume | $L^3$ | 2-11 |
| | Electric potential | $FL/q$ | A-1 |
| $w$............... | Work transfer per unit mass | $FL/M$ | 4-2, 5-7 |
| $W$............... | Work, in general | | |
| | Work, system | $FL$ | 2-13, 4-2, 5-5 |
| | Hemispherical emissive power | $F/\tau L$ | 19-11 |
| | Humidity ratio | None | 14-1 |
| $x$............... | Coordinate direction | $L$ | 5-1 |
| | Mole fraction | None | 11-9 |
| | Quality | None | 9-5 |
| $y$............... | Coordinate direction | $L$ | 5-1 |
| $z$............... | Coordinate direction | $L$ | 5-1 |
| | Compressibility factor | None | 10-3 |
| $Z$............... | Height above the earth; not necessarily the $z$ coordinate of the cartesian coordinate system | $L$ | 2-2 |

| Symbol | Concept | Dimension | Introductory or explanatory article |
|--------|---------|-----------|-------------------------------------|

### Greek Characters

| Symbol | Concept | Dimension | Introductory or explanatory article |
|--------|---------|-----------|-------------------------------------|
| $\alpha$................ | Absorptivity | None | 19-10 |
| | Number given by thermometer | None | 3-4 |
| | Thermal diffusivity | $L^2/\tau$ | 20-3 |
| $\beta$................ | Coefficient of thermal expansion | $1/T$ | 19-7 |
| $\gamma$................ | Specific weight | $F/L^3$ | 1-8 |
| $\Gamma$................ | Runoff | $M/\tau L$ | 20-5 |
| $\Delta Q_p$............ | Heating value at $p$ = constant | $FL/M$ | 16-4 |
| $\Delta Q_V$............ | Heating value at $V$ = constant | $FL/M$ | 16-4 |
| $\delta$................ | Half-thickness | $L$ | 20-2 |
| $\epsilon$................ | Emissivity | None | 19-13 |
| | Extent of reaction | None | 16-6 |
| | Mechanical strain | None | 6-8 |
| | Roughness factor | $L$ | 12-5 |
| $\eta$................ | Combustion-engine efficiency | None | 17-1 |
| | Efficiency, First Law | None | 4-4, 15-3 |
| | Fin efficiency | None | 20-2 |
| $\eta_c$................ | Compression efficiency | None | 12-2 |
| $\eta_e$................ | Expansion efficiency | None | 12-2 |
| $\eta_f$................ | Furnace efficiency | None | 8-4 |
| $\eta_m$................ | Mechanical efficiency | None | 17-4 |
| $\eta_n$................ | Nozzle efficiency | None | 13-8 |
| $\eta_r$................ | Regenerator efficiency | None | 17-8 |
| $\eta_t$................ | Thermal efficiency | None | 4-4 |
| $\eta_v$................ | Volumetric efficiency | None | 17-4 |
| $\theta$................ | Coordinate direction | None | 19-4 |
| $\Theta$................ | Temperature, scale unspecified | $T$ | 3-2, 7-2 |
| $\lambda$................ | Latent heat of condensation | $FL/M$ | 20-5 |
| | Mean free path | $L$ | 19-6 |
| | Wavelength | $L$ | 19-9 |
| $\mu$................ | Chemical potential | $FL/M$ | 8-2 |
| | Degree of saturation | None | 14-1 |
| $\mu_f$................ | Dynamic viscosity, force units | $F\tau/L^2$ | 1-8 |
| $\mu_m$................ | Dynamic viscosity, mass units | $M/L\tau$ | 1-8 |
| $\nu$................ | Frequency | $1/\tau$ | 19-9 |
| | Stoichiometric number of moles | None | 16-7 |
| $\rho$................ | Density | $M/L^3$ | 1-8 |
| | Reflectivity | None | 19-10 |
| $\sigma$................ | Stefan-Boltzmann constant | $F/\tau LT^4$ | 19-1 |
| | Stress | $F/L^2$ | 6-8 |
| $\Sigma$................ | Sigma function | $FL$ | 14-3 |
| $\tau$................ | Time (a dimension) | $\tau$ | 1-2, 1-3 |
| | Time (a variable) | $\tau$ | 4-1 |
| | Transmissivity | None | 19-10 |

| Symbol | Concept | Dimension | Introductory or explanatory article |
|---|---|---|---|
| | Greek Characters | | |
| $\phi$............... | Coordinate direction | None | 19-12 |
| | Relative humidity | None | 14-1 |
| $\psi$............... | Coordinate direction | None | 20-3 |
| $\omega$............... | Rate of heat generation | $F/\tau L^2$ | 20-1 |
| | Solid angle | None | 19-12 |

| Symbol | Concept | Introductory or explanatory article |
|---|---|---|
| | Subscripts | |
| $a$............... | Air | |
| | Arithmetic | 19-21 |
| $aw$............... | Adiabatic wall | 19-8 |
| $A$............... | Added | |
| $b$............... | Black-body surface | 19-11 |
| $c$............... | Critical | 9-2, 10-2, 20-4 |
| $D$............... | Piston displacement | 6-7, 17-4 |
| $f$............... | Film or fluid | 19-7, 19-6 |
| | Flow | 5-1, 5-2 |
| | Force unit | 1-2 |
| | Saturated liquid state | 9-5 |
| $\mathfrak{F}$............... | Friction | 12-4 |
| $g$............... | Gray-body surface | 19-14 |
| | Saturated vapor state | |
| $h$............... | High | 16-4 |
| $i$............... | Indicated | 6-7 |
| | Inner | 19-8 |
| $i, j$............... | Constituents $i, j$ | |
| id............... | Ideal gas | 3-5 |
| irrev............... | Irreversible process | 6-3 |
| $l$............... | Low | 16-4 |
| $m$............... | Mass unit | 1-2 |
| | Mean | 19-4, 19-21 |
| | Mixture | |
| $n$............... | Normal | 19-12 |
| | Polytropic | 3-5, 11-3 |
| nf............... | Closed (nonflow) system | |
| $o$............... | Outer | 19-8 |
| $p$............... | Constant pressure | |
| $Q$............... | Heat | |
| $r$............... | Reduced | 10-3 |
| | Relative | 5-8e |
| $r'$............... | Pseudo-reduced property | 10-3 |
| rev............... | Reversible process | |
| $R$............... | Rejected | |

| Symbol | Concept | Introductory or explanatory article |
|--------|---------|-------------------------------------|

### Subscripts

| Symbol | Concept | Introductory or explanatory article |
|--------|---------|-------------------------------------|
| $s$ . . . . . . . . . . . . . . . | Entropy | 19-8 |
| | Saturated water vapor | 14-1 |
| | Sensible | 2-3 |
| | Surface | 19-6, 19-8 |
| sf . . . . . . . . . . . . . . . | Steady-flow system | 6-5 |
| $v$ . . . . . . . . . . . . . . . | Constant specific volume | |
| $V$ . . . . . . . . . . . . . . . | Constant total volume | 16-4 |
| $w$ . . . . . . . . . . . . . . . | Steam | 14-1 |
| $x$ ($y$ or $z$) . . . . . . . . . | In the $x$ ($y$ or $z$) direction | |
| | At quality $x$ | 9-5 |
| 0 . . . . . . . . . . . . . . . | Atmospheric value | 6-1, 7-8, 12-2 |
| | Standard value | 1-2 |
| | Stagnation state | 13-1 |
| 1 . . . . . . . . . . . . . . . | Entrance to flow system | 2-7, 5-1 |
| 2 . . . . . . . . . . . . . . . | Exit from flow system | 2-7, 5-1 |

### Superscripts

| Symbol | Concept | Introductory or explanatory article |
|--------|---------|-------------------------------------|
| ° . . . . . . . . . . . . . . . | Degree of temperature | 1-8 |
| | Standard-state designation | 8-3 |
| * . . . . . . . . . . . . . . . | Reference value | 19-8 |
| | State of adiabatic saturation | 14-3 |
| | State at Mach 1 (sonic) | A-5, 13-1 |
| $n$ . . . . . . . . . . . . . . . | Polytropic exponent | 3-5, 11-3 |

| Symbol | Concept |
|--------|---------|

### Independent Variables and Functions*

*Closed system* (independent variable is *time* $\tau$)

| Symbol | Concept |
|--------|---------|
| $Q(\tau)$ . . . . . . . . . . . . . | Heat through surface of closed system from time zero |
| $q(\tau)$ . . . . . . . . . . . . . | $Q(\tau)/m$, where $m$ is mass of closed system |
| $W(\tau)$ . . . . . . . . . . . . . | Work through surface of closed system from time zero |
| $w(\tau)$ . . . . . . . . . . . . . | $W(\tau)/m$, where $m$ is mass of closed system |
| $P(\tau)$ . . . . . . . . . . . . . | Value of a property $P$ of system at time $\tau$ |

*Open system* (independent variables are *time* $\tau$ and *position* $L$)

| Symbol | Concept |
|--------|---------|
| $Q(\tau,L)$ . . . . . . . . . . . . | Heat through surface of open system from entrance $L_1$ to position $L$ and from time zero |
| $q(L)$ . . . . . . . . . . . . . | Heat through surface of *steady-flow* system from entrance $L_1$ to position $L$, per unit of mass flowing |
| $W(\tau,L)$ . . . . . . . . . . . | Work through surface of open system from entrance $L_1$ to position $L$ and from time zero |
| $w(L)$ . . . . . . . . . . . . . | Work through surface of *steady-flow* system from entrance $L_1$ to position $L$, per unit of mass flowing |
| $m_f(\tau,L)$ . . . . . . . . . . . | Mass which has crossed a coordinate $L$ from time zero |
| $m(\tau,L)$ . . . . . . . . . . . | Mass in system between entrance $L_1$ and coordinate $L$ at time $\tau$ |
| $P(\tau,L)$ . . . . . . . . . . . | Value of a specific thermostatic or of a mechanical property $P$ at position $L$ at time $\tau$ |

* Reference articles: 2-13, 4-1, 4-2, 5-2 to 5-7.

# CHAPTER 1

# SURVEY OF DIMENSIONS AND UNITS

*The measure of a man's real character is what he
would do if he knew he would never be found out.*
*Thomas Macaulay*

This chapter surveys the dimensions and units of primary importance
in engineering.   Such dimensions, after a system of units is established,
become the variables of engineering mathematics.

**1-1. Dimensions.**   A concept, if fundamental, cannot be clearly
explained.   For example, what is mass?   Or length?   Or time?   If mass
is defined as that which has inertia, what is inertia?   The problem is
avoided by assigning names which serve as definitions, and these descriptive names are called *dimensions*.   More precisely:

**A dimension is a name given to a measurable quality or characteristic of an entity.**

To reduce the number of dimensions, certain descriptions can be expressed
in terms of others.   For example, length, area, and volume are dimensions
describing certain characteristics of an object.   But since an area can be
conceived and measured as a length squared and a volume as a length
cubed, in place of these dimensions all these descriptions can be stated in
terms of some fundamental dimension, in this case, length.   For example,

$$[A] = [L^2]$$

The bracket is used to identify a dimension, and the above equation
should be read:   The dimension of area is equivalent to the dimension of
length squared.   By following this procedure, *secondary*, or *derived*,
dimensions can be obtained from a small number of *fundamental, basic*, or
*primary* dimensions.

Once the primary dimensions are accepted (without definition), other
concepts can be defined:   *A force is that phenomenon that can accelerate a
mass.*   This definition presupposes three primary dimensions: mass,
length, and time.   On the other hand, it is convenient to consider force
to be a primary dimension since its derived dimension is clumsy (Art. 1-4):

**In this text, the fundamental dimensions are mass, force, length,
time (and temperature).**

1

**1-2. Engineering Units for the Fundamental Dimensions.** While a dimension is a descriptive word picture, a *unit* is a definite standard or measure of a dimension:

**A unit is an arbitrary amount of the quantity to be measured with assigned numerical value of unity.**

For example, foot, yard, inch, rod, and meter are all different units but with the common dimension of length.

The units commonly used in engineering are defined as follows:

*Time.* The *second* (sec) is the fundamental unit of time and is defined as 1/86,400 part of a mean solar day.

*Length.* The *foot* (ft) is the fundamental unit of length and is defined as $\frac{1}{3}$ *yard* (yd), where the *yard* (international) is 0.9144 of the *meter* (m).

The *meter*† is the distance between two lines, measured at 0 degrees Celsius, on a platinum-iridium bar that is kept at the International Bureau of Weights and Measures at Sèvres, France. A meter is equivalent to 100 *centimeters* (cm).

*Mass.* The *pound* (international) ($lb_m$) is the fundamental unit of mass and is defined as 453.59237 *grams* (g). The *kilogram* (kg) by international agreement is a certain basic mass of platinum-iridium located at Sèvres, France. The *slug* is defined as 32.1739 pounds mass ($lb_m$). The *mole* is defined as the mass equal in numerical amount to the molecular weight $M$ of the substance. Any unit of mass can be included in this definition: the *pound mole* is $M$ pounds; the *gram mole* is $M$ grams.

*Force.* The *standard pound force* ($lb_f$) is the fundamental unit of force and is defined as the force necessary to accelerate the pound mass ($lb_m$) at the rate of 32.1739 ft/sec², or the slug at the rate of 1 ft/sec². The *poundal* is defined as the force required to accelerate the pound mass ($lb_m$) at a rate of 1 ft/sec². It is not in common use.

*Acceleration of Gravity.* The *standard acceleration of gravity* (by international agreement) is $g_0 = 32.1739$ ft/sec² $= 980.665$ cm/sec². The *local* acceleration of gravity will be designated by the symbol $g$.

Upon examination a serious fault appears in the above list of units. The name *pound* is assigned to the fundamental units of both mass and force; yet obviously *a pound of mass is an entirely different type of thing from a pound of force.* To overcome this difficulty, the terms *force pound* (or *pound force*) and *mass pound* (or *pound mass*) will be used and abbreviated $lb_f$ and $lb_m$ for identification.

---

† At the Tenth International Conference on Weights and Measures, Paris, France, 1954, it was agreed that the standard unit of length will have to be redefined in terms of the wavelength of the most suitable atomic transition of an isotopically pure element; mercury, cadmium, xenon, and krypton were being studied for action by the Conference in 1960.

**1-3. The Engineering Dimensional System.** In engineering a system involving four fundamental dimensions is in common use, called the *FMLτ system* of dimensions. In this system, the unit of force is the pound force [F]; the unit of mass is the pound mass [M]; the unit of length is the foot [L]; and the unit of time is the second [τ]. The procedure in constructing the dimensional system can be illustrated by the Newtonian equation

$$\text{Force} \sim \text{mass} \times \text{acceleration}$$

This proportionality can be written as an equation,

$$\frac{F}{F'} = \frac{ma}{m'a'} \tag{a}$$

and

$$F = \frac{1}{\dfrac{m'a'}{F'}}\, ma \tag{b}$$

Equations (a) and (b) are *dimensionally homogeneous* because each term of the equation has the same dimension [F].

Recall that, by definition, 1 lb$_f$ will accelerate 1 lb$_m$ at the rate of 32.1739 ft/sec², and substitute these values in Eq. (a),

$$\frac{F}{1\ \text{lb}_f} = \frac{ma}{32.1739\ \text{lb}_m\, \dfrac{\text{ft}}{\text{sec}^2}}$$

Upon rearranging, as in Eq. (b),

$$F = \frac{1}{32.1739\ \dfrac{\text{lb}_m\, \text{ft}}{\text{lb}_f\, \text{sec}^2}}\, ma$$

which is usually written

$$F = \frac{1}{g_c}\, ma \tag{c}$$

Here $g_c$ is a *dimensional constant,*

$$g_c = 32.1739\ \frac{\text{lb}_m\, \text{ft}}{\text{lb}_f\, \text{sec}^2} \qquad \left[\frac{ML}{F\tau^2}\right]$$

Note that $g_c$ is not the acceleration of gravity because it has different dimensions (although, unfortunately, the number 32.1739 is equal to that of the standard acceleration of gravity).

**Example 1.** Determine the force to accelerate 10 $\text{lb}_m$ at the rate of 10 ft/sec².
*Solution*

$$F = \frac{1}{g_c}\, ma = \frac{1}{g_c}\, 10\ \text{lb}_m\ 10\ \text{ft/sec}^2 = \frac{100\ \text{lb}_m\ \text{ft/sec}^2}{32.1739\ \dfrac{\text{lb}_m\ \text{ft}}{\text{lb}_f\ \text{sec}^2}} = 3.105\ \text{lb}_f \qquad Ans.$$

**Example 2.** Determine the force exerted by 10 $\text{lb}_m$ because of the attraction of gravity at a location where $g = g_0$.
*Solution*

$$F = \frac{1}{g_c}\, ma = \frac{1}{g_c}\, 10\ \text{lb}_m\ g_0 = \frac{321.739\ \text{ft}\ \text{lb}_m/\text{sec}^2}{32.1739\ \dfrac{\text{lb}_m\ \text{ft}}{\text{lb}_f\ \text{sec}^2}} = 10\ \text{lb}_f \qquad Ans.$$

Inspection of Eq. (*c*) shows that the dimensional constant $g_c$ can be used as a dimensional conversion factor:

**The dimension of mass can be eliminated from a quantity by dividing by $g_c$; the dimension of force can be eliminated from a quantity by multiplying by $g_c$.**

**Example 3.** Eliminate the dimension of force from 5.00(10⁻⁹) $\text{lb}_f$ hr/ft².
*Solution.* The easiest solution is to multiply by $g_c$ and 3,600 sec/hr,

$$\left[5.00(10^{-9})\,\frac{\text{lb}_f\ \text{hr}}{\text{ft}^2}\right]\left(\frac{3,600\ \text{sec}}{\text{hr}}\right)\left(\frac{32.1739\ \text{lb}_m\ \text{ft}}{\text{lb}_f\ \text{sec}^2}\right)$$

$$(5.00)(3,600)(32.1739)(10^{-9})\,\frac{\text{lb}_m}{\text{ft sec}} = 57.9(10^{-5})\,\frac{\text{lb}_m}{\text{ft sec}} \qquad Ans.$$

[Of course, the answer could have been found in units of $\text{lb}_m/(\text{ft})(\text{hr})$.]

It has been demonstrated that a four-fundamental-dimension system can be constructed for the engineering units of pound mass, pound force, foot, and second. The same reasoning can be used to devise a system of

TABLE 1-1. DIMENSIONAL SYSTEMS AND UNITS

| Quantity | Name of system | | | |
|---|---|---|---|---|
| | Absolute English, $ML\tau$ | Absolute metric, $ML\tau$ | Technical English, $FL\tau$ | Engineering English, $FML\tau$ |
| Length................. | foot | centimeter | foot | foot |
| Time................. | second | second | second | second |
| Mass................. | pound mass | gram | slug | pound mass |
| Force................. | poundal | dyne | pound force | pound force |
| Power................. | foot poundal/sec | erg/sec | foot $\text{lb}_f$/sec | foot $\text{lb}_f$/sec |
| Energy................. | foot poundal | erg | foot $\text{lb}_f$ | foot $\text{lb}_f$ |
| Dimensional constant $g_c$... | 1 $\text{lb}_m$ ft poundal sec² | 1 g cm dyne sec² | 1 slug ft $\text{lb}_f$ sec² | 32.1739 $\text{lb}_m$ ft $\text{lb}_f$ sec² |

four fundamental dimensions employing as units the slug, pound force, foot, and second (or a system employing the poundal, pound mass, foot, and second). It will be found that the dimensional constants for the English systems of units are

$$g_c = 32.1739 \, \frac{\text{lb}_m \, \text{ft}}{\text{lb}_f \, \text{sec}^2} = 1 \, \frac{\text{slug ft}}{\text{lb}_f \, \text{sec}^2} = 1 \, \frac{\text{lb}_m \, \text{ft}}{\text{poundal sec}^2}$$

**1-4. Dimensional Systems with Three Fundamental Dimensions.** It would appear that the dimensions of time $[\tau]$, length $[L]$, mass $[M]$, and force $[F]$ would be the minimum number of basic dimensions necessary to form a complete dimensional system. But for one of these four descriptions it is possible to find derived dimensions by using Newton's law, namely,

$$\text{Force} \sim \text{mass} \times \text{acceleration}$$

This proportionality can be written in equational form as

$$F = Cma$$

where $C$ is a dimensional constant. If it is desired to express the equation without the constant, for dimensional equality

$$[F] = [M][a] \tag{a}$$

Now acceleration need not be a primary dimension because it can be measured in terms of length $[L]$ and time $[\tau]$,

$$[F] = [M][L\tau^{-2}] = [ML\tau^{-2}]$$

Thus, the fundamental dimension, force $[F]$, can be given the derived dimension of $ML\tau^{-2}$, and thereby a system of dimensions can be constructed involving only mass $[M]$, length $[L]$, and time $[\tau]$. This is called the $ML\tau$ *system* of dimensions.

In the $ML\tau$ system, the English units are the pound mass $[M]$, the foot $[L]$, and the second $[\tau]$. With these units substituted in the Newtonian equation

$$F = ma$$

$$(1 \text{ unit force } [ML\tau^{-2}]) = (1 \text{ pound mass } [M]) \left( \frac{1 \text{ ft } [L]}{1 \text{ sec}^2 \, [\tau]^2} \right) \tag{b}$$

$$[ML\tau^{-2}] = [ML\tau^{-2}]$$

This equation is *dimensionally homogeneous* because each term of the equation has the dimension $ML\tau^{-2}$. Here the unit of force has derived dimensions, and it is called the *poundal*. The size of the poundal must satisfy Eq. (b): *The poundal is the force required to accelerate 1 pound mass at the rate of 1 ft/sec².* It must be remembered in using this system that

$$1 \text{ poundal} = \frac{1 \text{ lb}_m \, \text{ft}}{\text{sec}^2}$$

However, the poundal is rarely used.

An alternative system can be constructed by allowing the Newtonian equation to define the dimensions of mass,

$$F = ma$$

$$[M] = \frac{[F]}{[a]} = \frac{[F]}{[L\tau^{-2}]} = [FL^{-1}\tau^2]$$

Here the fundamental dimension of mass is given the derived dimension of $FL^{-1}\tau^2$, and a system is constructed that involves only force $[F]$, length $[L]$, and time $[\tau]$. This is called the $FL\tau$ system.

In the $FL\tau$ system, the English units are the pound force $[F]$, the foot $[L]$, and the second $[\tau]$. These units can be substituted in the equation

$$F = ma$$

$$\text{(1 pound force } [F]) = \text{(1 unit mass } [FL^{-1}\tau^2]) \left( \frac{1 \text{ ft } [L]}{1 \text{ sec}^2 [\tau^2]} \right) \qquad (c)$$

$$[F] = [F]$$

Here the unit of mass has derived dimensions of $FL^{-1}\tau^2$, and it is called the *slug*. The size of the *slug* is determined by Eq. (c): *The slug is the mass that can be accelerated 1 ft/sec² by a force of 1 standard pound force.* In using this system, it must be remembered that

$$1 \text{ slug} = 1 \frac{\text{lb}_f \text{ sec}^2}{\text{ft}}$$

**1-5. Weight.** Because the word *weight* is used in two different senses, some confusion has arisen: (1) Weight is the term used to indicate the mass or quantity of matter in a body (an unfortunate usage). (2) *Weight is the force exerted on a given mass by the gravitational effect of the earth* (a better usage). When a body is "weighed" on a beam scale, it is directly balanced by a known and presumably calibrated mass; this "weighing operation" is a measure of the mass in the body because the attraction of gravity on both the known and unknown masses is equal. On the other hand, when a pound mass is weighed on a spring scale, the deflection of the scale will be governed by the local value of the attraction of gravity. The weight will be

$$F = \frac{1}{g_c} ma = \frac{1}{g_c} (1 \text{ lb}_m)g = \frac{g}{g_c} (1 \text{ lb}_m)$$

(and this weight is in standard pound force units). When this method of weighing is made at a location where $g = g_0$, the mass of 1 pound will exert a force of 1 standard pound force (by definition). If $g$ is greater than $g_0$, a greater force will be exerted and it will be so indicated if the scales have been calibrated at a region where $g = g_0$.

In most engineering work, the problem of spring scales will not be a factor, but it is not unusual to use a mass as a weight. For example, consider a heavy piston free to descend in a vertical cylinder filled with gas; the force exerted by the mass of the piston on the gas will vary as the location or altitude of the cylinder is varied. This force in *standard* pound force units is $mg/g_c$.

In some instances the gravitational force exerted by a mass of 1 pound at a location where the acceleration of gravity is $g$ is called a *gravitational* pound force. This unscientific unit must be multiplied by the ratio $g/g_0$ if the force is to be reported in standard pound force units.

Note again that a mass of 1 pound measured on a conventional beam scale is *precisely* 1 pound mass, but the force exerted (the weight) is usually not 1 standard pound force because of the multiplying factor of $|g/g_c| = |g/g_0|$.

**1-6. Nonunitary Homogeneous Equations and the Conversion of Units.** Equations are often used to show the relative size of different units·

$$5{,}280 \text{ feet} = 1 \text{ mile} \qquad 3{,}600 \text{ seconds} = 1 \text{ hour}$$
$$[L] = [L] \qquad\qquad [\tau] = [\tau]$$

Inspection of these equations shows that they are dimensionally correct because all terms have the same dimensions. However, the equations do not have *unitary homogeneity* because the terms are expressed in different units. Because of this nonhomogeneity, there are some who prefer to show the equality in the form of a dimensionless ratio that has a value of unity:

$$5{,}280 \frac{\text{ft}}{\text{mile}} = 1 \qquad 3{,}600 \frac{\text{sec}}{\text{hr}} = 1$$

These unit conversion factors can be used as multiplying factors because it is always permissible to multiply (or divide) by unity.

**Example 4.** Convert $88 \dfrac{\text{ft}}{\text{sec}}$ to miles per hour (mph).

*Solution.* From Table B-1 select the appropriate unit conversion factors,

$$5{,}280 \frac{\text{ft}}{\text{mile}} \qquad 3{,}600 \frac{\text{sec}}{\text{hr}}$$

Now arrange these factors of unity to cancel the units of feet and seconds.

$$88 \frac{\text{ft}}{\text{sec}} \frac{\text{mile}}{5{,}280 \text{ ft}} \frac{3{,}600 \text{ sec}}{\text{hr}} = \frac{88(3{,}600)}{5{,}280} \frac{\text{mile}}{\text{hr}} = 60 \text{ mph} \qquad Ans.$$

**Example 5.** Repeat Example 4, but use the defining equations.
*Solution.* Substituting

$$5{,}280 \text{ ft} = 1 \text{ mile} \qquad 3{,}600 \text{ sec} = 1 \text{ hr}$$

in $88 \dfrac{\text{ft}}{\text{sec}}$ results in

$$88 \frac{\text{ft}}{\text{sec}} = 88 \frac{1 \text{ ft}}{1 \text{ sec}} = 88 \frac{\frac{1}{5{,}280} \text{ mile}}{\frac{1}{3{,}600} \text{ hr}} = 88 \frac{3{,}600}{5{,}280} \frac{\text{mile}}{\text{hr}} = 60 \text{ mph} \qquad Ans.$$

Both the poundal and the slug can be expressed as unit conversion factors; consider first the slug. By definition, 1 standard pound force will accelerate 1 slug mass at the rate of 1 ft/sec²; it will also accelerate 1 pound mass at the rate of 32.1739 ft/sec² (Art. 1-2). It follows that the slug is a definite quantity of matter of size 32.1739 times that of the pound mass:

$$1 \text{ slug} = 32.1739 \text{ lb}_m$$
$$[M] = [M]$$

and also
$$32.1739 \frac{\text{lb}_m}{\text{slug}} = 1 \qquad (a)$$

Recall that, by definition (Art. 1-2), 1 standard pound force will accelerate 1 pound mass at the rate of 32.1739 ft/sec²; a poundal will accelerate 1 pound mass at the rate of 1 ft/sec². Consequently, the poundal is a definite force of size 1/32.1739 that of the standard pound force:

$$32.1739 \text{ poundals} = 1 \text{ lb}_f$$
$$[F] = [F]$$

and
$$32.1739 \frac{\text{poundals}}{\text{lb}_f} = 1 \qquad (b)$$

Equations (a) and (b), like the factors of feet and mile, second and hour, are *unit conversion factors*.

**1-7. The Dimensional Constant as a Unit Conversion Factor.** Although the dimensional constant $g_c$ acknowledges the existence of four fundamental dimensions, still Newton's law shows that only three basic dimensions are necessary for a complete dimensional system (Art. 1-4). Because of this fact, the dimensional constant can always be treated as a unit conversion factor. In other words, the units for the dimensional system bear a definite relationship to each other, and this relationship is established by Newton's law whether or not a fourth basic dimension is premised. For example, there is 32.1739 lb$_m$ (units) in 1 slug (unit) of mass; this is shown by the unit conversion factor,

$$32.1739 \frac{\text{lb}_m}{\text{slug}} = 1 \qquad (a)$$

The equivalent units for the slug, by Newton's law, are

$$1 \text{ slug} = 1 \frac{\text{lb}_f \text{ sec}^2}{\text{ft}}$$

When this factor is substituted in Eq. (a), the result is

$$32.1739 \frac{\text{lb}_m \text{ ft}}{\text{lb}_f \text{ sec}^2} = 1$$

and this is recognized to be the dimensional constant $g_c$. Thus, the philosophy of this article shows that $g_c$ can be considered to be a unit conversion factor with absolute value of unity. For this reason, $g_c$ can be used at will as a multiplying or dividing factor irrespective of the origin of the quantities that are to be operated upon. In Example 3, $g_c$ was considered to be a dimensional constant that was used to eliminate an undesired *dimension;* it could also be considered that $g_c$ was merely a unit conversion factor that eliminated an undesired *unit.* Whether $g_c$ is regarded as a dimensional or unit conversion factor is relatively unimportant, for the same result will be obtained from either viewpoint. It must be admitted, however, that those who consider $g_c$ to be a unit conversion factor gain the advantage of simpler equations. However, in this text, $g_c$ is used as a dimensional constant (1) because the student should recognize that mass and force are two different phenomena and (2) because mass is invariably measured on balances that are calibrated (and engraved) in pound mass units (in engineering) and never in slug units or $F\tau^2/L$ units.

**1-8. Derived Dimensions and Units.** Derived, or secondary, dimensions will be constructed as the need for such descriptions becomes necessary.

*Density* is defined as the mass contained in unit volume:

$$[\rho] = \frac{[M]}{[L^3]} = [ML^{-3}]$$

Density is measured in units of pounds mass per cubic foot ($lb_m/ft^3$) and frequently in units of slugs per cubic foot.

*Specific gravity* is defined as the ratio of the density of a substance to the density of a selected reference material under prescribed conditions. The reference density is usually that of water at 39°F or at 60°F. Note that a ratio is dimensionless.

*Specific volume,* the volume of unit mass of material, is the name given to the inverse dimension of density. The symbol of designation is $v$, and the dimensions are

$$[v] = \frac{[L^3]}{[M]} = \left[\frac{1}{\rho}\right] = [L^3 M^{-1}]$$

Usually the engineering units are in cubic feet per pound ($ft^3/lb_m$).

**Example 6.** Density ($\rho$) is defined as the mass contained in unit volume. Specific weight ($\gamma$) is defined as the weight of the mass contained in unit volume. Investigate the relationship of these terms in the $FML\tau$ and $FL\tau$ systems.

*Solution. FML$\tau$ System*

$$F = \frac{1}{g_c} ma \qquad \text{and} \qquad \gamma = \frac{1}{g_c} \rho g$$

and at locations where $g = g_0$

$$\gamma = \frac{g_0}{g_c} \rho$$

or numerically $\gamma$ equals $\rho$ (at the specified location: $g = g_0$). This equality is convenient; for example, water at 68°F has a density of 62.305 $lb_m/ft^3$ and weighs 62.305 $lb_f/ft^3$ (unless $g \neq g_0$). These units and dimensions are of everyday experience.

*FL$\tau$ System*

$$F = ma \qquad \gamma = \rho g_0$$

under standard conditions, or numerically $\gamma$ is 32.1739 times $\rho$. Using the same figures as before, the density of water in this system is

$$\rho = \frac{62.305}{32.1739} = 1.936 \text{ slugs/ft}^3 \text{ or } \left[\frac{\text{lb}_f \text{ sec}^2}{\text{ft}^4}\right]$$

and the specific weight is

$$\gamma = 62.305 \text{ lb}_f/\text{ft}^3$$

*Pressure* is defined as the force exerted on unit area. The symbol to designate pressure in any units will be $p$, and dimensionally

$$[p] = \frac{[F]}{[A]} = [FL^{-2}]$$

The engineering unit for pressure is the standard pound of force per square inch ($\text{lb}_f/\text{in.}^2$, or psi). In most equations, pounds per square foot will be used ($\text{lb}_f/\text{ft}^2$). The usual pressure-measuring instruments measure the pressure either above or below the atmospheric condition, and such pressures are called *gauge pressures*. The *absolute pressure* is the algebraic sum of gauge and atmospheric pressures. If a *vacuum* is reported as 28 in. of mercury with a barometric pressure of 30 in. of mercury, the absolute pressure is 2 in. of mercury. For some work, the unit of pressure is taken as the *standard atmosphere*, which is defined as the pressure exerted by a column of mercury 760 millimeters in height and having a density of 13.595 grams per cubic centimeter when located in a region where the acceleration of gravity is standard.

$$1 \text{ atm} = 760 \text{ mm Hg} = 29.92 \text{ in. Hg} = 14.696 \text{ psi}$$

*Viscosity.* A *fluid* is a substance which deforms continuously under an applied shear stress, however small. The *viscosity* of a fluid is a measure of its resistance to deformation. Consider two parallel plates (Fig. 1-1) with one plate moving with

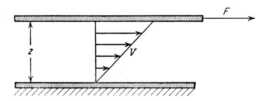

FIG. 1-1. Velocity gradient from viscosity (parallel plates, laminar flow, true fluid).

velocity $V$, while the other plate is stationary. Fluid adheres† to both plates; consequently, fluid in contact with the upper plate has the plate velocity $V$, while fluid touching the stationary plate has zero velocity. If the velocity is low or if the plates are close together, definite elements of the fluid will move parallel to the boundaries in fixed layers (on a macroscopic scale). This is *laminar*, or *streamline*, flow. To maintain the motion, a force $F$ must be applied to the moving plate, while an equal and opposite force must be exerted to hold the lower plate. The shearing stress is the force per unit area of plate ($F/A$). For this laminar-flow condition, Isaac Newton

† A no-slip condition is assumed between the fluid and its boundary. While this condition is usually satisfied, a notable exception occurs in highly rarefied gases, such as in the stratosphere.

defined the *coefficient of viscosity* (or the *absolute viscosity* or the *dynamic viscosity*) as

$$\mu_f = \frac{\text{shear stress}}{\text{shear rate}} = \frac{F/A}{dV/dz} \qquad \left[\frac{F/A}{1/\tau}\right] = \left[\frac{F\tau}{L^2}\right]$$

The subscript $f$ denotes force units. In units of mass,

$$\mu_m = \mu_f g_c$$
$$[\mu_m] = \left[\frac{F\tau}{L^2}\right]\left[\frac{ML}{F\tau^2}\right] = \left[\frac{M}{L\tau}\right]$$

The viscosity of a fluid arises from cohesion of the molecules and also from transfer of momentum as molecules diffuse from one layer to the next. Cohesion effects are dominant in liquids; hence, the *absolute viscosities of liquids decrease with temperature increase*. Molecular-activity effects are dominant in gases; hence, the *absolute viscosities of gases increase with temperature increase*.

Consider Fig. 1-2. A *perfect*, or *ideal, fluid*† has zero viscosity (and therefore a shear

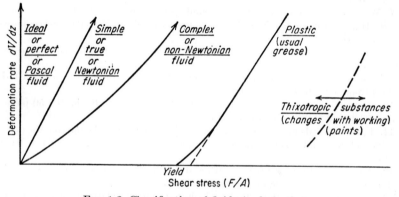

FIG. 1-2. Classification of fluids (and plastics).

stress cannot be imposed, nor can internal friction be induced). A *simple, true,* or *Newtonian fluid* has a coefficient of viscosity independent of the shear stress or rate of deformation (but dependent on the temperature and, to a lesser degree, on the pressure). *Greases* or *plastics* exhibit a yield stress before a deformation rate is established. Some substances have a decreased viscosity at high shear rates (*molecular slip*). At high shear rates when turbulence sets in, the linear relationship in Fig. 1-2 for Newtonian fluids disappears. The fluid remains Newtonian with the same value for $\mu$ as before, but $\mu$ can no longer be evaluated by the velocity gradients in the turbulent region (and a concept called the *eddy viscosity* is sometimes introduced).

**In this text discussion is limited either to ideal or else to Newtonian fluids.**

† The concept of the perfect, or ideal, *fluid* is not that of the perfect, or ideal, *gas* (Art. 3-5).

## PROBLEMS

**1.** Set up a dimensional system using as fundamental units the poundal [$F$], pound [$M$], foot [$L$], and second [$\tau$]. What are the value and dimensions for the dimensional constant? Eliminate the dimensional constant by giving the poundal derived dimensions, and compare the two systems. Which system is used?

**2.** Repeat Prob. 1, but use the slug [$M$], pound force [$F$], foot [$L$], and second [$\tau$] as the fundamental units.

**3.** A man weighs 180 lb. Explain the exact meaning of this sentence.

**4.** Explain the difference between a gravitational and a standard pound force.

**5.** A pound mass is "weighed" on a beam balance at a location where $g = 31$ ft/sec². What will be the reading? Discuss. The same mass is weighed on a spring scale originally calibrated in a region where $g \approx 32.2$ ft/sec². What will be the indicated weight? If calibration had been made at a location where $g = 31$ ft/sec², what would be the indicated weight and the weight in standard pounds force?

**6.** The value of $g$ at the equator and sea level is 32.088 ft/sec², and this value decreases about 0.001 ft/sec² for each 1,000 ft of ascent. At what height would $g$ equal 31 ft/sec²? At this location, how much mass must be used on a vertical piston to exert 5 lb$_f$?

**7.** What is the absolute pressure in pounds per square inch absolute if (a) vacuum is 2 in. Hg, (b) vacuum is 3.5 psi, (c) gauge pressure is 5.3 psi? (Barometric pressure is 750 mm Hg.)

**8.** Convert 60 ft/sec to miles per hour; 2,000 lb$_f$/ft² to pounds per square inch; 14 psi to atmospheres.

**9.** Convert 500 ft-lb$_f$ to horsepower-hours; $6.00(10^{-9})$ lb$_f$ hr/ft² to lb$_m$/(ft)(hr); 100 centipoises to lb$_m$/(sec)(ft) and lb$_f$ sec/ft².

**10.** Convert 15 slugs to pounds mass; 0.06 slug/(ft)(hr) to lb$_f$ hr/ft²; 50 poises to slugs/(ft)(hr); 2 moles of air to pounds.

**11.** Fifteen pounds mass is contained in a volume of ½ ft³. Determine the density, specific volume, and specific gravity. (Specific volume of water at 39°F is 0.01602 ft³/lb$_m$.)

**12.** A cylindrical tank 10 ft in length and 3 ft in diameter contains 5 lb$_m$ of fluid. Determine the specific volume and the density of the fluid.

**13.** An engine has a Prony brake which rests on the platform of a beam scale. The tare of the brake is 50 lb$_m$ (as shown by the scale). When the engine is running under full load, the scale reading is 150. Should a correction be made for the effect of gravity in computing the torque of the engine? (Let $g = 31$ ft/sec².) Discuss.

## SELECTED REFERENCES

1. Comings, E. W.: English Engineering Units and Their Dimensions, *Ind. Eng. Chem.*, **32**(7): 985 (July, 1940).
2. Bridgman, P. W.: "Dimensional Analysis," Yale University Press, New Haven, Conn., 1931.
3. Clemence, G.: Time and Its Measurement, *Am. Scientist*, May, 1951, p. 260.

# To the Student

*Thermodynamics* is the science which deals with energy and its transformations and with relationships between the properties of substances. The subject may also be called *physical chemistry* by the chemist or *heat* by the physicist. In these fields, study is almost invariably confined to the chemical and physical behavior of a fixed, quiescent mass of substance. Such studies could more properly be called *thermostatics* since motion (or gradients of temperature, pressure, etc.) is denied. The engineers have extended the subject of thermodynamics to include the study of substances in dynamic motion. This extension was made possible by evaluating, not a fixed mass of moving substance, but, rather, the *stationary region* which the moving substance penetrates. In the terminology of thermodynamics, the engineer studies relatively quiescent *closed systems* (as do the chemist and the physicist), and, in addition, he studies dynamic *open systems* (through which, for example, a high-velocity stream of fluid may be passing). In both *chemical-* and *mechanical*-engineering thermodynamics, the open system with or without chemical reactions is the primary theme. The difference between these two engineering subjects lies in the emphasis given by the mechanical engineer to the production of *power*.

Thermodynamics deals only with the macrostructure of matter and does not concern itself with events happening at the molecular level. For example, pressure, a macroscopic concept, is defined as the force per unit area at a "point," but the point must remain relatively large compared with molecular dimensions, and the force exerted by the matter must be continuous. Thus the laws and concepts of thermodynamics are independent of either present or future theories on the ultimate nature of matter (although molecular models or theories will be introduced, occasionally, to help explain the macroscopic behavior). The microstructure of matter is studied in subjects such as *kinetic theory* or *statistical mechanics* (which includes *quantum statistics*), and these subjects deal with large numbers of molecules (particles) and predict the *average* behavior of the group. Although thermodynamics includes the study of systems undergoing chemical (and atomic) reactions, no information is obtained on *reaction rates*, the subject of courses called *chemical kinetics*.

13

# CHAPTER 2

# FUNDAMENTAL CONCEPTS

Habit is a cable; we weave a thread of it
every day, and at last we cannot break it.
*Horace Mann*

The concepts and definitions in this chapter are indispensable in the analysis of thermodynamic problems. For this reason, the material should be reviewed habitually by the student so that the full significance of each definition or concept can be thoroughly understood.

**2-1. Introductory Axioms.** One of man's earliest observations was that a change could be made by exerting a force and that the product of force and distance was proportional to the expended effort. Thus, *force* was recognized to be the driving factor for change, but the magnitude of the change depended upon a capacity for supplying force. This *capacity* is called *energy:*

**Energy is the capacity, either latent or apparent, to exert a force through a distance.**

The presence of energy is indicated by the *properties* of matter, that is, by macroscopic characteristics of the physical or chemical structure of matter such as its pressure, density, or *hotness.* The concept of *hot* versus *cold* arose in the distant past as a consequence of man's sense of touch or feel. Observations show that, when a hot and a cold substance are placed together, the hot substance gets colder as the cold substance gets hotter. *The conclusion follows that energy is being transferred from the hot to the cold substance.* Since the bodies are in intimate contact, the process is called *conduction.* Observations also show that the transfer of energy continues even though the bodies are not in contact and even when the space between the two bodies is evacuated. Descriptively, the transmission of energy through space is called *radiation.* But whenever energy flows from a hot to a cold body, changes occur in one or more properties of each body, such as the pressure, volume, electrical resistance, etc. However, it is soon found by experiment that *none* of these properties can serve as an *explanation* of the energy transfer. *But the existence of the transfer of energy is, in itself, a proof that a difference exists in some fundamental property which is common to both bodies.* This fundamental dimension is called *temperature:*

**Temperature is the property which gauges the ability of matter to transfer energy by conduction or radiation.**

Thus, two bodies at the same temperature cannot change each other by the processes of either radiation or conduction. Since *equilibrium* means *balance* and *temperature* implies *thermal,* two bodies at the same temperature are said to be in *thermal equilibrium.* It logically follows that:

*Two bodies, each in thermal equilibrium with a third body, are in thermal equilibrium with each other.*

This axiom is sometimes called the *Zeroth Law of Thermodynamics;* it can as well be expressed by the observation that temperature is a fundamental dimension—a fundamental property.

On the other hand, if a Zeroth Law is necessary, it should include not only temperature but *all* intensity factors of energy. Or, from a broader viewpoint, a Zeroth Law should announce that functional relationships tie together the properties of matter:

**The properties of matter are functionally related.**

Although the *equivalence* of mass and energy is well established, it is convenient in thermodynamic calculations to use the older concept that mass is conserved in ordinary processes since the change in mass in other than nuclear reactions is too small to be detected. Thus, energy in various forms is considered to be a *companion* of *mass,* for example, energy of mass motion, called *kinetic energy;* energy of mass position, called *potential energy;* and energy of mass composition, called *internal energy.* Internal energy, like pressure, density, and temperature, is a property of matter:

**A fundamental axiom of modern thermodynamics is that internal energy is a property of matter.**

This axiom† is sometimes reserved to be a part of the First Law of Thermodynamics (Chap. 4).

All the forms of energy *accompanying* mass will be designated by the letter‡ $E$ (or $e$ for energy per unit mass). The derived dimension for energy of all kinds is

$$[E] = [LF]$$

and the basic unit in this text is the foot pound force.

---

† Although this axiom was not the first in the long history of thermodynamics, today it can serve as a logical starting point since it includes the pioneering work of Joule as well as the modern relativity laws of Einstein.

‡ Capital letters designate extensive quantities; lower-case letters designate quantities per unit mass; see p. xix.

When *changes* in energy occur without mass transfer, the energy is pictured as "flowing" to or from the mass under consideration (for example, radiation) and the transitory forms of energy are called *heat* and *work* (Art. 2-13) (with the symbols $Q$ and $W$ to warn that these concepts are not tied to mass, as is the symbol $E$).

**2-2. Potential and Kinetic Energy.** The external forms of energy are named *potential energy* ($PE$), or energy of position, and *kinetic energy* ($KE$), or energy of motion.

Potential energy is restricted to gravitational energy, that is, energy arising from the elevation of a mass with respect to the earth. The gravitational attraction of the earth on the mass is the source of a force (the weight, Art. 1-5) which is proportional to the mass,

$$F = \frac{m}{g_c} g$$

This force can be exerted through a distance which is the elevation $Z$ of the mass (since $g$ is practically constant for small vertical displacements). Thus, energy of position equals

$$PE = m \frac{g}{g_c} Z \tag{2-1}$$

$$[PE] = [LF]$$

Whenever a mass undergoes a change in velocity, a force is exerted,

$$F = \frac{m}{g_c} a = \frac{m}{g_c} \frac{dV}{d\tau}$$

When the force is exerted over a distance $\Delta L$, the change in energy is

$$\Delta E = \int_{L_1}^{L_2} F \, dL = \int_{\tau_1}^{\tau_2} FV \, d\tau = \frac{m}{g_c} \int_{V_1}^{V_2} V \, dV$$

The kinetic energy associated with the velocity $V$ and the mass $m$ is found by integrating the above equation from a velocity of zero to $V$,

$$KE = \frac{m}{2g_c} V^2 \tag{2-2}$$

$$[KE] = [LF]$$

Before potential energy can be calculated, an elevation with an assigned value of zero potential energy must be *arbitrarily* selected. From this elevation, the height of the mass is measured, and the potential energy *relative to this datum* is calculated. In measuring kinetic energy, a similar procedure is indicated, although, usually, the datum is automatically selected by measuring velocities relative to the earth. The same datum must be retained throughout a given problem for each form of energy.

**Example 1.** Calculate the potential energy of 32.1739 $lb_m$ at an elevation of 100 ft above an arbitrary datum of zero potential energy.

*Solution.* From the $FML\tau$ system,

$$\text{Potential energy } PE = (\text{gravitational force})(\text{height } Z)$$

$$PE = m \frac{g}{g_c} Z = (32.1739 \text{ lb}_m) \frac{g}{g_c} (100 \text{ ft}) = 3{,}217.39 \frac{g}{g_0} \text{ ft-lb}_f \qquad Ans.$$

(Note that $g/g_0$ is dimensionless but that $g/g_c$ is not.) Checking the dimensions,

$$PE = m \frac{g}{g_c} Z$$

$$[PE] = [M] \frac{\left[\dfrac{L}{\tau^2}\right]}{\left[\dfrac{ML}{F\tau^2}\right]} [L] = [LF]$$

and this is the correct dimension. In some cases $mZ$ is incorrectly taken to be the measure of potential energy (and $mV^2/2g$ for kinetic energy),

$$PE = mZ = 32.1739(100) = 3{,}217.39 \text{ ft-lb}_m$$

and this is numerically correct if $g/g_0 = 1.0$ but dimensionally wrong because foot pound mass is not the dimension for energy.

**2-3. Internal Energy.** All matter has energy arising from the motions and from the configurations of its internal particles. Such energy is called, quite descriptively, *internal energy,* and the amount of internal energy is reflected by properties such as pressure, temperature, and chemical composition. The symbol to designate internal energy is $U$ or, for unit mass, $u$.

Consider a mixture of air and gasoline vapor held under pressure and confined by a piston in a horizontal cylinder. Let the piston be connected by some means to an external load such that expansion of the mixture (but without ignition) will lift the load. Here internal energy of the mixture is transformed through the medium of pressure into potential energy of the load. The change in internal energy can be measured by the change in potential energy experienced by the external load. Examination of the mixture before and after the expansion would show no change in composition but a definite change in characteristics such as pressure (and temperature). Since chemical composition remained constant, the change in internal energy is sometimes called a change in *sensible internal energy.*

Let a small spark be used to ignite the gas-air mixture. A violent explosion will occur with the release of *chemical internal energy* far out of proportion to the energy of the electrical discharge, and a greater load than before can be lifted.

Suppose that the gas mixture is confined in the cylinder but with the piston locked in place. Suppose, too, that the temperature of the mixture is 600°F while the pressure is 200 psia. If this combination is surrounded by a water bath at 60°F, it is soon apparent that the water is increasing in temperature while the temperature (and pressure) of the mixture is decreasing. Here internal energy of the mixture is decreased by transfer of energy through the walls of the cylinder to the water bath because of a temperature difference.

Consider, as the next example, the familiar lead storage battery. The current from the battery is called *electrical energy* to distinguish it from mechanical energy. But when the battery delivers energy, no matter the name, its "stored energy"—its internal energy—decreases. Chemical changes occur in the battery corresponding to this decrease in internal energy. Some evidence of the change is shown by a hydrometer whereby the specific gravity of the acid solution is evaluated.

Thus, whenever energy is withdrawn from a piston-cylinder combination, from a battery, or from any other object under scrutiny, corresponding increases in energy appear in other objects.

**2-4. Units of Energy.** One object of thermodynamics is to provide tools for evaluating energy of all kinds in terms of the more outward manifestations of energy, such as pressure and temperature (but coupled with a knowledge of the chemical composition). A datum can be selected (say 14.7 psia, 60°F) and the internal energy of a selected substance can be arbitrarily assigned a value of zero internal energy per pound mass. Then, by measurements of the energy that need be transferred to change the temperature and pressure to new values, relative values of internal energy (sensible) can be obtained. Similarly, the energy released or absorbed when a chemical reaction occurs can be measured, and the internal energy of the products relative to that of the mixture is obtained. Tables of data are thus compiled for the substances in common use, with, in general, pressure, temperature, and/or volume serving as parameters (see, also, Art. 8-1).

The values of internal energy, relative to the arbitrarily selected datum of zero internal energy, could be recorded in units of foot-pounds, but a larger measure is more convenient. The *International Steam Table British thermal unit†* (IT Btu) is defined as

$$778.16 \ \frac{\text{ft-lb}_f}{\text{IT Btu}} \qquad (\text{symbol}, \ J)$$

† In this text, the symbol Btu designates the IT Btu (and the prefix IT will rarely be attached).

This *unit conversion factor* is called *Joule's equivalent*† and assigned, unfortunately, the symbol $J$.

A still larger unit of energy is the *international kilowatthour*,

$$3{,}412.76 \frac{\text{IT Btu}}{\text{int kwhr}}$$

The assignment of a symbol $J$ to a unit conversion factor is a pedagogical blunder, since the symbol implies a *dimensional* conversion factor. Unit conversion factors should not be indicated in equations. For example, internal energy ($u$) is usually given in Btu per pound mass units and $p$ and $v$ in pound mass, pound force, foot, and inch units. Then, when the addition of $u$ and $pv$ is indicated in a problem, the student may insist that $J$ should be included,

$$u + \frac{pv}{J} \qquad \left[\frac{\text{Btu}}{\text{lb}_m}\right]$$

But, with this reasoning, since $p$ is usually given in units of pounds force per square inch, another conversion factor, call it $B$, would also be required,

$$144 \frac{\text{in.}^2}{\text{ft}^2} \qquad (B)$$

or
$$u + \frac{Bpv}{J} \qquad \left[\frac{\text{Btu}}{\text{lb}_m}\right]$$

If this procedure were to be followed, each equation might well be cluttered with many unit conversion factors! [On the other hand, dimensional conversion factors should always be included, in particular, our chameleon friend $g_c$ (Art. 1-7).]

**2-5. Relativity Effects.** With the development of the theory of relativity, it became evident that mass and energy were different forms of the same fundamental phenomenon. Thus, the mass of a body is a measure of its energy content, and changes in mass accompany changes in energy from any cause whatsoever. The so-called *absolute*‡ energy of mass is given by the *Einstein mass-energy equation*,

$$E = \frac{m}{g_c} c^2 \qquad (a)$$

where $m$ is the mass§ and $c$ is the velocity of light. *Thus, theoretically, a means is available for calculating the energy of a substance on an absolute, rather than on a relative, basis.*

---

† In honor of James Joule (1818–1889), an English scientist who, by his experimental work, helped to establish the principle of conservation of energy (the First Law of Thermodynamics, Chap. 4).

‡ *Absolute* must be tempered somewhat since velocities, for example, are measured relative to the earth. Thus, Eq. ($a$) evaluates energy *relative to the datum of the measuring devices*, or *relative to the observer.*

§ And mass is defined here as that which has inertia. (Linus Pauling: "Matter is anything or everything that cannot travel at the speed of light.")

However, the mass in Eq. (a) is not a constant but, rather, changes with its velocity $V$ in accordance with the Lorentz-Fitzgerald equation

$$m = m_0 \frac{1}{\sqrt{1 - V^2/c^2}} \qquad (b)$$

where $m_0$ is the mass at rest (the rest† mass) and $V$ is the velocity relative to the observer. Upon substituting Eq. (b) into (a) and expanding,

$$E = \frac{m_0}{g_c} c^2 + \frac{1}{2} \frac{m_0}{g_c} V^2 + \frac{3}{8} \frac{m_0}{g_c} \frac{V^4}{c^2} + \frac{5}{16} \frac{m_0}{g_c} \frac{V^6}{c^4} + \cdots \qquad (c)$$

The energy of a moving body equals the rest energy plus a term which corresponds to the usual expression for kinetic energy, plus higher-order terms—all arising from the increase of mass with velocity. These latter terms, however, are insignificant in engineering calculations wherein $V$ is quite small relative to $c$ (and therefore the mass is not measurably affected by velocity, and separate accountings of mass and energy are permissible).

When a change in energy occurs, by Eq. (a),

$$\Delta m = \frac{\Delta E}{c^2} g_c \qquad (d)$$

Because of the size of $c^2$, the change in mass from heating a substance or from chemical reaction is very small, not detectable by the finest balance (but detectable in nuclear reactions).

*In following pages relativity effects will be ignored since the minute changes in mass are beyond the precision of engineering measurements.*

**Example 2.** (a) One mole of hydrogen and ½ mole of oxygen are in a bomb at 77°F. What is the absolute energy of this mixture if the total mass is exactly 18.016 $lb_m$? (b) A chemical reaction occurs and the temperature and pressure increase. Calculate the change in mass. (c) The bomb is cooled back to its original temperature and $94.4(10^6)$ ft-$lb_f$ of energy is transferred away from the bomb. Calculate the change in mass.

*Solution.* (a) By Eq. (a),

$$E = U = \frac{m_0}{g_c} c^2 = \frac{18.016 \, lb_m}{32.1739 \, \dfrac{lb_m \, ft}{lb_f \, sec^2}} \left(186,000 \, \frac{miles}{sec}\right)^2 \left(5,280 \, \frac{ft}{mile}\right)^2$$

$$U = 59.3(10^{16}) \, \text{ft-}lb_f \qquad Ans.$$

(b) No change in mass since the internal energy has only changed in form (electron bonding energy converted into primarily molecular motion).

† In theory, the operational measurement of internal energy is simply to measure the rest mass of substance on a precise scale (and then multiply the value by a constant); the fact that modern scales are not sufficiently precise to make the measurement does not destroy the logic.

(c) By Eq. (d) (note that $\Delta U$ is negative since energy was taken from the bomb),

$$\Delta m = \frac{\Delta U}{c^2}\, g_c = \frac{-94.4(10^6)(32.1739)}{(186,000)^2(5,280)^2} = -3.12(10^{-9})\ \text{lb}_m \qquad Ans.$$

And the answers cannot be checked experimentally although the atomic bomb attests to the magnitudes involved.

**2-6. Conduction and Radiation.**† Energy, in response to a temperature difference, is transferred by two basic mechanisms, called *conduction* and *radiation*. When conduction of energy occurs, a mass is the medium that serves as a conductor of energy from the high- to the low-temperature region (and therefore a *temperature gradient* is the driving influence). If one end of a metal rod is thrust into a fire, the other end will gradually become hot; energy (but not mass) is transferred from the hotter end by conduction to the colder end. Substances also transfer energy by *radiation*. Radiation requires no intervening medium to serve as a conductor for the energy (and therefore a *temperature difference* is the driving influence). Energy is continuously radiated from all substances. A substance that is hot relative to its surroundings radiates more energy than it receives, while its surroundings absorb more energy than they radiate. The result is a *net* transfer of energy from the high- to the low-temperature region.

In thermodynamics, it is sometimes convenient, but never necessary, to use the name *convection* to describe the transfer of energy from a hot to a cold region by mass movement of a fluid. A house can be warmed by passing cold air over a hot surface in a furnace to receive energy by conduction (primarily), then by passing the hot air over the walls of the house to transfer energy again by conduction (primarily). Thus, convection is simply a descriptive name for a combination of events—it can be loosely defined as double conduction distinguished by the presence of a fluid carrier.

**2-7. The System.** Before a change can be analyzed, it is essential that the participants, mass and energy, be known and included in the analysis; therefore, the *region* under study must be defined:

**The system is a specified region, not necessarily of constant volume, where transfers of energy and/or mass are to be studied.**

Since both mass and energy may be added to the system, an especially important concept is the *boundary*, or limits, of the system:

**The actual or hypothetical envelope enclosing the system is the boundary of the system.**

† Article 19-1.

All mass and energy transfers are evaluated at the boundary. The boundary may be either fixed or elastic if the system is allowed to expand or contract:

**The region outside the system is the surroundings.**

A system without mass transfer is called a *closed system* (Fig. 2-1a):

**A closed system is a region of constant mass, and only energy is allowed to cross the boundaries.**

The closed system may be at rest or else moving relative to the observer. A closed system may be *isolated* from the surroundings:

**An isolated system cannot transfer either energy or mass to or from the surroundings.**

During the period of isolation, the system is therefore constrained to a fixed volume and a fixed energy content.

Most engineering systems are *open:*

**An open system has a mass transfer across its boundaries, and the mass within the system is not necessarily constant.**

In Fig. 2-1b, note that the boundaries where mass (and the energy accompanying mass) enters or leaves the open system are fixed to the datum of

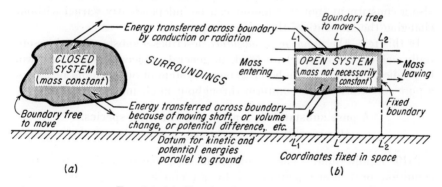

FIG. 2-1. (a) Closed and (b) open systems.

the observer and denoted by $L_1$ and $L_2$; other fixed-position coordinates $L$ can also be located, if desired. The enclosing walls *between* $L_1$ and $L_2$ are not necessarily stationary, and through this enclosing surface energy (but not mass) can pass by conduction or radiation, or by expansion of the walls, or by a rotating shaft, or by flow of electrical energy, etc. Through the stationary flow boundary at $L_1$ (or $L_2$) energy enters (or leaves) the system in direct proportion to the mass which crosses the boundary.

(Although energy, without regard for the mass flow, can also pass through the surfaces at $L_1$ and $L_2$ because of conduction or radiation or the presence of electric or magnetic fields, these effects will be ignored as being insignificant for engineering systems.)

**2-8. Homogeneous and Heterogeneous Systems.** A quantity of matter, homogeneous throughout in chemical composition and in physical structure, is called a *phase*. Homogeneity in chemical composition does not imply a single chemical species, for a mixture of gases or a solution is a phase. Homogeneity in physical structure specifies that the material is all gas, all liquid, or all solid. A system of water and its vapor (steam) contains two phases: a gas and a liquid phase. In all instances, *phase* signifies no abrupt change in either chemical or physical characteristics.

A system comprising a single phase is called a *homogeneous* system, while a *heterogeneous* system consists of more than one phase.

**2-9. Components, Constituents, and Pure Substances.** The chemical species making up a phase or a system are called either *components* or *constituents*. The names are synonymous when chemical changes are not present. Precisely, however, a *constituent* is a particular chemical species in molecular, atomic, ionic, or radical form. Since these forms of matter could all result (by dissociation, for example) from a single molecular substance (a single *component*), the amount of each constituent could bear a definite relationship to the amounts of other constituents—a constituent is not necessarily an independent variable. A *component* is also a constituent, but its amount can be independently varied without changing the amounts of the other components.

In this text, the name *pure substance* describes a substance with essentially one chemical structure (and, in general, molecular structures will be studied). Note that a heterogeneous system of one or more phases of a pure substance will be uniform throughout in chemical composition:

**A pure substance is a particular chemical species.**
*(One-constituent system of one or more phases)*

**2-10. Equilibrium.** The equilibrium of a mechanical system is a condition of balance maintained by an equality of opposing forces. *Thermodynamic equilibrium*, however, is a broader concept since it includes not only mechanical forces but also "forces" arising from thermal, electrical, chemical, and other influences. Each kind of influence on the system dictates a particular aspect of thermodynamic equilibrium: *thermal equilibrium* denotes an equality of temperature; *mechanical equilibrium* denotes an equality of pressure;† *electrical equilibrium* denotes an equality of electrical potential; *phase* and *chemical-reaction*

---

† Unless weight effects are present, as in a vertical column of fluid.

*equilibria* denote an equality of chemical potentials. Thus thermodynamic equilibrium is complete equilibrium:

**A system in thermodynamic equilibrium is incapable of spontaneous change even after being subjected to catalysts or to disturbances; it is in complete balance with the surroundings.**

Consider again the examples in Art. 2-3. The gasoline-vapor–air mixture in the cylinder is in *mechanical equilibrium* when its pressure is constant throughout and balanced by an opposing force in the surroundings. Similarly, the mixture is in *thermal equilibrium* when its temperature is constant throughout and balanced by an equal temperature of the surroundings. On the other hand, the gasoline-vapor–air mixture is not in *chemical equilibrium* since a small spark may cause a violent explosion. Here it seems logical to conclude that, since latent chemical energy was released by a small disturbance (the spark), a chemical "force" must also have been lying dormant in the mixture.

The absence of chemical equilibrium may be an unimportant detail of a particular problem. For example, suppose the problem is to find the variation of pressure with temperature (at constant volume) of a specified mixture of $CO$, $CO_2$, and $O_2$. If this system of fixed constituents is said to be in *equilibrium*, it is implicitly understood that only pressure and temperature equilibria are designated.

**2-11. Property, State, and Process.** Consider an isolated system in equilibrium. Here energy or mass transfers with the surroundings are absent, since the system is isolated, and gradients in temperature, pressure, composition, etc., are, in general, absent, since the system is in equilibrium. The system, however, would have many other characteristics, more or less observable, such as its mass, physical composition, pressure, temperature, volume, surface area, electrical potential, etc. In addition, a host of other characteristics could be defined† or else indirectly measured.‡ All these characteristics have different dimensions and are the variables called *internal*, or *thermostatic*, *properties:*

**An internal, or thermostatic, property is a characteristic of the matter within the equilibrium system.**

Thus, the class of variables called *internal*, or *thermostatic*, *properties* is restricted, arbitrarily, to those equilibrium characteristics of the physical and chemical structures of matter; gradients in these properties and energy or mass transfers with the surroundings are therefore not properties.

† For example, the temperature divided by the pressure is a characteristic.
‡ Chemical composition, internal energy, electrical and thermal conductivity, etc.

The concept of an open system (and also that of a closed system in motion relative to the observer) requires that another class of variables be defined—*external, or mechanical, properties:*

**An external, or mechanical, property is a characteristic of either the motion or the position of the system in a gravitational field.**

The *external, or mechanical, properties* are related to the *kinetic* and *potential energies* of the system (velocity, height, kinetic and potential energies); the *internal, or thermostatic, properties* are related to the *internal energy.* Both classes of properties are called *properties of the system.*

Several reference locations or datums may be required to evaluate the properties of the system:

**The thermostatic properties of the system are measured by an observer at rest relative to the substance.**

Thus, if motion is present, the observer is considered to be traveling with the substance (and therefore measuring quiescent matter).

**The mechanical properties of the system are measured relative to external datums.**

Thus, the height and velocity of the system or substance are measured by an observer located outside the system. The effects arising from the different reference locations are usually not important since differences in energy values are invariably calculated.

*Intensive* properties are independent of the mass of the system: pressure, temperature, viscosity, velocity, height, etc. *Extensive* properties are related to the mass of the system: volume, energies of all kinds, surface area, etc. *Specific*† values of extensive properties, that is, values per unit mass, can also be called intensive properties.

The system can be described or measured at each instant of time by its properties. Each unique condition of the system is called a *state:*

**State is the condition of the system (or a part of the system) at an instant of time as described or measured by its properties.**

Thus, state denotes not a change but particular values of the properties at one instant of time.

If the system is not in equilibrium or not homogeneous, no single state can be assigned. In the case of open systems, the "state of the system"

---

† Specific volume, specific internal energy, etc. In the following pages, specific values are most often implied, although the prefix *specific* will not always be shown (except, invariably, for *specific volume*).

is rarely implied; rather, the *intensive* state at each position coordinate of the region is specified by the values of the intensive properties.

The system can pass from one state to another state or undergo energy transfers at a *steady state* by a *process:*

**A process occurs whenever the system undergoes either a change in state or an energy transfer at a steady state.**

Since a *thermostatic* property is merely a parameter of the internal structure of matter, it is axiomatic (Art. 2-1) that functional relationships must exist between all *thermostatic* properties:

**A thermostatic property is a function of other thermostatic properties.**

On the other hand, the mechanical properties are *not* functionally related to each other or to the thermostatic properties. No function can be proposed, for example, that would enable the kinetic energy to be calculated from specified values of potential and internal energies. This difference between the external and internal properties of a system arises because the mechanical properties of velocity and elevation have been arbitrarily superimposed upon the system and declared to be *independent* by definition.

An *independent* property, as the name implies, is one that can be arbitrarily assigned a value. For example, water at constant pressure can be heated from the freezing point to the boiling point. Within this range, both temperature and pressure can be assigned values at will— each is an independent property. When boiling begins and two phases are present, only one of these two properties can be an independent variable (and the other is called a *dependent* variable), because the value of one fixes the value of the other. Also, some properties are *dependent* on other properties by definition. Thus the specific volume $v$ is defined as the reciprocal of the density $\rho$.

It follows from the concept of state that each *thermostatic* property can have but one value at each state (since *state* fixes the physical and chemical structure). The *mechanical* properties have this same quality (since *state* fixes specific values to both kinetic and potential energies). Therefore, *all* properties are *state* or *point functions:*

**A property has a single value at each equilibrium state—it is a function of the state.**

To recognize that a variable is a property, one of the following tests can be applied:

**A variable is a property if, and only if, it has a single value at each equilibrium state.**

A variable is a property if, and only if, its change in value between two equilibrium states is independent of the process (is single-valued).

A variable is a thermostatic property if, and only if, it is a function of other thermostatic properties.

To illustrate the concept of property, a number of new properties will be defined (and their usages left for future development). *Enthalpy* is defined as

$$H \equiv U + pV \qquad (2\text{-}3a)$$

or, per unit mass,

$$h \equiv u + pv \qquad (2\text{-}3b)$$

The properties $c_p$ (called $c$ sub $p$) and $c_v$ (called $c$ sub $v$) are defined only for chemically inert systems,

$$c_p \equiv \frac{\partial h}{\partial T}\bigg)_p \qquad c_v \equiv \frac{\partial u}{\partial T}\bigg)_v \qquad (2\text{-}4a)$$

The property $k$ is simply the ratio

$$k \equiv \frac{c_p}{c_v} \qquad (2\text{-}4b)$$

To illustrate that $c_v$ (or $c_p$) is a property, picture the function $u = u(T,v)$ as a smooth surface in three dimensions. Any point on this surface represents a state of the inert substance. Only one tangent line (lying within a plane $v = C$) can be drawn through a given point on the surface, and therefore $c_v$, which is the slope of this tangent, has a single value at the equilibrium state.

The reason for defining these properties is entirely for convenience—the notation can be simplified in certain types of problems.

**2-12. Fixing and Identifying the State.** Consider a single-phase thermostatic system of several constituents in temperature and pressure equilibria (but not in chemical equilibrium). The variables for the system are the mass, pressure, temperature, volume, and chemical composition, although other variables such as surface tension may be a factor, or the system may be charged or magnetized. To interrelate all the possible variables is a practical impossibility; hence, the number is reduced by the experimental procedure. Thus, the mass of the system can be held constant so that the properties of mass and volume can be replaced by specific volume, the chemical composition can be held constant, a large mass relative to the surface area can be used to make surface tension a negligible factor, and the experimental system can be guarded from electric or magnetic fields. With these provisos,

$$f(p,v,T) = 0 \qquad (a)$$

Equation (*a*) in its various forms is called the *equation of state* for liquids, gases, and solids. Observe that the *number of independent thermostatic properties has been arbitrarily reduced to two, for convenience.*

Despite its name, Eq. (*a*) is an *incomplete* description of the state since no knowledge of other properties is given. The internal energy, for example, must be experimentally determined and expressed in terms of two of the variables of Eq. (*a*) for each phase,

$$f(u,p,T) = 0 \qquad\qquad (b)$$

It will be apparent in Chap. 7 that for all states of a *pure* substance, without regard for phase, two independent properties can be *selected* (such as specific volume and internal energy, for example) such that the state is fixed. Since more complex systems are mixtures of pure substances, the same rule holds, in general, for all systems if the components are not allowed to vary in amount:

**The equilibrium states of a simple† system are fixed by two intensive properties. For the homogeneous system, any two independent properties will suffice. For the heterogeneous system, two selected independent properties are required.**

Also, a mechanical property is required for each external energy effect to be superimposed.

Consider, next, the number of variables required to *identify* a thermostatic system. For the equilibrium states (extensive) of a simple system, at *least* four properties are required:

**The chemical composition**
**The mass**
**Two thermostatic properties which fix the state**

If the two thermostatic properties are not carefully selected, the physical composition may also be required.

**2-13. Heat and Work.** Consider a system which contains, *within* itself, hot and cold regions. Here energy, because of the temperature difference, will transfer from the hot to the cold region by conduction, radiation, and/or convection. In the subject of heat transfer, such transitory forms of energy are called *heat*. In thermodynamics, however, the name heat is assigned only to energy, but not mass, passing to or from the *surface* of the system:

*Heat is energy transferred through the surface of the system by the mechanisms of conduction and radiation.*

† In the absence of gravitational, kinetic, surface, electrical, or magnetic effects and with constant components.

Note that the process of convection is not included since convection involves a mass flow and the energy accompanying mass will be evaluated separately. Thus, quite arbitrarily, heat is defined as a *surface* effect:

**Heat is energy transferred, without transfer of mass, across the boundary of a system because of a temperature difference between system and surroundings.**

Observe that the process of conduction (but not radiation) dictates a temperature *gradient* at the boundary of the system.

With this definition, it is wrong to speak of heat contained in a system— the correct phrase is *internal energy*. Nor can heat be carried by a mass flow since heat is a concept divorced from mass.

**Processes or systems that do not involve heat are called adiabatic.**

Work, like heat, is transitional in nature and cannot be stored in mass or in a system. Work exists or occurs only during a transfer of energy into or out of a system and, like heat, is a surface concept. After the work is done, no work is present, only the result of the work: energy. A general definition for all forms of work can be made by paraphrasing the definition for heat:

**Work is energy transferred, without transfer of mass, across the boundary of a system because of an intensive property difference other than temperature that exists between system and surroundings.**

The usual intensive property encountered in engineering problems is *pressure*. The pressure on the surface of the system gives rise to a force, and the action of the force through a distance is the concept called *mechanical work:*

**Mechanical work is energy alone crossing the boundary of a system in the form of a force acting through a distance.**

Since electrical energy can be completely converted into mechanical work by a perfect motor, *electrical work is simply electrical energy crossing the boundary of the system* (Art. 6-1).

The symbols† for heat and work will be $Q$ and $W$ and the dimension that of energy. Although heat, work, and energy have the same dimension, only energy is a property of a system. Heat and work are not prop-

---

† Page xix. The symbols $\Delta Q$ and $\Delta W$ were used by Clausius (1850), by Rankine (1849), and occasionally by Gibbs (1880).

erties because they appear only when a process occurs and disappear when the process is completed. The algebraic signs for heat and work are as follows:

$+\Delta Q$, heat *to* system $\qquad$ $+\Delta W$, work *from* system

$-\Delta Q$, heat *from* system $\qquad$ $-\Delta W$, work *to* system

## PROBLEMS

**1.** Calculate the potential energy (Btu) of 1 $lb_m$ at an elevation of 100 ft above a datum of zero potential energy if (*a*) $g = 31$ ft/sec$^2$, (*b*) $g = g_0$.

**2.** Determine the kinetic energy and the final velocity obtained by transformation of the potential energy of Prob. 1 without loss.

**3.** Convert 2,000 ft-lb$_f$ into IT Btu and international kilowatthours.

**4.** Are the following systems homogeneous or heterogeneous?

(*a*) A mixture of ice, water, and steam.

(*b*) The cooling fluid in the radiator of your car.

(*c*) Atmospheric air.

(*d*) A mixture of hydrogen and oxygen.

**5.** List the components and/or constituents in Prob. 4, and decide whether or not the name pure substance is applicable.

**6.** For each system of Prob. 4, list, to the best of your knowledge, the properties that could be used to fix the state and to identify the state. Are these intensive, extensive, thermostatic, independent, or dependent properties?

**7.** Let the *hot gas* (3 $lb_m$) in a balloon be the *system* which is moving through the atmosphere. Sketch and label the following properties: potential, kinetic, and internal energies; pressure, temperature, volume, velocity, and height of the balloon. Make lists of the thermostatic, independent, dependent, intensive, and extensive properties. Use the names intensive, extensive, and thermostatic state in a discussion.

**8.** Draw diagrammatic sketches of the following systems and emphasize the boundaries. Label open (or closed) system, boundary, surroundings, and show the directions of heat and work by arrows.

(*a*) Water pump

(*b*) Pressure cooker

(*c*) Water wheel

(*d*) Thermometer (system) and higher-temperature surroundings

(*e*) A baseball in flight

(*f*) Steam boiler in house, including all piping and radiators

(*g*) Automobile engine

(*h*) An ice-cream freezer (manual crank operation)

(*i*) A dashpot consisting of cylinder, piston, and contained air

(*j*) An automobile storage battery with leads

**9.** Discuss the concept of function (Art. A-6c). In Prob. 7, let $f(p,v,T) = 0$ be known for the gas. Will this function always hold for the gas as the balloon changes its state in any arbitrary manner? Why cannot a function $f(PE,KE,U) = 0$ be devised which would also hold?

**10.** Discuss why enthalpy is a property. Is it necessary to ascribe a physical meaning to this new property?

**11.** A substance has an internal energy of 100 Btu/lb$_m$, a pressure of 100 psia, and a specific volume of 5 ft$^3$/lb$_m$. Calculate the specific enthalpy.

## SELECTED REFERENCES

1. Morrison, J. L.: Fundamental Concepts of Thermodynamics, *Chem. in Canada*, July, 1956, p. 1.
2. Obert, E. F.: "Concepts of Thermodynamics," McGraw-Hill Book Company, Inc., New York, 1960.
3. Brønsted, J. N.: "Principles and Problems in Energetics," translated by R. P. Bell, Interscience Publishers, Inc., New York, 1955.
4. Bridgman, P. W.: "The Nature of Thermodynamics," Harvard University Press, Cambridge, Mass., 1941.

CHAPTER 3

# TEMPERATURE AND THE IDEAL GAS

For bragging time is over,
and fighting time has come.
*Sir Henry Newbolt*

Temperature has been defined as the property which measures the ability of the system to transfer energy by conduction or radiation. With this definition and the Zeroth Law (Art. 2-1), reproducible levels of temperature can be readily distinguished, such as the melting or boiling points of pure substances at specified pressures. The objective of this chapter is to devise a rational scale of temperature:

**A temperature scale is an arbitrary set of numbers and a method for assigning each number to a definite level of temperature.**

The *method* prescribed by the temperature scale requires that a *thermometer* be invented:

**A thermometer is a measuring device which yields a number at each thermal level, and this number is functionally related to the true temperature (another number).**

**3-1. Specifications for a Temperature Scale.** Since temperature is a property, it is a variable that can have but one value at each equilibrium state of the system. Therefore, the essential ingredient of a temperature scale is this:

**A function of the variables of the thermometer must be devised that will give a unique number at each and every temperature level of the system.**

Pedagogical objectives can also be included:

*The temperature scale should be independent of the properties of any substance.*

*The size of a unit of temperature, the degree, should have a rational interpretation.*

33

*A region of minimum temperature, an absolute zero, should be defined.*

That all these specifications can be fulfilled, however, will not be apparent until later (Chap. 7).

The size of an interval of temperature could be arbitrarily established without defining completely the temperature scale by assigning numbers to two *fixed points*.   Examples of fixed points are:

*Ice point:* the state of equilibrium between ice and air-saturated water at a total pressure of 1 atm

*Steam point:* the state of equilibrium between water and its vapor at a pressure of 1 atm

*Triple point:* the state of equilibrium between liquid, vapor, and solid phases

**3-2. Thermometers and the Properties of Materials.**   The usual thermometers are based upon the measurement of properties, such as the:

1. Volume expansion of gases, liquids, and solids
2. Pressure exerted by gases
3. Electrical resistance of solids
4. Vapor pressure of liquids
5. Thermoelectricity

Thus, the number $\alpha$ indicated by the usual thermometer depends upon:

1. The particular substance or substances used
2. The property or properties being measured
3. The design or construction of the instrument

Now, for *any* two (or more) thermometers $A$ and $B$ which yield numbers $\alpha_A$ and $\alpha_B$ at the same temperature level, a common temperature $\theta$ can be proposed, at least over a limited range,

$$\theta = f_A(\alpha_A) = f_B(\alpha_B)$$

But since temperature is a property, and therefore a function of **other** properties, the Zeroth Law announces that more general functions **exist**, valid over any range whatsoever,

$$T = f_{AA}(x_A, y_A, \text{ etc.}) = f_{BB}(x_B, y_B, \text{ etc.}) \tag{3-1}$$

wherein $x$, $y$, etc., are thermostatic properties of the substances.   The problem is to devise a temperature scale (and a thermometer) based upon the properties of substances to obey Eq. (3-1).   Such a scale can be called a *thermodynamic temperature scale based on the Zeroth Law.*

**3-3. Incomplete Temperature Scales.** Consider a mercury-in-glass thermometer which is calibrated to read temperatures of 32 and 212°F at the ice and steam points, respectively. Let the scale be arbitrarily divided into 180 parts, and suppose that the reading at an intermediate thermal level is $n = 32° + 38$ units of length. Here the property primarily in use is the volume expansion of mercury with temperature, but other factors are also present. Thus, various glasses have different expansion coefficients, and the construction of the thermometer also affects the reading. Also, pressure can squeeze the glass bulb and so cause the reading to increase with pressure increase. It can be concluded that a thermometer of this type is too arbitrary and capricious to be used to define a temperature scale [although mercury thermometers are quite precise today when corrections (from calibration tests) are applied to the readings].

Essentially the same comments, but in minor vein, can be made for all thermometers which are based upon a single property, such as the pressure of a gas at constant volume or the electrical resistance or thermoelectricity of metals at constant pressure and strain. Here, changes in properties that can affect the reading are not included in the definition of the scale—the scale is incomplete.

**3-4. The Gas Scale of Temperature.** Although incomplete scales can be made complete, it would seem that the resulting functional relationship among the independent properties would be quite complex and a temperature scale would necessarily be defined in terms of one particular substance. But when gases are investigated, a particularly simple solution becomes apparent and a fundamental relationship is found to exist among all gases.

Experience shows, when extraneous effects are eliminated (Art. 2-12), that the equation of state for a gas is of the form $f(p,v,T) = 0$. This function may be extremely complicated. To find cases where the relationship is simple, suppose that the values of pressure and specific volume are investigated at constant temperature for various gases. (And a constant *temperature* can be maintained at this stage even though the proper *number* to assign to it is unknown.) Let air be confined in a cylinder by a piston and the assembly placed in a water bath which is held at constant temperature. The piston can be moved to various positions so that a series of states at constant temperature and known (measured) pressure and mole volume can be visited. At first, the relationship between the values of $p$ and $v$ at each state is obscure. But as the pressure decreases, it becomes apparent that the product $pv$ is rapidly approaching a constant value. The limiting value is found by plotting the data and extrapolating to zero pressure (Fig. 3-1a). When the experiment is rerun at a higher temperature, the $pv$ product again approaches a constant value, but now the value is larger than before.   At each selected thermal level,† a unique value for the $pv$ product is obtained. Interestingly, the same behavior and the *same* numerical value of $pv$ (as $p \to 0$) at each thermal level are obtained when the air is replaced by another gas.   The experiments clearly show that the $pv$ product of real gases, when extrapolated to zero pressure, obeys Eq. (3-1):

$$T = f(\alpha) = f(pv) \Big]_{\substack{p \to 0 \\ \text{real gases}}} \tag{3-2}$$

The importance of Eq. (3-2) can be realized by noting its significance with

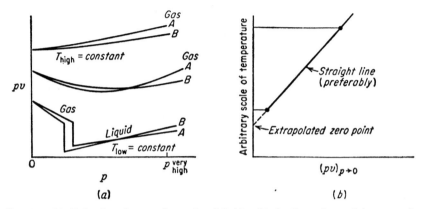

Fɪɢ. 3-1. (*a*) Behavior of $pv$ product of real fluids; (*b*) the "zero," or minimum, value of temperature on an arbitrary gas scale (mole specific volume).

respect to the specifications for a temperature scale (Art. 3-1).   In particular, note that:

1. A variable has been discovered which has a unique value at each temperature.

2. The variable has a wide but not unlimited range.

3. The range of the variable includes only positive numbers (if negative pressures or volumes of gases are conceded to be impossible).

4. The variable is independent of any one substance.

On the other hand, the solution offered by Eq. (3-2), while good, is not perfect.   An indication of the fault is that all gases condense to the liquid state at low temperatures and test pressures and, at high temperatures, gases dissociate, and experimental measurements are open to question. Once the range‡ of $f(pv)$ is exceeded, *temperature is again undefined*

---

† The various thermal levels can be distinguished by a mercury thermometer, for example.

‡ About 1 to 1850°K.

*and cannot be measured,* although the zero point† can be located (Fig. 3-1b). This collapse can be directly traced to the fact that the proposed temperature scale is tied to the specific properties of gases, and therefore a universal solution is not achieved (and this solution, found later in Chap. 7, includes and amplifies, but does not destroy, the gas scale of temperature).

The next step is to select a function of the product $pv$ in Eq. (3-2) and thereby define a temperature scale—a gas scale. An infinity of functions are available. One simple selection is

$$f(pv) = \frac{pv}{R_0}$$

where $R_0$ is the *universal gas constant.* Then the definition of what is called an *ideal-gas temperature scale* is

$$T \equiv \frac{pv}{R_0}\bigg]_{\substack{p \to 0 \\ \text{real gases}}} \tag{3-3}$$

Rather than specify $R_0$, the *absolute* temperature of the triple point of water is fixed by international agreement to be

$$T_{\substack{\text{triple} \\ \text{point}}} \equiv 273.16°\text{K}$$

on a temperature scale called the *Kelvin temperature scale.* [It will be shown, later, that the ideal-gas scale agrees with the Kelvin scale (Art. 7-2).] Then $R_0$ is evaluated from experimental values of $pv$ for real gases made at the temperature of the triple point of water,

$$R_0 = \frac{pv_{\text{triple point}}}{273.16}\bigg]_{\substack{p \to 0 \\ \text{real gases}}} \tag{3-4}$$

Values of $R_0$ for various units are listed in Table B-8.

The Celsius (formerly centigrade) scale is defined relative to the Kelvin scale,

$$t \, (°\text{C}) \equiv T \, (°\text{K}) - 273.15°$$

Thus, the zero of the Celsius scale (which is closely the temperature of the ice point) is, by definition, 0.01° below the triple point of water. Within the accuracy of experimental measurements, the steam-point temperature is 373.15°K. Thus, the interval between ice and steam points is 100.00°C, although this value is subject to change (slight) as measurements become more precise.

---

†With the stipulation of (3) above, the zero of the gas scale *could* be called a *true,* or *absolute, zero* even though temperatures from 0 to 1°K are not measurable.

A temperature scale in general use in engineering has degrees five-ninths the size of the Kelvin degree. It is called the *Rankine temperature scale*,

$$T_{\text{triple point}} \equiv 491.688°\text{R}$$

Values of $R_0$ to correspond to this scale are listed in Table B-8.

The Fahrenheit scale is defined in like manner to the Celsius scale,

$$t \text{ (°F)} \equiv T \text{ (°R)} - 459.67°$$

Thus, the ice point on the Fahrenheit scale is *closely* 32°F, the triple point is *exactly* 32.018°F, and the steam point is *closely* 212°F.

**3-5. The Ideal, or Perfect, Gas.** Since each gas has a unique internal structure, it also has a unique equation of state (Art. 2-12). But as pressure approaches zero, the $pv$ products of all gases at a given temperature approach the same value (Fig. 3-1a); therefore, all equations of state tend toward a common function. In addition, the effect of pressure on the internal energy of gases decreases rapidly with decrease in density and disappears as the density approaches zero. It follows that at the state of vanishing density, here called the *ideal* or *perfect state*, all gases behave similarly—behave ideally. This suggests a new concept—an ideal, or perfect, gas that will behave ideally at all densities:

**1. An ideal or perfect gas obeys the equation of state $pv = R_0T$.**
**2. The internal energy of an ideal gas is a function of temperature alone.**

Another thought can be added: *to create an ideal-gas model of each real gas.* Thus, the ideal-gas model of oxygen, for example, obeys the equation of state $pv = R_0T$ and its internal energy is a function of temperature alone while its molecular weight is 32.

The equations of an ideal gas lead to simple solutions, and therefore such solutions are convenient approximations for many engineering problems. Note that the mole and pound mass units are related by the molecular weight $M$, for example,

$$v \left[ \frac{\text{ft}^3}{\text{mole}} \right] = M \left[ \frac{\text{lb}_m}{\text{mole}} \right] v \left[ \frac{\text{ft}^3}{\text{lb}_m} \right]$$

Therefore, the ideal-gas equation of state (with $v$ in cubic feet per mole)

$$pv = R_0T \qquad (3\text{-}5a)$$

can also be shown as

$$pv = \frac{R_0}{M} T = RT \qquad (3\text{-}5b)$$

wherein $v$ has different units (cubic feet per pound mass) from those in Eq. (3-5a) and $R$ is called the *specific gas constant*,

$$R = \frac{R_0}{M} \quad \left[\frac{\text{ft lb}_f}{\text{lb}_m \, °\text{R}}\right]$$
(3-6a)

Most often, $R_0$ will be specified with the following units (Table B-8):

$$R_0 = 1,545 \frac{\text{lb}_f \text{ ft}}{\text{mole } °\text{R}}$$
(3-6b)

Equations (3-5a) and (3-5b) can also be written

$$pV = nR_0T$$
(3-5c)
$$pV = mRT$$
(3-5d)

(And the usual units will be $n$ lb moles, $m$ lb$_m$, $V$ ft$^3$.)

For processes of either real or ideal substances, the initial and final states can be connected by an equation of the form

$$p_1v_1{}^n = p_2v_2{}^n = C$$
(3-7a)

where in $n$ is a constant called the *polytropic exponent*. For an ideal gas, it follows from Eqs. (3-5a) and (3-7a) that

$$\frac{T_2}{T_1} = \left(\frac{p_2}{p_1}\right)^{\frac{n-1}{n}}$$
(3-7b)

and

$$\frac{T_2}{T_1} = \left(\frac{v_1}{v_2}\right)^{n-1}$$
(3-7c)

It may happen that all states encountered in the process obey Eqs. (3-7). Then, for an ideal gas,

$n = 0$    (constant-pressure process)
$n = 1$    (constant-temperature or isothermal process)
$n = \infty$    (constant-volume process)

A *Joule*, or *free, expansion* is the name given to a process devised by Joule in 1843. Joule took two containers, which were connected together by a valve, and immersed them in a water bath (Fig. 3-2). One of the containers was evacuated, while the second container was filled with air at 22 atm pressure and at the bath temperature. Now, when the valve was opened, the air rushed into the evacuated container, with a consequent radical drop in pressure. But when temperature and pressure equilibria of the air had been restored,

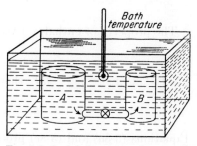

Fig. 3-2. Joule's free-expansion experiment.

Joule could detect little or no change in the bath temperature. He reasoned that the air had passed from state 1 $(p_1, T_1)$ to state 2 $(p_2, T_1)$ without change in internal energy, since no work was done and no heat was transferred, as shown by the constancy of the bath temperature. Joule concluded that, since pressure (or specific volume) could be changed without changing the internal energy, then the internal energy of air is a function of temperature alone.

Let the air remaining in the originally charged container be called system $A$ and the air displaced into the evacuated container be called system $B$. Then, toward the end of the process, system $A$ is at a low temperature since it had to do work to displace system $B$. Correspondingly, toward the end of the process, system $B$ is at a high temperature because work was done on this system. For both systems $A$ and $B$ considered together, no work was done, and the cooling of system $B$ was counterbalanced by the heating of system $A$, and therefore the bath temperature did not change. Thus, if $u = f(T,v)$ (Art. 2-12) and here $\partial u/\partial v)_T = 0$, then $u = f(T)$.

With more precise apparatus, it is found that a small drop in temperature is actually present for the conditions of Joule's experiment (a fraction of a degree). Thus, although the internal energy of gases is a function of both temperature and pressure, the effect of pressure is usually small. The important point is that, when the Joule experiment is repeated at lower and lower densities, the small effect of pressure becomes even smaller and disappears upon extrapolation to the limiting case of zero pressure. It can be concluded that for the ideal gas the internal energy is a pure temperature function. This statement is called *Joule's law:*

**The internal energy of an ideal gas is a function of temperature alone.**

Or, mathematically,

$$u = f(T) \Big]_{\text{ideal gas}} \tag{3-8a}$$

And from Eqs. (2-4a) and (3-8a), for inert gases,

$$c_v = \frac{du}{dT} \Big]_{\text{ideal gas}} \tag{3-8b}$$

The property of enthalpy (Art. 2-11) is defined as

$$h \equiv u + pv \tag{2-3b}$$

For the ideal gas, $pv = RT$; therefore,

$$h = u + RT \Big]_{\text{ideal gas}} \tag{3-8c}$$

Equations (3-8a) and (3-8c) proclaim:

**The enthalpy of an ideal gas is a function of temperature alone.**

It follows from Eqs. (2-4a) and (3-8c) that, for inert gases,

$$c_p = \left.\frac{dh}{dT}\right]_{\text{ideal gas}} \tag{3-8d}$$

And, with Eqs. (3-8b) to (3-8d),

$$c_p - c_v = R \text{ or } R_0 \Big]_{\text{ideal gas}} \tag{3-9}$$

Hence, it is not necessary to list temperature functions for both $c_p$ and $c_v$ since one differs from the other by a constant. (For real gases the difference between $c_p$ and $c_v$ is never less than $R$ but may be much greater because of pressure.)

For real substances, the properties of $h$, $u$, $c_p$, and $c_v$ are functions of both temperature and pressure, and extensive tables or graphs of functions are necessary for evaluation (Appendix B). However, in many instances the effect of pressure (density) can be neglected. In Fig. B-1 values of $c_p$, $c_v$, and $k$ for air are shown. Note that for temperatures above 400°F the effect of pressure on $c_v$ is almost imperceptible (at least for air). For temperatures above 1000°F and pressures less than 500 psi, the zero-pressure† values for $c_p$, in general, are less than 2 per cent low. On the other hand, $c_p$ for carbon dioxide at 100°F and 500 psi is 14 per cent higher than the zero-pressure value (Fig. B-2c). These comments suggest that the ideal-gas relationships can be used for approximate solutions of engineering problems and that, many times, the answers will be within the precision demanded by the problem.

It is usual to list temperature functions for the ideal-gas property of $c_p$ (Table B-2) since values of $c_v$ can be found from Eq. (3-9) and values of $u_{\text{id}}$ by integration of Eq. (3-8b),

$$\Delta u_{\text{id}} = \int_{T_1}^{T_2} c_v \, dT$$

Similarly, $\Delta h_{\text{id}}$ can be found from Eq. (3-8d). Arbitrary datums are selected where $u$ and $h$ are declared to be zero.

The conclusions of this article apply also to mixtures of ideal gases. In a mixture, the total pressure arises from the several constituents; therefore, each constituent is said to develop a *partial pressure*. In calculating the internal energy or enthalpy of a constituent of a mixture, its partial pressure is unimportant, since the internal energy of an ideal gas depends only upon temperature.

## PROBLEMS

Atmospheric pressure is 14.7 psi

**1.** What do you consider to be the most important specification for a temperature scale?

---

† The concept of zero pressure is highly unreal, and therefore the notation should be interpreted to mean limiting values: the ideal state.

**2.** Discuss what is meant by an incomplete temperature scale. How can such a scale be made complete?

**3.** Compare the ideal-gas temperature scale with the specifications listed in Art. 3-1.

**4.** Is the ideal-gas temperature scale an arbitrary scale? Discuss.

**5.** Suppose that only one gas in the entire world can be found that exhibits a different number for the $pv$ product at each thermal level but then only between the ice and sulfur points. Decide whether the absolute zero is affected. Can temperatures below 0°C be measured? Would you say that a temperature on this scale is not a universal property of all substances?

**6.** Decide whether the $pv$ product at the ice point for any gas ($v$ in mole units) is a universal constant. Discuss.

**7.** Why was the constant $R_0$ included in the ideal-gas equation of state?

**8.** Explain the difference between a *universal* and a *specific* gas constant. Are two symbols desirable?

**9.** Define a perfect gas. Why can gases such as air, steam, carbon dioxide, etc., be considered perfect?

In the following problems, the answers are to be based upon the assumption that the gases can be considered to behave ideally.

**10.** A volume of 3 ft³ of gas under atmospheric pressures has the pressure doubled while the temperature remains constant. What is the final volume of the gas?

**11.** A volume of 5 ft³ of gas at 200°F and 20 psia is expanded at constant pressure to a final volume of 10 ft³. Determine the final temperature.

**12.** If the gas in Prob. 11 is air, how many pounds (and moles) are present? Repeat, assuming that methane is the fluid.

**13.** A pound of air at a pressure of 100 psia and temperature of 60°F is to be stored in a tank. How large must the tank be? Repeat, assuming that methane is the fluid.

**14.** A tank containing nitrogen at 60 psig and 60°F is exposed to the sun and absorbs radiant energy until the temperature is 130°F. Calculate the final pressure.

**15.** A 10-ft³ tank contains hydrogen at 20 psia and 60°F. Hydrogen is pumped into the tank until the pressure is 100 psia and temperature 150°F. How much hydrogen was pumped into the tank?

**16.** A reservoir contains 100 ft³ of carbon monoxide at 1,000 psia and 100°F. An evacuated tank is filled from the reservoir to a pressure of 50 psia and temperature of 60°F, while the pressure in the reservoir decreases to 900 psia and the temperature to 80°F. What is the volume of the tank?

**17.** A tank contains air at 100 psia and 60°F. A pound of air is removed from the tank, and this causes the final conditions in the tank to be 50 psia and 50°F. Calculate the volume of the tank.

**18.** A pound mass of air at a temperature of 100°F is raised to 500°F at constant volume. Find the internal energy at both states if the datum of zero internal energy is 0°F (use Table B-2).

**19.** For the conditions of Prob. 18, find the enthalpy at both states. Locate the state of zero enthalpy.

**20.** Find the enthalpy of air at several temperatures between 0 and 100°F, and plot with $h$ as the ordinate and $T$ as the abscissa (use Table B-2). Construct a curve passing through these states and also a tangent to the curve at 50°F. Calculate $c_p$ at 50°F from the graphical data, and compare with the value in Table B-6.

**21.** A pound mass of carbon dioxide at 100° is raised to 1000°F at constant pressure. Calculate the change in internal energy.

**22.** Discuss why the internal energies of liquids and solids are relatively indifferent to pressures and also why the enthalpies are not.

**23.** Derive Eqs. (3-7$b$) and (3-7$c$).

**24.** Air undergoes a process from 100 psia and 260°F to 20 psia and −6°F. Calculate $n$ for the over-all process.

**25.** Justify the $n$ values indicated in the text for processes at constant temperature, constant volume, and constant pressure.

**26.** A process begins and ends at the same temperature, and therefore the $n$ value in Eq. (3-7$a$) is $n = 1$. Is this an isothermal process?

### SELECTED REFERENCES

1. American Institute of Physics: "Temperature: Its Measurement and Control in Science and Industry," Reinhold Publishing Corporation, New York, 1941.
2. Zemansky, M. W.: "Heat and Thermodynamics," 4th ed., McGraw-Hill Book Company, Inc., New York, 1957.
3. Roberts, J. K., and A. R. Miller: "Heat and Thermodynamics," 4th ed., Interscience Publishers, Inc., New York, 1954.
4. Jackson, L. C.: "Low Temperature Physics," 3d ed., John Wiley & Sons, Inc., New York, 1950.
5. Obert, E. F.: "Concepts of Thermodynamics," McGraw-Hill Book Company, Inc., New York, 1960.

## The First Law

The prerequisite for the First Law of Thermodynamics is contained in the axiom:

*The internal energy of a system is a property.*

In building upon this axiom, it became apparent that, whenever an energy increase appeared in one system, a corresponding energy decrease appeared in another system. This *conservation of energy* became the theorem known as *the First Law of Thermodynamics:*

**Energy can be neither created nor destroyed but only converted from one form to another.**

Thus, perpetual-motion machines of the *first kind* (or the *first class*) are declared to be impossible: *no machine can produce energy without corresponding expenditures of energy.*

The First Law dictates an energy balance between system and surroundings for *all* processes, real or ideal, perfect or imperfect:

$$\Delta E \Big]_{\text{surroundings}} + \Delta E \Big]_{\text{system}} = 0$$

(But the First Law gives no indication of whether or not the process was perfectly performed.) The change in energy of all types, $\Delta E$, is entirely a system term and is measured by changes in the properties of the system. The changes in energy can arise from three causes: transfers of mass, heat, and work. Energy while in transit in the form of heat or work is rarely measured because of the attendant difficulties (if not impossibilities) of instrumentation. Consider a water-cooled air compressor. The heat passed from the system (the cylinder of the compressor) to the surroundings (a water bath, for one element) is best apparent by noting the surroundings since the temperature of the water increased. But direct measurement of heat (which is energy in transit) is avoided by measuring $\Delta E$ for the system in the surroundings (the water bath) and then translating such data into the proper units, which well may be a rate (Btu per second). Although this indirect procedure will be followed in the laboratory, *it is convenient to think of heat and work as being measured "in*

45

*flight" and therefore equations will be developed that in appearance pertain only to the system.*†

Two distinct types of analyses will be developed, one for the closed system (Chap. 4) and one for the dynamic open system (Chap. 5). From the viewpoint of an energy balance, the closed system has the disadvantage that only very slow processes can be successfully evaluated. Thus, study of the closed system is, most often, a study of the equilibrium state and might better be called *thermostatics*. Conversely, flow processes can be successfully evaluated because of the extension of thermodynamics made possible by the definition of the dynamic open system.

It might appear that the theory of relativity (Art. 2-5) is not compatible with the principle of conservation of energy. But if mass is considered to be a form of energy, no inconsistency is present (although the First Law would be better named the law of conservation of mass-energy). Thus, the system that gains energy gains mass, and correspondingly, the system that loses energy loses mass. Since the slight changes in mass are too small to be measured in other than nuclear processes, separate balances for mass and for energy will be made, as before.

It is well to remark again that engineering thermodynamics has two distinct parts: (1) the study of energy transformations in both open and closed systems (our immediate problem in Chaps. 4 and 5), and (2) the study of the behavior—the property interrelationships—of quiescent substances (Chaps. 8 to 11, for examples).

---

† As a matter for discussion (or argument), note that heat cannot be measured without first assuming internal energy to be a property; thus "proofs" that internal energy is a property based upon hypothetical transfers of heat and work are invalid.

# CHAPTER 4

# THE FIRST LAW AND THE CLOSED SYSTEM

*The chessboard is the world, the pieces are the phenomena of the universe, the rules of the game are what we call the laws of nature.*
*Thomas Huxley*

In the fields of physics and chemistry, the energy balance of the First Law is merely a step in the development of the mathematical relationships between the equilibrium properties of quiescent matter.  Because of this end point, the independent variable *time*, which underlies classical thermodynamics, is not usually mentioned.  The engineer, however, is interested in energy and its conversion; therefore, the independent variable that he must use in the laboratory must be included in his theoretical analysis for clarity.

**4-1. Energy of the Closed System.**  The element of mass $\Delta m$ within the closed system of Fig. 4-1 has internal energy, and also it may have

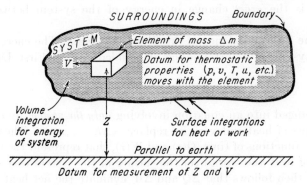

FIG. 4-1. The general closed system.

kinetic and potential energies.  The total specific energy at a point in the system at time $\tau$ is designated by the letter $e$ and the function $e(\tau)$:

$$e \equiv u + \frac{V^2}{2g_c} + \frac{g}{g_c} Z \qquad \left[ \frac{\text{energy}}{\text{mass}} \right] \qquad (4\text{-}1)$$

But each element may have different energy values from its neighbors,

47

especially when a process is under way. Therefore, at a specified time $\tau$, the energy of the entire system is indicated by a volume integral:

$$E\Big]_\tau = \int_{\substack{\text{volume} \\ \text{at } \tau}} e\rho \, dV = \int_{\substack{\text{volume} \\ \text{at } \tau}} e \, dm \qquad [\text{energy}] \qquad (4\text{-}2)$$

Notice that $e$ and $e\rho$ may depend on both time and position, while $E$ is a function of time alone. When $e$ varies with position in a chaotic manner (the nonequilibrium case found during the usual dynamic process), evaluation of Eq. (4-2) may not be possible. But if the system is in equilibrium or near equilibrium at the beginning and end of the process, or if appropriate average values are considered for $\rho$, $u$, $V$, and $Z$, Eq. (4-2) integrates to

$$E\Big]_\tau = U + \frac{mV^2}{2g_c} + m\frac{g}{g_c} Z \qquad (4\text{-}3)$$

In effect, this procedure reduces the independent variables of the system to that of time alone:

**The variables describing the closed system will be selected so that time is the only independent variable.**

Thus the function $E(\tau)$ represents the energy within the system at time $\tau$ (and $\Delta E$ is the *exact* change in energy of the system between times $\tau_1$ and $\tau_2$).

**4-2. The Energy Balance of the Closed System.** The energy balance between system and surroundings demanded by the First Law for *all* processes,

$$-\Delta E_{\text{surroundings}} = \Delta E_{\text{system}}$$

can be changed into an equation involving *only the system* by employing the concepts of heat and work (to replace $-\Delta E_{\text{surroundings}}$). The procedure is to define functions of time, $Q(\tau)$ and $W(\tau)$, that represent the integrated transfers of heat and work over the surface of the system (say from time zero). It then follows that $\Delta Q$ and $\Delta W$ evaluate the net heat and work transfers in the time period $\tau_1$ to $\tau_2$, and an *exact* energy balance can be proposed:

$$\Delta_r Q - \Delta_r W \Big]_{\substack{\text{surface} \\ \tau_1 \text{ to } \tau_2}} = \Delta_r E \Big]_{\substack{\text{system} \\ \tau_1 \text{ to } \tau_2}} \qquad (4\text{-}4a)$$

Here the algebraic signs of heat and work are dictated by convention:

| | |
|---|---|
| + for heat *to* system | + for work *from* system |
| − for heat *from* system | − for work *to* system |

Equation (4-4a) is an energy balance over the system alone: *surface* integrations for heat and work and *volume* integrations for energy of the system, all evaluated over the same time period $\tau_1$ to $\tau_2$; this is the *general energy equation* for analysis of the closed system.

Since time is the independent variable by dividing Eq. (4-4a) by $\Delta\tau$,

$$\frac{\Delta Q}{\Delta\tau} - \frac{\Delta W}{\Delta\tau} = \frac{\Delta E}{\Delta\tau}$$

and *then* allowing $\Delta\tau$ to approach zero, *by definition* of a derivative (Art. A-6h),

$$\frac{dQ}{d\tau} - \frac{dW}{d\tau} = \frac{dE}{d\tau} \tag{4-4b}$$

Thus the time rate of change of heat and work must equal the time rate of change of energy of the system. When Eq. (4-4b) is multiplied by the *increment* $\Delta\tau$,

$$\frac{dQ}{d\tau}\Delta\tau - \frac{dW}{d\tau}\Delta\tau = \frac{dE}{d\tau}\Delta\tau$$

*by definition* of the differential of a dependent variable [Eq. (A-4a)],

$$dQ - dW = dE \tag{4-4c}$$

*The meaning*† *of this differential equation is clear only if Eq. (4-4b) is kept in mind: Q, W, and E are functions of the single independent variable of time.*

Equation (4-4c) can be integrated back into the form of Eq. (4-4a); then, if the parts of $E$ can be evaluated,

$$\Delta Q - \Delta W \Big]_{\text{surface}} = \Delta E \Big]_{\text{system}} = \Delta U + \Delta\frac{mV^2}{2g_c} + \Delta m\frac{g}{g_c}Z \tag{4-4d}$$

All the work of the closed system may not be available for use since a part may be spent in pushing aside the surrounding atmosphere (which will be considered to exert a constant pressure $p_0$),

$$\Delta W_{\substack{\text{net}\\ \text{closed system}}} = \Delta Q - \Delta E - p_0(V_2 - V_1) \tag{4-4e}$$

However, if the system expands during one process and then contracts in succeeding processes so that the initial volume is restored, the $p_0\,\Delta V$ terms cancel.

Note, especially, that the equations in this article hold whether or not the process is perfect and whether or not the system is in equilibrium

---

† And the differentials $dQ$, $dW$, $dE$ have nothing to do with "infinitely small quantities." Nor is Eq. (4-4c) an energy balance for the process; rather, it is the equation which relates the differentials $dQ$, $dW$, $dE$ of the functions $Q(\tau)$, $W(\tau)$, and $E(\tau)$ (Art. A-6j).

(although, admittedly, unless equilibrium is present or closely approached at the beginning and at the end of the process, exact evaluation of the various quantities may be questionable).

The equations can also designate unit mass of system when shown with lower-case symbols; for example, Eq. (4-4c) becomes

$$dq - dw = de$$

**4-3. Processes of the Closed System.** The processes of a closed system, without kinetic or potential energy, are governed by Eq. (4-4d),

$$\Delta Q - \Delta W = \Delta U \Big]_{\substack{\text{closed system} \\ \Delta KE,\ \Delta PE\,=\,0}} \qquad (a)$$

Since the symbol $U$ designates all forms of internal energy (Art. 2-3), Eq. ($a$) holds whether or not a chemical reaction occurs during the process. When a property remains constant during the process, it is so indicated: *constant-volume* process, *constant-pressure* process, *constant-temperature* or *isothermal* process, *constant-internal-energy* process, etc.

By *system* is meant, usually, the *working substance* alone—the mechanical envelope is a part of the surroundings (Fig. 4-3b). However, the procedure to be followed in the laboratory demands that the containing structure or mechanism be included as a part of the system (Fig. 4-3a), and this practice will be occasionally indicated in the discussion.

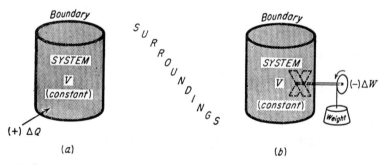

(a)                                         (b)

FIG. 4-2. Constant-volume processes of a closed system. (The boundary is a real or imaginary surface that separates system and surroundings.)

*a. Constant-volume Process.* Consider a constant-volume system of inert substance receiving heat of amount $\Delta Q$ (Fig. 4-2a). Since the volume is constant, compression (or expansion) work is zero and if electrical and shaft work are also zero,

$$\Delta Q = \Delta U = U_2 - U_1 \Big]_{V\,=\,C} \qquad (b)$$

Consider, next, the same system, but without heat transfer, and with mechanical work from the surroundings being dissipated by paddling the inert substance (Fig. 4-2b).   Let the same change† in internal energy be obtained as was evident from transfer of heat,

$$-\Delta W = \Delta U = U_2 - U_1 \Big]_{V=C} \qquad (c)$$

Equations (b) and (c) illustrate that heat and work are *by definition* surface effects and must be identified at the boundary because the effect of heat and work can be the same.   In other words, properties of the system cannot be used by themselves to identify either heat or work.

Suppose the constant-volume system is receiving work (by paddling) and heat is taken away at the same rate; conceivably, the properties of the system would remain essentially constant.   This would be a *steady-state* process.   Thus, a process does not necessarily involve a change in state.

*b. Friction.*   The result of a paddling process such as described above will be evidenced by a rise in temperature of the system (or by a phase change without change in temperature).   To obtain the same result, heat could have been substituted for the work.   The paddling process is a *frictional* process:

**Friction‡ is the dissipation of energy that otherwise could do work into a heating effect.**

Any process that dissipates mechanical, electrical, kinetic, or potential energy is a frictional process.   Examples of friction are quite common: *Mechanical friction* involves the rubbing together of two solids, thus causing a rise in temperature of the parts as if heat had been added; *fluid friction* involves the dissipation of either work or kinetic energy, so inducing a chaotic turbulence in a fluid, with the end result as if heat had been added.   The frictional process, of course, may be desirable (dissipation of electrical energy as in electric toasters, dissipation of kinetic energy by the brakes on a car).

*c. Adiabatic Process.*   An adiabatic process is a change in state without transfer of heat (Art. 2-13).   The work of the adiabatic system, $\Delta Q = 0$, equals

$$\Delta W = -\Delta U = U_1 - U_2 \qquad (d)$$

Equation (d) does *not* predict an optimum or maximum value for $\Delta W$; it merely specifies that the work must equal the change in internal energy of the system.   Thus the net work of the adiabatic expansion

† By definition, work done on a system is negative; hence, $-\Delta W$ is positive.
‡ A mild form of the Second Law.

process of a system similar to that in Fig. 4-3a may be zero or less than zero if mechanical and fluid friction losses are high, it may be greater than zero, and it may be far greater than zero if chemical internal energy is liberated during the process. Essentially the same comments can be made if the *system* is selected to be the *working substance* alone.

d. *Constant-internal-energy Process.* In the expansion process at constant internal energy, heat is added to compensate for the work delivered to the surroundings. Therefore, by Eq. (a), $\Delta U = 0$, and

$$\Delta W = \Delta Q \qquad\qquad (e)$$

Equation (e) specifies *a process of a closed system that will produce work of exact amount to the heat added.* However, this process *cannot* be continuously operated because expansion of a closed system has a definite limit.

e. *Constant-pressure Process.* Consider the *system* to be the *substance* alone in Fig. 4-3b. Let the system expand slowly, and suppose that, by

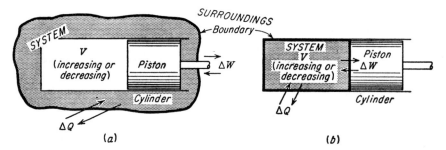

Fig. 4-3. (a) System of substance and mechanism; (b) system of substance alone.

some means, the pressure throughout the substance and on the piston (of area $A$) remains constant at its initial value. Then, by Eq. (a),

$$\Delta Q = \Delta U + \Delta W$$

The work done by the system is the force $pA$ times the distance $\Delta x$. Since $A\,\Delta x$ is the change in volume $\Delta V$,

$$\Delta Q = \Delta U + p\,\Delta V = \Delta U + \Delta pV$$

Here is a case where the property of enthalpy [Eq. (2-3a)] is convenient,

$$\Delta Q = \Delta H \Big]_{p=C} \qquad\qquad (f)$$

Consider, next, the means for holding the pressure constant. Let the system be an inert gas. Now, with expansion, the pressure will tend to fall, and heat must be added to hold the pressure constant. Thus, Eq. (f) for an inert gas is interpreted as the heat added in a constant-pressure expansion. The amount of heat required could be calculated if values of enthalpy for the gas were obtained for the end states (at $p_1$, $T_1$, and $p_1$, $T_2$). Or the approximate change in enthalpy could be calculated from ideal-gas relationships [Eq. (3-8$d$)].

If the substance were not inert and a chemical reaction occurred, conceivably the rate of the reaction could keep pace with the piston movement and Eq. ($f$) would still apply. In this instance, however, $\Delta Q$ could be positive, negative, or zero, depending upon the reaction and its degree of completion. The value of $\Delta Q$ as before, however, is equal to the change in enthalpy of the system. It is apparent in these discussions that *enthalpy is a property of interest in constant-pressure processes of a closed system.*

*f. $\Delta H$ and $\Delta U$ of Reaction.* The discussion in ($e$) suggests a method for finding the change in enthalpy with change in constituents, that is, the change in absolute† enthalpy for a chemical reaction. Suppose that the *system* is 1 *mole of carbon monoxide* and 10 *moles of oxygen*, a *piston and cylinder*, and a *large tank of water* (open to the atmosphere) which surrounds the gas mixture and cylinder. Let the piston be constrained in some manner such that it will yield very slowly to changes in pressure of the gas mixture. Now, with a spark, an explosion occurs, the pressure and temperature of the gases rise, and the piston moves slowly outward. Then, cooling of the gases takes place as energy passes to the water bath, and the piston moves slowly inward until the initial pressure of the gas mixture is regained; the initial temperature is also approached within a degree or so of temperature because of the size of the water bath. For the entire adiabatic system, by Eq. ($a$),

$$\Delta U = -\Delta W$$

Since the surrounding atmosphere was not accelerated, but only displaced at constant pressure $p_0$, the work done equals $p_0 \, \Delta V$,

$$\Delta U_{\text{system}} = -p_0 \, \Delta V$$

Or, for each part of the system,

$$\Delta U_{\text{gases}} + \Delta U_{\text{water}} + \Delta U_{\text{metal}} = -p_0 \, \Delta V_{\text{gases}} - p_0 \, \Delta V_{\text{water}} - p_0 \, \Delta V_{\text{metal}}$$

Hence the change in enthalpy of the reacting gases equals

$$\Delta H_{\text{gases}} = -\Delta H_{\text{water}} - \Delta H_{\text{metal}} \tag{g}$$

The right-hand side of Eq. ($g$) is readily evaluated since values of $c_p$ for the inert water‡ and $c_p$ for the inert metal are known, and the tempera-

---

† But the *absolute* enthalpy of one constituent alone cannot be found.

‡ Note that $\Delta H$ for the reacting gases cannot be calculated from $c_p$ values since such values are defined only for chemically inert substances (Art. 2-11).

ture rise (small) can be measured. Assume that the calculated value is $-121,664$ Btu at $77°F$: the absolute enthalpy of water and metal increased by 121,664 Btu, while the absolute enthalpy of the reacting gases decreased by 121,664 Btu.

*Thus, the property of enthalpy is of interest in processes of a closed system that begin and end at the same pressure.*

Consider the chemical reaction that took place,

$$CO + \tfrac{1}{2}O_2 + 9\tfrac{1}{2}O_2 \Big]_{p_1, T_1} \rightarrow CO_2 + 9\tfrac{1}{2}O_2 \Big]_{p_1, T_2 \approx T_1}$$

The excess oxygen ($9\tfrac{1}{2}$ moles) ensured that essentially all the carbon monoxide was converted into carbon dioxide. Note, too, that the initial and final states of the excess oxygen were closely equal [from a partial pressure $p_x$ (Art. 3-5) and temperature $T_1$ to a partial pressure $p_y$ and temperature $T_2 \approx T_1$]. Since the change in temperature is almost zero, and since the effect of pressure on the enthalpy of gases is slight (Art. 3-5), it can be concluded that $\Delta H$ for the excess oxygen is a small quantity relative to the over-all measurement of $-121,664$ Btu. Hence, the reaction, its direction, and the change in enthalpy are indicated as

$$CO + \tfrac{1}{2}O_2 \rightarrow CO_2 \qquad \Delta H_{p,77°F} = -121,664 \text{ Btu}$$

Thus, the enthalpy of 1 mole of $CO_2$ at $p_1$, $T_1$, relative to a mixture of 1 mole of CO and $\tfrac{1}{2}$ mole of $O_2$ at $p_1$, $T_1$, is $-121,664$ Btu (at $77°F$).

If the reacting gases are held at constant volume, the work of the adiabatic system arises only from changes in volume of the metal and the water bath. By Eq. ($a$),

$$\Delta U = -\Delta W$$

And here

$$\Delta U_{gases} + \Delta U_{water} + \Delta U_{metal} = -p_0 \, \Delta V_{water} - p_0 \, \Delta V_{metal}$$

Hence, the change in internal energy of the reacting gases equals

$$\Delta U_{gases} = -\Delta H_{water} - \Delta H_{metal} \qquad (h)$$

Laboratory measurements show that the right-hand side of Eq. ($h$) equals $-121,131$ Btu (at $77°F$). With the same assumptions as before,

$$CO + \tfrac{1}{2}O_2 \rightarrow CO_2 \qquad \Delta U_{V,77°F} = -121,131 \text{ Btu}$$

Thus, the internal energy of 1 mole of $CO_2$ at $V_1$, $T_1$, relative to a mixture of 1 mole of CO and $\tfrac{1}{2}$ mole of $O_2$ at $V_1$, $T_1$, is $-121,131$ Btu (at $77°F$).

It is not necessary in the laboratory to measure both $\Delta U$ and $\Delta H$ of reaction. By definition of enthalpy,

$$\Delta H \Big]_{p_1, T_1} = \Delta U \Big]_{p_1, T_1} + \Delta pV \Big]_{p_1, T_1} \qquad (4\text{-}5a)$$

If $\Delta U$, for example, is measured at $T_1$, $\Delta H$ at the same temperature $T_1$ can be closely calculated by adding $\Delta pV$ for the reaction to $\Delta U$ (with proper regard for the algebraic signs). This procedure ignores the fact that the pressure in the constant-volume experiment was, most probably, not equal to that in the constant-pressure experiment; however, the effect of pressure on the internal energy or enthalpy of gases is small and is insignificant relative to the large values attained for $\Delta U$ or $\Delta H$ of reaction. With these comments, Eq. (4-5$a$) can be shown as

$$\Delta H\Big]_{T_1} \approx \Delta U\Big]_{T_1} + \Delta pV\Big]_{T_1} \tag{4-5$b$}$$

Equation (4-5$b$) is closely valid for gas reactions and also for reactions wherein one or more of the substances is a solid or liquid (the internal energy of a solid or a liquid is also not affected greatly by pressure). Since the volume of a condensed phase is negligibly small, only the gas phase need be considered (and the ideal-gas law is adequate):

$$\Delta H\Big]_{T_1} \approx \Delta U + \Delta nRT\Big]_{T_1, \Delta n \text{ for the gases alone}} \tag{4-5$c$}$$

Consider the combustion of methane with pure oxygen,

$$CH_4 + 2O_2 \rightarrow CO_2 + 2H_2O$$

If the $H_2O$ product is a gas, then 3 moles of products are obtained from 3 moles of mixture, $\Delta n$ is zero, and $\Delta H$ equals $\Delta U$. And $\Delta H$ is larger (disregarding the algebraic sign) than $\Delta U$ if the reaction undergoes a decrease in volume since work is done on the system by the surroundings (and conversely); for example, if the $H_2O$ product is a liquid, $\Delta n$ is $-2$.

**4-4. Thermodynamic Cycles.** Several processes can be coupled together to form a *thermodynamic cycle*:†

**A thermodynamic cycle is a sequence of processes that eventually returns the working substance to its original state.**

Since the original state is regained,

*The thermodynamic cycle is a concept of the closed system.*

Consider that the constant-volume system of Fig. 4-2 can have work added in one process, and then have heat taken away in another process,

† The internal-combustion engine (such as the gasoline or diesel engine or the usual gas turbine) does *not* undergo a thermodynamic cycle since the working substance is not restored to its initial state (although the engine mechanism may undergo a *mechanical cycle*). Rather, such engines come under the class of processes called *steady-flow* (Art. 5-7).

to regain the original state (if chemical reaction is absent). The net effect of this *dissipation cycle* is

$$\Delta Q - \Delta W = 0$$
$$\Delta Q = \Delta W \qquad (a)$$

*In this manner, work can be completely and continuously converted into heat —observe that it is quite easy to devise cycles to convert work completely into heat.*

A primary interest of the engineer is *power cycles* to produce work from conversion of heat. To construct a power cycle, let the *system* be a *gas* confined in a cylinder by a piston (Fig. 4-3a). Arbitrarily select a number of processes, for example, an adiabatic compression process *ab*, a constant-volume heat-addition process *bc*, an adiabatic expansion process *cd*, and a constant-volume heat-rejection process *da* to the initial state. The sequence of states visited by the gas can be qualitatively shown on diagrams of properties by assuming ideal gases, $pv = RT$, and by recalling certain experimental facts: adiabatic compression of a gas raises its pressure and temperature, and the pressure and temperature can be lowered by adiabatic expansion. Figure 4-4 illustrates the cycle and also two invariable characteristics of all power cycles:

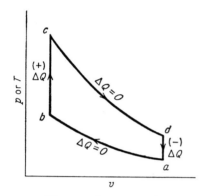

FIG. 4-4. A power cycle.

1. *A heat-addition process at a relatively high temperature*
2. *A heat-rejection process at a relatively low temperature*

Observe that work alone is transferred in processes *ab* and *cd* and heat alone in processes *bc* and *da* (characteristics of this arbitrary cycle). It can be surmised that the cycle could produce a net amount of work since the average pressure on the work-output process *cd* is much higher than the average pressure on the work-input process *ab*. The net amount of work can be found by noting that, for any cycle, $\Sigma \Delta U = 0$; therefore,

$$\Sigma \Delta Q - \Sigma \Delta W = \Sigma \Delta U = 0$$
$$\Sigma \Delta W = \Sigma \Delta Q \qquad (b)$$

If the net work of the cycle is to be positive—a work output—it can be concluded that more heat had to be added in process *bc* than was rejected in process *da*.

The *thermal efficiency* of the power cycle is defined as the fraction of the heat addition which was converted into work by the *closed system*,

$$\eta_t = \frac{\text{work output}}{\text{heat added}}\bigg]_{\text{cycle}} = \frac{\Delta Q_A + \Delta Q_R}{\Delta Q_A} = \frac{\Sigma\,\Delta W}{\Delta Q_A}\bigg]_{\text{cycle}} \qquad (4\text{-}6)$$

Obviously, from Eq. (*b*), the ratio of $\Sigma\,\Delta W$ to $\Sigma\,\Delta Q$ is always unity, for any substance and for any cycle. The thermal efficiency, however, is the ratio of the *net work* output to the *heat added* to the cycle of the closed system.

From experience it is found that the power cycle is characterized by a heat-addition process at a high temperature and an inevitable heat-rejection process at a lower temperature. Consequently, the thermal efficiency is *always* less than unity. No commercial power cycle has yet been devised with thermal efficiency of as much as 50 per cent, while values of 10 and 20 per cent are usual. The cycle of Eq. (*a*) completely converted work into heat. No problem attended this conversion, which could be operated continuously. The reasons for failure to convert heat into work completely and continuously† cannot be surmised from the First Law but must be left for the Second Law to explain.

If the power cycle of Fig. 4-4 is *reversed* so that the inverse sequence of states is visited, it becomes a *refrigeration cycle*. Here heat is absorbed in process *ad* and rejected in process *cb*; more work is required for process *dc* than that obtained from process *ba* (the net work of the refrigeration cycle is negative).

The excellence of a refrigeration cycle is judged by its *coefficient of performance*, which is the ratio of the *heat*‡ added to the *work* added in the cycle,

$$CP = \frac{\text{refrigeration}}{\text{work added}}\bigg]_{\text{cycle}} = \frac{\Delta Q_A}{-\Sigma\,\Delta W_A}\bigg]_{\text{cycle}} \qquad (4\text{-}7a)$$

Or if the refrigeration cycle is used as a *heat pump*, its coefficient of performance is the ratio between the heat rejected and the work added:

$$CP_{\text{heat pump}} = \frac{\text{heat rejected}}{\text{work added}}\bigg]_{\text{cycle}} = \frac{\Delta Q_R}{-\Sigma\,\Delta W_A}\bigg]_{\text{cycle}} \qquad (4\text{-}7b)$$

It follows that

$$CP_{\text{heat pump}} = CP_{\text{refrigeration}} + 1 \qquad (4\text{-}7c)$$

† Note that the constant-internal-energy process of Art. 4-3*d* could not be operated continuously; also, it would be improper to say that heat was converted into work in this process since the state of the working substance was changed.

‡ Note that $\Delta Q_A$ in Eq. (4-7*a*) refers to process *ad* of Fig. 4-4, while $\Delta Q_A$ in Eq. (4-6) refers to process *bc*.

## PROBLEMS

Draw a diagrammatic sketch of the system for each problem with the boundary emphasized; label *system, surroundings, boundary;* show *directions* of *heat* and *work* by arrows; designate *numerical values* of all *variables.*

**1.** A system receives 100 Btu of heat while work of amount 125 Btu is transferred to the surroundings. Is this possible?

**2.** A system transfers 100 Btu of heat to the surroundings while receiving 100 Btu of work. What is the name of the process?

**3.** The energy of a system increases by 60 Btu while 75 Btu of work is transferred to the surroundings. Is heat added or taken away from the system?

**4.** (a) The energy of a system increases by 50 Btu while the system is receiving 40 Btu of work. How much heat is transferred and in what direction?

(b) The energy of a system decreases by 25 Btu while the system is receiving 30 Btu of work. How much heat is transferred?

(c) During an expansion process, the work transfer is 10 Btu while the heat received by the system is 20,000 ft-lb$_f$. What is the change in energy for the system?

**5.** During a compression process, the work transfer is 8,000 ft-lb$_f$ while the heat received by the system is 25 Btu. What is the change in energy for the system?

**6.** A constant-volume system receives heat of amount 10 Btu and work of amount 5 Btu. Find the change in energy.

**7.** A closed system consists of a cylinder of water and ice stirred by a paddle wheel. For the process, the work was 17 Btu/hr; the initial internal energy was 133 Btu, and the final internal energy after $\frac{1}{2}$ hr of stirring was 126 Btu. Find the heat exchange in Btu per hour. Is the temperature of the system rising or falling?

**8.** A piston-cylinder arrangement has a gas in the cylinder space. During a constant-pressure expansion to a larger volume, the work effects for the gas are 1.60 Btu, the heat added to the gas and cylinder arrangement was 3.17 Btu, and the friction between the piston and cylinder wall amounted to 0.25 Btu. Determine the external useful work done by the process, and the change in internal energy of the entire apparatus (gas, cylinder, piston).

**9.** List a number of systems that will continuously and completely convert work into heat.

**10.** The energy of a constant-volume process increases by 10 Btu while only 5 Btu of heat is added. Was work transferred during this process? Can your statement be proved?

**11.** An adiabatic compression process involves the transfer of 10 Btu of work. (a) What are the value and direction for the change in energy of the system? (b) If the surrounding atmosphere has a constant pressure of 15 psia and the volume change of the system is 2 ft$^3$, show the division of work transfer and explain. (c) Repeat the above, but assume that the process is one of expansion.

**12.** An expansion process at constant energy involves the transfer of 10 Btu of heat. (a) Determine the value and direction for the transfer of work. (b) Devise an expansion process at constant energy that would not involve either heat or work.

**13.** A system receives 10 Btu of heat while expanding with volume change of 1 ft$^3$ against an atmosphere of 20 psia. A mass of 50 lb$_m$ in the surroundings is also lifted a distance of 20 ft. (a) Find the change in energy of the system. (b) The system is returned to its initial volume by an adiabatic process which requires 26 Btu of work. Find the change in energy of the system and the work, other than atmospheric.

(c) For the combined processes of (a) and (b), determine the change in energy of the system. Is it necessary to include the atmosphere in this calculation? Discuss.

**14.** In a closed system, it is found that the time relationships of heat and work flow can be expressed by $Q = a\tau^2 + b$ and $W = c\tau$ (a, b, c are constants). (a) Determine $dE$. (b) Determine $\Delta E$, and compare with $dE$. Should Eq. (4-4c) be interpreted as an approximation for the incremental equation? Is it an energy balance? Are the differentials $dQ$, $dW$, $dE$ infinitely small quantities?

**15.** Assume that the value of $\Delta H$ in Art. 4-3f is correct for the $CO$-$O_2$-$CO_2$ reaction, and check (calculate) the value for $\Delta U$. (Assume ideal gases.)

**16.** One-tenth gram of fuel is placed in a bomb calorimeter (a constant volume), which then is charged with oxygen. The mass of water in the water bath is 1,900 g, while the mass of metal, etc., is equivalent to 462 g of water. When the fuel is ignited (by an electric spark), the temperature rises from 77 to 77.51°F. Calculate $\Delta U$ for the reaction (Btu per pound mass fuel).

**17.** Assume that the fuel in Prob. 16 was pure carbon and that it was completely burned into carbon dioxide. Calculate $\Delta H$ for the reaction at 77°F.

**18.** In a thermodynamic cycle the heat and work are as follows:

| Process......... | 1 | 2 | 3 | 4 |
|---|---|---|---|---|
| $\Delta Q$.............. | +30 | −10 | −20 | +5 |
| $\Delta W$.............. | +3 | +10 | −8 | 0 |

Calculate the thermal efficiency.                                   *Ans.* 14.3%.

## SELECTED REFERENCES

1. Menger, Karl: The Mathematics of Elementary Thermodynamics, *Am. J. Phys.*, February, 1950, p. 89.
2. Keenan, J. H., and A. H. Shapiro: History and Exposition of the Laws of Thermodynamics, *Mech. Eng.*, November, 1947, p. 915.
3. Emmons, H. W.: Re-examination of Thermodynamic Fundamentals, *Mech. Eng.*, June, 1950, p. 475.
4. Osgood, W. F.: "Advanced Calculus," The Macmillan Company, New York, 1937 (in particular, p. 453).
5. Obert, E. F.: "Concepts of Thermodynamics," McGraw-Hill Book Company, Inc., New York, 1960.

CHAPTER 5

# THE FIRST LAW AND THE DYNAMIC OPEN SYSTEM

And for purposes of discipline—intellectual,
moral, religious—the most effective is science.
*Herbert Spencer*

The engineer's concept of the dynamic open system allows analyses to be made of continuous streams which are far from equilibrium in a part of their travels. *The procedure is to transform the problem from study of a fixed mass to study of a region, called by Prandtl a control volume.* Each flow boundary is established at a location of the stream where the assumption of equilibrium is valid (within the precision of the problem). Then, for example, energy values can be assigned to these selected position coordinates of the region so that the change in energy of the flowing stream between two (or more) coordinates is readily evaluated. The transformation, in effect, is to select *position* as well as *time* to be the independent variables:

**The variables describing the dynamic open system are functions of time and position.**

One other item should be emphasized:

**The position coordinates $L$ of the open system are at rest relative to each other.**

Once the open system is selected, the primary position coordinates (where a stream enters or leaves the system) are fixed, and the engineer may not be interested in employing position as an explicit variable. Thus, the engineering equations of analysis may often show time alone as the independent variable (but see Art. 5-7).

**5-1. One-dimensional Flow.** Three space variables $(x, y, z)$ may be required to locate and to describe accurately the properties of a stream (at time $\tau$), and therefore the flow is said to be *three-dimensional;* if two space variables are sufficient, the flow is called *two-dimensional* (Fig. 5-1a and c; each vector $V$ is a function of $x$ and $y$ for parallel-sided passages or of $x$ and $r$ for round conduits). When only one space variable is required, the flow, necessarily, has a set flow pattern and is called *one-dimensional* [for example, $V = f(x)$].

60

The assumption of one-dimensional flow is justified for most engineering analyses since the error can be made small by using appropriate average values of the properties at each cross section of the stream. Thus, at a particular cross section the velocity $V$ is the *average velocity* relative to the coordinate $L$, and the density $\rho$ is the *average density* so that, with the *true* cross-sectional area $A$, the *true* mass flow rate $\dot{m}_f$ can be calculated,

$$\dot{m}_f = A\rho V \qquad [L^2]\left[\frac{M}{L^3}\right]\frac{[L]}{[\tau]} = \left[\frac{\text{mass}}{\text{time}}\right] \qquad (a)$$

This procedure is followed even though the flow stream is in laminar flow or changes direction or converges or diverges by assuming that the average

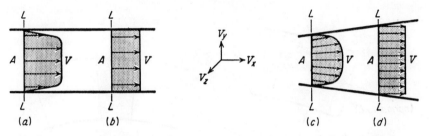

FIG. 5-1. (*a, c*) Two-dimensional flow; (*b, d*) hypothetical one-dimensional flow.

velocity vector remains perpendicular to the cross-sectional area $A$ (Fig. 5-1*d*):

**Most, if not all, of the flow streams in this text are considered to be one-dimensional.**

Since the one-dimensional stream can follow any configuration in space, each cross-sectional area is specified by the single coordinate $L$ $[A = A(L)]$. *The value assignable to the $L$ coordinate coorresponds to the length measured from the entrance of the system* $(L_1)$ *to the area under scrutiny (at $L$) along the center line* (Fig. 5-2*b*). Note, again, that the $x$ coordinate, for example, denotes a fixed direction, but not the coordinate $L$ $[L = L(x,y,z)]$.

**5-2. The Mass-flow Rate.** Consider the "moving" element of mass $\Delta m_f$ in Fig. 5-2*a* with frontal area $A$, velocity $V$, and length $V\,\Delta\tau$. The mass contained in $\Delta m_f$ as it crosses the coordinate $L$ is approximately

$$\Delta m_f\Big]_L \approx \rho A V\,\Delta\tau \qquad (a)$$

The instantaneous mass-flow rate $\dot{m}_f$ across a coordinate $L$ is defined as a function of time $\tau$ and position $L$,

$$\dot{m}_f = \lim_{\Delta\tau\to0} \frac{\Delta m_f}{\Delta\tau} = \frac{\partial m_f}{\partial\tau}\bigg)_L = \rho A V \qquad (5\text{-}1a)$$

On the other hand, the mass in the "stationary" element $\Delta m$ of Fig. 5-2$b$ is approximately equal to

$$\Delta m\bigg]_\tau \approx \rho A \,\Delta L \qquad (b)$$

and
$$\frac{\partial m}{\partial L}\bigg)_\tau = \lim_{\Delta L\to0} \frac{\Delta m}{\Delta L} = \rho A \qquad (5\text{-}1b)$$

Hence two mass functions are required to visualize correctly the open system: The physical significance of the function $m_f(L,\tau)$ is the amount of

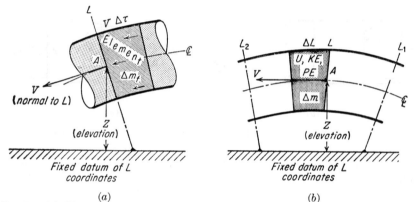

FIG. 5-2. ($a$) Concept of mass in motion: the infinitesimal $\Delta m_f$ crossing a coordinate $L$; ($b$) concept of contained mass: the infinitesimal $\Delta m$ at a time $\tau$.

mass which has crossed the coordinate $L$ from time zero to $\tau$—for the function $m(L,\tau)$, the amount of mass contained within the system between coordinate $L_1$ and position $L$ at time $\tau$ (and this is also the concept of $m$ for a closed system; hence the same symbol can be used).

The procedure in evaluation is the same as for the closed system: the density $\rho$ (and all other thermostatic properties) is evaluated by an observer traveling with the stream; the velocity $V$ (and height $Z$) is evaluated by an observer relative to the fixed coordinate system ($L$).

**5-3. Energy in Flow.** The energy of a flowing stream relative to a position coordinate $L$ can be found by considering the moving element of mass $\Delta m_f$ in Fig. 5-3. Here the energy† accompanying mass is made up

---

† And the properties $u$, $V$, $Z$ are evaluated exactly as for the closed system—two reference datums (Art. 4-1 and Fig. 4-1).

of internal, kinetic, and potential energies and, also, the work done by the substance following $\Delta m_f$ to push $\Delta m_f$ through the distance $V \Delta \tau$.

Then, approximately, since $p$, $\rho$, $u$, $V$, etc., are functions of position $L$ and time $\tau$,

$$\text{Energy of } \Delta m_f \approx \left( u + \frac{V^2}{2g_c} + \frac{g}{g_c} Z \right) \Delta m_f$$

$$\text{Work done on } \Delta m_f = (\text{force})(\text{distance}) \approx (pA)(V \Delta \tau)$$

Or, by Eq. ($a$) of Art. 5-2,

$$\text{Work done on } \Delta m_f \approx pv \, \Delta m_f$$

With these relationships, the *exact* energy flow† across $L$, as $m_{f_2} - m_{f_1}$ mass crosses the coordinate, is

$$\Delta_\tau E_f \bigg]_L = \int_{m_{f_1}(\tau_1)}^{m_{f_2}(\tau_2)} \left( u + pv + \frac{V^2}{2g_c} + \frac{g}{g_c} Z \right) dm_f \qquad (5\text{-}2a)$$

Equation (5-2a) suggests that the sum of the specific energy quantities of the flowing substance *assignable to the coordinate $L$ at time $\tau$ can be defined:*

$$e_f \equiv u + pv + \frac{V^2}{2g_c} + \frac{g}{g_c} Z \qquad \left[ \frac{\text{energy}}{\text{mass}} \right] \qquad (5\text{-}2b)$$

Comparison of Eqs. (4-1) and (5-2b) shows that the change in viewpoint (from watching a fixed mass as it moves in space to watching a position in space) has introduced a new energy term called *flow energy, pv.*

Note that the product $pv$ is a property and has a nonzero value for both open and closed systems, but only for a flow stream of the open system does it take on the significance of energy.

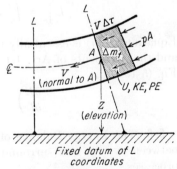

The energy $pv$ is sometimes called *flow work* since it originated from a work analysis. However, $pv$ is a property, and work is not; hence, the name *flow energy* is more descriptive. Moreover, our definition for work (Art. 2-13) excludes energy such as $pv$ being called *work* at a flow boundary. The question is one of definition: for those who define the system to be a *fixed mass*, $pv$ of a flow stream is *work;* for those who define the system to be a *region*, $pv$ of a flow stream is *energy.*

Fig. 5-3. Energy flow relative to a position coordinate $L$ (infinitesimal element).

† The physical significance of the function $E_f(L,\tau)$ is the energy accompanying mass which has flowed past $L$ from time zero to time $\tau$.

Since

$$\Delta E_f \Big]_L \approx e_f \, \Delta m_f$$

it follows, with Eq. (5-1a), that

$$\frac{\partial E_f}{\partial \tau}\bigg)_L = \lim_{\Delta\tau\to 0} \frac{\Delta E_f}{\Delta\tau} = \lim_{\Delta\tau\to 0} e_f \frac{\Delta m_f}{\Delta\tau} = e_f \dot m_f \qquad (5\text{-}2c)$$

Thus, in similarity with the concepts of $m_f$ and $m$, two specific energy functions are required for the open system: $e_f$ (the "moving energy") and $e$ (the "stationary energy") (Art. 5-4).

**5-4. Energy Stored within the Open System.** When the interior of the open system is analyzed, the procedure and the final equations are essentially the same as for the closed system. Consider the "stationary" element of mass in Fig. 5-2b, that is, the element at time $\tau$. The specific energy at a point in the element is

$$e = u + \frac{V^2}{2g_c} + \frac{g}{g_c} Z \qquad (4\text{-}1)$$

The mass of the element equals, approximately,

$$\Delta m \Big]_\tau \approx \rho A \, \Delta L \qquad (a)$$

Therefore, the energy stored within the system from $L_1$ to $L$ at a specified time $\tau$ is evaluated by a volume integral, designated by the function $E(\tau,L)$:

$$E \Big]_\tau = \int_{L_1}^{L} e \, dm \Big]_\tau = \int_{L_1}^{L} e\rho A \, dL \Big]_\tau \qquad (5\text{-}3a)$$

Equation (5-3a) is essentially the same as Eq. (4-2) for the closed system, and the comments of Art. 4-1 apply to both equations. It follows, with Eq. (5-1b), that

$$\frac{\partial E}{\partial L}\bigg)_\tau = \lim_{\Delta L\to 0} e \frac{\Delta m}{\Delta L} = e\rho A \qquad (5\text{-}3b)$$

**5-5. The Energy Balance of the Open System.** The energy balance between system and surroundings demanded by the First Law for all processes,

$$-\Delta E_{\text{surroundings}} = \Delta E_{\text{system}}$$

can be changed into an equation involving *only the system* by employing the concepts of heat, work, and energy in flow (to replace $-\Delta E_{\text{surroundings}}$). The procedure is to define functions of time and position for heat and work so that $Q(\tau,L)$ represents the integrated heat transfers over the surface of the open system from entrance $L_1$ to $L$ and from time zero [and similarly for $W(\tau, L)$]. Then with Eqs. (5-2a) and (5-3a), and noting that

$$\Delta_\tau E_f \Big]_L - \Delta_\tau E_f \Big]_{L+\Delta L} = -\Delta_\tau(\Delta_L E_f)$$

it follows as an *exact* energy balance that

$$\Delta_\tau(\Delta_L Q) - \Delta_\tau(\Delta_L W) - \Delta_\tau(\Delta_L E_f) = \Delta_\tau(\Delta_L E)$$

Here the algebraic signs of heat, work, and energy in flow are dictated by convention:

+ for heat and mass-energy *to* system     + for work *from* system
− for heat and mass-energy *from* system     − for work *to* system

Now, by dividing by $\Delta L$ and $\Delta \tau$, and with the appropriate limiting steps,

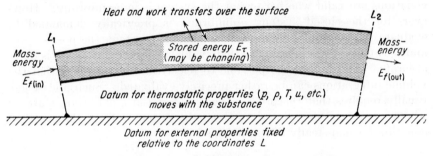

FIG. 5-4. Flow system with fixed entrance and exit.

*the general differential energy balance for an open system with a single one-dimensional flow stream* is obtained:

$$\frac{\partial^2 Q}{\partial \tau\, \partial L} - \frac{\partial^2 W}{\partial \tau\, \partial L} - \frac{\partial^2 E_f}{\partial \tau\, \partial L} = \frac{\partial^2 E}{\partial \tau\, \partial L} \qquad (5\text{-}4a)$$

Or, with Eqs. (5-2c) and (5-3b),

$$\frac{\partial^2 Q}{\partial \tau\, \partial L} - \frac{\partial^2 W}{\partial \tau\, \partial L} - \frac{\partial(\dot m_f e_f)}{\partial L} = \frac{\partial(\rho A e)}{\partial \tau} \qquad (5\text{-}4b)$$

To integrate Eqs. (5-4) for the real system is difficult since, for example, the surface integral for heat would be complex (and evaluation is part of the subject called heat transfer; Chaps. 19 and 20). Simplicity is achieved by restricting the system to a fixed entrance and exit (Fig. 5-4) so that time alone is the independent variable (the procedure followed for the closed system):

$$\Delta_\tau Q - \Delta_\tau W \quad + \int_{\tau_1}^{\tau_2} e_f\, dm_f \Big]_{L_1} - \int_{\tau_1}^{\tau_2} e_f\, dm_f \Big]_{L_2} = \int_{L_1}^{L_2} e\, dm \Big]_{\tau_2} - \int_{L_1}^{L_2} e\, dm \Big]_{\tau_1}$$

Surface of system     Surface     Surface     Volume     Volume
between $L_1$ and $L_2$     at $L_1$     at $L_2$     at $\tau_2$     at $\tau_1$

$$(5\text{-}4c)$$

Or, by definition of the functions $E_f$ and $E$,

$$\Delta_r Q - \Delta_r W = \Delta_r E + \Delta_r E_{f2} - \Delta_r E_{f1} \qquad (5\text{-}4d)$$

Equation (5-4d) can be treated in the same manner as was Eq. (4-4a) to obtain either time rate or differential equations (with time as the single independent variable). Also, when flow ceases, the $E_f$ terms become zero, and Eq. (5-4d) reverts to Eq. (4-4a).

The physical picture of Eqs. (5-4) is: Over a specified time period, the heat added plus the energy (mass) flow to the system minus the work done and minus the energy (mass) flow from the system equals the change in energy stored within the system.

The remarks in Art. 4-3 apply as well to the foregoing equations: the equations are valid whether or not the system is in equilibrium. However, for the closed system, equilibrium is practically demanded to evaluate the equations; in the case of the open system, the positions $L_1$ and $L_2$ can usually be selected so that evaluation of the flow terms is quite accurate. Also, it may be possible to select times $\tau_1$ and $\tau_2$ so that the volume integrations can be rather precisely evaluated. Such preciseness usually requires that the process begin and end at equilibrium states—where $e$ has but one value throughout the system. Thus, the practical equation for non-steady-flow problems is

$$\Delta_r Q - \Delta_r W = m_2 u_2 - m_1 u_1 + \int_{\tau_1}^{\tau_2} e_f \, dm_f \bigg]_{L_2} - \int_{\tau_1}^{\tau_2} e_f \, dm_f \bigg]_{L_1} \qquad (5\text{-}4e)$$

where $m_2$ = mass within system at time $\tau_2$ (with internal energy $u_2$)

$m_1$ = mass within system at time $\tau_1$ (with internal energy $u_1$)

Problems in nonsteady flow are discussed in Arts. 5-10 and 12-1.

**5-6. Continuity Equations.** The definitions (Art. 5-2) of the two mass functions $m$ and $m_f$ allow a simple and exact continuity equation to be directly proposed:†

$$-\Delta_L m_f \bigg]_\tau = \Delta_L m \bigg]_\tau \qquad (5\text{-}5a)$$

Therefore $\qquad -\dfrac{\partial m_f}{\partial L} = \dfrac{\partial m}{\partial L} \qquad$ and $\qquad -\dfrac{\partial^2 m_f}{\partial \tau \, \partial L} = \dfrac{\partial^2 m}{\partial \tau \, \partial L}$

With Eqs. (5-1),

$$-\dfrac{\partial \dot{m}_f}{\partial L} = \dfrac{\partial(\rho A)}{\partial \tau} \qquad (5\text{-}5b)$$

Any form of Eqs. (5-5) is *the general continuity equation for one-dimensional flow of a single flow stream.*

A simple but important class of flow problems is that of *steady flow.*

† Chapter 15 of Ref. 1.

Steady flow decrees that all variables of the stream be functions of position alone:

**A stream is in steady flow if all its variables are independent of time.**

Thus, steady flow demands that the channel area be fixed while the mass-flow rate at any section is constant (although, if the stream has several tributaries, the total mass-flow rate can change with position). With one flow stream, the mass-flow rate does not change with either time or position,

$$\dot{m}_f = A\rho V = \text{constant} \qquad (5\text{-}6a)$$

Or, from Eq. (5-5b),

$$\frac{d\dot{m}_f}{dL} = 0 = \frac{d(A\rho V)}{dL} = \frac{1}{\rho}\frac{d\rho}{dL} + \frac{1}{V}\frac{dV}{dL} + \frac{1}{A}\frac{dA}{dL} = 0 \qquad (5\text{-}6b)$$

Equation (5-6) in any of its various forms is called *the steady-flow continuity equation for a single stream with one-dimensional flow.*

**5-7. The Steady-flow Energy Equation.** When the restriction of steady flow is superimposed on Eq. (5-4b), so that all variables of the equation become functions of position $L$ alone (and $\dot{m}_f$ is a constant),

$$\frac{d\dot{Q}}{dL} - \frac{d\dot{W}}{dL} - \dot{m}_f \frac{de_f}{dL} = 0 \qquad (5\text{-}7a)$$

By dividing by $\dot{m}_f$,

$$\frac{dq}{dL} - \frac{dw}{dL} = \frac{de_f}{dL} \qquad (5\text{-}7b)$$

wherein $q$ represents the function $q(L)$—the heat through the surface $L_1$ to $L$ of the open system per unit mass of fluid flowing [and similarly for $w(L)$].

The foregoing equation could have been derived directly from the definitions of the new functions $q$ and $w$ (and this procedure may be preferable since the steps of Art. 4-2 are reviewed). Thus the *exact* energy balance on an element $\Delta L$ of a steady-flow system is

$$\Delta_L q - \Delta_L w = \Delta_L e_f$$

Dividing by $\Delta L$ (the only independent variable), and then taking the limit as $\Delta L$ approaches zero,

$$\frac{dq}{dL} - \frac{dw}{dL} = \frac{de_f}{dL}$$

Now, upon multiplying by $\Delta L$, by definition of the differential of a dependent variable [Eq. (A-4a)],

$$dq - dw = de_f \tag{5-7c}$$

Equation (5-7c) is the *differential form of the steady-flow energy equation with position L as the independent variable.*

It has become customary to solve flow problems on the basis of unit mass. Hence, Eq. (5-7c) is integrated from $L_1$ to $L_2$ of the system to yield

$$\Delta_L q - \Delta_L w = (h_2 - h_1) + \frac{V_2{}^2 - V_1{}^2}{2g_c} + \frac{g}{g_c}(Z_2 - Z_1) \tag{5-7d}$$

The *integrated* form of the steady-flow equation, Eq. (5-7d), is indifferent to events happening *within* the system since all terms pertain to measurements made at the boundaries. For this reason, the requirement of the differential equations that a *steady state* be maintained at each position coordinate *within* the system can be relaxed for the *practical system:*

**In engineering, an open system of constant mass is called a steady-flow system if all events at the boundaries are unaffected by time.**

An example of a system that can obey the integrated steady-flow equations is a multicylinder, reciprocating-piston engine. Here the properties of the working substance within the system periodically change from process to process, and internally neither a steady state nor a steady flow exists. But when the engine is run at constant speed and load with all operating temperatures stabilized, the conditions for steady flow at the boundaries are usually satisfied within the desired precision of the problem.

**Example 1.** A system has a flow rate of 5 $lb_m$/sec. The enthalpy, velocity, and height at the entrance are, respectively, 1000 Btu/$lb_m$, 100 ft/sec, and 100 ft. At the exit, these quantities are 1020 Btu/$lb_m$, 50 ft/sec, and 0 ft. Heat is transferred to the system at the rate of 50 Btu/sec. How much work can be done by this system?
  *Solution.* From Eq. (5-7d),

$$\Delta q - \Delta w = (h_2 - h_1) + \frac{V_2{}^2 - V_1{}^2}{2Jg_c} + \frac{g}{g_c}\frac{Z_2 - Z_1}{J}$$

$$\frac{50}{5} - \Delta w = (1{,}020 - 1{,}000) + \frac{50^2 - 100^2}{2(32.2)(778)} + \frac{32.2}{32.2}\left(\frac{0 - 100}{778}\right)$$

$$10 - \Delta w = 20 - 0.15 - 0.13$$

$$\Delta w = -9.7 \text{ Btu/lb}_m \qquad Ans.$$

This answer, when multiplied by the mass-flow rate (pounds mass per second), has the dimension of power,

$$\text{Power} = -48.5 \text{ Btu/sec} \qquad Ans.$$

Note that work (or power) must be supplied to the system, a condition indicated by the negative sign of $\Delta w$.

**Example 2.** For the data of Example 1, what must be the area of the inlet pipe if the specific volume of the fluid is 15 ft³/lb$_m$?

*Solution.* From Eq. (5-6a),

$$A = \frac{\dot{m}_f v}{V} = \frac{5(15)}{100} = 0.75 \text{ ft}^2 \qquad Ans.$$

**5-8. Steady-flow Devices.** The practical engineering application of the integrated steady-flow equations is to systems which include both working substance and mechanical structure. However, whether or not the mechanism or structure is included in the analysis, it will invariably influence the thermodynamic process undergone by the working substance.

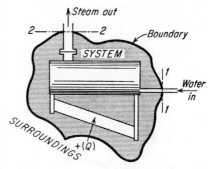

FIG. 5-5. System of boiler and single flow stream of water-steam. (Flow stream of hot gases is in the surroundings.)

*a. Boiler.* A steady-flow system, which also approaches steady-state operation throughout, is the boiler (Fig. 5-5). Here the boundary is located so that the combustion process and the hot gases of combustion are a part of the surroundings, and from these surroundings heat is received by the system. No work is transferred. Changes in potential and kinetic energies arising from the difference in elevation between inlet and outlet and from the velocities of flow are negligible relative to the flow of heat. Therefore, Eq. (5-7d) reduces to

$$\Delta q = h_2 - h_1 \qquad \left[\frac{\text{Btu}}{\text{lb}_m}\right] \qquad (a)$$

Equation (a) refers only to the properties of the flow stream, whether the system is the flow stream alone or includes the metal boiler (since it is assumed that the metal is in a steady state). The effect of the structure, however, is present in either case since the pressure at $p_2$ (and therefore the enthalpy $h_2$) is directly dependent on the design.

Note that, for a closed system, the heat added in a constant-volume process was $\Delta u$ and, in a constant-pressure process, $\Delta h$ (Art. 4-3); in this steady-flow process, with decreasing pressure in the direction of flow (because of friction), the heat also equals $\Delta h$.

*b. Nozzle.* Whenever a flow stream experiences a change in area, a change in velocity also occurs. A *nozzle* is a device designed to increase the velocity of the stream. Fluid enters the nozzle at high pressure and expands to a lower pressure maintained by some means at the exit. For the system of fluid and nozzle shown in Fig. 5-6, the heat transferred per pound of fluid is negligible (length short, mass-flow rate high), no work is done, and the change in potential energy is zero. By Eq. (5-7d),

$$V_2 = \sqrt{2Jg_c(h_1 - h_2) + V_1{}^2} \qquad \left[\frac{\text{ft}}{\text{sec}}\right] \qquad (b)$$

Is this an optimum velocity for the inlet and exit conditions? No, because a porous plug or roughened walls within the nozzle would cause a decided decrease in the velocity $V_2$. Equation (b) merely indicates the relationship between properties at the boundaries of the system that must be obeyed by any nozzle, real or ideal, perfect or imperfect.

c. *Throttling Process.* Whenever a substance expands from a region of relatively high pressure to a region at a lower pressure, either work

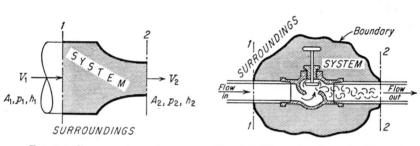

FIG. 5-6. Convergent nozzle.          FIG. 5-7. Throttling through globe valve.

can be done or changes in kinetic or potential energy can be produced. When such energy transformations are essentially absent, the process is described by the name *throttling*. In Fig. 5-7, a valve is used to throttle the flow. The fluid expanding through the valve acquires a high velocity, which is dissipated in aimless turbulence of the fluid. Here work is zero, and the heat flow per pound of fluid flowing is usually negligible. The over-all change in velocity is small, or it can be made zero by proper selection of pipe sizes. With these stipulations, Eq. (5-7d) reduces to

$$h_1 = h_2 \qquad (c)$$

A throttling process obeying Eq. (c) is called a *Joule-Thomson expansion.*

Observe that the fluid in expanding through the small opening of the valve in Fig. 5-7 acquired a high velocity with consequent drop in enthalpy [Eq. (b)]. Work could have been obtained by proper utilization of the kinetic energy. Instead, the kinetic energy was dissipated in restoring the enthalpy to its initial value [Eq. (c)]. Thus, throttling is a form of friction since it wastes the capabilities of the system to do work.

d. *Compression of Gases and Liquids.* The usual machine for compressing gases to high pressures† consists of a cylinder with double-acting

† Pressure ratios, in general, greater than 3: $p_{2,\text{outlet}}/p_{1,\text{inlet}} > 3$.

piston (Fig. 5-8). Here movement of the piston allows gas to be drawn into the cylinder at $A$ while gas trapped on the other side of the piston at $B$ is being compressed; hence, there are two delivery strokes per revolution of the crankshaft. Compression of the gas is regulated by the pressure of the receiver (a storage tank) because the outlet (and inlet) valves are merely check valves. When the gas being compressed reaches a pressure higher than that in the receiver, the delivery check valve opens and gas is discharged into the receiver. When the piston reaches the end of the compression stroke, not all the air in the cylinder has been delivered to the receiver, because a *clearance* must be allowed to prevent the piston from striking the cylinder head. Now, as the piston reverses its direction, the clearance gas expands, the pressure falls, and the receiver pressure

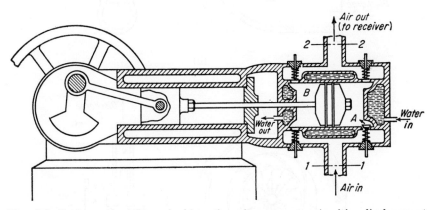

FIG. 5-8. Reciprocating-piston double-acting air compressor (positive-displacement machine).

closes the delivery check valve. The pressure within the cylinder continues to fall with movement of the piston until a pressure less than that in the intake pipe is obtained. Here the inlet check valve opens, and the remainder of the stroke inducts a new charge of gas. The reciprocating-piston compressor is an example of a *positive-displacement* machine.

Although flow in and out of the cylinder is intermittent, steady flow is approached at the outlet of the receiver (which acts as a surge suppressor); here a calibrated nozzle can be installed to measure the mass-flow rate.

Gases and liquids are also compressed and pumped by dynamic machines. For example, Fig. 5-9a illustrates a *diffuser*, or *turbine pump*, for liquids and Fig. 5-9b, a *centrifugal compressor* for gases. In both machines, the fluid enters the inlet at 1 and passes into the *impeller*, where part of the pressure rise occurs as a result of the radial flow of the fluid induced by centrifugal force. The fluid leaving the impeller has a high velocity, and when this velocity is reduced by the *diffuser* (or by the *scroll case* alone), the pressure is again increased. Thus, compression occurs without positive displacement of the fluid.

Relative to a piston-type compressor, the advantage of the dynamic machine is that flow is continuous, and therefore the size is small; the

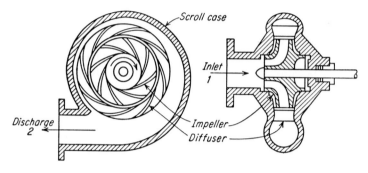

(a)

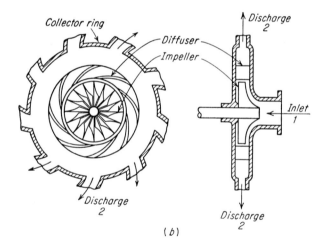

(b)

Fig. 5-9. (a) Diffuser, or turbine liquid pump; (b) centrifugal compressor (engine supercharger).

disadvantage is that high velocities lead to turbulence and consequent fluid-friction losses (lower efficiency). This disadvantage is serious for gases but not for liquids. To explain, note that a compressor does two things: it compresses the fluid and thereby increases its internal energy, and it displaces the fluid from the inlet to the discharge region and thereby increases its flow energy ($pv$). When a fluid is compressed (say by a piston in a cylinder of cross-sectional area $A$), the work is approximately

$$\text{Work} = (\text{force})(\text{distance}) \approx p_{av} A \, \Delta x$$

where $\Delta x$ is the piston movement and $p_{av}$ is the average pressure. For a gas, $\Delta x$ is relatively large; for a liquid, $\Delta x$ is almost zero! It follows that inefficiencies in the compression process (in any type of machine) are not serious in pumping liquids, and dynamic devices are universal. It also follows that the machine for liquids is best described as a *pump* (Fig. 5-9a) while the machine for gases is best described† as a *compressor* (Figs. 5-9b and 5-8).

For all types of compressors or pumps, the work must equal, by Eq. (5-7d),

$$-\Delta w = (h_2 - h_1) + \frac{V_2^2 - V_1^2}{2g_c} - \Delta q \qquad (d)$$

The change in potential energy is invariably negligible, the change in kinetic energy is also negligible for the positive-displacement compressor, and it *may* be negligible for the other types.

Equation (d) can be applied‡ to the working fluid alone or to the entire compressor. No assumptions need be made as to the perfections of the process or the mechanism. A part of the work may be dissipated in creating undesirable turbulence, and this energy reappears as an increase in the enthalpy of the fluid, or some of the work may be spent in overcoming mechanical friction, and the effect of this friction may be to increase the enthalpy of the fluid or else to increase the heat losses to the surroundings. In any event, the equation holds for real or for ideal processes or machines since it is simply an energy balance for the system under observation.

*e. Gas or Steam Turbine.* The elements of a small steam turbine (Fig. 5-10) are essentially similar to those of a gas turbine, and the same explanation will apply to either machine. Steam at high pressure passes through a regulating, or *throttle,*§ *valve* and enters the *steam chest.* Upon leaving the steam chest, the steam enters one or more nozzles that are located adjacent to a *wheel,* or *rotor.* Along the periphery of this wheel are the *blades,* or *buckets.* The steam passing through the nozzle expands

---

† Or as a *fan* (1 to 20 in. $H_2O$ pressure rise) or as a *blower* (1 to 30 psi rise).

‡ The difficulty in practical application is to measure accurately the heat quantities. For this reason, the work is usually measured directly by a *dynamometer* or, with piston-type compressors, by an *indicator* (Art. 6-7).

§ The line valve is also frequently called the *throttle,* even though this valve is not used to throttle the steam (except, of course, in starting); similarly, line pressure is unfortunately called the *pressure at the throttle.*

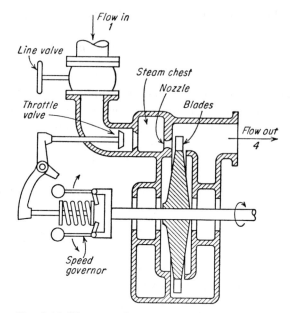

FIG. 5-10. Elements of steam or gas impulse turbine.

and attains a high velocity with correspondingly high kinetic energy. The high-velocity stream leaves the nozzle and impinges on the blading, causing it to move. (Accordingly, a turbine of this type is called an

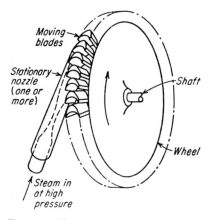

FIG. 5-11. Nozzle and blading of impulse turbine.

*impulse* turbine.) As each blade rotates out of the path of the high-velocity steam jet, another blade takes its place and a continuous rotational force is thus applied to the wheel (Fig. 5-11). The speed of the wheel is controlled by the *governor* (Fig. 5-10). If the load is decreased and the turbine overspeeds, centrifugal force will cause the governor weights to move outward and tend to close the throttle valve. As the throttle is closed, the pressure in the steam chest decreases, the drop in pressure through the nozzle also decreases, and lower velocity is attained by the steam and therefore less impulse is given to the turbine wheel.

It may be of interest to examine the mechanism whereby work can be done by a jet of high-velocity fluid. In Fig. 5-11, the fluid leaves the nozzle at high velocity and enters the curved-blade passage. A diagrammatic sketch of the vector velocities and blade shape appears in Fig. 5-12. The initial absolute velocity $V_2$ has a velocity relative to the blade of $V_r$ because of the blade speed $V_b$. The contour of the blade causes the direction of the fluid to be reversed, thus exerting a force on the blade. The fluid leaves the blade without change in value of the relative velocity if friction is absent. The absolute velocity of the fluid leaving the blade is the vector sum of $V_r$ and $V_b$, as shown. The work done can be found by calculating the force exerted on the blade or more simply by finding the decrease in energy of the fluid. First, assume that fluid friction is absent and that the process is adiabatic. Therefore, $h_3 = h_2$, and $\Delta q = 0$ (while $\Delta PE \approx 0$); hence, by Eq. (5-7d) applied to the system of Fig. 5-12, as represented by Fig. 5-13a,

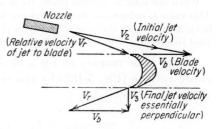

Fig. 5-12. Ideal velocities in turbine blade (no friction).

$$-\Delta w_{\text{blade}} = \frac{V_3{}^2 - V_2{}^2}{2Jg_c}$$

If fluid friction is present, $\Delta h$ is not zero and

$$-\Delta w_{\text{blade}} = h_{3'} - h_2 + \frac{V_{3'}{}^2 - V_2{}^2}{2Jg_c}$$

To evaluate the effect of friction, let the observer be located on the moving blade. This observer would see a stream of fluid with velocity $V_r$ entering his system and

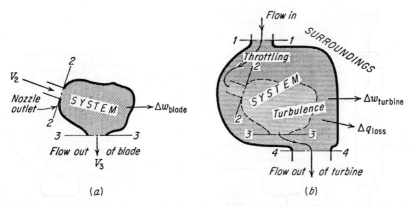

Fig. 5-13. (a) Adiabatic system for flow through turbine blade (nozzle is in the surroundings); (b) nonadiabatic system for flow through entire turbine.

leaving with velocity $V_{r'}$. To this observer, no work would be done since he would insist that his system was stationary (and the flow, as before, is assumed adiabatic). Hence, the observer would apply Eq. (5-7d) to his system and find that

$$h_{3'} - h_2 = \frac{V_r{}^2 - V_{r'}{}^2}{2Jg_c}$$

Thus, friction decreases the relative velocity of the fluid as it passes across the blade, and this dissipation of kinetic energy is reflected by an increase in enthalpy of the fluid.

When the fluid leaves the blade, its absolute velocity may still be relatively high. This value is reduced by fluid turbulence within the casing (with consequent increase in enthalpy) to a value compatible with the mass-flow rate through the exhaust pipe (and this leaving velocity is usually negligible).

In the same manner as for a blade, Eq. (5-7d) can be applied to the entire turbine (Fig. 5-13b) by ignoring entirely the changes taking place within the system,

$$-\Delta w_{\text{turbine}} = h_4 - h_1 + \frac{V_4{}^2 - V_1{}^2}{2Jg_c} - \Delta q$$

Although the heat radiated or conducted to the surroundings may be large, the amount per unit of mass flow is usually negligible; also, the velocities of $V_4$ and $V_1$ are essentially equal. Hence, the turbine equation is simply

$$-\Delta w_{\text{turbine}} = h_4 - h_1 \qquad \left[\frac{\text{Btu}}{\text{lb}_m}\right]$$

The question again arises: Is this the maximum work that can be done by the turbine? No decision can be made from the evidence at hand. The work output is merely shown to be the difference between the total energy put in and that taken out of the system, and this relationship will be true for real or ideal machines.

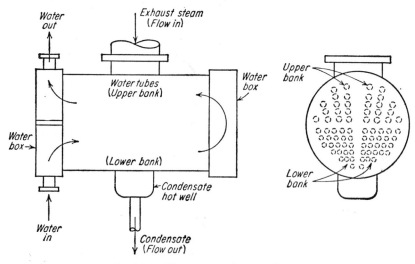

FIG. 5-14. Two-pass surface condenser.

f. *Condensers.* A *surface condenser* consists of a shell enclosing a number of tubes in the manner illustrated in Fig. 5-14. Cold water is admitted to the water box and then passes through the lower bank of tubes to the opposite water box. Here it passes upward and back through the upper bank of tubes. This is a *two-pass* condenser. Steam is admitted at the top of the condenser and passes downward over and between the tubes and condenses on the cold surface of the tubes. The condensate is collected in the *hot well* and pumped back to the boiler. An advantage of a surface condenser is that it permits reuse of the condensed steam as boiler feed-water. Since the feedwater in many localities must be distilled or treated to prevent deposits in boiler and turbine, the use of a surface con-denser will effect a considerable sav-ing in treating costs. Also, the water used for condensing the steam never comes in contact with the con-densate, and any cheap water supply can be used for cooling the con-denser.

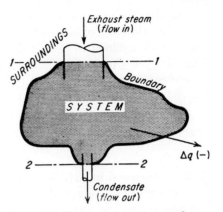

Fig. 5-15. System of surface condenser and single flow stream of steam and con-densate. (Flow stream of cooling water is in the surroundings.)

The system of surface condenser and flow stream of steam-condensate or else the flow stream of steam-condensate alone is diagrammatically shown in Fig. 5-15. Since work is zero and the changes in kinetic and potential energies are negligible relative to the amount of heat, Eq. (5-7d) reduces to

$$\Delta q = h_2 - h_1 \qquad \left[\frac{\text{Btu}}{\text{lb}_m}\right]$$

**5-9. The Steam Power Cycle.** A simple steam power plant is shown in Fig. 5-16. Here water, pumped into the boiler by the centrifugal pump, is evaporated into steam while heat is supplied at a high tempera-

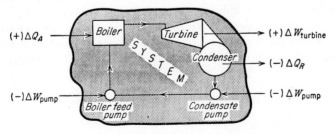

Fig. 5-16. Steam power cycle.

ture. The steam flows through the turbine, and work is produced. It then passes into the condenser, where it is condensed into water and heat is rejected at low temperature to the surroundings. The cycle is completed by pumping the condensate back to the boiler feed pump.

Note that the over-all system in Fig. 5-16 is closed, although each process within the system is steady-flow. Thus, a number of open systems have been combined to form a closed system, which executes a thermodynamic cycle (Art. 4-4). This was made possible by locating the furnace (where combustion took place) in the surroundings. For the system of Fig. 5-16, the thermal efficiency is defined by Eq. (4-6),

$$\eta_t = \frac{\Sigma \, \Delta W}{\Delta Q_A} \qquad (4\text{-}6)$$

And, as discussed in Art. 4-4, the thermal efficiency is *always* much less than 100 per cent, values of 5 to 40 per cent being usual.

Suppose that the furnace is included with the elements of Fig. 5-16 to form a new system (Fig. 5-17). This would be a steady-flow open system, and the work would equal (assuming $\Delta KE$, $\Delta PE$ to be zero)

$$-\Delta w = h_2 - h_1 - \Delta q_R$$

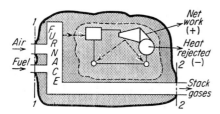

FIG. 5-17. Steady-flow system of furnace and cycle.

Observe that no heat is added to this open system and that the change in enthalpy $(h_2 - h_1)$ includes the chemical energy released by combustion (Art. 4-3f). Observe, too, that Eq. (4-6) does not apply since no thermodynamic cycle takes place [not to mention the fact that $Q_A$ is zero and therefore Eq. (4-6) would yield an infinite value].

**5-10. Non-steady-flow Problems.**† To solve Eq. (5-4e) in the general case is difficult, but for a restricted class of processes, simple time functions can be assigned (explicitly or implicitly) to allow an easy solution. A case of importance occurs when there is one flow boundary with $e_f$ being constant while the energy $e$ of the system is all in internal form $(u)$. Here Eq. (5-4e) reduces, with the help of the continuity equation [Eq. (5-5a)], to

$$\Delta Q - \Delta W = m_2 u_2 - m_1 u_1 - (m_2 - m_1)e_f \Big]_{e_f = C} \qquad (5\text{-}4f)$$

† Continued in Art. 12-1.

Also, the mass of fluid within the system equals

$$m = \frac{V}{v}$$

And if the system is of constant volume, the ratio of $m_2$ to $m_1$ is

$$\frac{m_2}{m_1} = \frac{v_1}{v_2}\bigg]_{V=C}$$

*Filling Processes.* Consider the system in Fig. 5-18. Here a *tank* (the system) is to be filled with fluid from a large *reservoir* (the surroundings). Work may be transferred by change in volume of the tank as well as by internal electrical or mechanical devices. The process is not necessarily adiabatic. However, if the reservoir is large relative to the system,

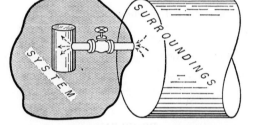

FIG. 5-18. The filling process from an infinite reservoir.

the properties of the entering fluid stream will be essentially constant and Eq. (5-4f) is applicable.

**Example 3.** A small evacuated container is connected to a large steam line that contains saturated steam at 100 psia. If the container is filled with steam to the line pressure, what will be the final state of the steam? (Assume that the heat capacity of the container and the heat transferred are zero.)

*Solution.* Since the container is empty at time zero, $m_1 = 0$ and $u_1 = 0$, while $e_f = h$, the enthalpy of the fluid in the reservoir. Hence, Eq. (5-4f) reduces to

$$m_2 u_2 = m_2 e_f$$

From Table B-4, for 100 psia saturated steam,

$$h = 1187.2 \text{ Btu/lb}_m \qquad t = 328°F$$

The final conditions in the tank are

$$p_2 = 100 \text{ psia} \qquad u_2 = e_f = 1187.2 \text{ Btu/lb}_m$$

Interpolating in Table B-4 for these values,

$$t_2 \approx 540°F \qquad (p_2 = 100 \text{ psia}) \qquad Ans.$$

The temperature rise from compression of the steam in the container is

$$540° - 328° = 212°F$$

Thus, the steam in the container is highly superheated although the steam in the reservoir is not.

**Example 4.** At the end of the exhaust stroke in an internal-combustion engine, the combustion chamber is filled with exhaust gas of mass $m_1$ and internal energy $u_1$. The piston then moves out on the intake stroke, and the system (the gases in the cylinder) enlarges from $V_1$ to $V_2$ while air and fuel are inducted. If all these processes occur at the constant pressure of the atmosphere, $p$, and if heat transfers are zero, determine an equation for the state of the mixture at the end of the intake stroke.

*Solution.* The work done by the expanding system is

$$\Delta W = \int_1^2 p \, dV = p(V_2 - V_1)$$

Also, by definition,

$$m_2 = \frac{V_2}{v_2} \qquad m_1 = \frac{V_1}{v_1}$$

or

$$\Delta W = p(m_2 v_2 - m_1 v_1)$$

With Eq. (5-4f),

$$-p(m_2 v_2 - m_1 v_1) = m_2 u_2 - m_1 v_1 - (m_2 - m_1)e_f$$

And this reduces to

$$m_2 h_2 = m_f e_f + m_1 h_1 \qquad Ans.$$

## PROBLEMS

Draw a diagrammatic sketch of the system for each problem with the boundary emphasized; label *system* and *surroundings;* show directions of heat and work by arrows; designate numerical values of all variables. Note that the continuity equation may be required for the solution.

**1.** A system has a mass-flow rate of 1 $lb_m$/sec. The enthalpy, velocity, and height at the entrance are, respectively, 100 Btu/$lb_m$, 100 ft/sec, 300 ft. At exit, these quantities are 99 Btu/$lb_m$, 1 ft/sec, $-10$ ft. Heat is transferred to the system of amount 5 Btu/sec. How much work is done by this system (a) per pound of fluid flow, (b) per minute, (c) in horsepower?

**2.** Repeat Prob. 1, changing the entrance and exit conditions to $h_1 = 1000$ Btu/$lb_m$, $V_1 = 10$ ft/sec, $Z_1 = 100$ ft, $h_2 = 1030$ Btu/$lb_m$, $V_2 = 10$ ft/sec, $Z_2 = 50$ ft.

**3.** On entering a system, the pressure is 100 psia, specific volume 3 ft³/$lb_m$, and internal energy 900 Btu/$lb_m$. On leaving, the pressure is 90 psia, specific volume 4 ft³/$lb_m$, and internal energy 850 Btu/$lb_m$. If the process is adiabatic and without change in kinetic or potential energy, how much work can be done?

**4.** A fluid enters a system with a velocity of 10 ft/sec through a 6-in.-diameter round pipe. The enthalpy is 1000 Btu/$lb_m$, internal energy is 900 Btu/$lb_m$, and the pressure is 100 psia. At exit, the enthalpy is 900 Btu/$lb_m$. If the process is adiabatic, while the changes in kinetic and potential energies are negligible, find the rate of work in horsepower.

**5.** Water enters a boiler with enthalpy of 60 Btu/$lb_m$, and steam leaves with enthalpy of 1200 Btu/$lb_m$. How much heat was transferred? Define the system.

**6.** From Fig. 5-5, define a system that will allow steam to be generated without flow of *heat* to your system. What is transferred in this double-circuit flow system? Set up an equation for the system of combustion gases alone in Fig. 5-5.

**7.** From Fig. 5-8, define a system that will be essentially adiabatic. How many flow circuits are involved?

**8.** An air compressor compresses 100 cfm of air with specific volume of 12 ft$^3$/lb$_m$; the enthalpy of the air increases 300 Btu/min, while the enthalpy of the cooling water increases 20 Btu/lb$_m$ of air delivered. Neglecting changes in kinetic or potential energy, find the horsepower required by the system.

**9.** A centrifugal air compressor compresses 5 lb$_m$/min of air from an initial pressure of 14.7 psia to a final pressure of 150 psia. The change in enthalpy of the air is 25 Btu/lb$_m$. How much work is required to drive the compressor?

**10.** A gas expands in a nozzle with the change of enthalpy equal to $-50$ Btu/lb$_m$. What will be the velocity of the fluid if the initial velocity is zero? If the initial velocity is 100 ft/sec, what will be the final velocity?

**11.** A fluid passes through a device wherein the velocity increases from 15 to 1,000 ft/sec without transfers of heat or work. What is the value for the change in enthalpy? The outlet velocity is dissipated by friction and turbulence until the initial value for velocity of 15 ft/sec is regained. What is the value for the change in enthalpy when this condition is reached?

**12.** As the flow jet passes across a turbine blade without friction, its velocity is reduced from 1,500 to 500 ft/sec. How much work can be done by the system of blade and fluid?

**13.** As the flow jet passes across a turbine blade with friction present, its velocity is reduced from 1,200 to 300 ft/sec while the enthalpy increases 10 Btu/lb$_m$. How much work can be done by the system?

**14.** The enthalpy of the fluid entering a reaction turbine is 1000 Btu/lb$_m$, while at the exit the enthalpy is 900 Btu/lb$_m$. How much work can be done by this system if changes in kinetic and potential energies are negligible and the system is essentially adiabatic? How can this problem be solved, since you do not know the sequence of processes or the mechanism within the turbine?

**15.** Steam enters a condenser with enthalpy of 1000 Btu/lb$_m$, and condensate leaves the condenser with enthalpy of 80 Btu/lb$_m$. The cooling-water circuit has a temperature increase from entrance to exit of 10°. If this temperature rise corresponds to an enthalpy increase of 10 Btu/lb$_m$ of cooling water, how many pounds of cooling water are needed to condense 1 lb$_m$ of steam?

**16.** Steam enters a four-pass condenser with enthalpy of 1000 Btu/lb$_m$ and velocity of 300 ft/sec, while the condensate leaves with enthalpy of 80 Btu/lb$_m$ and negligible velocity. How much heat must be transferred from the system of steam and condenser? Repeat, assuming a two-pass condenser.

**17.** The heat transferred to a power cycle is 1200 Btu/lb$_m$, while the heat rejected is 800 Btu/lb$_m$ and the pump work is 5 Btu/lb$_m$. What is the value for thermal efficiency?                                                     *Ans.* $\eta_t = 33\frac{1}{3}\%$.

**18.** The heat rejected from a power cycle is 400 Btu/lb$_m$, while the work done by the turbine is 200 Btu/lb$_m$ and the pump work is 5 Btu/lb$_m$. Calculate the thermal efficiency.

**19.** Fluid enters a turbine with velocity of 5 ft/sec and enthalpy of 900 Btu/lb$_m$ and leaves with enthalpy of 800 Btu/lb$_m$ and velocity of 300 ft/sec, while the heat loss is 30 Btu/min and flow rate is 1 lb$_m$/sec. What horsepower will be developed?

**20.** The absolute velocity of the fluid leaving the nozzle is 1,000 ft/sec, and the velocity of the blade is 500 ft/sec.

(*a*) If the angle of the nozzle to the blade is 12°, what angle will the relative velocity vector make to the axis of the blade (graphical solution)?

(*b*) Suppose the fluid leaves the blade with a relative velocity only 0.9 that of the entering relative velocity. What will be the value for the work done by the blade?

(*c*) If no friction is present, how much work will be done?

**21.** Water is drawn from a lake, pumped up to a city 486 ft above lake level, and forced through the nozzles of a fire hose at 64 ft/sec. Neglecting friction, what horsepower is theoretically needed to deliver 5,000 $lb_m$/sec of water?

**22.** An insulated, rigid tank with zero heat capacity contains 10 $lb_m$ of air (ideal gas) at 20 psia and 100°F. It is quickly filled with air from a large reservoir at 200 psia, 100°F. Determine the final state of the air in the tank and the amount added.

**23.** A bottle of 2 ft³ capacity is evacuated to 0.3 psia with the temperature that of the room, 70°F. The bottle is uncorked, and air (ideal gas) rushes in, restoring the pressure to atmospheric. What is the final temperature of the air in the bottle, assuming losses to be zero?

**24.** The intake valve on an automotive engine is not opened until the intake stroke is over, and then the cylinder fills with air (ideal gas) to atmospheric pressure. Assume that 0.0004 $lb_m$ of air is in the cylinder at $p = 4$ psia, $t = 20$°F when the valve opens. Calculate the final temperature if heat losses are neglected for outside air at 14.7 psia, 25°F.

**25.** Repeat Example 4, but (*a*) for a supercharged engine ($p_a > p_0$) and (*b*) for a throttled engine ($p_a < p_0$) with constant $p_a$.

## REFERENCE

1. Obert, E. F.: "Concepts of Thermodynamics," chaps. 5, 15, McGraw-Hill Book Company, Inc., New York, 1960.

CHAPTER 6

# THE REVERSIBLE PROCESS

Though the mills of the gods grind exceedingly
slow, yet they grind exceedingly fine.

*Von Logau*

In his search for the most efficient way to produce work from heat, the engineer must examine each process for imperfections. *From experience, he has found it easy to convert work into heat but impossible to convert heat into work completely and continuously.*† For this reason, he must avoid processes that allow work, or even a latent capability to do work, to be dissipated in friction or turbulence.

**6-1. Available Energy.** Energy can exist in many forms, but not all these forms can be converted completely into work, even under ideal conditions. Thus, energy is considered to have *available* and *unavailable* parts:

**Available energy is that portion of energy which could be converted into work by ideal processes which reduce the system to a dead state—a state in equilibrium with the earth and its atmosphere.**

Suppose that a closed system contains energy in several forms. This energy would be capable of doing work if the intensity factors for the various forms of energy were different from those of the ultimate surroundings for all systems, the earth and its atmosphere. (Here *atmosphere* implies the rivers and lakes as well as the air of the earth.) Consider, first, the sensible internal energy of a system. If this system has a pressure different from that of the surroundings, work can be obtained from an expansion process, and if the system has a different temperature, heat can be transferred to a cycle and work can again be obtained. But when the temperature and pressure become equal to those of the earth, transfer of energy ceases, and although the system contains internal energy, this energy is *unavailable*.

---

† A statement that can be called an axiom of the Second Law. This chapter discusses such axioms of experience that, in the next chapter, are formalized into a general theorem called the *Second Law*.

83

When the available energy of a flow system is considered, the defined form of energy called *flow energy* (Art. 5-3) must also be included even though the source of the flow energy lies in some other system. Nevertheless, the flow energy of the stream is available to do work and should properly be included as a part of the available energy of the system. Here, as in the case of expansion energy, the available part of flow energy is that over and above the energy necessary to overcome the pressure of the surroundings.

Potential and kinetic energies are considered to be entirely available energy because the properties of height and velocity are measured relative to the surface of the earth. If, for example, the datum of zero potential energy were to be located at the center of the earth, potential energy would be, to a great extent, unavailable energy. In close similarity to potential energy, electrical energy is also entirely available, since the intensive property of electrical potential is measured as a difference in potential (Art. A-1).

The concept of available energy is somewhat qualitative since the transformation of energy into work depends upon the type of system as well as on the surroundings and the allowed interactions between system and surroundings (Arts. 12-2 and 12-3).

Thus, *available energy* denotes the latent capability of energy to do work, and in this sense it can be applied to energy in the system or in the surroundings (since the surroundings are but another system); it is also a metonym for work in cases where the boundaries are not well delineated.

**6-2. The Ideal, or Reversible, Process.** Work must be a more precious form of energy than heat because only a fraction of the heat supplied to a cycle can be transformed into work. Work is more adaptable than heat. To accomplish the same objective, work can always be substituted for heat (turbulence and frictional effects), but the reverse is not true: heat cannot always be substituted for work. Evidently, the clue to perfection of a process is to ensure that work is not being used to effect a change that could be as readily attained by transfer of the cheaper commodity, heat:

> **A process is ideally performed when neither work nor available energy is used to cause a change that could have been accomplished in whole or in part by transfer of heat.**

It then follows immediately that the test of perfection is this: *If the process is stopped and made to retrace its steps by returning to the system or the surroundings the work and heat previously delivered, only the ideal process can do this and restore both system and surroundings to their initial states.* Hence, the ideal process must be a reversible process, and therefore this new name is adopted as a synonym for perfection:

> **The ideal process is called a reversible process.**

Note that many real processes can be stopped, reversed in direction, and the original state of the *system* regained, but these real processes are *not* reversible in the thermodynamic sense since examination will invariably show that either work or available energy *of the surroundings* has been dissipated in a heating effect.

On the other hand, the ability of the real process to undo itself—to reverse its procedures—is the initial guide in the search for perfection. For example, if a gas is adiabatically compressed by a piston in a cylinder, at any stage of the compression the process can be reversed in direction and *almost* all the work put in can be regained on the expansion process. The process approaches reversibility. But if the brakes on a car are applied, kinetic energy is dissipated into a heating effect of the brake drums. The process is highly irreversible since the available energy corresponding to the decrease in velocity was entirely dissipated in friction.

It should be carefully noted that *reversibility is a limit that cannot be attained by the real process.* For the real process, a difference must exist in some property, such as pressure or temperature, if energy is to be transferred. If such differences were to be made vanishingly small, the real process would require an infinite time for completion.† Certain processes, however, can approach reversible operation. *The task of the engineer is to recognize and evaluate the factors preventing reversibility and to select those processes which, while not reversible, at least have adequate justification for the irreversibilities present.*

**6-3. Reversible and Irreversible Processes.** In Fig. 6-1 is shown an adiabatic system of fluid alone. If the piston moves at a slow speed, essentially the same pressure will exist at any instant in all parts of the chamber. Now assume that the piston moves at a rate such that a pressure difference is present between chambers $A$ and $B$. A lower pressure will be exerted on the piston than was experienced before, and a lesser amount of work will be done. If no pressure gradients were present (and therefore no fluid turbulence), if the system were truly adiabatic (no temperature gradients), the piston could be stopped and reversed with the gas being compressed from state to state in the exact inverse order of the original expansion by returning to the system the work previously

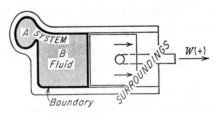

Fig. 6-1. Adiabatic system of fluid alone. (Similar to a precombustion-chamber diesel engine.)

---

† But the reversible process, being a hypothetical process, can be operated at any speed.

delivered to the surroundings. This would be a reversible process:

**A reversible process must pass through the same states on the reversed path as were initially visited on the forward path.**

Since the system *and* surroundings were returned to their initial states:

**A reversible process, when undone, will leave no history of the events in the surroundings.**

In the irreversible expansion process (Fig. 6-1), the fluid was highly turbulent because of the variation in pressure; therefore, definition of a single state for the system at any instant was impossible. It follows that

**A reversible process must pass through a continuous series of equilibrium states.**

Consider a reversible and an irreversible expansion process receiving heat from a single constant-temperature source. For the reversible isothermal process, the hypothetical transfer of heat takes place without temperature difference between source and system. For the irreversible process, the transfer of heat is taking place across a finite temperature difference. Then, at any stage of the expansion, the fluid in the irreversible process will be at a lower temperature than that in the reversible process, and therefore the pressure† will be lower. The work done by the irreversible process will be less than that of the reversible process. Similar reasoning shows that more work will be required for the irreversible compression process (relative to the reversible compression) to achieve the higher temperature necessary to establish a temperature difference between the system and surroundings.

**A process is irreversible if heat is transferred through a finite temperature difference.**

The examples have emphasized that reversible processes can be stopped at any instant and restored to their original states. This concept is more difficult to visualize when applied to open systems under steady flow, where kinetic effects usually lead to irreversibilities. Despite this difficulty, such processes can be examined to see how closely the original state can be regained by inverting or reversing the steps or stages of the process. For example, in flow through a nozzle the pressure decreases while velocity increases. Then, if the high-velocity fluid is led through a *diffuser* (Fig. 6-2), the velocity can be reduced to its original value and the pressure increased *toward* its original value. If expansion

---

† Let the substance be an ideal gas. Here $p = mRT/V$; hence, at equal stages of the expansion, the gas at the lower temperature has the lower pressure.

through the nozzle were reversible, then an ideal diffuser could restore the initial state.

Note that a diffuser is merely a reversed nozzle but of longer length. The combination of nozzle and diffuser is called a *venturi*.

It is helpful to divide the irreversibilities present in a system into two classes, internal and external. *An internal irreversibility occurs within the system and arises because of any unbalance, such as temperature and pressure gradients, that prevents attainment of an equilibrium state. An external irreversibility occurs at the boundary of the system and arises because of friction or because of any unbalance across the boundary, such as a temperature difference.* By redefining the system, external irreversibilities can be made a part of the surroundings, and the new system may approach reversible operation (if internal irreversibilities are negligible).

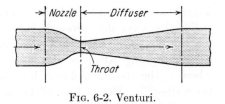

Fig. 6-2. Venturi.

With the concept of reversibility in mind, it would appear that the real process should be operated at an exceedingly slow pace. But consider the expansion of a hot gas. If the process is slowly executed, losses from turbulence will be low, but unavoidable heat losses, high. Thus the ideal expansion process for the production of power is universally considered to be adiabatic, and the adiabatic process is best approached by high-speed operation. When the transfer of heat is studied, a large temperature difference dictates a loss of available energy but the speed of heat transfer is increased by the temperature difference and therefore the size (and the cost) of the equipment is reduced. Hence, the engineer must weigh one advantage against the other and, in the final analysis, select the process and equipment which yield the greatest saving of money.

**6-4. Mechanical Work—Closed System.** The work of expansion or compression arises from the pressure of the system moving the boundary against a resistance offered by the surroundings,

$$\text{Work} = (\text{force})(\text{distance})$$

The force equals the product of the moving area $A$ and the *indicated pressure* $p_{im}$ at the boundary. Since $p_{im}$ is not necessarily constant, the approximate change in the work function for the system from time $\tau_1$ to $\tau_2$ is

$$\Delta W \approx p_{im} A \, \Delta x \qquad (a)$$

By dividing by $\Delta \tau$ and then allowing $\Delta \tau$ to approach zero (noting that

$A \, \Delta x$ is the change in total volume $\Delta V$ of the system),

$$\frac{dW}{d\tau} = p_{\mathrm{im}} \frac{dV}{d\tau} \tag{6-1a}$$

Or, by multiplying by $\Delta\tau$, by definition of $dW$ [Eq. (A-4a)],

$$dW = p_{\mathrm{im}} \, dV \tag{6-1b}$$

Equations (6-1) apply to both reversible and irreversible processes.

For the reversible process, $p_{\mathrm{im}}$ is not only the pressure (force) at the moving boundary but also the pressure (a thermostatic property) of the entire system. Also, for the reversible process, the volume equals $mv$, where $v$, the specific volume, has but one value throughout the system (at a time $\tau$). Hence Eqs. (6-1) for the reversible process become

$$\frac{dW_{\mathrm{rev}}}{d\tau} = mp \frac{dv}{d\tau} \tag{6-2a}$$

$$dW_{\mathrm{nf,rev}} = mp \, dv \tag{6-2b}$$

And integrating from $\tau_1$ to $\tau_2$ (corresponding to $v_1$ and $v_2$),

$$\Delta W_{\mathrm{nf,rev}} = m \int_{v_1}^{v_2} p \, dv \tag{6-2c}$$

Although Eqs. (6-1) and (6-2) are similar, note that $p$ and $v$ in Eqs. (6-2) are properties of the entire system; the corresponding variables $p_{\mathrm{im}}$ and $V$ 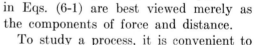 in Eqs. (6-1) are best viewed merely as the components of force and distance.

To study a process, it is convenient to show the change on a diagram that has properties as coordinates. The $pv$ diagram (Fig. 6-3) is important because the area under the pressure-volume path *may be* work, for this area is $\int p \, dv$. Since each state on the diagram presupposes an equilibrium state, a continuous path (1-2) can be rigorously interpreted only as an internally reversible process.[†] This is not as serious a restriction as it might seem. For fluids compressed or expanded within a cylinder by a piston, the internal irreversibilities may be quite negligible. However, for fluids compressed by centrifugal means, the $pv$ diagram should be used only to show the initial and final states. These two states can be located and connected by a dashed line (1-3) to indicate the uncertainty

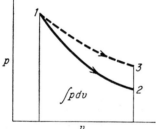

FIG. 6-3. Pressure–specific-volume diagram $(pv)$   1-2, reversible, and 1-3, irreversible expansion processes.

---

† Real processes, however, that approach reversibility may be shown as solid lines in subsequent figures.

of the path, for in such processes there may be finite differences in pressure, temperature, specific volume, and other properties, and a single state for the system is impossible to define. In other cases of irreversibilities, it may be possible to stop the process (if only in imagination) at frequent intervals and so attain an equilibrium state. The path can be approximated in this case by connecting the known (or imagined) equilibrium states and representing the process, as before, by a dashed line.

**The irreversible process is usually shown as a discontinuous line on the $pv$ diagram.**

**Example 1.** During a reversible and nonflow expansion process, the pressure and specific volume are observed to be related by the relationship $pv = C$ (constant). Derive a formula for the work done.

*Solution.* $\Delta w_{\text{rev}} = \int p\, dv$ and $p = C/v$.

$$\Delta w_{\text{rev}} = C \int_{v_1}^{v_2} \frac{dv}{v} = C \ln \frac{v_2}{v_1} = p_1 v_1 \ln \frac{v_2}{v_1} = p_2 v_2 \ln \frac{p_1}{p_2} \qquad Ans.$$

Work can also be obtained from changes in either kinetic or potential energies. Since these forms of energy are fully available (Art. 6-1), complete conversion is the reversible work. Hence, the ideal mechanical work of the closed system from expansion and from changes in $KE$, $PE$ is

$$dW_{\text{nf,rev}} = mp\, dv - d(KE) - d(PE) \qquad (6\text{-}2d)$$

$$\Delta W_{\text{nf,rev}}_{\tau_1 \text{ to } \tau_2} = m \int_{v_1(\tau_1)}^{v_2(\tau_2)} p\, dv - \Delta KE - \Delta PE \qquad (6\text{-}2e)$$

where, again, time is the independent variable.

**6-5. Mechanical Work—Steady-flow System.** For the simple case of steady flow, the ideal work can be readily deduced. Consider the flow of unit mass of substance through a steady-flow system. The work delivered to the surroundings equals

$$-\Delta w_{\text{sf}} = \Delta h + \Delta KE + \Delta PE - \Delta q \qquad (5\text{-}7d)$$

(and the increments are for functions of length). Now *follow* the *same* unit of mass (that is, consider it to be a closed system) as it passes from $L_1$ to $L_2$ in space. The work of this closed system involves the fluid behind and the fluid ahead of the element under observation as well as the net work delivered to the surroundings. The gross work, however, equals

$$-\Delta w_{\text{nf}} = \Delta u + \Delta KE + \Delta PE - \Delta q \qquad (4\text{-}4d)$$

(and the increments are for functions of time). Although the increments in Eqs. (4-4d) and (5-7d) are for different independent variables, the *numerical* values of $\Delta u$, $\Delta h$, $\Delta KE$, $\Delta PE$, and $\Delta q$ are the same for either system since the same element was observed. Thus the mechanical work of the steady-flow system differs from that of the closed system by the numerical difference between $\Delta u$ and $\Delta h$,

$$\Delta w_{sf} - \Delta w_{nf} = -\Delta pv \Big]_{\substack{\text{rev or} \\ \text{irrev}}} \qquad (a)$$

For the reversible process, $\Delta w_{nf}$ can be found from Eq. (6-2e), which must yield the same answer whether integrated from $\tau_1$ to $\tau_2$ or from $L_1$ to $L_2$; therefore, by Eq. (a),

$$\Delta w_{\substack{\text{steady-flow} \\ \text{reversible} \\ L_1 \text{ to } L_2}} = \int_{v_1(L_1)}^{v_2(L_2)} p \, dv - \Delta pv - \Delta KE - \Delta PE \qquad (6\text{-}3a)$$

Since all terms in Eq. (6-3a) are functions of length, it can be treated in the usual manner to yield

$$dw_{sf} \atop \text{rev} = p \, dv - d(pv) - d(KE) - d(PE) \qquad (6\text{-}3b)$$

$$dw_{sf} \atop \text{rev} = -v \, dp - d(KE) - d(PE) \qquad (6\text{-}3c)$$

Thus, the mechanical work of the steady-flow process is derived from expansion energy, flow energy, kinetic energy, and potential energy.

**6-6. Process Equations.** The sequence of states followed by a process of a closed system is dictated by the First Law energy equation,

$$dq - dw = de \Big]_{\substack{\text{closed system} \\ \text{rev or irrev}}} \qquad (4\text{-}4c)$$

If the work of the system is mechanical and reversible, by Eq. (6-2d),

$$dq_{rev} = du + p \, dv \qquad (6\text{-}4a)$$

If the pressure is constant,

$$dq_{rev} = dh]_{p=C} \qquad (6\text{-}4b)$$

If the volume is constant,

$$dq_{rev} = du]_{v=C} \qquad (6\text{-}4c)$$

When the derivations are repeated but for a steady-flow process (Probs. 16 and 17), the same equations are obtained; therefore, *Eqs.* (6-4) *hold for the reversible processes of either closed or steady-flow systems* (Art. 7-4) (but with different independent variables).

For the adiabatic process, $dq$ is zero in Eq. (6-4a), and

$$du = -p \, dv \qquad (a)$$

Observe that Eq. (a) involves only thermostatic properties that are evaluated by an observer at rest relative to the substance; therefore, Eq. (a) holds for either closed or steady-flow systems without inquiring as to the independent variables (Art. 2-11).†

† Problem 20.

Equation ($a$) can be solved for an ideal gas by substituting Eqs. (3-5$a$), (3-8$b$), and (3-9) and rearranging,

$$-\frac{1}{k-1}\frac{dT}{T} = \frac{dv}{v} \tag{b}$$

Integration yields, with $k$ considered constant,

$$\frac{T_2}{T_1} = \left(\frac{v_1}{v_2}\right)^{k-1} \tag{6-5a}$$

And, with the ideal-gas equation of state,

$$\frac{T_2}{T_1} = \left(\frac{p_2}{p_1}\right)^{\frac{k-1}{k}} \tag{6-5b}$$

Equations (6-5) are the relationships among temperature, specific volume, and pressure for the reversible adiabatic process of an ideal gas. Comparison of these equations with Eqs. (3-7) shows that the path of the reversible adiabatic process can be designated by $k$,

$$pv^k = C \tag{6-5c}$$

Equations that involve only heat or only work must always be examined to see whether or not the restriction of reversibility is present. For example, Eqs. (4-4) hold for all processes, but not Eqs. (6-2) or (6-4). *Once both heat and work disappear from the equation, the question of reversibility is left unanswered.* Thus Eqs. ($a$), ($b$), and (6-5) definitely hold for the reversible adiabatic process of an ideal gas because of the restrictions of the derivation, but they may also hold for certain (but not all) irreversible processes.† This is because the distinction between heat and work is a matter of definition and the properties of the system cannot distinguish, for example, between a reversible flow of heat and a frictional dissipation of work.

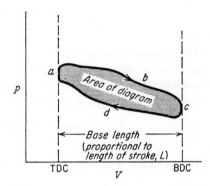

Fig. 6-4. Indicator diagram or pressure-volume diagram ($pV$). TDC, top dead center (inner limit of piston stroke); BDC, bottom dead center (outer limit of piston stroke).

**6-7. Indicated Work.** Reciprocating-piston machines are used for many purposes (Art. 5-8$d$). Since such machines are continuously operated, *a mechanical cycle* is completed and the atmospheric work on one stroke of the piston is canceled by that on the return stroke (Art. 4-2). The *indicated work* of the mechanical cycle can be calculated from the

† Article 11-3.

*indicator diagram* (Fig. 6-4). This diagram is a record of the pressures in the cylinder at each stage of the cycle drawn to scale by an *indicator*. Whether the real engine process approaches reversibility or not, the indicator, if precise, will yield a diagram with area proportional to the work done *on* or *by the fluid* if the pressure recorded by the indicator is essentially that experienced by the piston (since the indicated area is merely an integration of force through distance). For the usual locations where the indicator is tapped into the cylinder, even the fastest engines will not create a significant pressure difference between indicator and piston.

The indicated work of the real steady-flow system for one cycle of operation is found by integrating Eq. (6-1b) over the cycle of Fig. 6-4,

$$W_{\substack{\text{indicated} \\ \text{rev or irrev}}} = \oint p_{\text{im}} \, dV \qquad (6\text{-}6a)$$

The integration of Eq. (6-6a) is graphically made by measuring the enclosed area of the indicator diagram and calculating an average pressure. The net average pressure is called the *indicated mean effective pressure* (imep) for the diagram, and it is the difference of the average pressures for each stroke of the piston,

$$p_{\text{im}} = \text{imep} = \frac{\text{closed area of diagram}}{\text{length of base}} \times \text{ordinate scale} \qquad (6\text{-}6b)$$

The indicated mean effective pressure can be defined as that constant pressure which, by acting alone throughout one stroke, can perform the same amount of work as the varying pressure does in the complete cycle of strokes. Thus, the indicated work of one diagram is

$$W_{\text{im}} = (\text{imep})(\text{area of piston})(\text{length of stroke})$$
$$= p_{\text{im}} A L$$

The product $AL$ is called the *piston displacement* $V_D$, and for one diagram

$$W_{\text{im}} = p_{\text{im}} V_D \qquad (6\text{-}6c)$$

(The usual units are $p_{\text{im}}$ in pounds force per square inch, $V_D$ in cubic inches, and $W_{\text{im}}$ in foot-pounds force.)

The indicated work of an engine is greater than the work delivered to the shaft because of friction within the mechanism. Conversely, if the machine does not deliver work but absorbs work (an air compressor, for example), the indicated work is less (without regard for sign) than the shaft work. In either case, the difference is the friction work (let the algebraic sign of $W_{\text{friction}}$ be always positive while that of $W_{\text{shaft}}$ is positive or negative),

$$W_{\text{im}} = W_{\text{shaft}} + W_{\text{friction}} \qquad (6\text{-}6d)$$

The *indicated horsepower* is defined as

$$\text{ihp} \equiv \frac{p_{\text{im}}LAN}{33,000} \qquad (6\text{-}6e)$$

where $N$ = number of mechanical cycles per minute.

**6-8. Other Work Problems.** Although thermodynamics is primarily concerned with chemical systems, the thermodynamic method is applicable to all systems, the main objective, however, being to arrive at property interrelationships rather than to study energy transformation. For example, consider a bar or wire of length $L_0$ and cross-sectional area $A_0$ at time $\tau_0$. Let an axial force $F$ be applied to cause an axial displacement $\Delta L$. Then for reversible or irreversible processes with or without heat transfer,†

$$dW_{\substack{\text{rev or irrev} \\ \tau_1 \text{ to } \tau_2}} = -FdL \qquad (a)$$

If internal friction is absent (elastic bodies), then for pure tension or compression the stress ($\sigma = F/A_0$) and the unit strain ($\Delta\epsilon = \Delta L/L_0$) are single-valued throughout the bar at time $\tau$. By dividing ($a$) by $A_0 L_0$,

$$dw_{\substack{\text{rev} \\ \tau_1 \text{ to } \tau_2}} = -\sigma\, d\epsilon \qquad \left[\frac{\text{work}}{\text{unit volume}}\right] \qquad (6\text{-}7a)$$

To integrate Eq. (6-7$a$), the relationship between stress and strain for the process must be known. Although an infinity of relationships are theoretically possible, only the adiabatic and constant-temperature processes are practical. A *modulus of elasticity* is defined as

$$E \equiv \frac{\partial\sigma}{\partial\epsilon}$$

although the usual practice in engineering is to express the modulus as a constant (implying an exact linearity between stress and strain). For example, Young's isothermal modulus is generally shown as

$$E_T = \frac{\sigma}{\epsilon}$$

The solid system also has a work term arising from changes in volume of the bar against the pressure of the surroundings. But practically, this term is extremely small, and, more important theoretically, to bring it into the discussion would introduce an inconsistency: The system was

† Tensile forces, stresses, and strains are defined positive in sense, and compressive forces, stresses, and strains are defined negative, so that the conventions for the algebraic signs of work are obeyed (Art. 2-13).

restricted to pure tension or compression. Therefore, Eq. (6-7a) alone is substituted into Eq. (4-4c) to obtain the energy equation for an idealized solid system:

$$dq_{rev} = du - \sigma\, d\epsilon \qquad \left[\frac{\text{energy}}{\text{volume}}\right] \qquad (6\text{-}7b)$$

The stress $\sigma$ depends on the strain, which is also influenced by the temperature. Hence the "equation of state" for the idealized solid system is

$$f(\sigma,\epsilon,T) = 0 \qquad (6\text{-}7c)$$

Observe that the number of independent thermostatic properties has been decreed to be two [Eq. (6-7c)]. Hence the internal energy (and other thermostatic properties) is a function of any two of the properties in Eq. (6-7c).

The *electrical potential* $V$ is defined (Art. A-1) as a definite quantity of available energy per unit charge; therefore electrical work is the product of the potential difference $V_{ab}$ and the number of charges. Since $V_{ab}$ is not necessarily constant, the work of the reversible or irreversible system equals

$$dW_{\substack{\text{rev or}\\\text{irrev}}} = V_{ab}\, dq \qquad \left[\frac{\text{joule}}{\text{coulomb}}\right][\text{coulomb}] \qquad (6\text{-}8a)$$

If the system is reversible, the potential difference $V_{ab}$ is equal in magnitude to the emf and reversibility is "signaled" by the symbol $\mathcal{E}$:

$$dW_{\substack{\text{rev}\\\tau_1 \text{ to } \tau_2}} = \mathcal{E}\, dq \qquad [\text{joule}] \qquad (6\text{-}8b)$$

The work of the closed system may arise from mechanical as well as electrical effects:

$$dW_{rev} = p\, dV + d(KE) + d(PE) + \mathcal{E}\, dq \qquad (6\text{-}8c)$$

The electrical work, $\mathcal{E}\, dq$, may be independent of other work [similar to $d(KE)$, $d(PE)$], as, for example, in a system containing an ideal electric motor and an uncharged gas. When the electrical work is not independent, the problem becomes involved since three independent thermostatic properties are required to fix the state. But for a galvanic cell, the electric charge need not be a variable (for the usages of thermodynamics), and a relatively simple energy equation can be obtained.

When fuel and air react in a combustion process, available energy is dissipated since the process is irreversible. In a number of cases, an essentially reversible chemical reaction can be approached by the use of a galvanic cell. For example, suppose that two different metals (the electrodes) are inserted into an electrolyte. The atoms of the metals have a tendency to form positive ions in the solution by leaving behind their negative charges (electrons). Since the process depends on the materials, one of the electrodes will be at a higher potential than the other, and a flow of current can therefore be obtained in an external circuit. A gas can be the electrode if an inert metal, such as porous platinum, is used as a holder, for example, hydrogen and oxygen electrodes in an aqueous sulfuric acid solution. When the cell discharges, the chemical reaction is in the direction

$$H_2 + \tfrac{1}{2}O_2 \rightarrow H_2O$$

And, if the cell is charged (the flow of current reversed), the opposite reaction takes place. The test of reversibility parallels that for mechanical work. An opposing potential, controlled by a potentiometer, is applied to the cell so that the flow of current is very small, and it is observed whether or not a change in direction of the current (corresponding to a change in direction of the chemical reaction) can be obtained without significant change in the applied potential. The galvanic cell approaches reversibility if it can pass this test. The balancing potential at the point of zero current (equilibrium) is equal, by definition, to the emf $\mathcal{E}$ of the chemical reaction. It is the maximum work that can be done per unit charge when the chemical reaction is proceeding reversibly (Art. A-1). Since $\mathcal{E}$ depends on the states of reactants and products (which may include the concentrations of solutions) as well as the temperature and pressure, it is a property.

If the galvanic cell is operated reversibly (that is, a very large cell is imagined to be operated at constant $T$, $p$, and $\mathcal{E}$), the electric charge concentration on the electrodes does not change; therefore, the net effect is to convert reactants from their initial states into products at their final states and push aside the atmosphere. Hence Eq. (6-8c) reduces to

$$\Delta W_{\substack{\text{rev} \\ \tau_1 \text{ to } \tau_2 \\ \text{net}}} \Big]_{T,p,\mathcal{E}} = \mathcal{E}\,\Delta q \qquad (6\text{-}8d)$$

Every ion with the same valence carries the same basic charge, either positive or negative, although the masses of the various ions are proportional to their respective atomic weights. The passage of 1 faraday† of charge will transfer 1 gram equivalent of substance (1 gram-atomic weight of univalent ion; $\tfrac{1}{2}$ gram-atomic weight of bivalent ion, etc.). If $n$ is the number‡ of faradays for the reaction, $\Delta q = n\mathbf{F}$, and Eq. (6-8d) reduces to

$$\Delta W_{\substack{\text{rev} \\ \tau_1 \text{ to } \tau_2 \\ \text{net}}} \Big]_{T,p,\mathcal{E}} = n\mathbf{F}\mathcal{E} \qquad (6\text{-}8e)$$

$$[\text{joule}] = [\text{coulomb}] \left[ \frac{\text{joule}}{\text{coulomb}} \right]$$

With the foregoing comments and procedures in mind, an equation of state§ for a system containing a reversible galvanic cell could be proposed of the form

$$f(p, T, \mathcal{E}, m_1, m_2, \text{ etc.}) = 0 \qquad (6\text{-}9a)$$

† Article A-1.
‡ Equal to the valence of the ions being transferred.
§ Another version of an *"incomplete"* equation of state (Art. 2-12).

wherein $m_1$, $m_2$, etc., are variables of the cell (such as concentrations) that affect $\mathcal{E}$. However, when $m_1$, $m_2$, etc., are held constant, for the hypothetical reversible operation, the "equation of state" is simply

$$f(p,T,\mathcal{E}) = 0 \tag{6-9b}$$

In some cases, the effect of pressure is negligible, and temperature and emf are the only variables.

By substituting Eqs. (6-2e) and (6-8e) into Eq. (4-4d),

$$\Delta Q_{rev}\Big]_{T,p,\mathcal{E}} = \Delta U + p\,\Delta V + nF\mathcal{E} \tag{6-10a}$$

$$\Delta Q_{rev}\Big]_{T,p,\mathcal{E}} = \Delta H + nF\mathcal{E} \tag{6-10b}$$

(Here capital letters are used to warn of the chemical reaction.) Equation (6-10b) is the energy equation, not for a galvanic cell being charged or discharged, but for the chemical reaction portrayed by the cell (continued in Art. 8-5).

## PROBLEMS

**1.** List, to the best of your knowledge, and explain all the possible irreversibilities that may be present in a system.

**2.** Divide the irreversibilities listed in Prob. 1 into internal and external irreversibilities.

**3.** A reciprocating-piston air compressor has a $100°$ difference in temperature between the air in the cylinder and that in the cooling water. If the *system* is the *air* enclosed in the compressor, can the compression process be considered to approach reversibility? Discuss.

**4.** For a reversible process of a closed system, the increase in total energy is 6 Btu while heat of amount 10 Btu is abstracted. Compute the work, and explain the algebraic sign.

**5.** An irreversible process that passes through the same series of states as the reversible process of Prob. 4 has 12 Btu of heat transferred to the surroundings. Compute the work.

**6.** A gas expands in a cylinder against a piston, and work is delivered to the surroundings. The gas is then compressed back to the exact initial state. Is this a reversible process? Discuss.

**7.** Potential, kinetic, and electrical energy are said to be entirely available energy. Discuss.

**8.** If the real process is to approach reversibility, must the process be slowly conducted? Discuss and list examples.

**9.** List a number of manufactured devices which have a highly irreversible process as the basis for their successful operation.

**10.** If an ideal gas is the fluid in Example 1, what is the name of the process?

**11.** A reversible process of a closed system has pressure and specific volume related by the equation $pv^n = C$ ($C$ and $n$ are constants). Derive a formula for the work.

**12.** Repeat Example 1, but assume an open system and steady flow.

**13.** Repeat Prob. 11, but assume an open system and steady flow.

**14.** Are Eqs. (3-7a) to (3-7c) valid for irreversible processes?

**15.** Convert the answers of Probs. 11 and 13 to other forms by means of Eqs. (3-7b) and (3-7c).

**16.** Calculate by means of the energy equations the heat transferred in the following processes: (a) a reversible constant-pressure process of a closed system; (b) a reversible steady-flow process; (c) an irreversible steady-flow process with $\Delta w$, $\Delta KE$, and $\Delta PE$ essentially zero. Discuss.

**17.** Calculate the heat transferred in reversible processes of a constant-specific-volume (a) closed system, (b) steady-flow system ($\Delta KE$, $\Delta PE \approx 0$). Discuss.

**18.** Determine the indicated work in foot-pounds force for a reciprocating-piston engine that has a bore (cylinder diameter) of 3 in. and a (piston) stroke of 4 in. (these data are usually abbreviated $3'' \times 4''$) if the area of the indicator diagram is 1.38 in.² The reducing motion that drives the drum of the indicator has a ratio of 2:1, and the ordinate scale is 100 psi/in.

**19.** Repeat Prob. 18, but assume that the engine dimensions are 4 by 5 in.

**20.** Show that Eqs. (a) and (6-5) of Art. 6-6 hold for a reversible adiabatic process of a steady-flow open system.

**21.** A Daniell cell has an emf of 1.0934 volts at 273°K for the reaction $ZnSO_4 + Cu \rightarrow Zn + CuSO_4$. How much work (Btu) is transferred to deposit 1 mole of copper? To deposit 1 mole of zinc? Discuss.

**22.** Compare meanings of potential difference and emf.

**23.** Discuss the units of each term in Eq. (6-8e).

**24.** The equation of state for the galvanic cell is sometimes shown as $f(\mathcal{E}, T, q) = 0$, where $q$ is the charge (and volume changes are negligible). Discuss.

**25.** Calculate the reversible isothermal work to stretch a wire from a unit strain of $\epsilon_1$ to $\epsilon_2$.

**26.** The equation of state for a solid in Art. 6-8 is "incomplete." What other data are necessary (to parallel the case of a chemical substance)?

## SELECTED REFERENCES†

1. Lewis, G. N., and M. Randall: "Thermodynamics and the Free Energy of Chemical Substances," McGraw-Hill Book Company, Inc., New York, 1961.

2. Guggenheim, E. A.: "Thermodynamics," North-Holland Publishing Co., Amsterdam, 1957.

3. Brønsted, J. N.: "Principles and Problems in Energetics," Interscience Publishers, Inc., New York, 1955.

4. MacDougall, F. H.: "Physical Chemistry," The Macmillan Company, New York, 1936.

5. Zemansky, M. W.: "Heat and Thermodynamics," 4th ed., McGraw-Hill Book Company, Inc., New York, 1957.

6. Bridgman, P. W.: "The Nature of Thermodynamics," Harvard University Press, Cambridge, Mass., 1941.

7. Darken, L. S., and R. W. Gurry: "Physical Chemistry of Metals," McGraw-Hill Book Company, Inc., New York, 1953.

8. Joffe, A. F.: "The Physics of Crystals," McGraw-Hill Book Company, Inc., New York, 1928.

9. Obert, E. F.: "Concepts of Thermodynamics," McGraw-Hill Book Company, Inc., New York, 1960.

† References 1, 2, 3, and 4 for galvanic cells; Refs. 1, 2, 3, 5, 7, 8 for thermodynamic systems other than chemical; Ref. 6 for reversibility in general; see also Ref. 9.

# The Second Law

The *First Law* is a theorem on the *conservation of energy:*

**The energy of the isolated system is conserved in real or reversible processes.**

This law announces that all forms of energy are equivalent in the sense that, when energy in one form disappears, an equal quantity in another guise must appear. The First Law makes no attempt to designate whether or not a system or a process is ideal or to insist that one direction of the process is preferable to the other. For real or for ideal systems, the First Law is a precise bookkeeper to ensure that energy is neither created nor destroyed but only changed in form.

The *Second Law* is a theorem on the *degradation of energy:*

**The entropy of the isolated system increases in all real processes (and is conserved in reversible processes).**

Since entropy has yet to be defined, the thought of the Second Law is more clearly expressed at the moment by the statement:

*The available energy of the isolated system decreases in all real processes (and is conserved in reversible processes).*

The Second Law recognizes that all forms of energy are *not* equivalent in their ability to do work and announces that the available energy of the system and its surroundings can never increase—only decrease. In this thought is the key experience of man: heat is not entirely available energy, and yet, invariably, man's machines degrade available energy so that only heat can be recovered. Thus friction (defined in Chap. 4) is a Second Law concept.

If heat were fully available, then no loss could be attributed to the irreversible process (and a new world would emerge). Nothing in the First Law implies that difficulty will be encountered in converting heat into work, especially since the converse process of converting work into heat is readily accomplished. The experience of man denies the possibility of complete conversion in the axiom underlying the work of Carnot:

*Heat cannot be converted completely and continuously into work.*

Consider the consequences if this axiom be false. A system could be devised to convert heat completely and continuously into work. Such a system would necessarily have to undergo a cycle since the sole desired effect is for the *surroundings* to furnish heat and receive only work in return. But this system would be independent of the temperature difference between the source of heat and the cycle—it would not matter whether the heat was received at a high or a low temperature since the work delivered would always equal exactly the heat received by the system. With a machine of this type, power could be obtained by cooling the earth, its oceans, and its atmosphere. Such a device was called by Ostwald *a perpetual-motion machine of the second kind* (or the *second class*). Note that perpetual motion of the second kind does not contradict the First Law since the work obtained is equal to the heat absorbed. But the Second Law would be violated since energy is upgraded, not degraded.

If such a cycle could be constructed, heat could be obtained from a low-temperature source, converted into work, and the work used to raise the temperature of a high-temperature sink. The net effect of source, cycle, and sink would be that *heat would pass from a cold to a hot region without work being supplied by the surroundings.* This conclusion, so contrary to experience, is denied by the Clausius (1850) axiom:

*Heat cannot, of itself, pass from a lower to a higher temperature.*

The thought of Clausius includes the obvious—that heat flows toward regions of lower temperature—and it also includes the principle that no device can be constructed with the *sole* effect being a transfer of heat from a cold to a hot region.

All these deductions are included in the axiom of Planck (1897):

*It is impossible to construct an engine which will work in a complete cycle and produce no effect except the raising of a weight and the cooling of a heat reservoir.*

The Second Law, unlike the First, can gauge the approach of a process to perfection by examining the changes in available energy experienced by the system and its surroundings. It follows that the direction of natural or spontaneous processes is also dictated by the Second Law: such processes must proceed from state to state of lesser available energy, and the difference of available energy is the driving influence toward an equilibrium state—a state of rest.

Any of the axioms is a statement of the Second Law or a corollary since, *if one fails, all must fail.* The validity of the Second Law, as well as the First, rests upon the fact that none of the axioms has ever been broken.

# CHAPTER 7

# THE SECOND LAW

*I cannot believe that God plays dice with the world.*
*Albert Einstein*

The foundations of the Second Law were laid in 1824 by Sadi Carnot in an essay† which introduced the concept of a cycle, the concept of reversibility, and the principle that the temperature of the source and that of the sink determined the thermal efficiency of a reversible cycle. The logic of Carnot enabled William Thomson (Lord Kelvin) to take the next step in 1852—the definition of a thermodynamic scale of temperature. And this, in turn, led Rudolf Clausius in 1865 to the concept of entropy.

**7-1. The Carnot Cycle.** The experimental evidence available to Carnot indicated that, when a source and a sink had different temperatures, a system could be devised to do work and the larger the temperature difference, the greater was the conversion of heat into work. Carnot decided that a prerequisite for the study of the "motive power" of heat was for the system to undergo a cycle, because if it did not, a part of the work might be derived from changes in the working substance of the system. From the experimental facts, Carnot reasoned as follows: First, every cycle must have a heat-rejection process; else the cycle would have a thermal efficiency of 100 per cent and therefore be indifferent to the temperature of the source. Second, since a temperature difference was necessary for the production of work, then, for fixed temperatures of source and sink, a reversible cycle would experience the largest possible difference in temperature and therefore should be the most efficient. But was temperature the only factor that controlled the thermal efficiency? Possibly different working substances or the pressure of the substance could change the thermal efficiency.

To answer this question, Carnot invented a *power cycle* that had four reversible processes (Fig. 7-1):

$ab$   A reversible adiabatic compression process from the lower temperature $t_R$ to the higher temperature $t_A$
$bc$   A reversible isothermal addition of heat at the constant high temperature $t_A$

† S. Carnot, "Reflections on the Motive Power of Heat," translated by R. H. Thurston, American Society of Mechanical Engineers, New York, 1943.

*cd*   A reversible adiabatic expansion process from the higher temperature $t_A$ to the lower temperature $t_R$

*da*   A reversible isothermal rejection of heat at the constant low temperature $t_R$

Mark well the simplicity of the Carnot cycle: heat is added at a high (but constant) temperature and rejected at a low (but constant) temperature while the intervening processes are heatless (adiabatic).

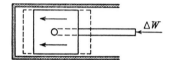

*ab* – Adiabatic compression from $t_R$ to $t_A$

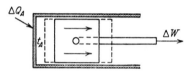

*bc* – Isothermal expansion at $t_A$

Carnot devised the test of reversibility and stipulated this restriction for each process of his cycle. And, when the cycle was reversed, it became a *heat pump* with heat being *added* to the cycle in process *ad* and *rejected* in process *cb* (Fig. 7-1).

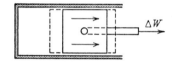

*cd* – Adiabatic expansion from $t_A$ to $t_R$

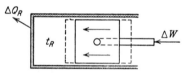

*da* – Isothermal compression at $t_R$

Consider the system illustrated in Fig. 7-2*a*. A Carnot power cycle receives heat from a source at $t_A$ and rejects heat to a sink at $t_R$ while driving a Carnot heat pump. The heat pump absorbs heat from the low-temperature sink and rejects heat to the high-temperature source. The net effect of the two cycles must be zero, since each quantity (of heat or work) for the power cycle is canceled by that for the heat pump. But let it be premised that some factor other than temperature could make the Carnot cycle more efficient, for example, a more efficient working substance in the power cycle. Then more work will be delivered by the power cycle than that required by the heat pump (Fig. 7-2*b*). A high-temperature source would not be required since the heat rejected by the heat pump exactly equals the heat absorbed by the power cycle. Carnot concluded that this result was absurd and therefore the premise that a change in the working substance could change the thermal efficiency of his cycle was false.

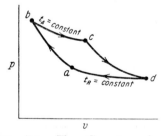

FIG.  7-1.   The  Carnot  cycle  of processes.

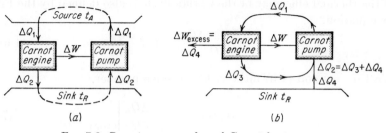

FIG. 7-2. Carnot power cycle and Carnot heat pump.

Carnot arrived at the following axioms:

**For fixed and constant temperatures of source and sink:**
1. **No cycle can be more efficient than a reversible cycle.**
2. **All reversible cycles have the same thermal efficiency.**
3. **The thermal efficiency of a reversible cycle is independent of the working substance† (which can be a solid, liquid, or gas, real or ideal).**
4. **A cycle not receiving work cannot transfer heat from the cold to the hot reservoir.**

These conclusions can be summarized in the thought:

**The thermal efficiency of a reversible cycle depends only upon the temperature of the source and that of the sink.**

Any of these axioms can be called *a statement of the Second Law.*

**7-2. The Thermodynamic Temperature Scales.** At the time of Carnot, the temperature scales in use were arbitrary scales that related the temperature of a system to some particular property of the substance of the thermometer. Lord Kelvin realized that the logic of Carnot opened the way for the definition of a *thermodynamic scale of temperature* which would be entirely divorced from the properties of matter since the thermal efficiency of Carnot's cycle was a function of temperature alone. Let $\Theta$ be a temperature on a new scale, not yet defined. Then the thermal efficiency of a Carnot cycle equals, by the Second Law, some function of $\Theta_A$, the temperature of the source, and $\Theta_R$, the temperature of the sink,

$$\eta_t = f(\Theta_A, \Theta_R) \qquad (a)$$

† Note that the pressure exerted by the working substance is only incidental to the production of work. In engineering, one goal is to use the widest possible difference of temperature between source and sink to obtain high thermal efficiency; another goal is to use pressures sufficiently low to obtain safety and ease of design but sufficiently high so that density is high, to obtain reasonable size and therefore cost.

But the thermal efficiency of the Carnot cycle is also defined by the First Law equation,

$$\eta_t = \frac{\Delta Q_A + \Delta Q_R}{\Delta Q_A} = 1 + \frac{\Delta Q_R}{\Delta Q_A} \qquad (4\text{-}6)$$

Since Eqs. (a) and (4-6) must yield the same number,

$$\eta_{t,\text{Carnot}} = f(\Theta_A,\Theta_R) = 1 + \frac{\Delta Q_R}{\Delta Q_A}\Big]_{\substack{\text{Carnot}\\ \text{cycle}}} \qquad (7\text{-}1)$$

This is *the general definition of all Second Law thermodynamic temperature scales* and satisfies the specifications of Art. 3-1, in particular: *A variable* $\eta_{t,\text{Carnot}}$ *has been found that yields a progressively larger number as the temperature difference increases, for any selected* $\Theta_A$ *(or* $\Theta_R$*), and therefore a unique number can be assigned to each thermal level.*

To fix the form of the function $f$ in Eq. (7-1), Kelvin decided that each degree drop in temperature experienced by a Carnot cycle should contribute equally to the work:

$$\text{Work from each } 1° \text{ drop in temperature} = \frac{\Delta Q_A}{\Theta_A}$$

$$\text{Work from } (\Theta_A - \Theta_R)° \text{ drop in temperature} = (\Theta_A - \Theta_R)\frac{\Delta Q_A}{\Theta_A}$$

Since $\eta_{t,\text{Carnot}} = \Delta W/\Delta Q_A$, then from the above,

$$\eta_{t,\text{Carnot}} = f(\Theta_A,\Theta_R) = \frac{\Theta_A - \Theta_R}{\Theta_A} \qquad (7\text{-}2a)$$

And with Eq. (7-1), noting that $\Delta Q_R$ is a negative number,

$$\frac{\Theta_R}{\Theta_A} = \frac{-\Delta Q_R}{\Delta Q_A}\Big]_{\text{Carnot cycle}} \qquad (7\text{-}2b)$$

Equation (7-2b) defines *a class of thermodynamic temperature scales with the following common characteristic: The ratio of two absolute thermodynamic temperatures equals the ratio of the heat quantities for a Carnot cycle over the same temperature interval.*

The final step is to assign a particular number to be the absolute thermodynamic temperature of a particular fixed point, and then the thermodynamic temperatures of other thermal levels could be found from Eq. (7-2b) and a "thermometer." Since the "thermometer" specified for the scale is a Carnot cycle, alternative methods for finding thermodynamic temperatures will be studied.

Recall that the efficiency of a reversible cycle between two temperature levels is independent of the properties of the working substance. Then an ideal gas can be selected as the working substance for a Carnot cycle, and the thermal efficiency for this substance must equal that for any other substance, real or ideal. The thermal efficiency of a Carnot cycle with an ideal gas equals (Example 1)

$$\eta_{t,\text{Carnot}} = \frac{T_A - T_R}{T_A} \tag{7-2c}$$

Upon comparing Eqs. (7-2a) and (7-2c), it is apparent that the temperatures $\Theta$ and $T$ can differ only by a constant multiplier. To avoid a new set of numbers, $\Theta$ is set equal to $T$, and

$$\frac{T_R}{T_A} = \left. \frac{-\Delta Q_R}{\Delta Q_A} \right]_{\text{Carnot cycle}} \tag{7-2d}$$

Thus, the symbol $T$ now denotes absolute temperatures on the Second Law scales defined by Eq. (7-2d), and these scales correspond exactly to the Zeroth Law scales defined by Eq. (3-3). Therefore, a gas thermometer (a precise instrument) can replace a Carnot-cycle thermometer (a relatively crude instrument) as the means for measuring thermodynamic temperatures over a wide interval.

Since the temperature of the triple point has been assigned a positive number (Art. 3-4), the range of temperatures on the scales defined by Eq. (7-2d) is thereby restricted to the range of positive numbers (from zero to infinity). That the scale cannot be extended to include negative temperatures can be realized from Eq. (7-2c): if $T_R$ is negative, the thermal efficiency is greater than unity and the First Law would be broken. The limiting efficiency of unity uniquely defines $T_R$ as zero for any $T_A$, and this *concept* is called *the absolute zero* since lower temperatures are thermodynamically inadmissible.

The question of whether or not the absolute zero can be attained cannot be answered by the Second Law and is denied by the Third Law [although, admittedly, in our choice of words we have, without consulting the Third Law, denied that a *cycle* can be constructed with the absolute zero as the sink temperature (page 100, for example)].

**Example 1.** Determine the thermal efficiency of a Carnot cycle between temperatures $T_A$ and $T_R$ with an ideal gas as the working substance.

*Solution.* Recall that an ideal gas is defined (1) by its equation of state, $pv = RT$, and (2) by the stipulation that its internal energy is a function of temperature alone. Therefore, the heat addition and rejection processes of Fig. 7-1 are not only at constant temperature but also at constant internal energy,

$$\Delta Q - \Delta W = \Delta U = 0$$

Thus, the heat and the work† for the isothermal process are equal in magnitude,

$$\Delta Q_{A,\text{rev}} = \Delta W_{\text{rev}} = mRT_A \ln \frac{v_c}{v_b} \qquad \Delta Q_{R,\text{rev}} = \Delta W_{\text{rev}} = mRT_R \ln \frac{v_a}{v_d}$$

Upon substituting in Eq. (4-6),

$$\eta_t = 1 + \frac{\Delta Q_R}{\Delta Q_A} = 1 - \frac{T_R \ln \ (v_d/v_a)}{T_A \ln \ (v_c/v_b)}$$

which reduces, with Eq. (6-5a), to

$$\eta_{t,\text{Carnot}} = \frac{T_A - T_R}{T_A} \qquad Ans.$$

**7-3. Entropy.** Consider two (or more) reversible adiabatic processes. Can these processes have a common state? Suppose that this were possible (Fig. 7-3). Here the reversible adiabatic processes $ab$ and $ca$ are premised to have the common state $a$. If this were true, a nonadiabatic process could be devised to link states $b$ and $c$ and so form a reversible cycle. The net effect of the cycle would be to convert completely heat into work—a consequence denied to be possible by the Second Law.‡ It follows that

**Two (or more) reversible adiabatic processes cannot have a common state.**

But this is not only an axiom of the Second Law; it is also a criterion§ of

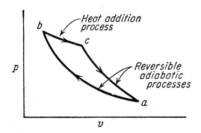

a property of the *state*. For example, two (or more) constant-temperature (or constant-volume or constant-pressure, etc.) processes cannot have a common state; else the state could be assigned two numbers—two temperatures. It follows that a new property is proclaimed by the behavior of the reversible adiabatics. Then *each state of matter*

FIG. 7-3. Failure of Second Law if two reversible adiabatic processes had a common state.

*can be assigned a value of this new property such that all states designated by the same number can be visited by a reversible*

† The work of the reversible isothermal process is found as in Art. 6-4, Example 1.
‡ Although Fig. 7-3 presupposes that two independent variables determine the state, the argument holds for any number of independent variables. Thus, the loci of states accessible to the reversible adiabatic process become a surface for three or more independent variables, and a family of nonintersecting surfaces maps all states of the system.
§ The criterion of single-valuedness (Art. 2-11).

*adiabatic process.* The new property is named *entropy*† and the symbol for entropy is $S$:

**Entropy is the property of matter held constant in a reversible adiabatic process.**

The problem of defining entropy by an equation was solved by Clausius, who noted that Eq. (7-2d) yielded a new variable,

$$\frac{\Delta Q_{1,\text{rev}}}{T_1} = \frac{\Delta Q_{2,\text{rev}}}{T_2} = \frac{\Delta Q_{j,\text{rev}}}{T_j} \qquad (7\text{-}2e)$$

He reasoned from Eq. (7-2e):

1. To pass reversibly from a selected datum adiabatic to other reversible adiabatics requires heat to be transferred; therefore, the variable $\Delta Q_{\text{rev}}/T$ will yield a set of numbers relative to the datum such that all reversible adiabatics can be assigned an identifying number.

2. To pass reversibly from state to state of an adiabatic process requires $\Delta Q_{\text{rev}}$ to be zero; therefore, the variable $\Delta Q_{\text{rev}}/T$ will remain constant, and each state in the sequence can be assigned the same value of $\Delta Q_{\text{rev}}/T$ as that which identifies the adiabatic.

3. Most important, *the change in the variable* $\Delta Q_{\text{rev}}/T$ *is independent of the isotherm selected for evaluation;* therefore, *any* reversible path between the initial and final state, or between the initial and final adiabatic, can be selected for evaluation.

Clausius concluded that the entropy of each state of matter, relative to a selected datum‡ state, could be calculated by

$$\Delta S = \frac{\Delta Q_{\text{rev}}}{T} \qquad \left[\frac{\text{energy}}{\text{temperature}}\right] \qquad (7\text{-}3a)$$

An isothermal process, however, is not demanded by the conditions of Eq. (7-2e) since the variable $\Delta Q_{\text{rev}}/T$ is an infinitesimal which is indifferent to the isotherm selected for evaluation. Therefore, let $\Delta Q_{\text{rev}}$ and $T$ be considered functions of time $\tau$, and divide Eq. (7-3a) by $\Delta\tau$. By allowing $\Delta\tau$ to approach zero,

$$\frac{dS}{d\tau} = \frac{1}{T}\frac{dQ_{\text{rev}}}{d\tau} \qquad (7\text{-}3b)$$

---

† The student invariably tries to picture entropy in the same light as pressure $p$ or temperature $T$, that is, as a single physical phenomenon. But $T/p$ is a property that needs no more explanation than that it is a function of other properties. Entropy belongs to this same class of properties; for each phase of matter, for example, a function $S(T,p)$ can be found which is called *entropy.*

‡ Although the definition of entropy is for a *change,* and not for an absolute value, note that this is the same procedure previously followed for internal energy and enthalpy.

And by multiplying by $\Delta\tau$,

$$dS \equiv \frac{dQ_{\text{rev}}}{T} \tag{7-3c}$$

Equation (7-3c) is considered to be *the fundamental definition of entropy. It is rarely used for the direct evaluation of entropy since confusion arises because of the restriction of reversibility.* The importance of the equation is that it establishes the variable $dQ_{\text{rev}}/T$ to be equal to an exact differential of a function of properties, and therefore the independent variable of time can be eliminated from the First Law energy equation (Art. 7-4).

**7-4. Entropy and the Process Equations.** Since entropy is a thermostatic property, it is a function of other thermostatic properties. To find the functional relationships, consider the differential energy equation for a closed system,

$$dQ = dE + dW\Big]_{\text{rev or irrev}} \tag{4-4c}$$

If the work of the system is all mechanical and reversible, by Eq. (6-2d)

$$dQ_{\text{rev}} = dU + p\,dV \tag{6-4a}$$

Recall that the independent variable of each term in Eq. (6-4a) is time $\tau$. Since $\tau$ does not appear in Eq. (6-4a), the variable $Q_{\text{rev}}$ may be also, perchance, a function of the properties $U$ and $V$. That this is incorrect can be recognized by noting that $\Delta Q_{\text{rev}}$ between two states is multivalued since the integral of $p\,dV$ is multivalued.† But let Eq. (6-4a) be divided by the temperature $T$,

$$\frac{dQ_{\text{rev}}}{T} = \frac{1}{T}\,dU + \frac{p}{T}\,dV$$

The line integral of the left-hand side of this equation between two states is single-valued since the integral equals $\Delta S$, by Eq. (7-3c). Therefore, the line integral of the right-hand side of the equation is also single-valued. Hence, the *inexact*‡ differential equation [Eq. (6-4a)], when divided by the temperature $T$, becomes an *exact* differential equation,

$$T\,dS = dU + p\,dV \tag{7-4a}$$

Equation (7-4a) is called *the first T dS equation.*

In Art. 2-12 it was stated that two independent intensive properties could be *selected* to identify the state. From Eq. (7-4a), it is evident that this selection is best made of two of *the three fundamental thermodynamic coordinates: specific volume, entropy, and internal energy.*

---

† Article A-6c.
‡ Inexact for the variables $U$ and $V$.

By adding $d(pV)$ to both sides of Eq. (7-4a), *the second T dS equation* is obtained,

$$T \, dS = dH - V \, dp \qquad (7\text{-}4b)$$

*Equations (7-4) are the means for calculating the change in entropy for reversible processes, or for irreversible processes with constant chemical composition* (Art. 8-2).

For example, at constant volume of an inert system Eqs. (7-4a) and (2-4a) yield

$$ds\bigg]_{v = C} = c_v \frac{dT}{T} \qquad (7\text{-}5a)$$

Or at constant pressure, with Eqs. (7-4b) and (2-4a),

$$ds\bigg]_{p = C} = c_p \frac{dT}{T} \qquad (7\text{-}5b)$$

If a complete functional relationship for the entropy of a substance is desired, an equation of state and an equation relating internal energy with temperature and pressure or volume are necessary. The procedure will be illustrated for the simple case of an ideal gas. By substituting Eqs. (3-5b) and (3-8b) into Eq. (7-4a),

$$ds = c_v \frac{dT}{T} + R \frac{dv}{v} \qquad \text{(ideal gas)} \qquad (7\text{-}6a)$$

which integrates to

$$s_2 - s_1 = c_v \ln \frac{T_2}{T_1} + R \ln \frac{v_2}{v_1} \qquad \text{(ideal gas)} \qquad (7\text{-}6b)$$

By selecting a datum state of zero entropy at $T_1$ and $v_1$,

$$s = c_v \ln T + R \ln v + C \qquad \text{(ideal gas)} \qquad (7\text{-}6c)$$

wherein $C$ is a constant. Equation (7-6c) illustrates a typical form of the functional relationship among the properties $s$, $T$, and $v$.

Equations (7-4) to (7-6) are valid for either reversible or irreversible processes since the value of a property is fixed by the state—not by the process—and these equations involve only properties. The question of reversibility need be raised only when heat and work are to be recognized. For example, in an irreversible process the states encountered may frequently be traced on both the $Ts$ and $pv$ diagrams with fair precision (and therefore large gradients in property values are not present). Here Eqs. (7-4) are quite valid, but the integral of $T ds$ is not heat nor is the integral of $p \, dv$ work unless the process is reversible (Art. 8-2).

The equations are also valid for unit mass of substance in a flow process since the properties are evaluated by an observer at rest with respect to the mass under consideration. Thus, each element of mass in a flow process is a closed system to an observer traveling with the system. Then the variations of the properties in this closed system are shown by Eqs. (7-4) quite independently of the conclusions of another observer who is evaluating the work and heat for the region of the flow process from a station at rest with respect to the system. Note that, when the velocity of the flow stream is used to do reversible work, the properties in Eqs. (7-4) are not affected. If the velocity changes reversibly without transfer of work, the effect on Eqs. (7-4) is exactly equivalent to that of work reversibly added to the closed system. The properties in Eqs. (7-4) cannot differentiate between a reversible change in velocity in a flow process and a reversible compression or expansion in a nonflow process. Also, if the velocity changes irreversibly, the properties in Eqs. (7-4) are affected in the same manner as if heat were added to the system.

**7-5. The $Ts$ Diagram.** Consider that a $Ts$ diagram, similar to that in Fig. 7-4, is to be constructed for the simple case of an ideal gas. The

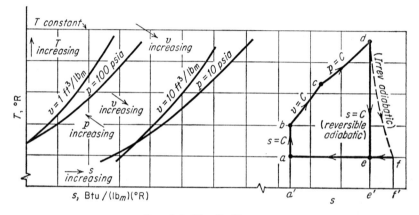

Fig. 7-4. The $Ts$ diagram.

first step is to select data which reflect the heat measurements of the research laboratory over the range of states of the diagram. If data for $c_p$ are selected, values for $c_v$ can be found from Eq. (3-9). From an arbitrary datum state, values of entropy are calculated for assigned values of temperature for both constant-pressure and constant-volume processes by Eqs. (7-5). Then, values of temperature are assigned so that states at various pressures and various volumes can be calculated by Eqs. (6-5). In this manner, a diagram such as that in Fig. 7-4 is constructed. Note that one set of "heat" data (that for $c_p$) *and* the equation of state were the prerequisites for the problem.

**All $Ts$ diagrams in this text are drawn for unit mass of substance.**

Figure 7-4 and Eqs. (7-5) show that the slope $dT/ds$ of the constant-volume (or constant-pressure) process is dictated by the temperature and by $c_v$ (or $c_p$). Since $c_p$ is always larger than $c_v$ [Eq. (3-9)],

**At each state on the $Ts$ diagram, the slope of the constant-volume process is greater than that of the constant-pressure process.**

Lines of constant entropy mark the path of the reversible adiabatic process in Fig. 7-4:

## A reversible adiabatic process is an isentropic process.†

All states (points) on a property diagram such as Fig. 7-4 are for states of equilibrium. Then a continuous line marks a succession of equilibrium states and therefore denotes a reversible process (within the precision of measurement of the represented data). The area bounded by the path of the process and the abscissa is the integral of $T\ ds$. For the reversible process, by comparing Eqs. (6-4a) and (7-4a) an identity is established,

$$\Delta q_{rev} = \int_{\tau_1}^{\tau_2} dq_{rev} = \int_{s_1}^{s_2} T\ ds \tag{7-7}$$

Hence, the area (in proper units) can be interpreted as a quantity equal to the reversible heat. For example, consider the reversible cycle $abcdea$ in Fig. 7-4,

$$\Delta q_{rev,bcd} = \text{heat reversibly added} \quad = \text{area } a'bcde' = \int_{b}^{d} T\ ds$$

$$\Delta q_{rev,ea} = \text{heat reversibly rejected} = \text{area } e'eaa' \quad = \int_{e}^{a} T\ ds$$

$$\overline{\Sigma\ \Delta q \quad\quad = \text{work of cycle} \quad\quad\quad = \text{area } abcdea}$$

Note that, for reversible or irreversible cycles, the First Law demands that the difference between the heat added and the heat rejected (without regard for algebraic sign) must equal the work of the cycle. For the *reversible* cycle, this difference is also equivalent to a difference between two areas on the $Ts$ diagram: area $abcdea$. This area is clockwise generated; therefore, the heat reversibly added is greater than that rejected, and work is produced by the cycle.

Suppose that the reversible adiabatic process $de$ in Fig. 7-4 is replaced by an irreversible adiabatic process $df$ and that the new cycle is $abcdfa$. The heat-addition processes remain unchanged, while less work will be done by the irreversible cycle; therefore, more heat must be rejected. This requires area $a'aff'$ to be larger than area $a'aee'$, and thus state $f$ lies to the right of state $e$. Here

$$\Delta q_{rev,bcd} = \text{heat reversibly added} \quad = \text{area } a'bcde' = \int_{b}^{d} T\ ds$$

$$\Delta q_{rev,fea} = \text{heat reversibly rejected} = \text{area } f'faa' \quad = \int_{f}^{a} T\ ds$$

$$\overline{\Sigma\ \Delta q \quad = \text{work of cycle} \quad\quad\quad < \text{area } abcdfa}$$

† In this text and almost invariably in scientific articles, the names *reversible adiabatic* and *isentropic* are synonymous, for simplicity of expression (Art. 11-3).

Unlike the reversible case, the enclosed area on the $Ts$ diagram is *not* equivalent to the work (but is *greater* than the work). Also, the integral of $T\,ds$ in this case is finite (area $e'dff'$); yet the process was adiabatic ($\Delta q = 0$), and this emphasizes that the integral of $T\,ds$ equals the heat only for the reversible process. Because of these differences between the reversible and irreversible processes, the latter are shown as dashed lines on the $Ts$ diagram ($df$ in Fig. 7-4), and this practice warns, too, that the exact path is probably nebulous since gradients in temperature and pressure are probably present:

**The irreversible process is usually shown as a discontinuous line on the $Ts$ diagram.**

These comments parallel those made for the $pv$ diagram and the integral of $p\,dv$ (Art. 6-4).

**Example 2.** For the reversible cycle of Fig. $A$, check the values of entropy and

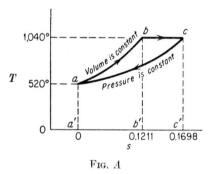

Fig. $A$

compute the thermal efficiency. [$c_v = 0.175$ Btu/(lb$_m$)(°R), ideal gas with molecular weight of 29.]

*Solution.* By Eq. (3-9),

$$c_p = c_v + \frac{R_0}{M} = 0.175 + \frac{1.986}{29} = 0.244 \text{ Btu/(lb}_m)(°R)$$

Integrating Eqs. (7-5) with $c_p$ and $c_v$ constants,

$$\Delta s_{ab} = c_v \ln \frac{T_2}{T_1} = 0.175 \ln \frac{1,040}{520} = 0.1211 \text{ Btu/(lb}_m)(°R) \qquad Ans.$$

$$\Delta s_{ac} = c_p \ln \frac{T_2}{T_1} = 0.244 \ln \frac{1,040}{520} = 0.1698 \text{ Btu/(lb}_m)(°R) \qquad Ans.$$

By Eqs. (6-4a), (3-8b), and (3-8d),

$$\Delta q_{\text{rev},ab} = c_v(T_b - T_a) = 0.175(1,040 - 520) = 91 \text{ Btu/lb}_m$$
$$\Delta q_{\text{rev},ca} = c_p(T_a - T_c) = 0.244(520 - 1,040) = -127 \text{ Btu/lb}_m$$

By Eq. (7-3c),

$$\Delta q_{\text{rev},bc} = T_{bc}(s_c - s_b) = 1,040(0.0487) = 50.7 \text{ Btu/lb}_m$$

And

$$\eta_t = \frac{\Sigma\,\Delta w}{\Delta q_A} = \frac{\Sigma\,\Delta q}{\Delta q_A} = \frac{91 + 50.7 - 127}{91 + 50.7} = 0.1022, \text{ or } 10.22\% \qquad Ans.$$

**7-6. Principle of the Increase in Entropy.** In Art. 7-5 it was decided that the entropy must increase during an *irreversible* adiabatic process if the reversible cycle is to be the criterion of perfection. Another type of irreversibility arises when heat flows through a temperature difference from a source to a sink. How will this affect the entropy? Since the temperature difference is the item of interest, let source and sink be simple subsystems such that the volume and temperature of each can be assumed constant. Then, by integrating Eq. (7-4a),

$$\Delta S = \frac{\Delta U}{T}\bigg]_{\text{source or sink}}$$

Since $\Delta U$ arises from transfer of heat alone, it equals $\Delta Q$ numerically and for the isolated system of source and sink together

$$\Delta S_{\text{source}} + \Delta S_{\text{sink}} = \frac{\Delta Q}{T}\bigg]_{\text{source}} + \frac{\Delta Q}{T}\bigg]_{\text{sink}} \qquad (T_{\text{source}} > T_{\text{sink}}) \qquad (a)$$
$$(-) \qquad (+) \qquad (-) \qquad (+)$$

Observe that $\Delta Q_{\text{sink}} = -\Delta Q_{\text{source}}$ and that the positive $\Delta Q$ of the sink is accompanied by the lower temperature. Therefore,

$$\Sigma\,\Delta S_{\substack{\text{isolated or} \\ \text{adiabatic} \\ \text{system}}} \geq 0 \qquad\qquad (7\text{-}8)$$

If the transfer of heat approached reversibility, both temperatures would approach each other and entropy would be conserved, but when heat flows through a finite temperature difference, the entropy of the isolated or the adiabatic† system must increase (and entropy is said to be created).

Consider, next, that a cycle is placed *between* the source and sink of Eq. (a). When the cycle of processes is completed, Eq. (a) applies but now $\Delta Q_{\text{sink}}$ is a smaller number since a part of the heat was converted into work. Observe that, if the cycle is reversible, the sum of the terms on the right-hand side of Eq. (a) is zero by Eq. (7-2d). If one or more processes of the cycle are irreversible, its efficiency is *less* than that of the reversible cycle and hence *more* heat must be rejected to the sink and $\Delta Q/T$ for the sink must also be *a greater positive number*. It follows that the sum of the terms on the right-hand side of Eq. (a) is positive (greater than zero). *Thus, the inequality of (7-8) holds for all irreversibilities within the isolated or adiabatic† system.*

For the general case, assume that a process of an isolated or adiabatic system could take place with *decrease* in entropy. Then the isolation

† The argument holds for complete isolation or adiabatic isolation.

could be set aside while heat was added to regain the initial entropy and work transferred to regain the initial state. The net effect of the cycle of processes would be dictated by the First Law,

$$\Delta Q - \Sigma \Delta W = \Delta E = 0$$

and
$$\Sigma \Delta W = \Delta Q \text{ (positive)}$$

Thus, work must be *delivered* by the cycle of amount equal to the heat added (and a perpetual-motion machine of the second kind created) to obey the premise. On the other hand, if the entropy of the isolated system increased, heat could be taken away (to regain the initial entropy) and work could be added (to regain the initial energy) without violating

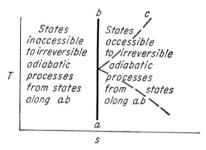

FIG. 7-5. Accessible and inaccessible states.

either the First or Second Law. The conclusion follows that for all isolated systems:

**The entropy of the isolated system increases in all real processes (and is conserved in reversible processes).**

Planck calls this *"the most general statement of the Second Law."* Note that the *real* isolated system can experience either a *spontaneous* process or else no change at all. This is because the reversible process has no unbalance, no driving force, to make it operate—it is a hypothetical process. Thus

**The real processes of an isolated system are spontaneous (and irreversible).**

The Second Law supplies a rigorous method for identifying an irreversible process:

**A process is irreversible if the entropy of the isolated system increases during the process.**

The foregoing discussions are illustrated in Fig. 7-5. For the reversible adiabatic process, all states along *ab* are accessible states; for the irreversible adiabatic process, only states with greater entropy than the initial state (such as $c$) are accessible:

## States of lesser entropy are inaccessible by adiabatic processes.

**Example 3.** A pound of ice at $32°F$ is placed in $5\ lb_m$ water at $80°F$ with the system open to the atmosphere, which has a pressure of 14.7 psia. Compute the change in entropy for the spontaneous process of the adiabatic system.

*Solution.* For the system of ice and water, by the First Law,

$$\Delta Q - \Delta W = \Delta U \tag{4-4$d$}$$

Since the system is assumed adiabatic and work is done only to push aside the atmosphere, by Eq. ($f$) of Art. 4-3,

$$\Delta H\bigg]_{p=C} = 0 \tag{$a$}$$

Since the effect of modest changes in pressure on the enthalpy of solids and liquids is small, values of $c_p$ for water and the latent enthalpy of fusion for ice can be selected from Table B-3 and the equilibrium temperature computed,

$$(m\ \Delta h)_{ice} + (m\ \Delta h)_{water} = 0$$
$$[144 + (t - 32)]1.0 + (t - 80)(1.0)(5) = 0$$
$$-288 + 6t = 0$$
$$t = 48°F$$

Then, with Eqs. (7-4$b$) and (7-5$b$),

$$\Delta S_{ice} = \Delta S_{\substack{\text{melting ice at} \\ \text{constant temperature}}} + \Delta S_{\substack{\text{heating melted ice} \\ \text{from 32 to 48°F}}}$$
$$= \frac{144}{32 + 460} + c_p \ln \frac{460 + 48}{460 + 32}$$
$$= 0.292 + 0.032 = +0.324$$
$$\Delta S_{water} = \Delta S_{\text{cooling from 80 to 48°F}}$$
$$= 5c_p \ln {}^{508}\!/_{540} = -0.3055$$
$$\Sigma\ \Delta S_{system} = \Delta S_{ice} + \Delta S_{water} = +0.0185\ Btu/°R \qquad Ans.$$

Thus, the entropy of the adiabatic system has increased because of the irreversible process.

**7-7. The Irreversible Process.** Consider a process, not necessarily adiabatic, and let the system together with a heat reservoir be isolated. By the inequality of (7-8),

$$dS_{system} + dS_{reservoir} \geq 0 \tag{$a$}$$

Since the system is the item of interest, let the heat processes of the reservoir be reversible,

$$dS_{reservoir} = \frac{dQ_{rev}}{T} \tag{$b$}$$

For either reversible or irreversible processes, the heat function for the system is equal to that of the reservoir (but of opposite algebraic sign),

$$dQ_{rev}\bigg]_{reservoir} = -dQ\bigg]_{system} \tag{$c$}$$

By substituting Eqs. (*b*) and (*c*) into (*a*),

$$dS_{\text{system}} - \frac{dQ_{\text{system}}}{T_{\text{system}}} \geq 0$$

$$dS \geq \frac{dQ}{T} \tag{7-9a}$$

If the process is reversible, the equality sign in (7-9a) holds and Eq. (7-9a) reverts to Eq. (7-3c). Thus, the change in entropy of the system equals the integral of $dQ/T$ for the reversible process and is greater† than the integral of $dQ/T$ for the irreversible process.

The integral of the left-hand side of (7-9a) for a cycle is zero, since entropy is a property; therefore, the cyclic integral of the right-hand side of (7-9a) is less than zero if the cycle is irreversible,

$$\oint \frac{dQ}{T} \leq 0 \tag{7-9b}$$

Principle (7-9b) is called *the inequality of Clausius.*

Recall that mechanical and fluid friction were defined qualitatively as the heating effects arising from the dissipation of work or available energy. Let the symbol $\mathfrak{F}$ denote all forms of dissipations of available energy (and therefore be a function restricted to positive‡ values). To define $\mathfrak{F}$ quantitatively, (7-9a) can be converted into an equality,

$$dS = \frac{dQ}{T} + \frac{d\mathfrak{F}}{T} \tag{7-10a}$$

and

$$d\mathfrak{F} \equiv T \, dS - dQ \tag{7-10b}$$

When the path of the irreversible process can be approximated, Eq. (7-10b) is the means for evaluating the frictional heating. Note that friction may not be an entire loss since the frictional heating may allow a portion of the dissipated available energy to be regained.

The time rate of change of entropy equals

$$\frac{dS}{d\tau} = \frac{1}{T}\left(\frac{dQ}{d\tau} + \frac{d\mathfrak{F}}{d\tau}\right) \tag{7-10c}$$

Although the rate $dS/d\tau$ can be positive or negative, since $dQ$ can be positive or negative, the rate $d\mathfrak{F}/d\tau$ can only be positive, since both friction and time are restricted to positive variations.

**7-8. The Availability of Heat.** Consider that whenever a cycle occurs, the system is eventually restored to its initial state; therefore, the work of a power cycle is obtained entirely from transformation of heat. But even

---

† And $-20$, for example, is less than $-10$.
‡ Friction or irreversibilities are always a heating effect—never a cooling effect.

under ideal conditions, not all the heat supplied to a cycle can be converted into work, and therefore heat can be considered to have *available* and *unavailable* parts. The available part is called *the availability of heat* and equals *the maximum amount of work that can be obtained when* (1) *the cycle is reversible and when* (2) *the heat-rejection process of the reversible cycle is made at the temperature of the atmosphere*, $T_0$. The unavailable part equals the heat rejected at $T_0$ since no means can be devised to obtain work from heat alone at the temperature of the atmosphere. (And a sink at a temperature less than $T_0$ requires work.)

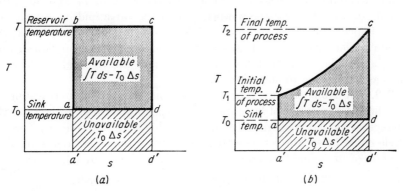

FIG. 7-6. (*a*) Availability of heat from an infinitely large reservoir; (*b*) availability of heat from a finite reservoir.

Consider, first, heat transferred at a constant temperature $T$ and of amount $\Delta Q_A$. Here the maximum work could be produced by a Carnot cycle with $T_0$ as the sink temperature,

$$\Delta W_{\text{Carnot}} = \Delta Q_A + \Delta Q_R \qquad (a)$$

By the identity of Eq. (7-7), Eq. (*a*) equals (note Fig. 7-6*a*),

$$\Delta W_{\text{Carnot}} = \int_b^c T\, ds + \int_d^a T_0\, ds \qquad (b)$$

This work is, by definition, the availability $\Delta \mathcal{Q}$ of the heat. Hence, Eq. (*b*) can be shown as

$$\Delta \mathcal{Q} = \int_1^2 T\, ds - T_0\, \Delta s_{12} \Big]_{\text{rev}} \qquad (7\text{-}11a)$$

and also as

$$\Delta \mathcal{Q} = \Delta Q_{\text{rev}} - T_0\, \Delta S \qquad (7\text{-}11b)$$

Consider, next, a reversible transfer of heat not at constant temperature, such as $bc$ in Fig. 7-6*b*. To obtain the maximum work from this heat, a reversible cycle must be devised so that the heat-rejection process occurs only at the temperature $T_0$. The solution is simple. Two heatless processes (reversible adiabatics) are used (*ab* and *cd*) to complete the cycle: *abcda*. The work of this cycle equals the availability of the heat, $\Delta \alpha$, which is evaluated, as before, by Eqs. (7-11).

Note that the system furnishing the heat suffers a decrease in available energy while the system receiving the heat gains in available energy (if $T$ is greater than $T_0$). The gain of one system will exactly offset the decrease of the other system if the process is reversible. Thus, Eqs. (7-11) can be used to evaluate either the system which receives a heat transfer or the system which supplies the heat.

The availability of heat is influenced by the value of $T_0$, and this value is not constant because the temperature of the earth and its atmosphere will change from hour to hour and from season to season. Thus, the steam power plant is aided in winter by the low-temperature water that can be circulated through the condenser, and the thermal efficiency is increased (relative to summer values).

**7-9. Available Energy.** In passing from one state of a system to another, the transfers of heat and work depend upon the process, but the change in available energy does not. If this were not true, a reversible cycle could be set up between the two states with the net effect the conversion of unavailable energy into available energy. It follows that

### The available energy of a system is a property.

With this axiom of the Second Law, the loss of available energy arising from an irreversible process can be calculated. Let the system and its heat reservoir, if any, be isolated during the irreversible process; the entropy of the isolated system, by (7-8), will increase. Then the isolation can be lifted and the system and/or heat reservoir changed back to their initial entropies. The net effect of this reduction in entropy must be a cooling effect, and the net available energy removed by the cooling is given by Eq. (7-11*b*),

$$\Sigma \, \Delta \alpha = \Sigma \, \Delta Q_{rev} - T_0 \Sigma \, \Delta S \qquad (a)$$

During the same interval, work is reversibly added to restore the energy removed by the cooling,

$$-\Sigma \, \Delta W_{rev} = -\Sigma \, \Delta Q_{rev} \qquad (b)$$

The available energy required to restore the system reversibly from the final to the initial state is obtained by adding together Eqs. (*a*) and (*b*); and therefore, since available energy is a property, this same quantity must equal the loss in available energy in the irreversible process between the same† two states,

$$\Delta \mathcal{G}_{\substack{\text{isolated or} \\ \text{adiabatic} \\ \text{system}}} \equiv - T_0 \Sigma \, \Delta S \qquad (7\text{-}12)$$

† And the algebraic sign is automatic by the sign of $\Delta S$.

Observe that available energy is conserved in reversible processes of an isolated system and decreases in irreversible processes:

**The available energy of the isolated system decreases in all real processes (and is conserved in reversible processes).**

**Example 4.** Water in a constant-pressure container is agitated by a paddle until the temperature rises from 60 to 100°F. Compute the change in available energy of the water and the decrease in available energy of the isolated system if the temperature of the atmosphere is 32°F.

*Solution.* The water could be restored to its initial state by serving as a heat source for a reversible cycle. The heat reversibly transferred would then be

$$\Delta q_{\text{rev}} = \int T \, ds = \Delta h \bigg]_{p=C} = c_p(T_2 - T_1) = 1.0(60 - 100) = -40 \text{ Btu/lb}_m$$

The change in entropy would equal

$$\Delta s = c_p \ln \frac{T_2}{T_1} = 1.0 \ln \frac{520}{560} = -0.075 \text{ Btu/(lb}_m)(°R)$$

And, by Eq. (7-11b), the available energy that could be regained is

$$\Delta \alpha = \Delta q_{\text{rev}} - T_0 \Delta s = -40 - 492(-0.075) = -40 + 36.9 = -3.1 \text{ Btu/lb}_m$$

Therefore, the *increase* in available energy from paddling the water must have been +3.1 Btu/lb$_m$.

The isolated system consists of a work reservoir, which furnished −40 Btu/lb$_m$ of available energy, and the water, which increased in available energy by 3.1 Btu/lb$_m$. Hence, the available energy of the isolated system *decreased* in amount −36.9 Btu/lb$_m$. And this answer agrees, of course, with Eq. (7-12).

Only a reversible cycle can produce work equal in amount to the availability of the heat that it receives; if irreversibilities arise in the cycle, the work will be *less* and the heat rejected *greater* than in the ideal case. Consider, for example, the Carnot cycle in Fig. 7-7a, wherein the heat reversibly added is represented by area $b'bcc'$, the heat reversibly rejected by area $b'adc'$, and therefore the availability of the heat supplied

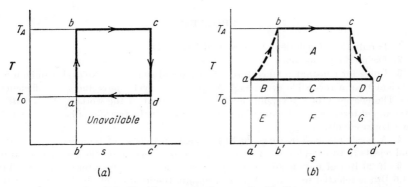

Fig. 7-7. Significance of areas on the $Ts$ diagram.

by area *abcd*.  Now suppose the compression and expansion processes are irreversible (but adiabatic), while a temperature difference prevents rejection of heat at the temperature $T_0$ of the atmosphere, as shown in Fig. 7-7*b*.  Here the heat added is premised to be the same as before (area *b'bcc'*), but now the heat rejected is much larger (area *a'add'*).  The various areas in Fig. 7-7*b* have the following significance:

Area $A + C$    Available energy supplied to the cycle by the reversible heat addition

Area $F$    Unavailable energy accompanying the reversible heat addition

Area $E$    Loss of available energy in the irreversible adiabatic compression process

Area $G$    Loss of available energy in the irreversible adiabatic expansion process

Area $B + C + D$    Loss of available energy because of the temperature difference between cycle and sink

The work of the cycle is the difference between the available energy supplied and that "lost" in the various processes.  If the paths *ab* and *cd* are well defined, the dissipations of available energy for these two processes can also be shown [Eq. (7-10*b*)]:

$$\text{Area } a'abb' = \int_a^b T\, ds \qquad \text{friction in adiabatic process } ab$$
$$\text{Area } c'cdd' = \int_c^d T\, ds \qquad \text{friction in adiabatic process } cd$$

Note that the friction is greater than the loss of available energy for a process.  Note, too, that no area in Fig. 7-7*b* is proportional to the work of the irreversible cycle.

However, it is not always possible to represent friction and changes in available energy on the *Ts* diagram so nicely, since the usual process involves both friction and heat transfer and the path of the process is usually indeterminate.

### PROBLEMS

**1.** Is radiation a violation of the Second Law?  Discuss.

**2.** Define the terms *source*, *sink*, and *heat reservoir*.

**3.** Prove for constant temperatures of source and sink:  (*a*) No cycle can be more efficient than a reversible cycle.  (*b*) All reversible cycles have the same efficiency. (*c*) The efficiency of a reversible cycle is independent of the working fluid.  (*d*) A system not receiving work cannot have as a net result the transfer of heat from a cold to a hot reservoir.

**4.** A reversible cycle receives 20 Btu of heat at a constant temperature of 180°F and rejects 15 Btu at a lower constant temperature.  Determine the sink temperature.

**5.** If 20 Btu of heat is added to a Carnot cycle at a temperature of 212°F while 14.6 Btu is rejected to a sink at 32°F, compute the location of absolute zero on the Fahrenheit scale.

**6.** Devise an absolute temperature scale that would read 100° at the boiling point of sulfur, and determine the temperature on this scale for the freezing point of water (boiling point of sulfur is 444.6°C).

**7.** A Carnot cycle has the same efficiency between 1000°R and 500°R and between $x$°R and 1000°R (the sink). Determine $x$.

**8.** An inventor claims that a new cycle will develop 5 hp for a heat-addition rate of 300 Btu/min. The highest temperature in the cycle is 3000°R, while the lowest is 1000°R. Are his claims possible?

**9.** A reversible power cycle is used to drive a reversible refrigeration cycle. The power cycle takes in heat $\Delta Q_1$ at $T_1$ and rejects heat $\Delta Q_2$ at $T_2$. The heat pump abstracts heat $\Delta Q_4$ from the sink at $T_4$ and discharges $\Delta Q_3$ at $T_3$. Derive a general expression for the ratio $\Delta Q_4/\Delta Q_1$ in terms of temperature alone.

**10.** A Carnot cycle operates between 300 and 100°F while 100 Btu of heat is supplied. Determine the thermal efficiency, work, heat rejected, and change in entropy for each process. Draw the cycle on the $Ts$ diagram.

**11.** What is the independent variable underlying Eq. (7-3c)? Are the variables $Q_{rev}$, $S$, and $T$ properties? If they are properties, how was each established as a property?

**12.** Is it feasible to measure the heat reversibly transferred, say in an isothermal process of a gas? If it is not feasible, of what use is Eq. (7-3c)? (Note Art. 7-4.)

**13.** Show that Eq. (6-4a) is an inexact differential equation for the variables $u$ and $v$ by applying the test of exactness (Art. A-6k).

**14.** Can the test of exactness show Eq. (7-4a) to be exact?

**15.** Construct a portion of a $Ts$ diagram for air considered to be an ideal gas. Show two isentropic, two constant-volume, and two constant-pressure processes. Use $c_v = 0.17$ and $c_p = 0.24$ Btu/(lb$_m$)(°R), $M = 29$.

**16.** A gas cycle consists of three reversible processes: $ab$, isothermal compression; $bc$, constant-pressure expansion; and $ca$, reversible adiabatic. Draw the cycle on the $Ts$ diagram, and indicate areas for the heat added, heat rejected, and the work of the cycle.

**17.** Repeat Prob. 16, but assume that all processes are irreversible although the same sequences of states are encountered. Discuss.

**18.** A gas cycle has the following reversible processes: $ab$, isentropic compression; $bc$, constant-pressure addition of heat; $cd$, constant-volume addition of heat; $de$, adiabatic expansion; $ea$, constant-pressure compression. Draw the cycle on the $Ts$ diagram, and label areas that are proportional to heat or work quantities.

**19.** Express the change in entropy from $v_1$, $T_1$ to $v_1$, $T_2$ if $c_v = A + BT + CT^2$.

**20.** A closed system undergoes a reversible process from 500 to 600°R with the heat transferred per degree increase in temperature being given by $dQ/dT = 0.5$ Btu/deg and the work by $dW/dT = 1 - (10^{-4})T$ Btu/deg. Find the change in internal energy and in entropy for the process.

**21.** During a reversible process, the temperature and entropy are related by the equation $T = As + Bs^2$, wherein $A$ and $B$ are constants. Express the heat added between the temperatures $T_1$ and $T_2$.

**22.** An iron rivet at 1800°F and weighing 1 lb$_m$ is dropped into an insulated bucket containing 10 lb$_m$ of water and 5 lb$_m$ of ice in temperature equilibrium. Determine the final temperature and change of entropy for the adiabatic system (Table B-3).

**23.** A pound of ice at 32°F is dropped into 6 lb$_m$ of water at 80°F. Calculate the change in entropy of the adiabatic system.

**24.** A pound of ice at 32°F is dropped into 3 lb$_m$ of water at 70°F. Calculate the change in entropy of the adiabatic system.

**25.** A stone weighing 50 $lb_m$ falls from a height of 1,000 ft *in vacuo* and strikes a 100-$lb_m$ iron plate. For the adiabatic system of stone and plate, compute the change in entropy. Assume that the heat capacities of stone and plate are equal, with a value of 0.2 Btu/($lb_m$)(°R), and that both objects experience the same temperature rise from 60°F.

**26.** A tub contains 30 $lb_m$ of water at 150°F. It is desired to cool the water by adding 15 $lb_m$ of cold water at 60°F. (*a*) Compute the entropy change for the adiabatic mixing process of cold and hot water. (*b*) Sketch the processes followed by the hot and cold water (considered to be separate entities) on a $Ts$ diagram.

**27.** A rigid container filled with an ideal gas is equipped with a paddle wheel. The state of the gas before rotation of the paddle wheel is $p = 15$ psia, $V = 10$ ft³, $t = 60°F$. After paddling, $p = 22$ psia, $V = 10$ ft³, and $t = 300°F$. During the process, 5 Btu of heat was irreversibly transferred to the surroundings, which are at 40°F. Calculate the change in entropy per pound of gas if $c_p = 0.25$ and $c_v = 0.20$ Btu/($lb_m$)(°R).

**28.** A reversible cycle consists of an isentropic compression from initial temperature to 1000°R, a constant-volume process from 1000 to 1500°R, a reversible adiabatic expansion to 1000°R, a constant-pressure expansion from 1000 to 1500°R, followed by a constant-volume process to the initial temperature. Draw the cycle on $Ts$ and $pv$ diagrams, and calculate the initial temperature if $k = c_p/c_v = 1.40$.

**29.** For the data of Prob. 28, calculate the thermal efficiency of the cycle if $T_1 = 566°R$. Calculate the availability of the heat added to the cycle and rejected from the cycle if the atmosphere is at 60°F and $c_p = 0.24$ Btu/($lb_m$)(°R). (The difference in availability should check the product of thermal efficiency and heat added.)

**30.** Consider the iron rivet in Prob. 22 to be a heat reservoir, and calculate the availability of the heat transferred in cooling the rivet to 32°F (which is $T_0$). Since the cooling of the rivet does not melt all the ice, the temperature of the ice water remains at 32°F. What happened to the available energy originally in the rivet?

**31.** For the data of Prob. 23, calculate the change in availability of the ice and of the water and the net change. Illustrate the results of the adiabatic process on the $Ts$ diagram for each constituent considered alone.

**32.** For the data of Prob. 25, calculate the increase in availability of the two masses from temperature increase and the loss of available energy of the stone in falling.

**33.** From a substance which holds a constant temperature of 3000°R, 200 Btu of heat is transferred to another substance, which accepts the heat at a constant temperature of 800°R. If the atmosphere is at 60°F, calculate the loss of available energy arising from the heat transfer.

**34.** An ideal gas at 600°R with $k = 1.4$ is isentropically compressed from 60 to 10 ft³/$lb_m$. Heat of amount 1200 Btu/$lb_m$ is then reversibly added at constant temperature, followed by an adiabatic expansion to the original temperature. The cycle is closed by a reversible isothermal rejection of heat. The cycle efficiency is 40 per cent; calculate the work done. Draw the cycle on the $Ts$ diagram. Does the cycle have a name?

**35.** (*a*) For an internally irreversible process, the entropy change (1) is always zero; (2) is always positive; (3) is always negative; (4) may be positive, negative, or zero. (*b*) If heat exchange occurs between two systems because of an appreciable temperature difference, the entropy change of the adiabatic system comprised of the two systems is (1) positive, (2) zero, (3) negative, (4) indeterminate. (*c*) For an isentropic process, the heat capacity is 0, ∞, $c_p$, $c_v$, or indeterminate. (*d*) When heat is reversibly added to a gas while it expands and its temperature drops, then $\Delta s$ for the gas is +, −, ∞, or indeterminate.

## SELECTED REFERENCES

1. Planck, M.: "Treatise on Thermodynamics," Dover Publications, Inc., New York, 1926.
2. Keenan, J. H.: "Thermodynamics," John Wiley & Sons, Inc., New York, 1941.
3. Schmidt, E.: "Thermodynamics," Oxford University Press, New York, 1949.
4. Lewis, G. N., and M. Randall: "Thermodynamics and the Free Energy of Chemical Substances," McGraw-Hill Book Company, Inc., New York, 1923 or 1961.
5. Bridgman, P. W.: "The Nature of Thermodynamics," Harvard University Press, Cambridge, Mass., 1941.
6. Carnot, S.: "Reflections on the Motive Power of Heat," translated by R. H. Thurston, American Society of Mechanical Engineers, New York, 1943.
7. Keenan, J. H.: Adventure in Science, *Mech. Eng.*, May, 1958, p. 79.
8. Denbigh, K. G.: "Chemical Equilibrium," Cambridge University Press, New York, 1955.
9. Sears, F. W.: "Thermodynamics," Addison-Wesley Publishing Company, Reading, Mass., 1953.
10. Rushbrooke, G. S.: "Introduction to Statistical Mechanics," Oxford University Press, New York, 1955.

# CHAPTER 8

# SECOND AND THIRD LAW TOPICS

The rank is but the guinea's stamp,
The man's the gowd for a' that.
*Robert Burns*

It has been emphasized that mechanical engineering thermodynamics has two objectives: energy analyses and property interrelationships. In this chapter the latter topic will be primarily considered.

**8-1. Differential Relationships for Systems of Constant Chemical Composition.** The fundamental equation to describe processes of a closed system of unit mass was derived in Art. 7-4:

$$du = T \, ds - p \, dv \qquad (8\text{-}1a)$$

Other properties are arbitrarily defined:

$$h \equiv u + pv \qquad \text{(enthalpy)} \qquad (2\text{-}3b)$$
$$a \equiv u - Ts \qquad \text{(Helmholtz free energy)} \qquad (8\text{-}2a)$$
$$g \equiv h - Ts \qquad \text{(Gibbs free energy)} \qquad (8\text{-}2b)$$

Then by differentiating Eq. (2-3b),

$$dh = du + p \, dv + v \, dp$$

and by substituting Eq. (8-1a),

$$dh = T \, ds + v \, dp \qquad (8\text{-}1b)$$

Similarly,

$$da = -p \, dv - s \, dT \qquad (8\text{-}1c)$$
$$dg = v \, dp - s \, dT \qquad (8\text{-}1d)$$

Equations (8-1) are the means for obtaining property relationships. For example, Eq. (8-1a) declares that $u = f(s,v)$; it follows that

$$du = \frac{\partial u}{\partial s}\bigg)_v ds + \frac{\partial u}{\partial v}\bigg)_s dv \qquad (A\text{-}5b)$$

And by comparing Eqs. (A-5b) and (8-1a),

$$\frac{\partial u}{\partial s}\bigg)_v = T \qquad \frac{\partial u}{\partial v}\bigg)_s = -p \qquad (8\text{-}3a)$$

The second cross partials of a continuous function with continuous derivatives are equal (Art. A-6$k$). Hence differentiate Eqs. (8-3$a$):

$$\frac{\partial^2 u}{\partial s\, \partial v} = \frac{\partial T}{\partial v}\Big)_s = \frac{\partial^2 u}{\partial v\, \partial s} = -\frac{\partial p}{\partial s}\Big)_v$$

and obtain the

*Maxwell from u*
$$\frac{\partial T}{\partial v}\Big)_s = -\frac{\partial p}{\partial s}\Big)_v \tag{8-3b}$$

Or, by operating on Eq. (8-1$c$), obtain the

*Maxwell from a*
$$\frac{\partial p}{\partial T}\Big)_v = \frac{\partial s}{\partial v}\Big)_T \tag{8-3c}$$

Equations (8-3) illustrate one form of the relationships between properties that can be found from mathematical manipulation.

**Example 1.** Check the data for superheated steam at 500 psia, 700°F, by the Maxwell from $u$.

*Solution.* Select $p$, $s$ data, all for the same specific volume (that at 500 psia, 700°F), from the Steam Tables, and plot; measure the slope of the graph at 500 psia. This slope measures about

$$\frac{\partial p}{\partial s}\Big)_v = 1,490\ \frac{\text{lb}_f/\text{in.}^2}{\text{Btu/lb}_m\ °\text{F}}\ \frac{144\ \text{in.}^2/\text{ft}^2}{778\ \text{ft-lb}_f/\text{Btu}} = 273\ \frac{°\text{F}}{\text{ft}^3/\text{lb}_m} \qquad Ans.$$

Next, select $t$, $v$ data, all for the same entropy (that at 500 psia, 700°F), from the Steam Tables, and plot; measure the slope of the graph at 700°F. This slope measures about

$$-\frac{\partial T}{\partial v}\Big)_s = 273\ \frac{°\text{F}}{\text{ft}^3/\text{lb}_m} \qquad Ans.$$

The two values are equal, as predicted by Eq. (8-3$b$).

Another objective is to change the differential equation into one involving a heat capacity and a $pvT$ relationship. For example, since $u = f(T,v)$,

$$du = \frac{\partial u}{\partial T}\Big)_v dT + \frac{\partial u}{\partial v}\Big)_T dv \tag{A-5b}$$

and since the first coefficient is recognized to be $c_v$ [Eq. (2-4$a$)],

$$du = c_v\, dT + \frac{\partial u}{\partial v}\Big)_T dv$$

Half of the objective is now fulfilled. To complete the task, select Eq. (A-6$b$) (because one of us, fortunately, has found the right path by trial):

$$\frac{\partial u}{\partial v}\Big)_T = \frac{\partial u}{\partial s}\Big)_v \frac{\partial s}{\partial v}\Big)_T + \frac{\partial u}{\partial v}\Big)_s \tag{A-6b}$$

And by substituting Eqs. (8-3a) and (8-3c),

$$\left(\frac{\partial u}{\partial v}\right)_T = T\left(\frac{\partial p}{\partial T}\right)_v - p \qquad (8\text{-}4a)$$

(Thus the change of internal energy with isothermal compression can be calculated without heat measurements from the equation of state alone.) The complete differential equation is

$$du = c_v\,dT + \left[T\left(\frac{\partial p}{\partial T}\right)_v - p\right]dv \qquad (8\text{-}4b)$$

Since $du$ is an exact differential, any path between two states can be selected for integration. An "easy" path is to let temperature changes occur at constant volume and to let volume changes occur at constant temperature:

$$\Delta u_{v_1,T_1 \text{ to } v_2,T_2} = \int_{T_1}^{T_2} c_v\,dT\Big]_{v_1 = C} + \int_{v_1}^{v_2}\left[T\left(\frac{\partial p}{\partial T}\right)_v - p\right]_{T_2 = C} dv \qquad (8\text{-}4c)$$

To calculate $\Delta u$, note that heat measurements must be made to obtain $c_v$ values at $v_1$ (and at various temperatures), and the equation of state must be established [so that substitutions can be made for $\partial p/\partial T$ and $p$ in Eq. (8-4c)].

By similar procedures, it can be shown that

$$dh = c_p\,dT + \left[v - T\left(\frac{\partial v}{\partial T}\right)_p\right]dp \qquad (8\text{-}5)$$

And, by other procedures,

$$c_p - c_v = T\left(\frac{\partial p}{\partial T}\right)_v\left(\frac{\partial v}{\partial T}\right)_p \qquad (a)$$

Hence it is not necessary to measure both $c_p$ and $c_v$, since one of these properties can be found from Eq. (a) and the equation of state.

The slope of the vapor-pressure curve (Fig. 9-2) is dictated by the Clapeyron equation

$$\frac{dp}{dT} = \frac{\Delta h_{\text{phase change}}}{T\,\Delta v_{\text{phase change}}}\bigg]_{T,p}$$

Or conversely, by measuring the slope and the volume change with change in phase, the latent heat of phase change can be calculated.

**8-2. Extensive Property Equations.** Consider Eq. (8-1a) in the form

$$dU = T\,dS - p\,dV\Big]_{n_a,n_b,\,\ldots\,=\,C} \qquad (8\text{-}1a)$$

Recall that Eq. (8-1a) was derived for a system of fixed mass in equilibrium. But suppose that the moles of constituents $n_a$, $n_b$, etc., are considered to be independent variables. Then Eq. (8-1a) implies that

$$U = f(S,V,n_a,n_b, \ldots)$$

and therefore

$$dU = \frac{\partial U}{\partial S}\,dS + \frac{\partial U}{\partial V}\,dV + \frac{\partial U}{\partial n_a}\,dn_a + \frac{\partial U}{\partial n_b}\,dn_b + \cdots \qquad (a)$$

By comparing (a) and (8-1a), it follows that

$$dU = T\,dS - p\,dV + \frac{\partial U}{\partial n_a}\,dn_a + \frac{\partial U}{\partial n_b}\,dn_b + \cdots \qquad (8\text{-}6a)$$

By repeating the foregoing steps for the properties $H$, $A$, and $G$,

$$dH = T\,dS + V\,dp + \frac{\partial H}{\partial n_a}\,dn_a + \frac{\partial H}{\partial n_b}\,dn_b + \cdots \qquad (8\text{-}6b)$$

$$dA = -S\,dT - p\,dV + \frac{\partial A}{\partial n_a}\,dn_a + \frac{\partial A}{\partial n_b}\,dn_b + \cdots \qquad (8\text{-}6c)$$

$$dG = -S\,dT + V\,dp + \frac{\partial G}{\partial n_a}\,dn_a + \frac{\partial G}{\partial n_b}\,dn_b + \cdots \qquad (8\text{-}6d)$$

Equation (8-6a) can be converted into an equation for $dH$ by adding $d(pV)$ to both sides of the equation [and $dU$ into $dA$ by subtracting $d(TS)$ and similarly for $dH$ into $dG$]. Then by comparing the converted equations with (8-6), the following equalities must hold:

$$\mu_j \equiv \frac{\partial G}{\partial n_j}\bigg)_{T,p,n_a,\ldots} = \frac{\partial U}{\partial n_j}\bigg)_{S,V,n_a,\ldots} = \frac{\partial H}{\partial n_j}\bigg)_{S,p,n_a,\ldots} = \frac{\partial A}{\partial n_j}\bigg)_{T,V,n_a,\ldots} \qquad (8\text{-}7)$$

Since $\mu_j$ is a function of properties, it too is a property (Art. 2-11) and is called *the chemical potential of constituent j*. The value of the chemical potential is unchanged when the size of the homogeneous system is increased or decreased at constant temperature, pressure, and chemical composition; hence it is an intensive property [note Eq. (8-6d)].

As the name implies, the chemical potential can be shown to be the force function for chemical or phase equilibrium, just as temperature is the potential for thermal equilibrium and pressure the potential for mechanical equilibrium. Thus two phases of a pure substance are in equilibrium when their chemical potentials are equal; a constituent will diffuse from one phase into another if the chemical potential of the constituent is higher in one phase than in the other; or chemical change will occur unless the chemical potentials of reactants and products are in balance.[1]

For example, if Eq. (8-6a) is applied to a closed system of water and steam in equilibrium,

$$dU = T\,dS - p\,dV + \mu_f\,dn_f + \mu_g\,dn_g$$

Since the change in amount of water must be exactly compensated by an opposite change in the amount of steam,

$$dU = T\,dS - p\,dV + (\mu_f - \mu_g)\,dn_f$$

And by accepting that the chemical potentials of water and steam in equilibrium are equal ($\mu_f = \mu_g$), just as $p_f = p_g$ and $T_f = T_g$,

$$dU = T\,dS - p\,dV \tag{7-4a}$$

An equivalent type of argument can be made for chemical equilibrium. *Thus Eq. (7-4a) holds for closed systems doing only mechanical work in phase or chemical equilibrium.*

Suppose that the closed system contains inert constituents. Then $dn_a$, $dn_b$, etc., in Eq. (8-6a) are zero and Eq. (7-4a) is again obtained. *Thus Eq. (7-4a) also holds for closed systems of inert constituents* (doing only mechanical work).

**8-3. The Reaction Equation.**  Consider the chemical reaction

$$\nu_a a + \nu_b b = \nu_c c + \nu_d d$$

When pure reactants $a$, $b$ are converted into pure products $c$, $d$, the change in the extensive property $X$ for the reaction is *defined* as

$$\Delta X \equiv X_c + X_d - X_a - X_b = \nu_c x_c + \nu_d x_d - \nu_a x_a - \nu_b x_b$$

wherein $x_a$, $x_b$, etc., and $X_a$, $X_b$, etc., are necessarily *absolute values* of the particular property.  For example, the complete conversion of 1 mole of CO and $\frac{1}{2}$ mole of $O_2$, each at a reference or *standard state*, into 1 mole of $CO_2$ at a standard state is shown as

$$CO(g) + \tfrac{1}{2}O_2(g) \rightarrow CO_2(g) \qquad \Delta H^\circ = -121{,}664 \text{ Btu}$$

Here
$$\Delta H^\circ = H^\circ_{CO_2} - H^\circ_{CO} - H^\circ_{O_2}$$

The gaseous phase of each constituent of the reaction is shown by the letter $g$ (and $l$ and $s$ would be used if the *liquid* and *solid* phases were present).  The change in *absolute* enthalpy for the reaction is $-121{,}664$ Btu for each mole of $CO_2$ that is formed; thus, the absolute enthalpy of $CO_2$ at its standard state is $-121{,}664$ Btu *less* than that of 1 mole of CO and $\frac{1}{2}$ mole of $O_2$, each at its standard state (or the enthalpy of 1 mole of $CO_2$ *relative* to 1 mole of CO and $\frac{1}{2}$ mole of $O_2$ is $-121{,}664$ Btu).  Although usable values (see Art. 2-5) of the absolute internal energy or enthalpy of a single substance are not available, note that $\Delta H$ (or $\Delta X$) is indeed a difference of absolute quantities.

The use of the *standard state* for recording data reduces the number of values of $\Delta H$, for example, that can be found in the literature for a particular chemical reaction. Thus, for gases the degree sign usually signals that each constituent of the reaction was evaluated at 1 atm and at a temperature of 25°C (77°F) (18°C has also been used). Although the standard state implies a definite pressure, this requirement is not essential for the properties of internal energy and enthalpy if the gases behave ideally because the internal energy and enthalpy of ideal gases are dependent upon temperature alone (and the internal energies of solids and liquids are little affected by pressure). For properties such as volume and entropy, however, pressure is a strong variable.

Since the value of a property is fixed by the state, the change in a property in passing from reactants to products is entirely independent of the intermediate products that may be formed or the transfers of heat and work that may accompany the change. When the enthalpy of reaction of solid carbon is found by experiment to be

$$C(s) + O_2(g) \rightarrow CO_2(g) \qquad \Delta H^\circ_{77°F} = -169,182 \text{ Btu} \qquad (a)$$

and also for carbon monoxide for the same standard states

$$CO(g) + \tfrac{1}{2}O_2(g) \rightarrow CO_2(g) \qquad \Delta H^\circ_{77°F} = -121,664 \text{ Btu} \qquad (b)$$

then $(b)$ can be subtracted from $(a)$ to yield

$$C(s) + \tfrac{1}{2}O_2(g) = CO(g) \qquad \Delta H^\circ_{77°F} = -47,518 \text{ Btu} \qquad (c)$$

This procedure was possible because the $CO_2$, for example, was in the same state in either equation and therefore its enthalpy in each equation had the same value. In this manner, the enthalpy of reaction can be calculated for reactions [such as $(c)$] that cannot be experimentally performed (since both CO and $CO_2$ would be formed in burning C).

To facilitate the calculation of enthalpies (or free energies) of reaction, a table of relative enthalpies (or relative free energies) can be constructed. The procedure is to set equal to zero, arbitrarily, the enthalpy (or free energy) of each element in its standard state. Then, when a compound is formed from the elements, the enthalpy (or free energy) of the compound is equal to the $\Delta H$ (or $\Delta G$) for the reaction. For example,

$$C(s) + O_2(g) \rightarrow CO_2(g) \qquad \Delta H^\circ_{77°F} = -169,182 \text{ Btu}$$

and, arbitrarily, at a datum of $t = 77°F$, $p = 1$ atm,

$$h_{C(s)} = 0 \qquad h_{O_2(g)} = 0 \qquad h_{CO_2(g)} = -169,182 \text{ Btu/mole}$$

*The enthalpy of a compound relative to its elements is called the heat of formation or enthalpy of formation.* The standard states must always be specified.

The enthalpy of reaction can be calculated from the enthalpies of formation. For example,

$$CO(g) + H_2O(l) \rightarrow CO_2(g) + H_2(g)$$

and for this reaction

$$\Delta H° = h°_{CO_2} + h°_{H_2} - h°_{CO} - h°_{H_2O(l)}$$

Values can be substituted from Table B-13,

$$\Delta H° = -169,182 + 0 + 47,518 + 122,891$$
$$\Delta H°_{77°F} = 1227 \text{ Btu}$$

**8-4. The Power Process with Combustion.** The creation of a heat source for the commercial power cycle is accomplished by burning a fuel, such as coal or fuel oil, with atmospheric air. A flow process is invariably used because of the large quantities of heat demanded by commercial systems. Figure 8-1 illustrates the usual furnace. Air and fuel at atmospheric temperature and pressure enter the system, and an irreversible mixing process takes place. The mixture is not in chemical equilibrium, but the inertness of the substances prevents reaction until a spark or a flame is present. Then an irreversible chemical reaction occurs with the attainment of a high temperature. As heat is transferred from this heat source, the temperature of the hot gases falls until the limiting temperature is approached—that of the atmosphere.

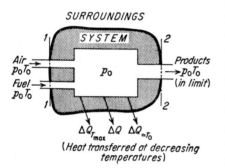

FIG. 8-1. System of combustion furnace.

Let the steady-flow energy equation be applied to the furnace of Fig. 8-1,

$$\Delta q = h_2 - h_1 \Big]_{\Delta KE, \Delta PE = 0} \qquad (5\text{-}7d)$$

Here $h_2$ is for the products of combustion and $h_1$ for the reactants. Assume combustion is complete (complete conversion of reactants into products), that the process occurs at constant pressure, and that the products are cooled back to the initial temperature. Then, for $M$ units of mass† of reactants and products, the heat received from the furnace equals

$$\Delta Q \Big]_M = H_2 - H_1 \Big]_{p,T} \qquad (8\text{-}8)$$

† In general, equations of processes involving a chemical reaction will be shown with extensive values for the properties to warn of the change in composition.

Thus, the limiting amount of heat that can be obtained from the furnace equals the enthalpy of reaction (Art. 4-3$f$). The usual furnace can supply about 85 per cent of this maximum to a power cycle, while the remainder is "lost" to the surroundings; the efficiency of the furnace, then, is 85 per cent. The efficiency is thus defined as

$$\eta_{\text{furnace}} \equiv \frac{\Delta Q_{\text{useful}}}{\Delta H_{p,T}} \tag{8-9}$$

wherein $-\Delta H$ is called the *heating value* of the fuel.

The heat passes from furnace to power cycle but here the Second Law intervenes and the work obtained can be no greater than the availability of the heat. Hence, *the work of the commercial power cycle is far less than the heating value of the fuel because of the Second Law restriction on a cycle, because of the inefficiencies of the furnace, and because of the inefficiencies of the cycle.*

**Example 2.** Methane and air are burned at atmospheric pressure in a furnace and a temperature of 3000°F attained. Determine the thermal efficiency of a reversible cycle (the system) which uses the hot gases (in the surroundings) as a source of heat. [Assume that $c_p = 0.25$ Btu/(lb$_m$)(°R) for the gases and that the atmospheric temperature is 60°F.]

*Solution.* The irreversible combustion process is not included in the system, and hence it is not a part of the problem. Let the hot gases be reversibly cooled at constant pressure to $T_0$,

$$\Delta q_{\text{rev}} = \Delta h \bigg]_{p=C} = c_p(T_0 - T_1) = 0.25(520 - 3{,}460) = -735 \text{ Btu/lb}_m \text{ hot gases}$$

The change in entropy of the gas is

$$\Delta s = c_p \ln \frac{T_0}{T_1} = 0.25 \ln \frac{520}{3{,}460} = -0.475 \text{ Btu/(lb}_m)(°R)$$

Substituting these values into Eq. (7-11$b$),

$$\Delta \alpha = \Delta q_{\text{rev}} - T_0 \Delta s = -735 + 520(0.475) = -489 \text{ Btu/lb}_m$$

(The minus sign applies to the heat reservoir.) Thus, the work of a reversible cycle which reversibly cools the combustion gases to $T_0$ is $+489$ Btu/lb$_m$. The thermal efficiency of the reversible cycle equals

$$\eta_t = \frac{\Delta \alpha}{\Delta Q_A} = \frac{489}{735} = 0.665, \text{ or } 66.5\% \quad Ans.$$

**8-5. The Reversible Power Process.** The power process of Art. 8-4 produced work because of a chemical reaction that began with the reactants in temperature and pressure equilibria with the surrounding atmosphere and ended with the products of the reactions in a similar state. The question to be answered is this: If the chemical reaction had been reversibly accomplished, what would be the maximum work?

The work of a closed system is dictated by its energy equation,

$$-\Delta W = \Delta U - \Delta Q \Big]_{\Delta KE, \Delta PE = 0} \qquad (4\text{-}4d)$$

If $\Delta W$ is to be a maximum, the process must be reversible and by Eq. (7-7)

$$-\Delta W_{\substack{\text{rev} \\ \text{closed} \\ \text{system}}} = \Delta U - \int T\, dS$$

Since the surroundings have a constant temperature, reversible transfers of heat can be only at this temperature (which is also the temperature of the end states),

$$\Delta W_{\substack{\text{rev} \\ \text{closed} \\ \text{system}}}\Big]_T = -\Delta U + T\,\Delta S \qquad (a)$$

Since the surroundings have a constant pressure, the work to displace the surroundings equals

$$p(V_2 - V_1) = \Delta pV$$

When this amount is subtracted from Eq. ($a$), the *net* or *useful* work that can be delivered by the reversible process is obtained,

$$\Delta W_{\substack{\text{rev net} \\ \text{closed} \\ \text{system}}}\Big]_{T,p} = -\Delta U - \Delta pV + T\,\Delta S = -\Delta H + T\,\Delta S \qquad (b)$$

The foregoing steps can be repeated for a steady-flow open system,†

$$-\Delta W = \Delta H - \Delta Q \Big]_{\substack{\Delta KE, \Delta PE = 0 \\ M \text{ units of mass}}}$$

And for reversible transfers of heat at the constant temperature $T$ of the surroundings and the end states,

$$\Delta W_{\substack{\text{rev} \\ \text{steady-flow}}}\Big]_T = -\Delta H + T\,\Delta S \qquad (c)$$

If the steady-flow process begins and ends at the pressure $p$, Eqs. ($b$) and ($c$) become identical.

The interesting aspect of Eqs. ($a$) to ($c$) is this: The work of a reversible process that uses only the constant-temperature surroundings as the medium for reversible heat transfers is independent of the path and depends only on the end states. Thus, *high temperatures or high pressures during the power process are not demanded.*

---

† And the steady-flow process is evaluated for $M$ units of mass, so that the capitalized symbols can warn of the change in composition.

With Eqs. (8-2a) and (8-2b), it follows that

$$\Delta W_{\substack{\text{rev} \\ \text{closed} \\ \text{system}}} \Big]_T = -\Delta A \tag{8-10a}$$

$$\Delta W_{\substack{\text{rev net} \\ \text{closed} \\ \text{system}}} \Big]_{T,p} = -\Delta G \tag{8-10b}$$

$$\Delta W_{\substack{\text{rev} \\ \text{steady-flow}}} \Big]_T = -\Delta G \tag{8-10c}$$

Most if not all of the commercial power processes begin and end at essentially the same temperature and pressure as the surroundings, and therefore Eqs. (8-10) can be the criteria of the maximum work from such processes. Consider the general process in Fig. 8-2, which can represent any one of various commercial systems. Equation (8-10c) predicts the maximum work to be

$$\Delta W_{\substack{\text{rev} \\ \text{steady-flow}}} \Big]_{T,p} = -\Delta G$$
$$= -\Delta H + T\,\Delta S$$

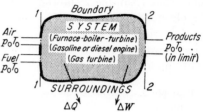

FIG. 8-2. General heat power system.

But this work is greater or less than the negative of the enthalpy of reaction (the heating value) according as the sign of $\Delta S$ is positive or negative. Since the process is reversible, a positive $\Delta S$ can arise only if heat flows into the system from the surroundings and, by this aid, the work output is increased to a value greater† than the heating value of the fuel. Typical values of $\Delta H°$ and $\Delta G°$ are as follows:

|  | $-\Delta H°$, Btu/lb$_m$ | $-\Delta G°$, Btu/lb$_m$ |
|---|---|---|
| Coal (carbon)................... | 14,087 | 14,118 |
| Carbon monoxide (gas)......... | 4,344 | 3,942 |
| Methane (gas)................. | 21,502 | 21,069 |
| Octane (liquid) (gasoline)....... | 19,256 | 19,647 |

Recall (Art. 8-4) that, for the system of furnace and cycle, the Second Law restricted *the work of a reversible cycle* to be much less than the heating value of the fuel, but for *the work of a reversible process* the Second Law makes no such restriction!

For certain chemical reactions a galvanic cell can be constructed that will allow the chemical reaction to be conducted under conditions

† And less if $\Delta S$ is negative.

approaching reversibility (Art. 6-8). Then, by Eqs. (8-10$b$) and (6-8$e$),

$$\Delta W_{\substack{\text{rev net}\\\text{closed}\\\text{system}}}\bigg]_{T,p} = -\Delta G\bigg]_{T,p} = n\mathbf{F}\varepsilon \qquad (8\text{-}11)$$

In this manner, values of $\Delta G$ for a chemical reaction can be calculated from experimentally measured values of the reversible potential $\varepsilon$ (obtained in an isothermal process!).

When the galvanic cell principle is used as a power source with a flow system, the device is called a *fuel cell*. Here the maximum work is also predicted by Eq. (8-11), but now reversibility is not approached because speed of operation is demanded. Even so, an irreversible fuel cell may be more efficient than a combustion system that includes a cycle. Consider that Example 2 showed that only about 66 per cent of $-\Delta H$ could be obtained as work from a certain combustion process and reversible cycle. Since the real cycle is inefficient, the over-all efficiency would be much lower, of the order of 30 to 40 per cent. Hence a fuel cell can be quite inefficient and yet outproduce the system of Example 2.

**Example 3.** Calculate the loss of available energy for the combustion process of Example 2.

*Solution.* The maximum work obtainable in a steady-flow process from the methane-oxygen reaction is

$$\Delta W_{\substack{\text{rev}\\\text{steady-flow}}}\bigg]_{T_0} = -\Delta G = 21{,}069 \text{ Btu/lb}_m \text{ methane}$$

The work obtained in Example 2 was

$$\Delta W = \eta_t(-\Delta H) = 0.665(21{,}502) = 14{,}299 \text{ Btu/lb}_m \text{ methane}$$

Hence the loss of available energy because of the irreversible combustion process (the only irreversibility in Example 2) was

$$\Delta \mathcal{I} = -6{,}770 \text{ Btu/lb}_m \qquad Ans.$$

(Although $\Delta G$ at 77°F is not necessarily equal to $\Delta G$ at 60°F, the difference is quite negligible since the change in $G$ with temperature for the products is offset by the change for the reactants.)

**8-6. The Third Law.** The entropy of a pure substance was defined in Arts. 7-3 and 7-4 only up to an arbitrary constant,

$$s = s(T,p) + s_0$$

This procedure was satisfactory since, for changes in state without chemical reaction, only the differences in entropy were of interest and the value of $s_0$ canceled from the calculation. However, an absolute value for entropy can be determined from the theorem known as the *Third Law of Thermodynamics:*

**The entropy of a pure substance in complete thermodynamic equilibrium becomes zero at the absolute zero of temperature.**

The entropy of a *mixture* of substances can *not* be zero at the absolute zero since an entropy of mixing is present (Art. 11-13). Also, a *glass* (supercooled fluid) is not in internal equilibrium at the lowest test temperature and probably retains a positive entropy even as the absolute zero is approached. For these reasons, it is not uncommon to state the Third Law as an unattainability principle. In the words of Fowler and Guggenheim:[3]

**It is impossible by any procedure no matter how idealized to reduce any assembly to the absolute zero in a finite number of operations.**

The Third Law enables the absolute entropies of pure substances to be calculated from the fundamental definition of entropy [Eq. (7-3c)], with $s_0$ set equal to zero,

$$s = \int_0^T \frac{dQ_{rev}}{T}\bigg]_{\substack{\text{pure substance} \\ \text{in equilibrium}}} \tag{8-12}$$

Equation (8-12) can be evaluated if the complete thermal history of the equilibrium states of a substance is known from a temperature close to zero to the temperature $T$. This requires experimental measurements of heat capacities, heats of transition, fusion, and vaporization. The data are then extrapolated to the absolute zero by theoretical considerations. Also, the absolute entropies of many substances are found today from theoretical calculations resting upon the methods of statistical mechanics, for example, the Sackur-Tetrode equation for the translational contributions to the entropy of an ideal gas:

$$s^\circ = \tfrac{3}{2}R \ln M + \tfrac{5}{2}R \ln T - R \ln p - 2.298$$

Table B-13 lists values of absolute entropies for various pure substances found from both experimental and theoretical relationships.

Simon's remarks on the Third Law are illustrated in Fig. 8-3. Classical theory dictates that the heat capacities should tend toward constancy as the temperature approaches zero. Then for an ideal gas

$$s_2 - s_1 = c_v \ln \frac{T_2}{T_1} + R \ln \frac{v_2}{v_1}$$

and $s_1$ becomes negatively infinite (Fig. 8-3a) as $T_1$ approaches zero (and the same conclusion is reached for any substance in any phase with finite heat capacity). The picture was changed by Einstein (1907), who predicted that the heat capacities would fall away to zero at the absolute zero. With this new concept, Fig. 8-3b might be proposed. But the Third Law demands not only that the heat capacities must approach zero but also that entropy differences at the absolute zero must disappear (Fig. 8-3c).

The unattainability of the absolute zero can also be demonstrated by the common intersection of the property lines in Fig. 8-3c. Note that lines of constant entropy cannot intersect; else the Second Law would be broken. Then $T = 0$ is the locus of $s = 0$, which cannot intersect another locus of constant $s$. Nor can $T = 0$ be attained by cooling since this implies the existence of a sink at least at zero temperature. On the other hand, if Fig. 8-3b were true, zero temperature could be reached by a reversible adiabatic expansion (path $bc$).

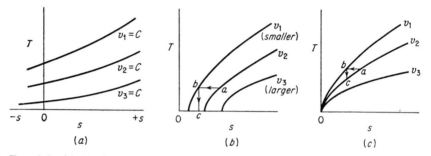

FIG. 8-3. (a) $Ts$ diagram for substance with constant (nonzero) heat capacity; (b) $Ts$ diagram for substance with zero heat capacity at the absolute zero but not obeying the Third Law; (c) $Ts$ diagram for Third Law substance. (After Simon.[2])

With the Third Law, experimental methods may often be replaced by theoretical calculations. For example,

$$\Delta G^\circ \Big]_{T,p} = \Delta H^\circ - T\,\Delta S^\circ \qquad (a)$$

1. $\Delta S^\circ$ can be found from theoretical data (Table B-13), $\Delta H^\circ$ can be found from calorimeter tests (Art. 4-3f and Table B-11), and $\Delta G^\circ$ can be calculated. The emf of a reversible cell can thus be predicted [Eq. (8-11)].

2. Or the emf can be experimentally measured and $\Delta G^\circ$ calculated. Then, from values of $\Delta S^\circ$ [Table B-13 and Eq. (a)], $\Delta H^\circ$ can be predicted (usually with better accuracy than the value found experimentally).

3. Or $\Delta G^\circ$ and $\Delta H^\circ$ can be found experimentally (galvanic cell and calorimeter) and $\Delta S^\circ$ predicted. (Tests of this type confirmed the Third Law in the days when the law was not universally accepted.)

## PROBLEMS

1. Derive Eqs. (8-1c) and (8-1d).
2. Obtain relationships similar to Eqs. (8-3a) from the properties $h$, $a$, and $g$.

**3.** Derive the Maxwells from $h$ and $g$.

**4.** Find $\partial h/\partial p)_T$.

**5.** Derive an equation for $\Delta h$.

**6.** Solve Eq. (8-4c) for an ideal gas and for a van der Waals gas $[p + a/v^2 = RT/(v - b)]$ (with $a$, $b$ constants).

**7.** Derive Eqs. (8-6b), (8-6c), and (8-6d).

**8.** Derive Eq. (8-7).

**9.** For the reaction $CO + \frac{1}{2}O_2 \rightarrow CO_2$, write an equation that shows precisely (by appropriate subscripts, etc.) the meaning of $\Delta H°$.

**10.** Calculate the enthalpy of reaction ($Btu/lb_m$) when burning ethane with oxygen (gaseous products) (Table B-13).

**11.** Calculate $\Delta S°$ for the reaction of ethane with oxygen from absolute entropy values, and check by $\Delta G°$, $\Delta H°$ data in Table B-13.

**12.** Given the three enthalpies of reaction for methane, carbon, and hydrogen reacting with oxygen (Table B-11), calculate the heat of formation of methane.

**13.** An inventor claims that a new power cycle has a thermal efficiency of 75 per cent. His heat source is the usual furnace, with the hot gases from the combustion process at 3260°R being cooled at constant pressure to 260°F at the chimney (stack) exit. Is his claim possible? Given $T_0 = 60°F$ and $c_p = 0.25\ Btu/(lb_m)(°R)$.

**14.** For the data of Prob. 13, assume that methane was the fuel, and calculate the loss of available energy from the combustion process. How can this loss be reduced?

**15.** An ideal gas is sometimes defined with the stipulation that $c_p$ and $c_v$ are constants. Does this defined gas conform to the Third Law?

**16.** Decide the units of the Sackur-Tetrode equation.

**17.** What would be the "efficiency" (work divided by the heat of combustion) of a reversible fuel cell that uses methane and oxygen? Discuss.

**18.** Discuss why a fuel cell can have an "efficiency" which is greater than the Carnot (Second Law) prediction.

**19.** Calculate $\Delta G°$, given $\Delta H°$ and $S°$ values (Tables B-11 and B-13), for the reaction of solid carbon with oxygen to form carbon dioxide at 77°F.

## SELECTED REFERENCES

1. Obert, E. F.: "Concepts of Thermodynamics," chaps. 7, 11, 14, McGraw-Hill Book Company, Inc., New York, 1960.

2. Simon, F. E., N. Kurti, J. Allen, and K. Mendelssohn: "Low Temperature Physics," Pergamon Press, Inc., New York, 1952.

3. Fowler, R. H., and E. A. Guggenheim: "Statistical Thermodynamics," The Macmillan Company, New York, 1939.

# CHAPTER 9

# PROPERTIES OF THE PURE SUBSTANCE

Mine honour is my life, both grow in one;
Take honour from me, and my life is done.
*Shakespeare* (Richard III)

To measure experimentally and to compute analytically the properties of substances are expensive tasks; therefore, relatively complete data are available only for substances that fulfill a commercial need. The development of the steam power plant was accompanied by extensive study of the properties of water; similarly, the growth of the refrigeration industry produced a corresponding increase in knowledge of the properties of refrigerants.

**9-1. The $pvT$ Surface for a Pure Substance.** The interrelationships among pressure, specific volume, and temperature can be shown on a three-dimensional surface such as that illustrated, in part, by Fig. 9-1. Although this drawing is for water, it can be viewed as a characteristic surface for all substances (of course, with different numerical values for the coordinates). The coordinates of a point *on* this surface represent the values for pressure, specific volume, and temperature that the substance must assume if it is to be in a stable equilibrium state. (Conversely, if the properties of the substance do not dictate a point *on* the surface, the substance is in a *metastable* state.) There are three regions, labeled $S, L$, and $G$, on this model, where the substance exists only in a single phase: *solid* (ice), *liquid* (water), and *gas* (steam). In each of these single-phase regions, the state of the substance is fixed by any two of the three properties of pressure, specific volume, and temperature, for all these properties are independent of each other.

Between the single-phase areas are the transitional, or two-phase, regions, where two phases exist in equilibrium: *liquid-vapor*, *solid-vapor*, and *solid-liquid*. The solid-liquid region extends (through $gh$) from $a$ to $b$, and the volume at $b$ is greater than that at $a$. This is because, when ice melts, the volume decreases. (Most substances have the opposite characteristic of increasing volume when the solid is liquefied. For these substances, the solid region will be moved back to $b'$.)

Within the two-phase regions the properties of temperature and pressure are interdependent, for one cannot be changed without changing the other. For this reason, the state of the substance, although fixed by two independent properties, cannot be fixed by temperature and pressure alone. Specific volume and either temperature or pressure can be used for this purpose.

The specific volume of the two-phase mixture is equal to the sum of the volumes (not specific volumes) of each phase because the *mixture* has unit mass. Hence, if the amount of one phase is increased, the amount of the other phase is correspondingly decreased and the specific volume of the

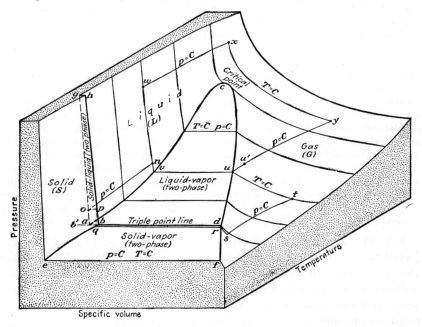

FIG. 9-1. The $pvT$ surface for water (not to scale).

mixture must change because for each phase the specific volumes have different values.

Three phases can exist in equilibrium along line *abd*, which is the locus of the *triple point*. Here, if the amount of one phase is increased, the amount of the *two* other phases is correspondingly decreased but the exact amount for each of these two phases is indeterminate. Therefore, the state of a mixture at the triple point cannot be completely fixed by the properties of pressure, temperature, and specific volume because an infinity of mixtures can exist at any point along line *abd*. However, inspection of the first $T\,dS$ equation [Eq. (7-4a)] shows that in the absence of potential and kinetic effects the state of the system is determined by the values assumed by two of the three properties of entropy, internal energy, and specific volume (Art. 8-2).

The liquid from $v$ to $u$ is in equilibrium with vapor and is therefore called *saturated liquid*. The locus of single-phase states of saturated liquid is $ac$. Similarly, the locus of single-phase states of *saturated vapor* is $dc$. *Saturated vapor* is vapor in equilibrium with saturated liquid. States in the two-phase area $acd$ contain mixtures of saturated vapor and saturated liquid.

A number of processes can be devised to allow a traverse to be made on the surface of Fig. 9-1. Consider a liquid at a given pressure and temperature, say point $n$ in Fig. 9-1. When the liquid at state $n$ is cooled at constant pressure, the volume and temperature decrease while the state shifts toward point $o$. Now, when the liquid at $o$ is cooled at constant pressure, ice appears, the volume increases, the temperature remains constant, and the state shifts toward point $p$ while more and more of the two-phase mixture of liquid and solid turns into ice. The solid at state $p$ is heated at constant temperature, while pressure is decreased, the volume increases, and the state shifts toward $q$ along the path $pq$ (lying on surface $S$). The ice at state $q$ is heated at constant pressure, vapor appears, the volume increases, the temperature remains constant, and the state shifts toward point $r$ through the two-phase mixture of solid and vapor. At $r$, the solid has been entirely converted into vapor without the appearance of a liquid phase (sublimation). State $u$ is attained by traversing the gas region: a constant-temperature (expansion) path from $r$ to $s$, a constant-pressure (expansion) path from $s$ to $t$, and a constant-temperature (compression) path from $t$ to $u$. The vapor at state $u$ is cooled at constant pressure; the temperature remains constant, liquid appears, and the state moves toward $v$ while the two-phase mixture of liquid and vapor shrinks into liquid. At state $v$, only liquid is present. The original state $n$ is restored by isothermal compression of the liquid from $v$ to $n$.

**9-2. The $Tp$ Diagram.** Suppose that all points on Fig. 9-1 are projected to the temperature-pressure plane. The resulting plot is called the *equilibrium*, or *phase, diagram*, Fig. 9-2. All two-phase areas are reduced to lines on a diagram of temperature and pressure because these properties are interdependent during a phase change (phase changes at constant pressure also occur at constant temperature).

The triple point[†] that was a line $abd$ on Fig. 9-1 appears as a point on Fig. 9-2. Limit $agh$ slopes to the left for water and to the right for other substances. At extremely high pressures, different phases may be assumed by the solid. Bridgman[‡][§] has identified seven different crystalline forms for ice; the regions where they exist are shown in Fig. 9-2. The pure substance $H_2O$ can exist in at least nine different and distinct phases.

† The *triple point* of water is often confused with the *ice point*. At the triple point, the pressure of 0.089 psia is exerted entirely by the $H_2O$, and the temperature is 32.018°F; at the ice point, the pressure of 14.696 psia is exerted primarily by *air* on a mixture of ice, water, and vapor, and the temperature is 32.000°F.

‡ P. W. Bridgman, High Pressure and Five Kinds of Ice, *J. Franklin Inst.*, **177** (3): 315–332 (1914).

§ P. W. Bridgman, The Phase Diagram for Water, *J. Chem. Phys.*, **5**: 964–966 (December, 1937).

Point $c$ is called the *critical point* and marks the termination of any distinction between liquid and gaseous phases.   If liquid-vapor phases in equilibrium at $uv$ and confined in a glass cylinder are maintained in equilibrium while the temperature and pressure are raised, the attainment of state $c$ will be marked by the disappearance of the meniscus that

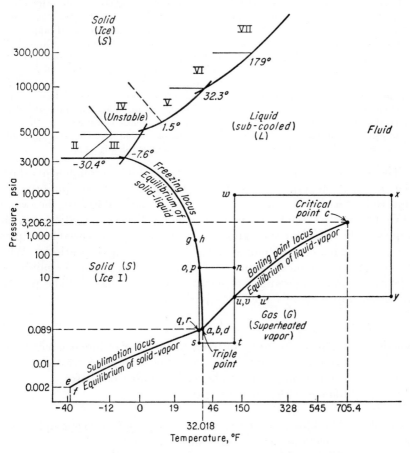

Fig. 9-2. Phase diagram for water (not entirely to scale).

identified the presence of two phases.   In the vicinity of state $c$, the properties of the liquid phase and the vapor phase approach each other in similarity; for example, the specific volumes of liquid and gas approach the same value.   At state $c$, all properties of both phases become identical. Note that a single phase of a fluid when confined in a glass container could not be visually recognized to be a solid, liquid, or gas (unless a change in color occurs with change of phase).

The $Tp$ diagram shows that the liquid and the gas phase cannot be separately identified when the temperature and pressure are above the critical values.   In this region, these two phases merge into one phase without forming a two-phase mixture, and the general name *fluid* is more appropriate than the phase names *liquid* and *gas*.

It is interesting to traverse the fluid region with the glass cylinder and piston. Consider two phases in equilibrium at state $uv$.   If the pressure is raised while temperature is held constant, the vapor phase disappears, as evidenced by disappearance of the liquid meniscus, and the compressed liquid will pass to state $w$.   Now, if the temperature is raised while pressure is held constant, the state $x$ can be attained without the appearance of a phase change.   Similarly, the state can be changed from $x$ to $y$ and $y$ to $u'$ without the appearance of two phases.   But with the attainment of $u$, a liquid phase definitely appears from what was presumably all liquid.   Hence, somewhere along the path $uvwxyu'$ the liquid phase must have changed into the gas phase. Since the change occurred without evidence of two phases being present, the exact point cannot be stated where the describing names of *gas* and *liquid* became applicable.

The path $nopqrstuvn$ followed in Fig. 9-1 is reproduced in Fig. 9-2.

The diagram of Fig. 9-2 for a pure substance (or one component) shows the values of temperature and pressure which allow phase equilibrium, in this case, for water.   If other components are added to the water, obviously the $pT$ relationships of Fig. 9-2 will be changed; hence, the values in the equilibrium diagram are controlled by the variables of pressure, temperature, and composition.   The question then arises:   What effect will there be on the number of phases that can exist in equilibrium when pressure, temperature, or composition is changed?   The possible variations are shown by the Gibbs phase rule

$$F = C - P + 2 \tag{9-1}$$

where $F$ = number of degrees of freedom: number of independent variations of *phase* intensive properties or of concentrations of the components without change in number of phases

   $C$ = number of components (Art. 2-9)

   $P$ = number of phases

The rule can be illustrated by referring to Fig. 9-2 for the simplest case of a one-component system.   For a single phase,

$$F = C - P + 2 = 1 - 1 + 2 = 2$$

Thus, a single phase of a pure substance has two degrees of freedom: any two intensive properties can be independently varied without forming additional phases.   Similarly, if two components are present, $F = 3$ for a single phase, and temperature, pressure, and composition can all be independently varied without the appearance of a new phase.   If two phases of a pure substance are to coexist, $F = 1$; here only temperature (or any other *phase* intensive property) is independent.   In the case of three phases of a pure substance, $F = 0$; hence, no intensive property of a phase can be varied. A more impressive deduction is that the maximum number of phases of a pure substance that can coexist in equilibrium is three, a result evident in Fig. 9-2. The importance of the phase rule is mainly apparent when more than one component is present; thus, the phase rule predicts that four phases can exist in equilibrium for a system of two components.

**9-3. Definitions.** At the risk of some duplication a number of new conditions will be defined here.

Limit *ef, abd, c* of Fig. 9-2 is the *vapor-pressure* curve.

Liquid in equilibrium with vapor is *saturated liquid.*

Vapor in equilibrium with liquid is *saturated vapor.*

Liquid at a lower temperature than the saturated liquid at the same pressure is called *subcooled liquid* (state *n*).

Vapor at a higher temperature than the saturated vapor at the same pressure is called *superheated vapor* or *gas* (states *s, t, u', and y*).

When a phase change occurs from the solid directly to the gas phase, the process is called *sublimation.* The heat required to effect this change (at constant pressure and temperature) is called the *latent heat of sublimation.* Similarly, the heat required for a phase change from solid to liquid at constant pressure and temperature is called the *latent heat of fusion,* while the heat necessary to vaporize liquid into gas at constant pressure and temperature is called the *latent heat of vaporization.* The term *latent heat* is used because no rise in temperature accompanies the transfer of heat during a phase change.

**9-4. The *pv* Diagram.** A diagram of pressure and specific volume (Fig. 9-3) can be obtained by projecting all points of Fig. 9-1 over to the *pv* plane. The regions *S, L,* and *G* correspond to the phases indicated by the same letters in Fig. 9-2. The triple point on the pressure-volume diagram is a line, while the phase boundaries of Fig. 9-2 become areas. Since the phase change from solid to liquid is marked by a decrease in volume (for water), the liquid region in a plane diagram lies to some extent under the solid-liquid region (*abgh*).

Figure 9-3 illustrates that, for temperatures above the critical temperature, compression of a gas (without cooling) will not cause the appearance of two phases signifying condensation to the liquid state. Because of this, it is often stated that a gas with temperature above the critical temperature cannot be liquefied by compression. It would be better to say that a liquid and a gas cannot be distinguished from each other under these conditions.

The paths *nopqrstuvn* and *uvwxyu'u* are reproduced on this diagram.

**9-5. The Two-phase Mixture.** The relative amounts of each phase that are present in a two-phase mixture depend on the *quality* of the mixture. The *quality* of a two-phase mixture is equal to the ratio of the mass of vapor to the total mass of mixture. Although defined as a ratio, it is frequently used as a percentage. Saturated liquid is zero quality or 100 per cent moisture. Saturated vapor is 100 per cent vapor and 0 per cent moisture. A quality of 0.60 signifies that the mixture consists of

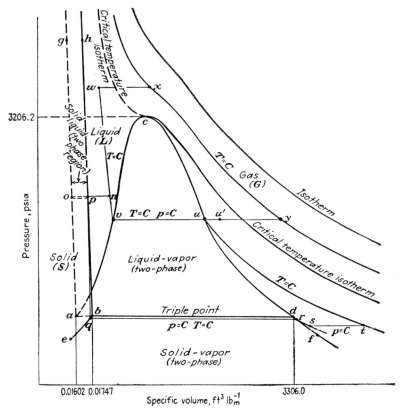

Fig. 9-3. *pv* diagram for water (not to scale).

60 per cent vapor and 40 per cent liquid. The letter $x$ will be used to designate the quality,

$$x = \frac{\text{mass of vapor}}{\text{mass of mixture}}$$

Since the pound mass will be the usual unit,

$$1 \text{ lb}_m \text{ mixture of } x \text{ quality} = x \text{ lb}_m \text{ vapor} + (1 - x) \text{ lb}_m \text{ liquid} \quad (a)$$

The *specific*† values (that is, values for 1 lb$_m$) for the extensive properties

† These *specific* values should not be called *intensive* properties; else the phase rule (for one matter) becomes unintelligible.

of volume, enthalpy, entropy, and others can be determined by Eq. (a) for the mixture. (The state of saturated liquid is denoted by the subscript $f$ and that for the saturated vapor by the subscript $g$.)

$$v_{\text{at quality } x} = v_x = xv_g + (1 - x)v_f$$
$$h_x = xh_g + (1 - x)h_f \tag{b}$$
$$s_x = xs_g + (1 - x)s_f$$

The change in these properties during the phase change is shown by the subscript $fg$, or

$$h_g - h_f = h_{fg} \qquad v_g - v_f = v_{fg} \qquad s_g - s_f = s_{fg} \tag{c}$$

Equations (b) [by the use of Eqs. (c)] can be converted to

$$v_x = v_f + xv_{fg} \qquad h_x = h_f + xh_{fg} \qquad s_x = s_f + xs_{fg} \tag{9-2}$$

Thus by introducing specific values for the mixture of phases, continuous functions can be established for the specific properties of the heterogeneous system; therefore, such systems can be treated as if they were homogeneous.

**9-6. The $Ts$ Diagram.** A diagram of temperature and entropy for a pure substance would resemble Fig. 9-4. Here the solid, solid-liquid, and liquid regions are quite compact since these phases are relatively incompressible. Note that the state of a subcooled phase (for example, subcooled liquid $g$, Fig. 9-4) will apparently be on a saturation line because of the proximity of the lines of constant pressure. When the subcooled liquid is heated at constant pressure, its temperature rises and the state changes at constant pressure ($gh$) to the saturation temperature ($h$). Continued heating at constant pressure causes a change from the saturated liquid state ($h$) to a two-phase state ($i$), then to the saturated vapor state ($j$) and, finally, to a gas or superheated vapor state ($k$).

On the other hand, if the liquid has a pressure greater than the critical, it can be raised in temperature without the appearance of two phases to achieve a state such as $l$. An interesting feature of water is that all the constant-pressure lines cross at 39°F. Thus, the properties of temperature and entropy cannot identify the state at this temperature, and also isentropic compression of the liquid does not change the temperature.

The behavior of the constant-pressure lines can be predicted from two of the Maxwell equations (Art. 8-1),

$$\left(\frac{\partial s}{\partial p}\right)_T = -\left(\frac{\partial v}{\partial T}\right)_p \qquad \left(\frac{\partial T}{\partial p}\right)_s = \left(\frac{\partial v}{\partial s}\right)_p$$

Water is one of the few substances that expand upon freezing.† Thus, when water is cooled at constant pressure, it reaches a state of maximum density at 4°C; continued

† Because this characteristic is displayed by a few substances and not, say, by half of all substances, the saying of Einstein on statistical methods is often stated as: "I do not believe in a dice-throwing God."

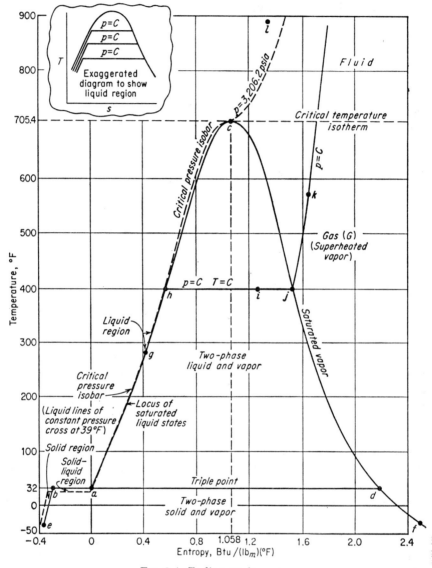

FIG. 9-4. $Ts$ diagram for water.

cooling results in expansion. The state at 4°C of maximum density requires that $\partial v/\partial T$ be zero; then $\partial s/\partial p$ is also zero, and the lines of constant pressure cross each other.

If the volume of the solid or liquid phase increases with increase in temperature at constant pressure (usual),

$$\frac{\partial s}{\partial p}\bigg)_T = \text{negative}$$

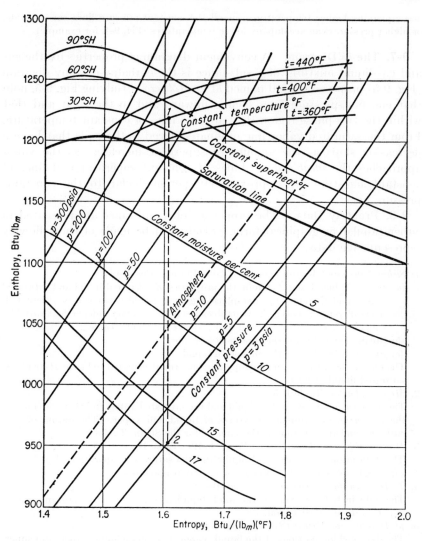

FIG. 9-5. *hs* (Mollier) diagram for steam.

Hence, higher pressures are at lesser entropy (for the same temperature). (From 4°C down to the triple point, the inverse characteristic is displayed by a few substances, such as water.)

Upon melting at constant pressure, the volume can increase (usual), be zero, or decrease (water, for example); it follows that

$$\frac{\partial T}{\partial p}\Big)_s = \text{positive, zero, or negative}$$

Hence, the constant-pressure lines in the solid-liquid region can have higher pressures accompanying higher temperatures, or all the constant-pressure lines can coincide, or higher pressures can accompany lower temperatures (Fig. 9-4, for example).

**9-7. The *hs* Diagram.** A convenient diagram of properties for the gas and gas-liquid mixtures of a substance is the enthalpy-entropy diagram (Fig. 9-5) proposed by and named after *Mollier*. Studying Fig. 9-5, note that constant-pressure lines incline upward and to the right and that within the two-phase region they are also lines of constant temperature. From the saturated-vapor line originate curved isothermals that have a decreasing slope as the amount of superheat increases (as the vapor approaches the *ideal state*). Lines of constant superheat and lines of constant moisture are approximately perpendicular to the constant-pressure lines.

**9-8. Property Data.** The properties of a substance are presented in either tabular or graphical form. Study will be made of the following representative data:

1. *Steam Tables and Mollier Diagram* (Table B-4 and Fig. B-6):
   (a) Steam Tables 1 and 2 include the properties of saturated liquid and saturated vapor, and therefore between the two-phase region of liquid-vapor can be calculated. The difference between the two tables lies in the primary independent variable: *temperature* for Steam Table 1 and *pressure* for Steam Table 2.
   (b) Steam Table 3 is for the superheated vapor.
   (c) Steam Table 4 is for the subcooled liquid.
   The datum for enthalpy and entropy is arbitrarily designated to be the state of the saturated liquid at 32°F.
2. *Mercury Tables* (Table B-5):
   The data are for the saturated liquid and the saturated vapor and therefore include the properties of two-phase mixtures of liquid and vapor. The datum of zero enthalpy and zero entropy is the same as for the Steam Tables.
3. *Dichlorodifluoromethane Tables* (*Freon*, F-12) (Table B-14):
   The data are for the saturated liquid and the saturated vapor and, also, for two superheat states at each of the given saturated pressures.
4. *Ts Diagram for Carbon Dioxide* (Fig. B-3):
   The data include portions of the solid, liquid, two-phase solid-vapor and liquid-vapor, and superheat region.
5. *ph Diagram for Methyl Chloride* (Fig. B-4):
   The data include portions of the liquid, vapor, two-phase liquid-vapor, and superheat region.

It should be carefully noted that the number of significant figures in the various tables and graphs may be an indication of the *internal consistency* of the data, but it usually is not an indication of the *accuracy* of the data. Note that the assignment of an over-all accuracy or a percentage deviation to a chart or a table is not possible since the data in certain regions may be *remarkably precise* and in other regions *extremely inaccurate* (and the *true* values of the data are unknown). In general, property data from reputable laboratories are accurate to within 1 per cent. The main body of the Steam Tables because of their long history is probably more accurate than this figure, possibly of the order of one-fifth of 1 per cent.

In the examples below, the calculations are performed without regard to the accuracy of the data, and the number of significant values depends upon the particular table or graph.

**Example 1.**   Find the internal energy of 1 $lb_m$ of saturated water at 32°F.
*Solution.*   From Steam Table 1 (Table B-4),

$$h_f = 0.00 \text{ Btu/lb}_m \qquad p = 0.08854 \text{ psia} \qquad v_f = 0.01602 \text{ ft}^3/\text{lb}_m$$

And by definition of enthalpy (Art. 2-11),

$$u = h - \frac{pv}{J} = 0.00 - \frac{0.08854(144)(0.01602)}{778.16} = -0.0002625 \text{ Btu/lb}_m \qquad Ans.$$

The minus sign is a consequence of the arbitrary datum assigned to enthalpy.

**Example 2.**   Determine the change in entropy of methyl chloride from saturated liquid to saturated vapor at a constant pressure of 40 psia.
*Solution.*   Figure B-4 does not show values of entropy for the saturated liquid. However, with Eq. (7-4b) and Fig. B-4,

$$\Delta s = \frac{h_{fg}}{T} \bigg]_{T,p} = \frac{200.8 - 27}{495} = 0.351 \text{ Btu/(lb}_m)(°R) \qquad Ans.$$

**Example 3.**   Determine the specific enthalpy of a liquid-vapor mixture of Freon, F-12, if the quality is 35 per cent and temperature 18°F.
*Solution.*   From Table B-14

$$h_f = 12.12 \qquad h_g = 80.27 \qquad \text{or} \qquad h_{fg} = 68.15 \text{ Btu/lb}_m$$

Hence, by Eqs. (9-2),

$$h = h_f + xh_{fg} = 12.12 + 0.35(68.15) = 36.0 \text{ Btu/lb}_m$$

**Example 4.**   Calculate the heat required to raise the temperature of the mixture in Example 3 to 55.5°F in a reversible constant-pressure process ($\Delta KE$, $\Delta PE = 0$).
*Solution.*   A temperature of 55.5°F is equivalent to a superheat of 37.5°F.   Linear interpolation of Table B-14 yields $h = 85.68$ Btu/$lb_m$.   Then, by Eq. (6-4b),

$$\Delta q_{rev} = h_2 - h_1]_{p=C} = 85.68 - 36.00 = 49.68 \text{ Btu/lb}_m \qquad Ans.$$

**Example 5.**  Carbon dioxide at 1,460 psia and 135°F expands isentropically to 700 psia.   Determine the enthalpy and quality.
*Solution.*   Locate the initial state on Fig. B-3, and proceed at constant entropy to $p = 700$ psia (vertically downward).   Read from the chart

$$h = 120 \text{ Btu/lb}_m \qquad x = 0.83 \qquad Ans.$$

**Example 6.** How much heat must be transferred to 1 $lb_m$ of steam in a constant-volume container to raise the pressure from 144 to 150 psia?   The initial temperature is 360°F.
*Solution.*   From Steam Table 3 for the initial conditions,

$$p_1 = 144 \text{ psia} \qquad t_1 = 360°F \qquad v_1 = 3.160 \text{ ft}^3/\text{lb}_m \qquad h_1 = 1196.5 \text{ Btu/lb}_m$$

The heat added at constant volume is equal to the change in internal energy [Art. 4-3a and Eq. (4-4$d$)]. Since this property is not listed in Steam Table 3, it must be computed in the manner illustrated in Example 1,

$$u_1 = h_1 - \frac{p_1 v_1}{J} = 1196.5 - \frac{144(144)3.160}{778.16} = 1112.2 \text{ Btu/lb}_m$$

The volume and pressure at the end of the process are known; hence, interpolation can be made in Steam Table 3 between temperatures of 380 and 390°F at a pressure of 150 psia to find $h_2 = 1211.4$ Btu/lb$_m$. Solving for internal energy gives

$$u_2 = 1211.4 - \frac{150(144)3.160}{778.16} = 1123.6 \text{ Btu/lb}_m$$

and      $\Delta q_{\text{rev}, v=C} = u_2 - u_1 = 1123.6 - 1112.2 = 11.4$ Btu/lb$_m$      *Ans.*

**Example 7.** Water is isentropically compressed in a flow process from the saturated state at 100°F to a pressure of 1,000 psia. How much work is required? How much work is required if the efficiency of the pump is 60 per cent (of the reversible value)?

*Solution.* An approximate method can be used for problems of this type. The work of a reversible steady-flow process ($\Delta KE$, $\Delta PE = 0$), by Eq. (6-3$c$), is

$$\Delta w_{\text{rev}} = -\int v \, dp = \int p \, dv - \Delta FE$$

Since the volume does not change greatly,

$$\Delta w_{\text{rev}} \approx -v(p_2 - p_1) = -\frac{0.01613(144)(1000 - 0.9)}{778.16} = -3 \text{ Btu/lb}_m \qquad Ans.$$

But this same equation can be obtained by assuming that the work is necessary only to increase the flow energy of the fluid. The small amount of work spent in compressing the liquid was noted in Art. 5-8$d$,

$$\Delta w_{\text{comp}} = \int p \, dv \approx p_{\text{av}} \, \Delta v = \left(\frac{1,000.9}{2}\right)(0.000051)\left(\frac{144}{778.16}\right) = 0.005 \text{ Btu/lb}_m$$

Hence, the work supplied to the usual liquid pump is used to increase the flow energy of the fluid, and only a negligible portion of the work is used to compress the fluid. It can be concluded that the temperature change from isentropic compression is also negligible in most cases. The efficiency of the actual pump is 60 per cent; then the work required is

$$\eta_{\text{pump}} = \frac{\text{ideal work}}{\text{actual work}} = \frac{\Delta w_{\text{isen}}}{\Delta w_{\text{actual}}}$$

or      $$\Delta w_{\text{actual}} = \frac{-3}{0.60} = -5 \text{ Btu/lb}_m \qquad Ans.$$

Hence, for the actual pump the enthalpy of the liquid will also be increased because of friction and turbulence. This heating effect will cause a pronounced rise in temperature when compared with the temperature rise of the reversible pump. Noting that the heat capacity is closely 1 Btu/(lb$_m$)(°F) and that 2 Btu/lb$_m$ of energy is dissipated in turbulence,

$$\Delta t \approx 2°F \qquad \text{or} \qquad t_{2\,\text{actual}} \approx 102°F$$

**Example 8.** Water at a temperature of 100°F and a pressure of 1,000 psia is to be heated in a steady-flow process to the state of dry saturated steam at the same pressure. How much heat must be transferred?

*Solution.* For the liquid phase, Steam Tables 1 and 2 give only saturated values.

However, the enthalpy of subcooled water at the state given in this problem corresponds to that of Example 7. From this example,

$$p_a = 1,000 \text{ atm} \qquad t_a \approx 100°F$$

$$h_a \approx h_f \Big]_{100°F} + |\Delta w_{s-C}| = 67.97 + 3 = 71 \text{ Btu/lb}_m$$

From Steam Table 2,

$$p_c = 1,000 \text{ psia} \qquad t_c = 544.61°F \qquad h_c = 1191.8 \text{ Btu/lb}_m$$

For this steady-flow process, Eq. (5-7d) shows that

$$\Delta q_{rev} = h_c - h_a = 1120.8 \text{ Btu/lb}_m \qquad Ans.$$

The area representing this quantity of heat is laid out in Fig. $A$ as $a'abcc'$.

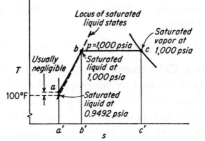

Fig. A. Constant-pressure heating process (at $p = 1,000$ psia).

**Example 9.** Steam at a pressure of 144 psia and a temperature of 400°F enters a turbine and leaves at a pressure of 3 psia under conditions of steady flow. How much work is delivered (a) if the process is adiabatic and reversible and entering and leaving kinetic energies are negligible, (b) if the turbine is irreversible and only 70 per cent of the isentropic work is obtained?

*Solution.* (a) The work done by a steady-flow system under these conditions is (Art. 5-8e)

$$\Delta w_{rev} = (h_1 - h_2)_{isen}$$

The properties of the steam entering the turbine are obtained from Steam Table 3,

$$h_1 = 1220.4 \text{ Btu/lb}_m \qquad s_1 = 1.6050 \text{ Btu/(lb}_m)(°R)$$

At exit, the known properties are

$$s_2 = 1.6050 \text{ Btu/(lb}_m)(°R) \qquad p_2 = 3 \text{ psia}$$

Inspection of Steam Table 2 shows that for this value of entropy the steam must be a two-phase mixture. The quality must be calculated from Eqs. (9-2),

$$s_2 = s_f + x s_{fg}$$

(Values of $s_f$ and $s_{fg}$ are obtained from Steam Table 2, while $s_2$ is known.)

$$1.6050 = 0.2008 + x(1.6855) \qquad x = 0.835$$

Since the quality is known, the enthalpy can be calculated,

$$h_2 = h_f + xh_{fg} = 109.37 + 0.835(1,013.2) = 956.4 \text{ Btu/lb}_m$$

and the work can be found from the change in enthalpy,

$$\Delta w_{rev} = h_1 - h_2 = 1220.4 - 956.4 = 264 \text{ Btu/lb}_m \qquad Ans.$$

This problem is greatly simplified by use of a Mollier diagram. The initial conditions can be located on this chart as shown in Fig. 9-5 for the data of this example. From the initial state to the end state is a process at constant entropy, and therefore following a vertical path from point 1 down to the known pressure line enables the enthalpy of state 2 to be read directly from the ordinate without calculation for the quality.

(b) The irreversible process will deliver

$$\Delta w = 0.70(264) = 184.8 \text{ Btu/lb}_m \qquad Ans.$$

The final state of the steam can also be located on the Mollier diagram since the pressure and enthalpy are known,

$$h_{2'} = h_1 - \Delta w = 1220.4 - 184.8 = 1035.6 \text{ Btu/lb}_m \qquad \text{and} \qquad p = 3 \text{ psia}$$

**9-9. The Throttling Calorimeter.** Although the state of a one-phase system is completely described by measuring the independent properties of temperature and pressure, for a two-phase system temperature and pressure are no longer independent and some other property such as enthalpy or internal energy must therefore be measured before the state can be determined. However, measurements of the intensive properties of temperature and pressure are more easily and more precisely made than measurements of extensive properties; hence, it is preferable to devise a means to use these intensive properties as indicators of the state of a two-phase system. In Fig. 9-6 are shown the elements of a Mollier diagram, and point $a$ represents the state of a two-phase system. The pressure and temperature of $a$ do not determine the state because any other state such as $x$ would have the same values for these properties. Now, if the fluid at state $a$ passes to state $b$ by a process that begins and ends at the same enthalpy, state $b$ is defined by pressure and temperature alone because only one phase will be present. Then state $a$ can be found from calculation because the enthalpy at $a$ is equal to that of $b$, and this property along with the independent property of pressure (or temperature) determines the state of a two-phase system.

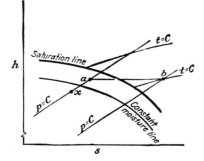

FIG. 9-6. Throttling (calorimeter) process on the $hs$ diagram.

The procedure is to use a *throttling calorimeter* (Fig. 9-7). Here a sampling tube *s* is inserted into a vertical pipe wherein (preferably) the flow is downward.† A representative sample of the fluid enters the tube and the calorimeter with negligible velocity and expands through an orifice *o* to the outlet pressure (usually atmospheric). In passing through the orifice, enthalpy decreases and kinetic energy increases.

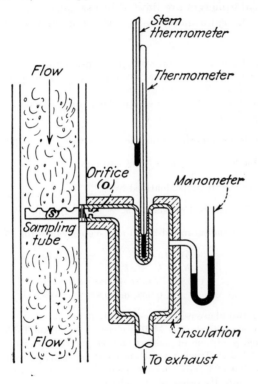

FIG. 9-7. Throttling calorimeter.

But when the kinetic energy is dissipated in turbulence, the initial enthalpy is regained, as can be seen from the steady-flow equation,

$$\Delta q - \Delta w = (h_b - h_a) + \frac{V_b{}^2 - V_a{}^2}{2g_c} + \frac{g}{g_c}(Z_b - Z_a) \quad (5\text{-}7d)$$

For the over-all process, heat and work are zero while changes in potential and kinetic energies are negligible. Therefore,

$$h_a = h_b \quad (9\text{-}3)$$

† This does not disturb the distribution of the water particles in the vapor to as great a degree as other positions of the pipe.

If the heat loss is not zero, the calorimeter must be calibrated (by test at a known quality) and a *normal correction* calculated. The normal correction is expressed as the number of degrees that must be *added* to the observed exit temperature to compensate for the heat *loss*. With this correction, the data from the calorimeter are $t_b$ and $p_b$ (Fig. 9-6), and therefore $h_b$ can be found from the superheat tables for the substance.

Throttling calorimeters are limited in use since it is essential that the vapor leaving the calorimeter be superheated (at least 5° to avoid misleading results).

**Example 10.** Steam under a pressure of 150 psia enters a throttling calorimeter and emerges as superheated steam at atmospheric pressure with a temperature of 300°F. What is the quality of the wet steam?

*Solution.* The state of the steam at the outlet from the calorimeter can be determined from Steam Table 3.

Measured:     $p_b$ = 14.696 psia (atmospheric pressure)     $t_b$ = 300°F

From Steam Table 3,         $h_b$ = 1192.8 Btu/lb$_m$

The initial state of the steam is identified by

$$p_a = 150 \text{ psia} \qquad h_a = 1192.8 \text{ Btu/lb}_m$$

Steam Table 2 yields the saturated-liquid and -vapor values for 150 psia,

$$h_f = 330.51 \text{ Btu/lb}_m \qquad h_{fg} = 863.6 \text{ Btu/lb}_m$$

and
$$h_a = h_f + x h_{fg}$$
$$1192.8 = 330.51 + x(863.6)$$
$$x = 0.998, \text{ or } 99.8\% \qquad Ans.$$

The steam in the two-phase region is essentially dry.

The easier solution is to use the Mollier chart of the Steam Tables (Fig. B-6). On this chart, the final conditions of pressure and temperature can be located in the vapor region (as in Fig. 9-6). Tracing a path of constant enthalpy from this point to the constant-pressure line of 150 psia locates the initial state in the two-phase region, and the quality can be directly read from the chart.

## PROBLEMS

The following points should be borne in mind in solving the problems:

1. A diagrammatic sketch of the system must be drawn showing the boundaries and the following:

   (*a*) Whether a closed or a steady-flow system
   (*b*) Initial and final properties
   (*c*) All transfers of heat, work, and mass
   (*d*) All other given data ($v = C$, $s = C$, $\Delta Q = 0$, etc.)

2. The appropriate First Law energy equation, valid for reversible or irreversible processes [Eq. (4-4*d*) or (5-7*d*)], should be used as the initial step in solutions for work or heat transfers.

Determine from the data whether or not the process is reversible, and use this knowledge in the solution of the energy equation.

3. The $T\,ds$ equations [Eqs. (7-4)] should be used as the initial step in solutions not involving work or heat transfers.

4. Note especially whether the initial and final states can be identified (recall that two properties usually identify the state). In most problems, only one property is apparently given for either the initial or the final state. Here the wording of the problem must be carefully studied to find the missing data. For example, a reversible adiabatic process dictates that $s_1 = s_2$.

5. Always illustrate the process on the appropriate property diagram ($Ts$, $hs$, $ph$ or $pv$, etc.) since in constructing the diagram the solution may become obvious.

**1.** Why do not the properties of pressure and specific volume define the state of a mixture at the triple point?

**2.** Explain the differences between the triple point and ice point.

**3.** For the traverse or cycle made in Fig. 9-1, list the proper name for each process.

**4.** Between states $v$ and $u$ in Fig. 9-1, the amount of saturated vapor increases from 0 to 100 per cent. Do the characteristics of this vapor change as the amount increases?

**5.** (a) If steam has a quality of 0.8 and pressure of 100 psia, find the specific values for volume, enthalpy, and entropy.

(b) Repeat, but use mercury as the fluid.

**6.** One pound of steam at a pressure of 100 psia is confined in a closed tank with a volume of 4 ft³. Determine the quality, enthalpy, and entropy.

**7.** If the steam in Prob. 6 is to be heated until the temperature is 500°F, how much heat must be added?

**8.** If the steam in Prob. 6 is heated in a cylinder at constant pressure to 500°F, how much heat must be added?

**9.** Find the *mean* heat capacity of superheated steam at constant pressure from the saturated state at 160 psia to 500°F. Repeat, using the range from 400 to 500°F.

**10.** (a) Saturated water at 100°F enters a pump and is reversibly pumped to 600 psia without transfer of heat. How much work is done by the pump?

(b) If the pump is only 60 per cent as efficient as a reversible pump, how much work is done?

(c) What is the approximate enthalpy of the fluid leaving the irreversible pump?

**11.** If water from the pumps of Prob. 10 enters a boiler where it is evaporated into dry saturated steam that leaves at a pressure of 600 psia, find the heat transferred for both cases.

**12.** Saturated water is isentropically pumped in a flow process from 100°F to a pressure of 800 psia. Find the work necessary for this process.

**13.** Feedwater enters a steam boiler at 120°F while the steam leaves at 1,000 psia and 98 per cent quality. Find the heat transferred.

**14.** (a) A pound of saturated steam at 250 psia pressure is cooled at constant volume until its pressure falls to 100 psia. Determine the amount of heat transferred and the change in entropy.

(b) Repeat, using carbon dioxide as the fluid.

**15.** Three pounds of steam is processed in a closed and rigid container from $p_1 = 14.7$ psia and $t_1 = 300°F$ to $t_2 = 100°F$. Find $p_2$, $\Delta Q$, and $\Delta W$, and sketch $pv$ and $Ts$ diagrams for the change.

**16.** Water, at the rate of 1,000 lb$_m$/min, is to be heated from 60 to 200°F in a feed-water heater (water passes through coils surrounded by steam). The exhaust steam passing round the heater coil is at a pressure of 15 psia and contains 10 per cent moisture and leaves as saturated water.

(a) How much steam is required per hour?

(b) What is the change in entropy for the steam and for the water?

(c) Sketch $Ts$ diagrams to illustrate the processes experienced by the liquid and the vapor.

**17.** In the inlet pipe to an ideal turbine, the steam is at a pressure of 160 psia and a temperature of 400°F. The exhaust is at 80°F, while the rate of steam flow is 1,200 $lb_m$/hr.

(a) Determine the work done per pound of steam.

(b) Calculate the horsepower developed.

**18.** (a) Saturated steam at 150 psia enters a long pipe and leaves with a pressure of 144 psia. If no heat is lost by unavoidable radiation, etc., find the quality of the steam at exit.

(b) If 10 Btu of heat is lost per pound of steam by radiation, etc., what will be the quality?

**19.** (a) Steam at 100 psia and 60 per cent quality receives 350B tu/$lb_m$ of heat, while the pressure remains constant. Determine the resultant properties of the steam.

(b) Repeat, but use methyl chloride as the fluid and add only 50 Btu/$lb_m$ of heat.

**20.** (a) A container with volume of 10 ft³ contains saturated steam at 100 psia. Determine the mass of steam within the tank.

(b) Repeat, but use methyl chloride as the fluid.

**21.** (a) Saturated steam at 100 psia enters a perfect turbine and expands isentropically to 5 psia. Find the work done by the turbine. Check the calculations using a Mollier chart.

(b) Determine the final state of the steam if the work obtained from an adiabatic but irreversible expansion process is only 80 per cent of that determined in (a).

**22.** Repeat Prob. 21, assuming that the limits are 160 to 5 psia and 500 to 60°F.

**23.** (a) Repeat Prob. 21, assuming that the limits are 100 psia (saturated) to 1 psia.

(b) Repeat, using mercury as the fluid.

**24.** (a) Saturated steam expands in a perfect nozzle from 150 psia to 60°F. Find the exit velocity if the initial velocity is negligible. Check the computations, using a Mollier chart.

(b) Repeat, using Freon, F-12, as the fluid.

**25.** Determine the heat that must be transferred to water initially at 60°F if the final temperature is to be 300°F and the process passes through a continuous series of saturated states (graphical solution).

**26.** (a) Steam at 100 psia and 96 per cent quality is throttled to atmospheric pressure. Find the temperature after throttling. Check the computation, using a Mollier chart.

(b) Repeat, using methyl chloride as the fluid.

**27.** Steam at 150 psia is throttled to atmospheric pressure with consequent temperature of 230°F. What was the quality of the steam before the expansion? (Check computations, using a Mollier diagram.) If the normal correction is 5°, what will be the quality?

**28.** Repeat Prob. 27, but assume that the initial pressure is 100 psia.

**29.** Repeat Prob. 27, but assume that the initial pressure is 50 psia.

**30.** If saturated water at 150 psia were throttled to atmospheric pressure, what would be the final quality?

**31.** (a) Dry (saturated) steam at 100 psia enters a turbine and is throttled to 50 psia in the steam chest. Find the properties of the steam in the steam chest.

(b) Repeat, but use Freon, F-12, as the fluid.

**32.** The pressure in a steam calorimeter similar to that in Fig. 9-7 is 4 in. Hg above the atmospheric pressure. For steam initially at 100 psia, find the quality if the temperature in the calorimeter is 225°F.

**33.** For conditions similar to those of Prob. 32, the normal correction is 5°F, while the flow of steam is 1 $lb_m$/sec. How much heat is lost per pound of steam by radiation, etc.?

**34.** Steam at 150 psia and 500°F is isothermally and reversibly expanded in a cylinder until the pressure is 100 psia. Determine the work and heat transfers for this process of a closed system.

**35.** (a) Steam at 150 psia and 500°F is expanded in a perfect nozzle to 3 psia. Find the final velocity if the initial velocity is 30 ft/sec.

(b) The high-velocity stream is slowed down in a perfect diffuser until the pressure is 100 psia. Determine the condition of the steam leaving the diffuser.

**36.** Saturated steam at 180 psia and 30 ft/sec velocity undergoes an irreversible but adiabatic steady-flow process to a pressure of 100 psia, quality of 98 per cent, and final velocity of 10 ft/sec. (a) Determine the work transferred and the change in entropy. (b) Repeat, using mercury as the fluid.

**37.** Repeat Prob. 36, but assume that the process is reversible and that the final conditions are 100 psia and unknown quality.

**38.** Carbon dioxide at 100 atm pressure and 660°R is isentropically expanded to 10 atm pressure. Find the change in enthalpy, the work, and the quality.

**39.** Carbon dioxide at 30 atm pressure and 500°R expands in a throttling calorimeter to a pressure of 1 atm. What will be the final temperature?

**40.** Steam at 5.3 psig and 300°F enters a steam-heating radiator in a room. The liquid leaving the radiator is saturated. Determine the steam required in pounds per hour if the heat exchange between the radiator and room is 15,000 Btu/hr.

**41.** Methyl chloride flows through a superheater at the rate of 1,000 $lb_m$/hr. At entrance to the heater, $p = 200$ psia, $x = 100$ per cent; at exit from the heater, $p = 180$ psia, and $t = 250$°F. Determine the heat transferred, and show the process on $pv$ and $Ts$ diagrams.

**42.** Steam flows through a nearly closed valve in a horizontal insulated pipe of 10 in. ID. Upstream from the valve the pressure is 400 psia; downstream the pressure is 50 psia, and the temperature is 300°F. The flow rate is 6.97 $lb_m$/hr. Determine the transfers of heat and work, the initial temperature, and the change in entropy, and show the process on $pv$ and $Ts$ diagrams.

*Complementary problems for the integrated course*

**43.** Determine the thermal and over-all efficiencies for a Carnot-cycle power plant. Heat is added to saturated liquid at 500 psia with saturated steam from the boiler delivered to the turbine. Heat rejection is at 80° while the furnace efficiency is 75 per cent (Art. 8-4).

**44.** Upon test of the cycle of Prob. 43, the real adiabatic turbine delivered only 85 per cent of the isentropic work, while the real adiabatic pump required double the isentropic input work (the flow from the pump is saturated liquid, as before). Calculate the thermal and over-all efficiencies for the power plant.

**45.** Repeat Prob. 43 but for a Rankine cycle and therefore with an improved furnace efficiency, say 85 per cent. (Heat is now added to subcooled liquid at about 80°F.)

**46.** Repeat Prob. 44 for the Rankine-cycle data of Prob. 45, noting that the beginning of heat addition is now at a slightly higher temperature.

**47.** Summarize the data of Probs. 43 to 46 in a table.

**48.** Calculate the loss of available energy for the turbines and pumps of Probs. 44 and 46 (see Art. 12-2).

**49.** Calculate the loss of available energy for the process of Prob. 16.

**50.** Calculate the loss of available energy for the process of Prob. 18.

## SELECTED REFERENCES

1. Keenan, J. H.: "Thermodynamics," John Wiley & Sons, Inc., New York, 1941.
2. Lee, J. F., and F. W. Sears: "Thermodynamics," Addison-Wesley Publishing Company, Reading, Mass., 1955.
3. Zemansky, M. W.: "Heat and Thermodynamics," 4th ed., McGraw-Hill Book Company, Inc., New York, 1957.
4. Dodge, B. F.: "Chemical Engineering Thermodynamics," McGraw-Hill Book Company, Inc., New York, 1944.
5. Roberts, J. K.: "Heat and Thermodynamics," Interscience Publishers, Inc., New York, 1954.
6. Obert, E. F.: "Concepts of Thermodynamics," McGraw-Hill Book Company, Inc., New York, 1960.

# CHAPTER 10

# THE $pvT$ RELATIONSHIPS

When the state is most corrupt
then laws are most multiplied.
*Tacitus* (100 A.D.)

In formulating a relationship among pressure, volume, and temperature, the microstructure of matter and the macroscopic behavior should
be considered.   By so doing, a theoretical form can be proposed for the
equation of state and the theory tested by observing the agreement of the
equation with experimental data.   In general, a separate equation of
state is required for each phase of matter (Art. 2-12).

The atoms of the solid are closely packed (*order* rather than *disorder*)
and held together by attraction forces; the liquid phase allows more
freedom (more disorder) for molecular displacement (and therefore
liquids are slightly more compressible than are solids); the molecules
of a gas are much farther apart (most disordered state) (and therefore the
compressibility of gases is many times that of solids or liquids).   The
molecules of the gas have kinetic forms of energy arising from motion
and also potential energy arising from intermolecular forces.   When the
molecules are apart, the intermolecular force is a weak attraction which
rises slowly to a maximum as the molecules approach each other; with
further approach, the attraction decreases and turns into a repulsion;
the repulsion increases rapidly as the molecules come together.   As the
density of the gas is decreased, the effects of the intermolecular forces are
minimized and therefore a limiting condition is approached by real
gases where the behavior is unaffected by the presence of other molecules.   Since all gases can approach this common limit, it is considered
to be an *ideal* or *perfect* state (Art. 3-5).

**10-1. Requirements for the Equation of State.**   An equation of state
serves two purposes:

1. The equation is the means for directly calculating one of the three
properties of pressure, specific volume, and temperature.

2. The partial derivatives of the equation are the means for calculating
other property data (Chap. 8).

Recall that the partial derivatives of pressure, specific volume, and
temperature correspond to certain slopes on the $pvT$ surface.   Thus, the

159

equation of state should portray the experimental $pvT$ data, and it should also portray the contours of the experimental $pvT$ surface. Since equations are never exact, the equation of state may be quite precise in predicting, say, a value of pressure and yet be far in error in predicting a partial derivative of pressure. Consider that the effect of isothermal compression on internal energy is given by the equation (Art. 8-1)

$$\Delta u \bigg]_{v_1 T_1 \text{ to } v_2 T_1} = \int_{v_1}^{v_2} \frac{\partial u}{\partial v}\bigg)_T dv = \int_{v_1}^{v_2} \left[ T \frac{\partial p}{\partial T}\bigg)_v - p \right] dv$$

Hence, the evaluation of internal energy depends upon precise values of the pressure and also on precise values of the partial derivative of pressure with respect to temperature.

A three-dimensional surface of the $pvT$ properties is not a convenient tool; hence, the behavior of the real fluid is shown by plotting values of the ratio $pv/RT$ on a diagram in the manner of Fig. 10-1. Observe that, as pressure approaches zero, the gas approaches the ideal state ($pv/RT \to 1$) at all temperatures. Consider the critical isotherm $T_c$; starting from $p = 0$, $pv/RT = 1$, the isotherm drops rapidly to a value of about $pv/RT = 0.3$ at the critical point. It can be surmised that, when $pv/RT$ is less than unity, attraction or cohesive forces between molecules are predominant and thus the volume of the gas is *less* than the ideal value. As the pressure is increased above $p_c$, however, the critical isotherm rises and at high pressures the close proximity of the molecules brings into play repulsion forces to cause the volume to be *greater* than the ideal value ($pv/RT > 1.0$).

At temperatures below the critical ($T = 0.9T_c$, for example), the isotherms descend from $pv/RT = 1$ and pass through the two-phase region of liquid and vapor, and then the liquid isothermals rise and cross the critical isotherm. At temperatures above the critical, such as $T = 1.2T_c$, the isotherms must also cross the critical isotherm since the numerical order of the isotherms at high pressures is the progression shown in the upper right of Fig. 10-1. Hence, the critical isotherm is not a lower boundary to the diagram. At rather high pressures the orderly progression of the isotherms shows that raising the temperature causes the fluid to approach the ideal state, $pv/RT \to 1$.

At low temperatures, the isotherms approach zero pressure with negative slopes, and as the temperature is raised, the slopes decrease toward zero. The isotherm of zero slope at zero pressure is called the *Boyle isotherm* ($T \approx 2.5T_c$). At temperatures above the Boyle isotherm, the slopes of the isotherms at zero pressure are positive and increase numerically to a maximum near $5T_c$. At temperatures greater than this, the slopes of the isotherms progressively decrease but never become negative.

*Thus, the ratio pv/RT is always greater than unity at temperatures above the Boyle isotherm.* The isotherm $T = 5T_c$ is not an upper boundary since it is progressively crossed by isotherms of lower temperature to achieve the order shown in the upper right of Fig. 10-1 (nor is the critical isotherm the lower boundary since it is crossed by higher- and lower-temperature isotherms).

Because of the crossings of the isotherms, an increase in temperature at constant pressure does not always bring the gas closer to that of the ideal state. Consider process $ab$ of Fig. 10-1. Here, as temperature increases, $pv/RT$ also increases to exceed the value of unity. This trend continues to about $4T_c$ and then reverses: raising the temperature

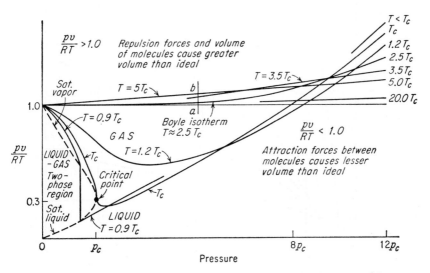

Fig. 10-1. Deviations of a real gas from ideal behavior (not to scale).

causes $pv/RT$ to decrease and approach 1 as its limit (note the isotherm $T = 20T_c$).

The equation of state in portraying the behavior of Fig. 10-1 should yield isometrics (constant volume or density) similar to those in Fig. 10-2. For an ideal gas, the isometrics would be straight lines originating from the origin and with slope proportional to the density. For the real gas, the isometrics are displaced from the origin and are either straight or slightly curved:

1. The isometrics are straight at low densities, at or near the critical density, and also at about twice the critical density.

2. The isometrics are straight at high temperatures.

3. The isometrics have negative curvature (convex upward) at medium densities and at very high densities.

4. The isometrics have positive curvature in the region from critical to about twice the critical density.

Figures 10-1 and 10-2 illustrate that, if the ideal-gas equation of state were to be used for the real gas, it would be quite exact if the density were low and, most probably, quite inexact at high densities. It is evident that an exact equation of state is highly desirable but that an equation able to follow precisely the entire complex patterns of Figs. 10-1 and 10-2 will be difficult, if not impossible, to find. Fortunately, most problems involve only a small region of the $pvT$ domain, and therefore a relatively

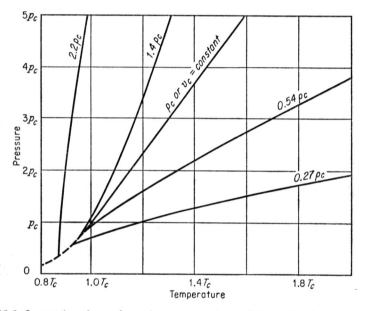

FIG. 10-2. Isometrics of a real gas (curvatures of $\rho = C$ lines slightly exaggerated).

simple equation of state *may* be satisfactory. (And the automatic computer may change this entire chapter by quickly handling elaborate equations of state, of form not yet known.)

**10-2. The van der Waals Equation.** An equation of historic importance is that proposed by van der Waals in 1873 to overcome, in part, the deficiencies of the ideal-gas equation. A perfect gas has point molecules; therefore, the length of path between collisions with the enclosure is not changed by the presence of the molecules themselves. Actually, in the real gas as more and more molecules are present in a volume (as the density is increased), the number of times a molecule will strike the walls will be increased and a greater pressure will be exerted. Van der Waals premised that the increased pressure would be inversely proportional to the actual volume, that is, to the free space, when allowance was made for the finite size and number of molecules:

$$\frac{p_{\text{perfect}}}{p_{\text{actual}}} = \frac{\text{actual volume}}{\text{total volume}} = \frac{v - b}{v}$$

$$p_{\text{perfect}} = p \frac{v - b}{v}$$

But also

$$p_{\text{perfect}} = \frac{RT}{v}$$

and therefore

$$p = \frac{RT}{v - b}$$

where $b$ is a constant for each gas.

The second basic characteristic of the perfect gas is that the molecules are neither attracted nor repulsed by the other molecules present. In the real gas, a force field exists around the molecule. As the density increases, the molecules approach closer to each other, and near collision, short-range forces of repulsion come into play. It can be considered that the constant $b$ includes repulsion forces since it simulates the same effect on pressure (an increase).

An opposing effect is present before collision since the molecules are attracted to each other by long-range cohesive forces. Van der Waals premised that the pressure of the gas is decreased, because of these attraction forces, by an amount proportional to the square of the density:

Total pressure = (kinetic and repulsion pressure) − (cohesive pressure)

$$p = \frac{RT}{v - b} - \frac{a}{v^2} \tag{10-1a}$$

where $a$, like $b$, is a constant for each gas. When $a$ and $b$ are zero, the equation is that of the ideal gas.

The calculation of temperature or of pressure can be readily made from the van der Waals equation, but calculations of specific volume are best made by successive approximations. For this reason, the equation is said to be *explicit* only for temperature and pressure.

The constants $a$ and $b$ in the van der Waals equation, the *specific constants*, should be found by empirically fitting the equation to experimental data for one particular region, and thus several sets of constants will be necessary to include all regions. An easier and less accurate solution is to calculate the constants from certain conditions which are known to be valid at the critical point:

1. The equation of state.
2. The critical isotherm has a horizontal tangent.
3. The critical isotherm has a point of inflection.

The last two conditions are expressed mathematically as

$$\left(\frac{\partial p}{\partial v}\right)_{T_c} = 0 \qquad \left(\frac{\partial^2 p}{\partial v^2}\right)_{T_c} = 0 \qquad \text{(at critical point)} \qquad (10\text{-}2)$$

When the three equations, Eqs. (10-1a) and (10-2), are solved simultaneously, the three constants $a$, $b$, and $R'$ can be evaluated† in terms of the critical constants $p_c$, $v_c$, and $T_c$. For the van der Waals equation,

$$\left(\frac{\partial p}{\partial v}\right)_T = \frac{-R'T}{(v-b)^2} + \frac{2a}{v^3} = 0$$

$$\left(\frac{\partial^2 p}{\partial v^2}\right)_T = \frac{2R'T}{(v-b)^3} - \frac{6a}{v^4} = 0$$

$$p = \frac{R'T}{v-b} - \frac{a}{v^2}$$

And, upon solving,

$$a = 3v_c{}^2 p_c \qquad b = \frac{v_c}{3} \qquad R' = \frac{8}{3}\frac{p_c v_c}{T_c} \qquad (a)$$

But the van der Waals equation should reduce to the ideal-gas equation as pressure approaches zero, and therefore $R'$ should be the universal gas constant $R$ and not a specific constant for each gas. This reasoning allows one of the critical constants to be eliminated from Eqs. (a), and the critical volume is selected, since its value is not usually known within the same accuracy as the critical temperature and pressure,

$$a = \frac{27}{64}\frac{R^2 T_c{}^2}{p_c} \qquad b = \frac{1}{8}\frac{R T_c}{p_c} \qquad (10\text{-}3)$$

Values of $a$ and $b$ for several gases and calculated from Eq. (10-3) can be found in Table B-8. Since the constant $R$ is a universal constant, the van der Waals equation (and many others) is called a *two-constant equation of state.*

**Example 1.** Compute from the van der Waals equation the pressure exerted by 1 lb$_m$ of $CO_2$ at a temperature of 212°F if the specific volume is 0.193 ft³/lb$_m$.
    *Solution.* Since Table B-8 is in mole units, it will be easier to convert the specific volume to a mole basis (or change the constants in Table B-8 by use of the factors listed in this table). Then:

----

† And if $R'$ is arbitrarily specified, the equation of state, in general, can contain three other constants, $a$, $b$, and $c$.

Table B-6

$M_{CO_2} = 44$

$v = 0.193(44) = 8.492 \text{ ft}^3/\text{mole}$

$T = 672°R$

Equation (10-1a)

$p = \dfrac{RT}{v-b} - \dfrac{a}{v^2}$

$= \dfrac{0.73(672)}{8.492 - 0.685} - \dfrac{924.2}{8.492^2}$

$= 62.9 - 12.8 = 50.1 \text{ atm}$

Table B-8

$a = 924.2 \dfrac{\text{atm ft}^6}{\text{mole}^2}$

$b = 0.685 \text{ ft}^3/\text{mole}$

$R = 0.73 \dfrac{\text{atm ft}^3}{°R \text{ mole}}$

Equation (3-5a)

$p = \dfrac{RT}{v} = \dfrac{0.73(672)}{8.492}$

$p = 57.8 \text{ atm} \qquad Ans.$

Compare these two answers. Which answer is correct can be determined only by experiment, although the van der Waals answer should be, and usually is, more reliable.

**10-3. The Compressibility Factor.** The deviation of the real gas from ideal behavior is shown by the compressibility factor $z$,

$$z \equiv \frac{pv}{RT} \qquad pv = zRT \qquad (10\text{-}4a)$$

where $z$ is a complex (and unknown) function of any two of the variables $p$, $v$, and $T$. The factor $z$ can be calculated from the experimental data, and a diagram such as Fig. 10-1 constructed. Although only isotherms are shown in Fig. 10-1, isometrics ($v = C$) can also be superimposed, since $v = zRT/p$. Figure 10-1 can be called a *specific compressibility chart* since it is constructed from data for one substance. It represents a graphical equation of state which is not restricted to the gaseous phase of the substance.

In engineering practice, reasonable approximations to the $pvT$ relations of a substance are frequently required, especially for the case, more usual than unusual, where $pvT$ data are entirely nonexistent. To fulfill this need, a *generalized compressibility chart* can be constructed from known data for one or more substances, and this chart can then be used for all substances. To construct a generalized compressibility chart, one method is to reduce the properties by dividing by the appropriate critical constant:

$$p_r \equiv \frac{p}{p_c} \qquad T_r \equiv \frac{T}{T_c} \qquad v_r \equiv \frac{v}{v_c} \qquad (10\text{-}5)$$

The compressibility factor $z$ is then plotted versus any two of these reduced properties. For example, Fig. 10-3 shows the experimental data for 10 different gases on a generalized compressibility chart which has $z$, $T_r$, and $p_r$ as the coordinates, and the data correlate with an over-all average deviation of the order of 1 per cent. This excellent correlation suggests that Fig. 10-3 can be used for any gas with acceptable accuracy for most engineering problems (and, for $z$ values greater than 0.6, the generalized compressibility chart is almost invariably within 5 per cent of the correct answer *for all gases*).

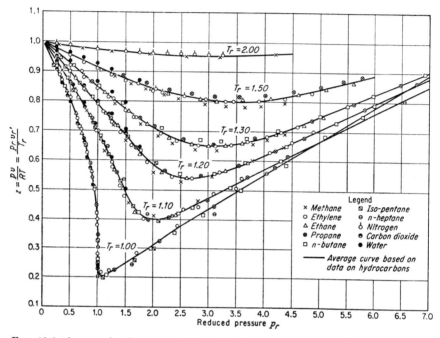

FIG. 10-3. A generalized compressibility chart for various gases. (Gouq-Jen Su; see footnote, p. 167.)

**Example 2.** Determine by means of Fig. 10-3 the specific volume of 1 $lb_m$ of $CO_2$ if the pressure is 50.9 atm and the temperature is 212°F.
*Solution.* From Table B-7,

$$p_c = 72.9 \text{ atm} \qquad\qquad t_c = 87.9°F$$

$$p_r = \frac{p}{p_c} = \frac{50}{72.9} = 0.685 \qquad T_r = \frac{T}{T_c} = \frac{459.7 + 212}{459.7 + 87.9} = 1.23$$

From Fig. 10-3, $z = 0.88$, and, by Eq. (10-4a),

$$v = \frac{zRT}{p} = \frac{0.88(0.73)(671.7)}{50.9} = 8.49 \text{ ft}^3/\text{mole} = 0.193 \text{ ft}^3/lb_m \qquad Ans.$$

The chart solution is almost invariably more accurate than that found by a two-constant equation of state.

Figure 10-3 does not show values for the reduced specific volume $v_r$. The reason for this omission is that $z$ has been correlated against two independent variables, $T_r$ and $p_r$, and therefore Fig. 10-3 represents a graphical function of

$$z = f(T_r, p_r) \qquad (a)$$

By substituting Eqs. (10-5) into Eq. (10-4a),

$$z = \frac{p_r v_r}{T_r} \frac{p_c v_c}{RT_c} = \frac{p_r v_r}{T_r} z_c \qquad (10\text{-}4b)$$

a new variable $z_c$ appears: *the compressibility factor at the critical point.* Unfortunately, $z_c$ is not a constant but varies widely from substance to substance (Table B-7). However, comparison of Eqs. (a) and (10-4b) shows that $v_r z_c$ must also be a function of $T_r$ and $p_r$. This variable, proposed by Kamerlingh Onnes, was called by Su† the *ideal reduced volume* $v_{r'}$,

$$z = \frac{p_r v_r}{T_r} z_c = \frac{p_r v_{r'}}{T_r} \qquad (10\text{-}4c)$$

Thus, the reduced properties of the compressibility chart are defined as

$$T_r \equiv \frac{T}{T_c} \qquad p_r \equiv \frac{p}{p_c} \qquad v_{r'} \equiv \frac{1}{\rho_{r'}} \equiv \frac{v}{RT_c/p_c} = z \frac{T_r}{p_r} \qquad (10\text{-}6)$$

wherein $RT_c/p_c$ could be called the *ideal-gas volume at the critical point.* Note again that reduced isometrics of the form $v/v_c$ cannot be added to Fig. 10-3 since $z_c$ is not a constant but that Eqs. (10-6) show that this disadvantage can be overcome by redefining the reduced volume to be a function of only $T_r$ and $p_r$.

Figure B-8 is a generalized compressibility chart with ideal reduced isometrics. Equations (10-6) allow additional isometrics and isotherms to be readily constructed on the compressibility chart, and thus interpolation can be avoided.

**Example 3.** Determine the pressure exerted by 1 $lb_m$ of $CO_2$ at 212°F if the specific volume is 0.193 ft³/$lb_m$.

*Solution.* From Table B-7,

$$t_c = 87.9°F \qquad z_c = 0.275 \qquad v_c = 1.51 \text{ ft}^3/\text{mole}$$

Hence, $\quad T_r = \dfrac{T}{T_c} = \dfrac{671.7}{547.6} = 1.23 \qquad v_{r'} = v_r z_c = \dfrac{0.193(44)}{1.51}(0.275) = 1.55$

From Fig. B-8, $z = 0.88$; hence,

$$p = \frac{zRT}{v} = \frac{0.88(0.73)(671.7)}{44(0.193)} = 50.9 \text{ atm} \qquad Ans.$$

Examples 1 to 3 are for the same state.

† Gouq-Jen Su, Modified Law of Corresponding States, *Ind. Eng. Chem.*, **38**: 803–806 (August, 1946).

**10-4. Compressibility and the Equations of State.** Although the specific compressibility chart is a convenient and precise method for obtaining the $pvT$ properties of a substance, an equation of state is more desirable when knowledge of other properties, such as internal energy and entropy, is required (Art. 8-1). This is because derivatives must be evaluated, and graphical differentiations and integrations are tedious and open to error.

A large number of equations of state have been derived from considerations based upon kinetic theory, upon thermodynamic relationships, and upon the known behavior of substances; some 56 equations are listed in Ref. 1. The equations are usually explicit in pressure, and this form of equation is an invariable consequence of a kinetic-theory background. The specific constants of the equations are found in various manners and without necessarily restricting the equation to have a point of inflection at the critical point. For precise work, the specific constants should be found by fitting the equation to the experimental data in the region under study. In this article, however, the specific constants are determined primarily by conditions at the critical point, and therefore the particular values are not necessarily the optimum values.

On the other hand, for many engineering applications it is neither expedient nor necessary to have a high degree of preciseness. *Thus, the validity of the equation of state for regions beyond the extent of the experimental data or for regions where the constants were determined by theoretical considerations can be shown by superimposing the equation on a generalized compressibility chart.* The procedure can be adapted to any equation of state, although tests only of reduced equations will be discussed here. The objective of this article is to study the $pvT$ predictions of the equation of state, while the question of property evaluation will be delayed until Art. 10-5.

*Van der Waals (1873)*

$$p = \frac{RT}{v - b} - \frac{a}{v^2} \tag{10-1a}$$

The constants dictated by Eqs. (10-2) are

$$a = \frac{27}{64} \frac{R^2 T_c^2}{p_c} \qquad b = \frac{1}{8} \frac{RT_c}{p_c} \tag{10-3}$$

A *reduced* equation is obtained by substituting Eqs. (10-3) and (10-6) into Eq. (10-1a):

$$p_r = \frac{T_r}{v_{r'} - \frac{1}{8}} - \frac{\frac{27}{64}}{v_{r'}^2} \tag{10-1b}$$

Since Eq. (10-1b) is entirely divorced from the critical properties of any one gas, it is *a reduced equation with generalized constants,* called also *a generalized equation.* This equation can be transposed into a compressibility equation by substituting $p_r = zT_r/v_{r'}$ for the explicit term and rearranging,

$$z = \frac{v_{r'}}{v_{r'} - \frac{1}{8}} - \frac{27/64}{T_r v_{r'}} \tag{10-1c}$$

Equation (10-1c) can be solved for selected values of $v_{r'}$ and $T_r$ and the $z$ network superimposed on the appropriate generalized compressibility chart to show the region or regions where the equation is valid.

Two-constant equations are not of much value if a large region of Fig. 10-1 is to be surveyed. But for the low-density region, such equations are usually quite accurate for many problems. The van der Waals equation can be expanded (Prob. 15) into the *virial* form (for use at low to medium densities),

$$z = 1 + \left(\frac{1}{8} - \frac{27/64}{T_r}\right)\rho_{r'} + \frac{1}{8^2}\rho_{r'}{}^2 + \frac{1}{8^3}\rho_{r'}{}^3 + \cdots \tag{10-1d}$$

or into a power series in $p$ (Prob. 16) (for use at low densities),

$$z = 1 + \left(\frac{1/8}{T_r} - \frac{27/64}{T_r{}^2}\right)p_r + \cdots \tag{10-1e}$$

These two equations reflect the data of Fig. B-8 quite accurately over a large region.

The compressibility factor at the critical point for the van der Waals equation is found from

$$z_c = \frac{p_c v_c}{RT_c} = 0.375 \tag{10-1f}$$

The value of Eq. (10-1f) is high relative to that of most substances (Table B-7); hence, the van der Waals equation is poor in the vicinity of the critical point and along the critical isotherm at high pressures.

Su[†] reported an *average* deviation of only 3 per cent or less from experimental data for 17 gases at densities almost up to the critical. An average deviation, however, is misleading; for example, at $T_r = 1.5$, $p_r = 2.5$, the deviation is of the order of 7 per cent at a density half that of the critical (Fig. 10-4a). The equation does show the foldback of the isotherms at the upper boundary of Fig. 10-1, and, at higher temperatures, it is quite accurate. The van der Waals equation, like many others, yields straight isometrics, and therefore it does not obey Fig. 10-2.

*Dieterici* (1899)

$$p = \frac{RT}{v - b}e^{-a/RTv} \tag{10-7a}$$

With the same procedures as for the van der Waals equations

$$a = \frac{4R^2 T_c{}^2}{p_c e^2} \qquad b = \frac{RT_c}{p_c e^2} \tag{10-7b}$$

$$p_r = \frac{T_r}{v_{r'} - 1/e^2}e^{-4/T_r v_{r'} e^2} \tag{10-7c}$$

$$z = \frac{v_{r'}}{v_{r'} - 1/e^2}e^{-4/T_r v_{r'} e^2} \tag{10-7d}$$

$$z = 1 + \frac{1}{e^2}\left(1 - \frac{4}{T_r}\right)\rho_{r'} + \frac{1}{e^4}\left(1 - \frac{4}{T_r} - \frac{8}{T_r{}^2}\right)\rho_{r'}{}^2 + \cdots \tag{10-7e}$$

[†] Gouq-Jen Su and C. H. Chang, A Generalized van der Waals Equation of State, *Ind. Eng. Chem.*, August, 1946, p. 800.

The factor $z_c$ can be calculated from its definition,

$$z_c = \frac{p_c v_c}{R T_c} = 0.271 \qquad (10\text{-}7f)$$

The critical value of 0.271 for $z_c$ is close to that for many substances (Table B-7), and therefore the Dieterici equation shows good agreement with the compressibility chart in the neighborhood of the critical point and also along the critical isotherm; in other regions, it is far in error (Fig. 10-4). (Of course, the chart itself is poor in the neighborhood of the critical point.)

The equation has positive-curvature isometrics.

*Berthelot* (1903)

$$p = \frac{RT}{v - b} - \frac{a}{T v^2} \qquad (10\text{-}8a)$$

With the same procedures as for the van der Waals equation,

$$a = \frac{27}{64} \frac{R^2 T_c{}^3}{p_c} \qquad b = \frac{1}{8} \frac{R T_c}{p_c} \qquad (10\text{-}8b)$$

$$p_r = \frac{T_r}{v_{r'} - \frac{1}{8}} - \frac{2\frac{7}{64}}{T_r v_{r'}{}^2} \qquad (10\text{-}8c)$$

$$z = \frac{v_{r'}}{v_{r'} - \frac{1}{8}} - \frac{2\frac{7}{64}}{T_r{}^2 v_{r'}} \qquad (10\text{-}8d)$$

Figure 10-4 shows that Eq. (10-8d) is inferior to the van der Waals equation. But if the constant $\frac{1}{8}$ is replaced by $\frac{9}{128}$ and the equation is expanded into a series in $p_{r'}$,

$$z = 1 + \left( \frac{9}{128} - \frac{2\frac{7}{64}}{T_r{}^2} \right) p_{r'} + \left( \frac{9}{128} \right)^2 p_{r'}{}^2 + \left( \frac{9}{128} \right)^3 p_{r'}{}^3 + \cdots \qquad (10\text{-}8e)$$

or in $p_r$,

$$z = 1 + \left( \frac{\frac{9}{128}}{T_r} - \frac{2\frac{7}{64}}{T_r{}^3} \right) p_r + \cdots \qquad (10\text{-}8f)$$

the resulting equations are superior to the similar van der Waals open equations. For example, many of the data of Fig. B-8 can be reproduced within 1 per cent deviation [versus about 2 per cent for Eqs. (10-1d) and (10-1e)]. Because of this excellent correlation, the Berthelot equation is widely used for low-pressure gas corrections.

The isometrics have negative curvature.

*Kamerlingh Onnes*† (1901)

$$z = \frac{pv}{RT} = 1 + \frac{B}{v_{r'}} + \frac{C}{v_{r'}{}^2} + \frac{D}{v_{r'}{}^4} + \frac{E}{v_{r'}{}^6} + \cdots \qquad (10\text{-}9)$$

wherein the coefficients $B$, $C$, $D$, etc., are of the form

$$B = b_1 + \frac{b_2}{T_r} + \frac{b_3}{T_r{}^2} + \frac{b_4}{T_r{}^4} + \frac{b_5}{T_r{}^6} \qquad (10\text{-}10)$$

The constants $b_1, b_2, \ldots, c_1, c_2, \ldots$, etc., represent the average values for hydrogen, nitrogen, oxygen, ethyl ether, and isopentane.

Excellent results are obtained from this equation. Note in Fig. 10-4 that the maximum deviation from experimental data is slightly greater than 2 per cent and, in most cases, considerably smaller.

† H. Kamerlingh Onnes, Expression of the Equation of State of Gases and Liquids by Means of a Series, *Communs. Phys. Lab. Leiden*, no. 71 (1901).

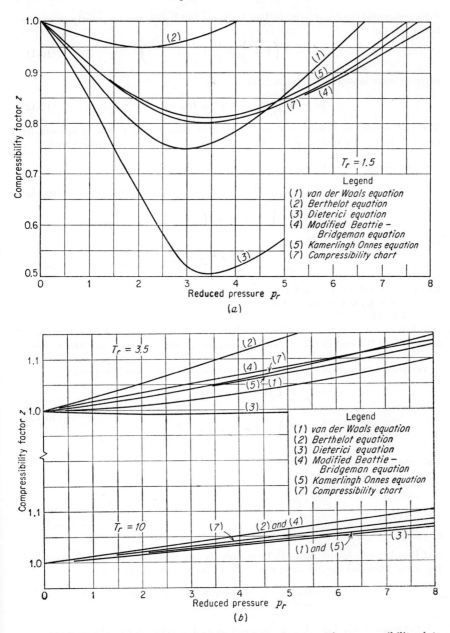

FIG. 10-4. Comparison of various reduced equations of state with compressibility data.

Experimental (and isothermal) data for gases and liquids are frequently presented in the form of

$$pv = A \left( 1 + \frac{B}{v} + \frac{C}{v^2} + \frac{D}{v^4} + \cdots \right) \tag{10-11}$$

or, from some laboratories, in the form

$$pv = A + B'p + C'p^2 + D'p^4 + \cdots \tag{10-12}$$

The odd powers of density and pressure (other than the first power) are usually omitted since fewer terms, practically, are required to fit the experimental data.

*Beattie-Bridgeman†* (1928)

$$p = RT \frac{1 - \epsilon}{v^2} (v + B) - \frac{A}{v^2} \tag{10-13a}$$

$$A = A_0 \left( 1 - \frac{a}{v} \right) \qquad B = B_0 \left( 1 - \frac{b}{v} \right) \qquad \epsilon = \frac{c}{vT^3}$$

Or, in virial form,

$$p = \frac{RT}{v} + \frac{\beta}{v^2} + \frac{\gamma}{v^3} + \frac{\delta}{v^4} \tag{10-13b}$$

where $\beta = RTB_0 - A_0 - Rc/T^2$
$\gamma = -RTB_0b + A_0a - RB_0c/T^2$
$\delta = RB_0bc/T^2$

The five specific constants for the equations have been determined for a number of gases (Table B-8 and Ref. 3) by curve fitting to the experimental data; the methods are discussed in detail in Ref. 5.

The reduced equation cannot be entirely divorced from critical properties,

$$p_r = \frac{T_r(1 - \epsilon')}{v_{r'}{}^2} (v_{r'} + B') - \frac{A'}{v_{r'}{}^2} \tag{10-13c}$$

$$A' = A_0' \left( 1 - \frac{a'}{v_{r'}} \right) \qquad B' = B_0' \left( 1 - \frac{b'}{v_{r'}} \right) \qquad \epsilon' = \frac{c'}{v_{r'}T_r{}^3}$$

since the new dimensionless ratios equal

$$a' = \frac{a}{RT_c/p_c} \qquad b' = \frac{b}{RT_c/p_c} \qquad c' = \frac{c}{RT_c{}^4/p_c}$$

$$A_0' = \frac{A_0}{R^2T_c{}^2/p_c} \qquad B_0' = \frac{B_0}{RT_c/p_c} \tag{10-13d}$$

Su‡ found the foregoing ratios to be essentially constants for a number of nonpolar§ gases and proposed the following generalized values (and $z_c = 0.27$):

$$a' = 0.1127 \qquad b' = 0.03833 \qquad c' = 0.05$$
$$A_0' = 0.4758 \qquad B_0' = 0.18764$$

† Reference 5; the behavior of gases is discussed in detail in this classic paper.

‡ Gouq-Jen Su and C. H. Chang, Generalized Beattie-Bridgeman Equation for Real Gases, *J. Am. Chem. Soc.*, **68**: 1080 (June, 1946).

§ Polar molecules have *dipole moments*—the electrical centers of positive and negative charges do not coincide ($CH_4$ nonpolar, $H_2O$ polar, for examples).

Generalized constants can also be calculated by Eqs. (10-13$d$) from the specific constants for a particular gas (and used for similar-structure gases).

Su† found that for each of 17 gases an average deviation of 2 per cent or less could be expected up to nearly the critical density. Figure 10-4 shows that the equation with the Su constants is quite precise (although, at $T_r = 1.2$, $p_r = 2.5$, the deviation is at least 6.5 per cent). At high reduced temperatures the deviation, although small, is greater than that of the van der Waals equation (Su's constants were fitted for the temperature range $T_r = 1$ to 5).

The equation has isometrics with negative curvature.

*Beattie‡* (1930)

$$v = \frac{RT}{p} + \frac{\beta}{RT} + \frac{\gamma}{R^2 T^2}\, p + \frac{\delta}{R^3 T^3}\, p^2 \tag{10-14a}$$

Beattie approximated Eq. (10-13$b$) by partially solving for the volume and then replacing $1/v$ by $p/RT$ in the higher-order terms. The Beattie equation is convenient in those cases where an equation explicit in $v$ facilitates computations [Eq. (8-5)]. At low pressures, Eq. (10-14$a$) is approximated by

$$z = 1 + \frac{\beta}{R^2 T^2}\, p \tag{10-14b}$$

or in reduced form

$$z = 1 + \left( \frac{B_0'}{T_r} - \frac{A_0'}{T_r^2} - \frac{c'}{T_r^4} \right) p_r \tag{10-14c}$$

Equation (10-14$c$) deviates from Fig. B-8 by about 1 per cent over most of the region (the generalized constants are those of the Beattie-Bridgeman equation).

*Benedict-Webb-Rubin§* (1940)

$$p = RT\rho + \left\{ RT(B_0 + b\rho) - (A_0 + a\rho - a\alpha\rho^4) - \frac{1}{T^2}\left[ C_0 - c\rho(1 + \gamma\rho^2)e^{-\gamma\rho^2} \right] \right\} \rho^2 \tag{10-15a}$$

$$z = 1 + \left( B_0 - \frac{A_0}{RT} - \frac{C_0}{RT^3} \right) \rho + \left( b - \frac{a}{RT} \right) \rho^2 + a\alpha\frac{\rho^5}{RT} + \frac{c\rho^2}{RT^3}(1 + \gamma\rho^2)e^{-\gamma\rho^2} \tag{10-15b}$$

The Benedict equation has eight specific constants, which have been evaluated for a number of hydrocarbons. The constants proposed by Benedict§ predict the pressure of gases with a maximum deviation of $1\frac{3}{4}$ per cent (and an average deviation of 0.35 per cent) up to densities of $1.8\rho_c$; methods for obtaining the vapor pressure of the pure substance and the phase behavior of hydrocarbon mixtures were also proposed.

The isometrics have negative curvature at low densities, positive at intermediate densities, and negative again at high densities, in agreement with Fig. 10-2.

† Su and Chang, *loc. cit.*

‡ J. A. Beattie, A New Equation of State for Fluids, Part IV, *Proc. Natl. Acad. Sci.*, **16:** 14 (1930).

§ M. Benedict, G. Webb, and L. Rubin, An Empirical Equation for the Thermodynamic Properties of Light Hydrocarbons and Their Mixtures, *J. Chem. Phys.*, **8:** 334 (1940).

*Martin-Hou*[6] (1955)

$$p = \frac{RT}{v - b} + \frac{A_2 + B_2 T + C_2 e^{-cT}}{(v - b)^2} + \frac{A_3 + B_3 T + C_3 e^{-cT}}{(v - b)^3} + \frac{A_4}{(v - b)^4}$$
$$+ \frac{B_5 T + C_5 e^{-cT}}{(v - b)^5} \qquad (10\text{-}16)$$

Most equations of state represent best a restricted class of substances: the Benedict equation, hydrocarbons. The Martin-Hou equation was designed to represent all gases by the use of 11 specific constants, which are functions of the critical properties $p_c$, $T_c$, $z_c$, and the slope of the critical isometric.

In general, it is claimed that the equation reproduces the experimental data within 1 per cent up to densities of about $1.5\rho_c$ and up to temperatures of about $1.5T_c$. Seven gases were studied, with various structures, polar as well as nonpolar: $CO_2$, $H_2O$, $C_6H_6$, $N_2$, $C_3H_6$, $H_2S$, and $C_3H_8$. Near the limiting density of $1.5\rho_c$, the equation is sensitive to the value of $b$ (because of the difference term, $v - b$) and becomes unusable.

*Liquids.* A simple equation for liquids is the *Buehler equation*,[7]

$$p_r = \frac{x_1}{x_2 - \rho_r} + x_3 \qquad (0.95 > T_r > 0.5) \qquad (10\text{-}17)$$

where $x_1 = 10T_r - 3.1$
      $x_2 = 8.284 - 18.07z_c - (4.482 - 14.1z_c)T_r$
      $x_3 = 15T_r - 20$

An advantage of the equation is that it can be solved explicitly for either pressure or density. Reduced equations for condensed phases, however, may be highly inaccurate since density changes are small.

*Plasma.* Phenomena such as atomic explosions and high-speed missiles require that consideration be given to the breakdown of gases, in particular, air. When the temperature of air at atmospheric pressure exceeds about 3000°K, intermolecular collisions become so violent that molecular oxygen ($O_2$) *dissociates* into atomic oxygen (O) and, to a lesser extent, molecular nitrogen ($N_2$) dissociates into atomic nitrogen (N), while various oxides of nitrogen appear. At still higher temperatures (say 6000°K), the atoms become appreciably *ionized*—the atoms break up into positively charged ions and free (unbound) electrons. Too, the bound electrons of the atom may move from inner orbits to outer orbits by absorbing energy so that the atom is in an "excited" state.

Consider that the effect of dissociation or ionization is to increase the moles of particles (for example, 1 mole of molecular oxygen forms 2 moles of atomic oxygen). Therefore, a gas resists a rise in temperature by dissociating and/or ionizing to absorb the added energy and resists a rise in pressure by recombining particles to reduce the volume. Thus the tendency for dissociation and ionization is greatest at high temperatures and low pressures (low densities). When the free-electron population of the gas is sufficient to make the gas an electrical conductor, it is called a *plasma*.

At 2000°K and 7 atm, air is essentially a nondissociated, nonionized gas with an approximate composition (in mole per cent) of 78 per cent $N_2$, 21 per cent $O_2$, and traces of A and $CO_2$ (Table 14-1); at the same temperature but $7 \times 10^{-6}$ atm, air becomes a partially dissociated but negligibly ionized gas with a composition of about 73 per cent $N_2$, 15 per cent $O_2$, 10 per cent O, 0.9 per cent A, 0.7 per cent NO, and traces of CO and $CO_2$; at 15,000°K and 1.5 atm, air becomes a fully dissociated and partially ionized gas—a plasma—with a composition of about 30 per cent N, 30 per cent electrons, 25 per cent $N^+$ (singly ionized nitrogen atom), 9.8 per cent O, 4.9 per cent $O^+$, 0.16 per cent A, 0.16 per cent $A^+$ (Hilsenrath[8]).

The equation of state for the mixture of particles called a *plasma* is usually shown in one of the two following forms (since density is low, the ideal-gas laws are adequate or $z$ is unity):

$$pv = Z^*RT$$
$$pv = (1 + \gamma)RT$$

where $v$ = specific volume (lb$_m$ basis)
$R$ = specific gas constant for the gas before breakdown
$Z^*$ = moles of particles $(n_p)$ per mole of gas before breakdown $(n)$
$\gamma$ = degree of dissociation = $(n_p - n)/n$

The symbol $Z^*$ (a dissociation-ionization factor) should not be confused with $z$ (the compressibility factor of undissociated, un-ionized gas; see Figs. 10-1, 10-3, and B-8).

**10-5. Property Evaluation and the Equation of State.** The primary purpose of the equation of state is to evaluate the thermodynamic properties other than pressure, volume, and temperature. Unfortunately, discussions in the literature of the ability of an equation of state to predict these properties are either absent or unsatisfactorily sparse. The problem is that slight errors in the $pvT$ or $zpT$ data can be greatly magnified in the partial derivatives of these data; also, the property equations usually involve a difference term between two relatively large numbers (with the difference approaching zero as density approaches zero) [Eq. (8-5), for example], thus leading to large and unforeseen errors. For example, the Beattie-Bridgeman operation portrays the compressibility of $CO_2$ to densities well above the critical (Fig. 10-4) but, in this same range, poorly represents the internal-energy deviation (Fig. 10-5). Note that, at very high densities, the results are far in error.

**10-6. Principles of Corresponding States.** A theorem of corresponding states was first proposed by van der Waals. He reasoned that all substances existed as a gas, liquid, or solid phase or as a mixture of these phases and that all displayed common characteristics such as the phenomenon of a critical point, a triple point, a Boyle isotherm, etc. Therefore, the $pvT$ relationship of all substances could be portrayed by a single model or three-dimensional surface if the parameters of this surface were properly selected. Since he was primarily interested in gases, he selected the critical point to be the base reference and the critical properties to be the reducing parameters,

$$p_r \equiv \frac{p}{p_c} \qquad T_r \equiv \frac{T}{T_c} \qquad v_r \equiv \frac{v}{v_c} \tag{10-5}$$

Therefore, the van der Waals principle of corresponding states has the form

$$f(p_r, v_r, T_r) = 0 \tag{10-18a}$$

Van der Waals illustrated his principle by substituting Eqs. (10-3) and (10-5) into Eq. (10-1a) along with the critical ratio of $\frac{3}{8}$,

$$p_r = \frac{8T_r}{3v_r - 1} - \frac{3}{v_r^2} \tag{10-19}$$

Equation (10-19) is a *reduced* or *generalized* or *universal* equation of state since it contains no constants specific to one gas. For that matter, many equations of state can be reduced in the same manner; thus, the van der Waals principle of corresponding states [Eq. (10-18a)] is a much broader generalization than a particular reduced equation based upon the van der Waals reduced properties [Eq. (10-19)].

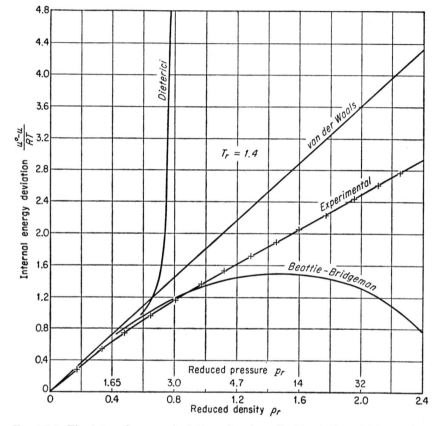

Fig. 10-5. The internal-energy deviation of carbon dioxide at $T_r = 1.4$ for various equations of state. (From Ref. 4.)

The van der Waals principle of corresponding states can be shown in a different form. Consider that $v_r$, by definition, equals

$$v_r \equiv \frac{v}{v_c} = \frac{zRT/p}{z_cRT_c/p_c} = \frac{z}{z_c}\frac{T_r}{p_r}$$

Hence, substituting for $v_r$ in Eq. (10-18a),

$$f\left(p_r, T_r, \frac{z}{z_c}\right) = 0 \qquad (10\text{-}18b)$$

Observe that, if reduced isotherms were to be constructed on a diagram with reduced pressure and reduced volume (or $z/z_c$) as coordinates, correlation for all substances would be *exact* at the critical point (since $v_r$, $T_r$, $p_r$, $z/z_c$ all would have the value of unity). *Thus, the van der Waals principle of corresponding states is an exact law for all substances at the critical point.* On the other hand, in other regions, for example, at low pressures, where $p_r$ isothermally approaches 0 as its limit, Fig. 10-1 shows that $z$ approaches 1 as its limit; therefore, the variable $z/z_c$ approaches a limit $1/z_c$ which has different values for different gases [since $z_c$ varies from about 0.2 to 0.3 for various substances (Table B-7)]. *Thus, the van der Waals principle of corresponding states is badly in error at low pressures if the substances being correlated have different $z_c$ values.*

The fault of the van der Waals principle at low pressures for substances with different $z_c$ values was corrected by Kamerlingh Onnes. Since $z$ *exactly* equals

$$z = \frac{pv}{RT} = \frac{p_r p_c v_r v_c}{RT_r T_c} = \frac{p_r v_r}{T_r} \frac{p_c v_c}{RT_c} = \frac{p_r v_r}{T_r} z_c \qquad (10\text{-}20)$$

and since all gases approach the ideal state (where $z = 1$) as pressure isothermally approaches zero (Fig. 10-1), a new set of reduced properties can be defined by combining $z_c$ with either $p_r$, $T_r$, or $v_r$. The combination of $v_r z_c$ was called by Su the *ideal reduced volume* (Art. 10-3). It follows that the new reduced properties are

$$T_r \equiv \frac{T}{T_c} \qquad p_r \equiv \frac{p}{p_c} \qquad v_{r'} \equiv \frac{1}{\rho_{r'}} \equiv \frac{v}{RT_c/p_c} = z \frac{T_r}{p_r} \qquad (10\text{-}6)$$

On the basis of Eqs. (10-6), Su proposed a *modified principle of corresponding states,*

$$f(p_r, T_r, v_{r'}) = 0 \qquad (10\text{-}21a)$$
or
$$f(p_r, T_r, z) = 0 \qquad (10\text{-}21b)$$

*Note that the Su principle of corresponding states approaches exactness for all substances as the pressure approaches zero but is badly in error at the critical point if the substances being correlated have different $z_c$ values* (the inverse behavior to the van der Waals law).

The generalized compressibility chart (Figs. B-8 and 10-3) represents graphically the function of Eq. (10-21b) with a precision adequate for many problems (and the fact that correlation is not perfect on the $zp_r$ chart shows that no exact function of the variables $z$, $T_r$, and $p_r$ exists). Thus, the $zp_r$ chart and all equations reduced with Eqs. (10-6) are examples of the modified principle of corresponding states. If the substances being correlated have the same $z_c$ values, the van der Waals and the Su principles are equivalent; if the substances have different $z_c$ values, the van der Waals principle is exact at the critical point and the Su principle approaches exactness as the pressure approaches zero, but in between these regions neither principle is exact (and the Su theorem is preferable since it is more exact in the low-pressure and medium-pressure regions and, also, quite accurate in all regions save that of the critical point, as demonstrated by the validity of the generalized compressibility chart).

## PROBLEMS

**1.** Determine the pressure exerted by 2 $lb_m$ air at 100°F if the volume is 30 ft³, using the (a) perfect-gas equation, (b) van der Waals equation, (c) compressibility factor.

**2.** Repeat Prob. 1, but for (a) methane, (b) carbon monoxide, (c) carbon dioxide, or (d) hydrogen.

**3.** Repeat Prob. 1, but use the Beattie-Bridgeman equation.

**4.** Locate the following states of $H_2O$ on the compressibility chart (Fig. B-8 or 10-3):

320 psia, 480°F; 590°F, $x = 0.5$; 32°F saturated liquid; 32°F, 6,000 psia.   (Compute $z$ from Steam Table data.)

**5.** A pound of carbon dioxide has a volume of 0.15 ft³ and a pressure of 100 atm. Compute the temperature by the (a) perfect-gas equation, (b) van der Waals equation, (c) compressibility factor.

**6.** Repeat Prob. 5, but for (a) methane, (b) carbon monoxide, (c) hydrogen, or (d) air.

**7.** A pound of carbon dioxide has a temperature of 100°F and a pressure of 1,000 atm.   Compute the volume by the (a) perfect-gas equation, (b) van der Waals equation, (c) compressibility factor.

**8.** Repeat Prob. 7, but for (a) methane, (b) carbon monoxide, (c) hydrogen, or (d) air.

**9.** Repeat Prob. 7, but for the Beattie equation.

**10.** A 10-ft³ tank containing 0.25 $lb_m$ of air at 60°F is connected to another tank of the same volume that is at the same temperature but contains 0.50 $lb_m$ of air.   Find the final pressure, assuming that the air behaves ideally.

**11.** Reduce the Beattie equation to a $z$ equation in $p_r$ and $T_r$.

**12.** Derive Eqs. (10-3), (10-1b), and (10-1c).

**13.** Are Eqs. (10-2) necessary and sufficient to dictate a point of inflection?

**14.** Why is $v_r$ rarely used?

**15.** Derive the virial form of the van der Waals equation.   [Equation (10-1d): Expand the $(v_{r'} = \frac{1}{8})^{-1}$ term in Eq. (10-1c) in a Maclaurin series.]

**16.** Change the virial form of the van der Waals equation into a power series in $p$.

**17.** Calculate the compressibility factor at the critical point for the van der Waals equation, and check by Eq. (10-1c).

**18.** Lay out isotherms for the van der Waals virial equation for $T_r = 1.05$, 1.20, 3.00 on Fig. B-8.

**19.** Investigate the curvature of the isometrics for the van der Waals equation.

**20.** Determine the constants $a$ and $b$ for the Berthelot equation.

**21.** Repeat Prob. 20, but for the Dieterici equation.

**22.** Repeat Probs. 15, 16, 17, and 19 for (a) the Berthelot equation or (b) the Dieterici equation.

**23.** Repeat Prob. 18, but for the Berthelot equation.

**24.** Calculate generalized constants for the Beattie-Bridgeman equation from the specific constants for nitrogen.

**25.** Transform Eq. (10-13a) into Eq. (10-13b).

**26.** Obtain the Beattie equation.

## SELECTED REFERENCES

1. Otto, J.: "Handbuch der Experimental Physik," vol. 8, p. 79, Akademische Verlagsgesellschaft m.b.H., Leipzig, 1929.
2. Dodge, B. F.: "Chemical Engineering Thermodynamics," chap. 5, McGraw-Hill Book Company, Inc., New York, 1944.
3. Taylor, H. S., and S. Glasstone: "A Treatise on Physical Chemistry," 3d ed., vol. 2, D. Van Nostrand Company, Inc., Princeton, N.J., 1951.
4. Hirschfelder, J. O., C. F. Curtiss, and R. B. Bird: "Molecular Theory of Gases and Liquids," John Wiley & Sons, Inc., New York, 1954.
5. Beattie, J. A., and O. C. Bridgeman: A New Equation of State for Fluids, *Proc. Am. Acad. Arts Sci.*, **63**: 229–308 (1928).

6. Martin, J. J., and Y. C. Hou: Development of an Equation of State for Gases, *J. AIChE*, **1**: 142 (June, 1955).
7. Hirschfelder, J., R. Buehler, H. McGee, and J. Sutton: Generalized Equation of State for Gases and Liquids, *Ind. Eng. Chem.*, **50**: 375 (March, 1958) (released in a technical report in 1956).
8. Hilsenrath, J., M. Klein, and H. Woolley: Tables of Thermodynamic Properties of Air, NBS, AEDC TR-59-20, December, 1959.
9. Obert, E. F.: "Concepts of Thermodynamics," McGraw-Hill Book Company, Inc., New York, 1960.

CHAPTER 11

# THE IDEAL GAS AND MIXTURE RELATIONSHIPS

*Ideals are like stars . . . choose them as your*
*guides, and, following them, reach your destiny.*

*Carl Schurz*

The concept of the ideal gas represents a limiting state that can be approached but not attained by any real gas. However, in mechanical engineering many of the processes are conducted at such pressures and temperatures that the ideal-gas laws can be used as simple and reasonably close guides to the behavior of the real gas.

**11-1. Equations for Real or Ideal Gases.** Because of the number of equations that will be developed in this chapter, basic relationships† will be summarized here. For reversible or irreversible processes of unit mass of substance,

$$\Delta q - \Delta w \Big]_{\substack{\text{nf} \\ \text{rev or} \\ \text{irrev}}} = \Delta u + \Delta KE \qquad (4\text{-}4d)$$

$$\Delta q - \Delta w \Big]_{\substack{\text{sf} \\ \text{rev or} \\ \text{irrev}}} = \Delta h + \Delta KE \qquad (5\text{-}7d)$$

and, for the change in value of a thermostatic property between two states,

$$dh \Big]_{p=C} = c_p \, dT \qquad du \Big]_{v=C} = c_v \, dT$$
$$du = T \, ds - p \, dv \qquad (7\text{-}4a)$$
$$dh = T \, ds + v \, dp \qquad (7\text{-}4b)$$

while, by arbitrary definition,

$$k \equiv \frac{c_p}{c_v} \qquad (2\text{-}4b)$$

† Changes in potential energy are ignored throughout the chapter. Also, the independent variables of time $\tau$ for the closed system and length $L$ for the steady-flow open system are implicit for all energy equations. Thus, $\Delta q$ for the closed system is an increment from time $\tau_1$ to time $\tau_2$, while $\Delta q$ for the steady-flow system is an increment from position $L_1$ to position $L_2$.

180

With the restriction that the process is reversible,

$$\Delta w_{\text{rev}\atop \text{nf}} = \int p \, dv \tag{6-2c}$$

$$\Delta w_{\text{rev}\atop \text{sf}} = \int p \, dv - \Delta FE - \Delta KE = -\int v \, dp - \Delta KE \tag{6-3a}$$

$$dq_{\text{rev}} = T \, ds \tag{7-3c}$$

**11-2. Property Relationships for the Ideal Gas.** Upon the structure of general equations in Art. 11-1, the restrictions of an ideal gas can be placed,

$$pv = RT \tag{3-5a}$$
$$du = c_v \, dT \tag{3-8b}$$

As a consequence,

$$dh = c_p \, dT \tag{3-8d}$$
$$c_p - c_v = R \tag{3-9}$$

By substituting Eq. (2-4b) into Eq. (3-9),

$$c_p = \frac{kR}{k-1} \tag{11-1a}$$

$$c_v = \frac{R}{k-1} \tag{11-1b}$$

Process equations for the perfect gas can be found by substituting Eqs. (3-8b) and (3-8d) into Eqs. (7-4a) and (7-4b),

$$T \, ds = c_v \, dT + p \, dv \tag{11-2a}$$
$$T \, ds = c_p \, dT - v \, dp \tag{11-2b}$$

From these equations and Eq. (3-5a),

$$ds = c_v \frac{dT}{T} + R \frac{dv}{v} \tag{11-3a}$$

$$ds = c_p \frac{dT}{T} - R \frac{dp}{p} \tag{11-3b}$$

The equation of state can be differentiated and divided by $RT$ to yield

$$\frac{dv}{v} + \frac{dp}{p} = \frac{dT}{T}$$

With this relationship and Eq. (3-9) substituted in either Eq. (11-3a) or (11-3b),

$$ds = c_p \frac{dv}{v} + c_v \frac{dp}{p} \tag{11-3c}$$

For simplicity of presentation, the heat capacities will be considered, most often, to be constants; with this restriction, Eqs. (11-3) integrate to

$$\Delta s = c_v \ln \frac{T_2}{T_1} + R \ln \frac{v_2}{v_1} \qquad (11\text{-}4a)$$

$$\Delta s = c_p \ln \frac{T_2}{T_1} - R \ln \frac{p_2}{p_1} \qquad (11\text{-}4b)$$

$$\Delta s = c_p \ln \frac{v_2}{v_1} + c_v \ln \frac{p_2}{p_1} \qquad (11\text{-}4c)$$

**11-3. The Polytropic Process.** The properties of the *end* states of the reversible or irreversible process of the ideal† gas can be related by equations of the form

$$p_1 v_1{}^n = p_2 v_2{}^n \qquad (3\text{-}7a)$$

$$\frac{T_2}{T_1} = \left(\frac{p_2}{p_1}\right)^{\frac{n-1}{n}} \qquad (3\text{-}7b)$$

$$\frac{T_2}{T_1} = \left(\frac{v_1}{v_2}\right)^{n-1} \qquad (3\text{-}7c)$$

If *all* states encountered in the process are represented by the foregoing equations, a general *path* equation for the ideal gas can be proposed,

$$pv^n = C \qquad (11\text{-}5)$$

wherein $n = \infty$      (constant volume‡)
       $n = k$      (constant entropy§)
       $n = 1$      (constant temperature‡)
       $n = 0$      (constant pressure‡)
       $n = $ constant      (polytropic process)

A process that follows the equation $pv^n = C$ is called a *polytropic process* unless some property remains constant during the process, and then it is assigned a name to show the constancy of the property—constant volume, constant pressure, isothermal, or isentropic.

Equation (11-5) does not attempt to trace every possible process on the $pv$ diagram between two states, for this would require the exponent $n$ to be a variable that would change in value throughout the process. Since Eq. (11-5) relates only properties of the system (and does not involve the surface effects of heat and work), it can be used for reversible and also for irreversible processes (*if* it is possible to assign property values to each state observed during the irreversible process). For

---

† And for a real gas the same equations could also be used, but then $n$ would have a different value for each equation.
‡ Article 3-5.
§ Article 6-6.

example, an irreversible expansion could be made to approach closely a constant-entropy process by transferring heat from the process to counter-balance exactly the increase in entropy from irreversibilities. However, the probability of encountering such a process is *extremely* remote; hence, *the terms constant entropy and isentropic and the exponent k will be used to denote only reversible processes* (Art. 7-5).

**11-4. Work and Heat Equations.** The reversible work of the nonflow process ($\Delta KE = 0$) is found by substituting $p = C/v^n$ into Eq. (6-2c),

$$\Delta w_{\substack{\text{rev nf} \\ (n \neq 1)}} = \int p \, dv = C \int \frac{dv}{v^n} = C \, \frac{v_2^{1-n} - v_1^{1-n}}{1 - n}$$

The constant $C$ equals $p_1 v_1^n = p_2 v_2^n$; hence,

$$\Delta w_{\substack{\text{rev nf} \\ (n \neq 1)}} = \frac{p_2 v_2 - p_1 v_1}{1 - n} \tag{11-6}$$

The reversible work of the steady-flow system differs from that of the closed system because of the presence of flow energy $\Delta pv$. Also, changes in kinetic energy may be of importance in flow systems. Hence, from Eq. (6-3a),

$$\Delta w_{\substack{\text{rev sf} \\ (n \neq 1)}} = \frac{n(p_2 v_2 - p_1 v_1)}{1 - n} - \Delta KE \tag{11-7a}$$

The foregoing equations are indeterminate for the isothermal process wherein $n = 1$. By substituting $p = C/v$ into either Eq. (6-2c) or (6-3a), the same form† of work equation is obtained for either the non-flow or the steady-flow process,

$$\Delta w_{\substack{\text{rev} \\ T=C}} = RT \ln \frac{v_2}{v_1} - \Delta KE = pv \ln \frac{p_1}{p_2} - \Delta KE \tag{11-8}$$

An equation for the heat transferred during a reversible polytropic process can be found from progressive substitutions of Eqs. (3-8b), (11-6), (3-5a), (3-9), and (2-4b) into Eq. (4-4d) [or Eq. (5-7d)],

$$\Delta q_{\text{rev}} = c_v \, \frac{k - n}{1 - n} \, (T_2 - T_1) \tag{11-9a}$$

Equation (11-9a) has the form of a heat-capacity equation; hence a *polytropic heat capacity*‡ is defined as

$$c_n = c_v \, \frac{k - n}{1 - n} \tag{11-10}$$

---

† The independent variables are different (p. xix).

‡ An illogical name since $c_p$ and $c_v$ are properties while $c_n$ is not.

Note that $c_n$ can be used to find the heat transfer only for a reversible process. On the other hand, for reversible or irreversible processes,

$$\Delta s = \int \frac{dQ_{rev}}{T} = \int \frac{c_n \, dT}{T} = c_n \ln \frac{T_2}{T_1} \qquad (11\text{-}9b)$$

**Example 1.** Air is reversibly compressed in a nonflow process along a polytropic path for which $n$ is 1.3. Find the work of the process if the initial pressure is atmospheric at a temperature of 60°F and the final pressure is 125 psia.

*Solution.* The first step in the solution is to draw a diagrammatic sketch of the system and indicate the initial and final states and the quantities to be evaluated. Here $p_1$, $T_1$, $p_2$, and $n$ are given. Since two properties are required to fix the state, the choice is guided by Eq. (3-7b) [or by Eq. (3-7c) if $v_1$ and $v_2$, rather than $p_1$ and $p_2$, are given],

$$T_2 = T_1 \left( \frac{p_2}{p_1} \right)^{(n-1)/n} = 520 \left( \frac{125}{14.7} \right)^{0.231} = 853°R$$

The reversible work of the system is given by Eq. (11-6),

$$\Delta w_{rev \atop nf} = \frac{p_2 v_2 - p_1 v_1}{1 - n}$$

*Now* the equation of state $pv = RT$ is used (molecular weight of air is 29),

$$\Delta w_{rev \atop nf} = \frac{R(T_2 - T_1)}{1 - n} = \frac{1{,}545(853 - 520)}{29(-0.3)} = -59{,}100 \frac{\text{ft-lb}_f}{\text{lb}_m} \qquad Ans.$$

**Example 2.** In a steady-flow isentropic process air enters the system at 440°F and 100 psia and leaves with pressure of 10 psia. Find the change in kinetic energy if the system does 50 Btu/lb$_m$ of work.

*Solution.* The procedure is the same as before: a diagrammatic sketch of the system is labeled to show the transfers of heat, work, and mass; also, the desired properties at the initial and final states are indicated. Here $p_1$, $p_2$, $T_1$, and $n = k$ are given; hence, Eq. (3-7b) is selected since $T_2$ can then be calculated. A value for $k$ could be obtained from Table B-6; however, Fig. B-1 enables a mean value of $k = 1.39$ to be selected for a more exact solution,

$$T_2 = T_1 \left( \frac{p_2}{p_1} \right)^{(k-1)/k} = 900 \left( \frac{10}{100} \right)^{0.275} = 477°R$$

Since for the reversible adiabatic process $\Delta q = 0$, Eq. (5-7d) is convenient to use with the help of Eq. (3-8d),

$$\Delta q - \Delta w \bigg]_{\substack{rev \ or \\ irrev}} = c_p(T_2 - T_1) + \Delta KE$$

The value of $c_p$ is also from Fig. B-1,

$$0 - 50 = -102.3 + \Delta KE$$
$$\Delta KE = 52.3 \text{ Btu/lb}_m \qquad Ans.$$

Since this was a reversible process, Eq. (11-7a) could have been used for solution, but it is better, when possible, to select the more general First Law equation [Eq. (5-7d)].

**11-5. Summary of Processes for the Perfect Gas.** Comparison of the processes for the perfect gas can be made by observing the relationships between the paths of these processes on the $Ts$ and $pv$ diagrams (Fig. 11-1). In reproducing Fig. 11-1, it will be found convenient to draw first the constant-pressure process for which $n$ is zero. Then, the other curves can be drawn, it being noted that $n$ increases from zero to infinity in the clockwise direction. Inspection of Fig. 11-1 shows that $n$ may have negative values. For a polytropic process with a negative number for the exponent $n$, Fig. 11-1 reveals that pressure and specific volume will both increase (or both decrease) during the process. This will occur if energy is transferred to an expansion process at such a rate that the pressure rises instead of falling. (Essentially, this occurs in an internal-combustion engine at the start of the power stroke, when chemical energy is liberated.) Similarly, for a hypothetical compression process, energy

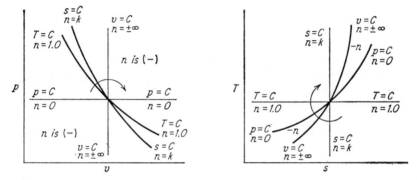

FIG. 11-1. Process paths on $pv$ and $Ts$ diagrams.

is taken away at a rate to cause the pressure to fall even though the volume is being decreased.

**11-6. The Irreversible Process.** For an irreversible process the value for $n$ cannot be used to compute either heat or work, but the process may still be represented in many cases on the $pv$ or $Ts$ diagram. The irreversible adiabatic process is often encountered. For this process, the value of $n$ will assume values dictated by the amount of irreversibility that may be present, and the direction of the change in $n$ can be readily deduced: if an adiabatic expansion process is irreversible, available energy is being degenerated into a thermal effect. This thermal effect on the $pv$ path will be exactly the same as if a transfer of heat had occurred, and $n$ will approach as a limiting† value $n = 1.0$. This will be the limit because, if all the available energy is wasted, no work will be done and the process will be at constant internal energy and this process is an isothermal process for the perfect gas. For the adiabatic com-

---

† In a flow process with high initial kinetic energy, even this value could be lessened, but, practically, such cases are rarely encountered.

pression process, $n$ is always greater than $k$ with no limiting value because there is no limit to the amount of work that can be added to the system. Accordingly, the effect of irreversibilities is to cause $n$ to *decrease* from the value for the reversible *expansion* process and to *increase* from the value for the reversible *compression* process. (These same trends will be evident for all irreversible processes.)

Although the transfers of heat and work for the irreversible process cannot be computed by evaluating the $\int T\,ds$ and the $\int p\,dv$, such transfers can always be measured by changes in the surroundings that are evaluated by the First Law. Consider an irreversible steady-flow process; for this process,

$$\Delta q - \Delta w \Big]_{\substack{\text{sf} \\ \text{rev or} \\ \text{irrev}}} = \Delta h + \Delta KE \qquad (5\text{-}7d)$$

This equation holds for reversible or irreversible processes since all energy terms are measured at the boundary of the system. Equation (5-7d) can be disguised by substituting Eq. (3-8d), and then Eqs. (3-7b) and (11-1a), to yield

$$\Delta q - \Delta w \Big]_{\substack{\text{sf} \\ \text{rev or} \\ \text{irrev}}} = \frac{kRT_1}{k-1}\left[\left(\frac{p_2}{p_1}\right)^{\frac{n-1}{n}} - 1\right] + \Delta KE \qquad (11\text{-}11)$$

Here the value of $n$ links together the end states but does not necessarily mark the path. In other words, if the end states of the irreversible process are known, it is always possible to find a value for $n$ that will satisfy Eq. (3-7b). The same substitutions can be made in Eq. (11-7a) to yield

$$\Delta w_{\substack{\text{rev sf} \\ (n\neq 1)}} = \frac{nRT_1}{1-n}\left[\left(\frac{p_2}{p_1}\right)^{\frac{n-1}{n}} - 1\right] - \Delta KE \qquad (11\text{-}7b)$$

But this equation is valid only for reversible processes since it is evaluated from the work integral [Eq. (6-3a)] (and then disguised by ideal-gas relationships).

Equations (11-11) and (11-7b) should not be used for problems since the steps outlined in Examples 1 and 2 are more instructive.

**11-7. The Compression and Expansion of Gases.** The *pressure ratio* for a process or cycle and for any working substances is defined as a number greater than unity,

$$r_p = \frac{p_{\text{high}}}{p_{\text{low}}}\Big]_{\text{process or cycle}} \qquad (11\text{-}12a)$$

The *compression* or *expansion ratio* of a process or cycle is also a number greater than unity and is best viewed as a ratio of specific volumes,

$$r_v = \frac{v_{\text{large}}}{v_{\text{small}}}\bigg]_{\text{process or cycle}} \qquad (11\text{-}12b)$$

It follows from Eqs. (3-7) that, for the usual processes of an ideal gas,

$$\frac{T_{\text{high}}}{T_{\text{low}}} = r_p^{(n-1)/n} = r_v^{n-1} \qquad (n \geq 1) \qquad (11\text{-}12c)$$

The problem in industry is cost, and cost dictates that the work of compression should be small while the work of expansion should be large. The $pv$ diagram will be used to illustrate the work of various processes. In Fig. 11-2a, the isothermal and isentropic compression processes are

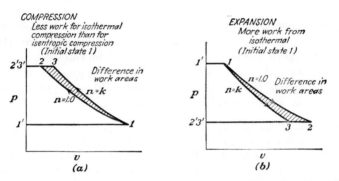

FIG. 11-2. Comparison of isothermal and isentropic steady-flow processes.

shown starting from the same initial state and proceeding over the same pressure ratio. Since most engineering machines use a steady-flow process, the reversible work equals $-\int v\, dp$ (if changes in kinetic energy are negligible). Inspection of Fig. 11-2a reveals that $-\int v\, dp$ will decrease as the exponent $n$ decreases; that is, less work is required when the gas is cooled during the compression process. However, if state 1 is at atmospheric temperature, the maximum cooling that can be attained without using a refrigerator (and a refrigerator would require an additional source of work) is the isothermal process. Hence, the isothermal process is the ideal† flow process for compressing fluids because the minimum amount of work is required (area 1'122' is less than area 1'133'). Figure 11-2b shows that, in steady flow, isothermal expansion will deliver

---

† Although the ability of the compressed fluid to do expansion work is reduced by the cooling (area 2'211' is less than area 2'311'), yet in most practical cases this is not a loss, for either the compressed fluid is stored or else it is transferred through pipes to other locations before use, and during this time the fluid is unavoidably cooled to the temperature of the surroundings.

more work than isentropic expansion if the fluid in both processes is at the same initial state (see Prob. 32).

It has already been demonstrated that the amount of work required to compress a fluid can be determined without inquiring into the mechanism of the compressor. If the reversible compression (or expansion) is isothermal, Eq. (11-8) shows that, for gases such as air,

$$\Delta w_{\substack{\text{rev sf} \\ \text{isothermal}}} = RT \ln \frac{p_1}{p_2} - \Delta KE \qquad (11\text{-}8)$$

while, for polytropic (and isentropic) compression (or expansion),

$$\Delta w_{\substack{\text{rev sf} \\ \text{polytropic}}} = \frac{nRT_1}{1-n} \left[ \left(\frac{p_2}{p_1}\right)^{\frac{n-1}{n}} - 1 \right] - \Delta KE \qquad (11\text{-}7b)$$

Inspection of these equations reveals that the work depends not only upon the pressure ratio but also upon the initial temperature: the lower the initial temperature of the air, the less will be the work required for the compression. Interestingly, for ideal gases the amount of work is the same for reversible pressure changes of 100 to 10 psia, 1,000 to 100 psia, or 1 to 0.1 psia, for example.

In industry, large quantities of air (for example) must be compressed, and the compressor must be driven at high speed to supply the quantity of air demanded by a process. For this reason, it is difficult to cool the fluid at a fast enough rate to maintain constant temperature. Therefore, the compression process in commercial machines approaches isentropic rather than the more desirable isothermal compression. In water-cooled reciprocating-piston air compressors, the polytropic exponent for the relationship $pv^n = C$ will have a value of 1.25 to 1.35. In centrifugal blowers, without cooling, values of $n$ above 1.4 are quite usual. It is best to use these values of $n$ in the general equation [Eq. (11-11)] since the real compression is an irreversible process and values of $n$ are usually determined from boundary conditions.

In some instances, water is injected into the air as a means of decreasing the work of compression. The saving in work arises from the high latent heat of the liquid; when vaporization of the water takes place, a reduction in temperature and therefore pressure of the air is obtained even though the vaporized water will contribute some part of the total pressure of the final mixture. This practice, however, must be used with care, since the presence of water may add to corrosion and erosion difficulties.

**11-8. Gas Tables.** The compression, expansion, and flow of gases are frequent problems in engineering, and solutions based upon the ideal-gas laws are usually adequate. To improve the accuracy of the computations, the variations of the heat capacities with temperature should be included. For example, the isentropic process with variable heat capacity can be solved by substituting the general form of $c_p$ from Table B-2,

$$c_p = A + BT + CT^2 + \cdots$$

into Eq. (11-3$b$) and integrating,

$$A \ln \frac{T_2}{T_1} + B(T_2 - T_1) + \frac{C}{2}(T_2{}^2 - T_1{}^2) = R \ln \frac{p_2}{p_1} \qquad (11\text{-}13)$$

Solution, however, of Eq. (11-13) is best made by trial. This tedious procedure can be avoided by the use of property tables such as those of Keenan and Kaye.[2]

Table 1 of the Gas Tables[2] lists the properties of dry air at low pressures (that is, in the ideal state) in the form illustrated in Table B-9. The values of enthalpy $h$, internal energy $u$, relative pressure $p_r$, relative volume $v_r$, and the entropy function $\phi$ are tabulated against temperature, as the only independent variable, for the range 100 to 6500°R. The zero values of enthalpy and $\phi$ are at zero degrees absolute to avoid negative values. The new quantities relative pressure $p_r$, relative volume $v_r$, and $\phi$ will be shown to be functions only of temperature.

The function $\phi$ is defined as

$$\phi = \int_0^T c_p \frac{dT}{T} \qquad [\text{Btu}/(\text{lb}_m)(°R)] \qquad \text{and} \qquad \phi = 0 \text{ at } 0°R \qquad (11\text{-}14)$$

This equation shows that $\phi$ is a function only of temperature. Values of $\phi$ were computed from heat-capacity data similar to those in Table B-2 except that adjustments were made to allow the heat capacities to approach zero at the absolute zero. Suppose that the change in entropy between two states is to be evaluated. Then, from Eq. (11-3$b$),

$$\Delta s = \int_1^2 c_p \frac{dT}{T} - R \ln \frac{p_2}{p_1}$$

And, by Eq. (11-14),

$$\Delta s = \phi_2 - \phi_1 - R \ln \frac{p_2}{p_1} \qquad (11\text{-}15a)$$

The values for $\phi$ in Table B-9 enable the change in entropy to be readily computed.

The entropy at any state is defined by rewriting Eq. (11-15$a$) in the form

$$s = \phi - R \ln p + s_0 \qquad (11\text{-}15b)$$

Two states have the same entropy when

$$\phi_2 - R \ln p_2 = \phi_1 - R \ln p_1$$

Therefore,

$$\frac{\phi_2 - \phi_1}{R} = \ln \frac{p_2}{p_1} \bigg]_{s=C} \qquad (a)$$

The form of Eq. ($a$) suggests defining† a *relative pressure* as

$$\ln p_r = \frac{\phi}{R} \qquad (11\text{-}16a)$$

since, then,

$$\frac{\phi_2 - \phi_1}{R} = \ln p_{r2} - \ln p_{r1} = \ln \frac{p_{r2}}{p_{r1}} \qquad (b)$$

† The values of $p_r$ in Table B-9 differ from the definition of Eq. (11-16$a$) by an arbitrary constant, used to reduce the magnitude of the numbers.

By equating Eqs. (a) and (b),

$$\left.\frac{p_2}{p_1}\right]_{s=C} = \frac{p_{r2}}{p_{r1}} \tag{11-16b}$$

Hence, *the ratio of two relative pressures is equal to the isentropic pressure ratio.* Equation (11-16b) is helpful in solving isentropic compression or expansion processes. Note, however, that the relative pressure $p_r$ is a pure temperature function [Eq. (11-16a)].

**Example 3.** A pound mass of dry air at $T_1 = 900°R$, $p_1 = 100$ psia undergoes a reversible adiabatic expansion to 10 psia. Calculate the final temperature.

*Solution.* From Table B-9,

$$T_1 = 900°R \qquad p_{r1} = 8.411$$

And, by Eq. (11-16b),

$$p_{r2} = p_{r1}\frac{p_2}{p_1} = 8.411(\tfrac{1}{10}) = 0.8411$$

With this value of $p_{r2}$, enter Table B-9, and find

$$T_2 \approx 468°R \qquad Ans.$$

As a matter of interest, use Eq. (3-7b) and $k = 1.39$ from Fig. B-1, and obtain $T_2 = 471°R$.

Equation (11-16b) can be converted into a volume ratio by substituting $p = RT/v$,

$$\left.\frac{T_2/v_2}{T_1/v_1}\right]_{s=C} = \frac{p_{r2}}{p_{r1}} \qquad \text{or} \qquad \left.\frac{v_1}{v_2}\right]_{s=C} = \frac{p_{r1}/T_2}{p_{r2}/T_1} \tag{c}$$

The form of Eq. (c) suggests defining a *relative volume* $v_r$ as

$$v_r = \frac{RT}{p_r} \tag{11-17a}$$

By substituting Eq. (11-17a) into Eq. (c),

$$\left.\frac{v_1}{v_2}\right]_{s=C} = \frac{v_{r1}}{v_{r2}} \tag{11-17b}$$

Hence, *the ratio of two relative volumes is equal to the isentropic volume ratio.*

**Example 4.** A pound of dry air is compressed in a reversible adiabatic nonflow process through a compression ratio (a volume ratio) of 6 from an initial state of 14.7 psia and 60.3°F. Determine the final state and the work required.

*Solution.* From Table B-9 at 520°R,

$$v_{r1} = 158.58 \qquad u_1 = 88.62 \text{ Btu/lb}_m \qquad p_{r1} = 1.2147$$

From Eq. (11-17b),

$$v_{r2} = \frac{v_{r1}}{v_1/v_2} = \frac{158.58}{6} = 26.43$$

For this value of $v_{r2}$, from Table B-9

$$T_2 = 1050°R \qquad u_2 = 181.47 \text{ Btu/lb}_m \qquad p_{r2} = 14.686 \qquad Ans.$$

From Eq. (11-16b),

$$p_2 = p_1\frac{p_{r2}}{p_{r1}} = 14.7\left(\frac{14.686}{1.2147}\right) = 178 \text{ psia} \qquad Ans.$$

From Eq. (4-4d),

$$\Delta w = u_1 - u_2 = 88.62 - 181.47 = -92.85 \text{ Btu/lb}_m \qquad Ans.$$

Also, by definition,

$$h_2 = u_2 + p_2 v_2 = u_2 + RT_2 = 181.47 + \frac{1.986(1,050)}{29} = 253.4 \qquad Ans.$$

And this value checks that in Table B-9.

**11-9. General Relationships between Mass and Mole Units.**† Consider a mixture made up of constituents $a$, $b$, and $c$. The mass of mixture must equal the sum of the masses of the constituents,

$$m = m_a + m_b + m_c \qquad (11\text{-}18)$$

The composition of a mixture is best evaluated if the mole is used as the mass unit. The total number of moles in the mixture is defined as

$$n \equiv n_a + n_b + n_c \qquad (11\text{-}19)$$

The *mole fraction* $x$ of constituents $a$, $b$, and $c$ is defined as

$$x_a \equiv \frac{n_a}{n_a + n_b + n_c} \qquad x_b \equiv \frac{n_b}{n_a + n_b + n_c} \qquad x_c \equiv \frac{n_c}{n_a + n_b + n_c} \quad (11\text{-}20a)$$

and therefore

$$x_a + x_b + x_c = 1.0 \qquad (11\text{-}20b)$$

Mole and mass units are related by

$$m = nM \qquad \text{and} \qquad n \equiv \frac{m}{M}$$

where $M$ is the molecular weight. Then, from Eq. (11-18),

$$nM = n_a M_a + n_b M_b + n_c M_c = m \qquad (a)$$

and the *molecular weight of the mixture is determined*,

$$M = x_a M_a + x_b M_b + x_c M_c \qquad (11\text{-}21)$$

It should be recalled (Art. 3-5) that the mole unit allows a *universal* gas constant $R_0$ to be used for all gases, while the pound or gram units of mass demand a *specific* gas constant that has different values for different gases. Hence, if a mass unit other than the mole is to be used for the mixture, a specific gas constant must be determined. The specific gas constant can be found from the relationship between the molecular weight and the universal gas constant,

$$R \equiv \frac{R_0}{M} \qquad (3\text{-}6a)$$

† It will be especially helpful in studying this article if the student will write out, in words, the meaning of each equation.

Equation (3-6a) can be used for either mixture or constituent.

**11-10. Mixtures of Ideal Gases.** In a mixture of gases, each constituent occupies the entire volume and displays the common temperature while exerting only a fraction of the entire pressure. The premise underlying ideal-gas mixtures is that each constituent is unaffected by the presence of other constituents. It follows that the equation of state for ideal gases can be substituted in Eq. (11-19),

$$n = n_a + n_b + n_c$$

$$\frac{pV}{R_0 T} = \frac{p_a V_a}{R_0 T_a} + \frac{p_b V_b}{R_0 T_b} + \frac{p_c V_c}{R_0 T_c}$$

But $\qquad V = V_a = V_b = V_c \qquad T = T_a = T_b = T_c$

and therefore $\qquad p = p_a + p_b + p_c \Big]_{T,V\dagger}$ $\qquad\qquad$ (11-22a)

Equation (11-22a) expresses *Dalton's law of partial pressures:*

**The pressure of a mixture of ideal gases is equal to the sum of the partial pressures which the constituent gases would exert if each existed alone in the mixture volume at the mixture temperature.**

For the conditions of Dalton's law applied to the mixture and to one of the constituents,

$$\frac{p_i V_i}{pV} = \frac{n_i R_0 T_i}{n R_0 T}$$

and here $\qquad V_i = V \qquad T_i = T$

Hence $\qquad \dfrac{p_i}{p}\bigg]_{V,T} = \dfrac{n_i}{n} = x_i \qquad p_i = x_i p \bigg]_{V,T}$ $\qquad$ (11-22b)

Equation (11-22b) shows that the partial pressure and the mole fraction of the constituent are proportional.

A slightly different and more imaginary concept can be applied to the mixture of gases. Suppose the mixture to be divided by imaginary partitions into spaces each occupied by a separate constituent. Each constituent, then, would exert the mixture pressure and temperature while occupying only a *partial volume* of the mixture. With this concept in mind, the perfect-gas equation can be again substituted in Eq. (11-19),

$$n = n_a + n_b + n_c$$

$$\frac{pV}{R_0 T} = \frac{p_a V_a}{R_0 T_a} + \frac{p_b V_b}{R_0 T_b} + \frac{p_c V_c}{R_0 T_c}$$

But here $\qquad p = p_a = p_b = p_c \qquad T = T_a = T_b = T_c$

and therefore $\qquad V = V_a + V_b + V_c \Big]_{p,T\ddagger}$ $\qquad\qquad$ (11-23a)

---

† These subscripts indicate that the pressures are to be evaluated at the temperature and volume of the mixture.

‡ All volumes are measured at the pressure and temperature of the mixture.

Equation (11-23a) expresses *Amagat's law*, also called *Leduc's law:*

**The volume of a mixture of ideal gases is equal to the sum of the partial volumes which the constituent gases would occupy if each existed alone at the pressure and temperature of the mixture.**

The partial volume, like the partial pressure, is related to the mole fraction,

$$\frac{p_i V_i}{p V} = \frac{n_i R_0 T_i}{n R_0 T}$$

and here
$$p = p_i \qquad T = T_i$$

Hence
$$\frac{V_i}{V}\bigg]_{p,T} = \frac{n_i}{n} = x_i \qquad V_i = x_i V\bigg]_{p,T} \qquad (11\text{-}23b)$$

It is evident by Eq. (11-23b) that mixture and constituents have identical molar *specific* volumes.

**Example 5.**   A mixture of 10 lb$_m$ of oxygen and 15 lb$_m$ of nitrogen has a pressure of 50 psia and a temperature of 60°F.   Determine for the mixture (a) the mole fraction of each constituent, (b) the average molecular weight, (c) the specific gas constant, (d) the volume and density, (e) the partial pressures and partial volumes.

*Solution.*   (a) The molecular weights of the constituents are found in Table B-6. The moles of mixture are

$$n_O = \frac{10 \text{ lb}_m}{32} = 0.3125 \text{ mole}$$

$$n_N = \frac{15 \text{ lb}_m}{28.02} = 0.5350 \text{ mole}$$

$$n = \overline{0.8475 \text{ mole}}$$

The mole fractions are
$$x_O = \frac{0.3125}{0.8475} = 0.369 \qquad Ans.$$

$$x_N = \frac{0.5350}{0.8475} = 0.631 \qquad Ans.$$

$$\overline{1.000}$$

(b)   The molecular weight is obtained from Eq. (11-21),

$$M = x_O M_O + x_N M_N = 0.369(32.0) + 0.631(28.02) = 29.5 \qquad Ans.$$

or by   $$M = \frac{m}{n} = \frac{25}{0.8475} = 29.5 \qquad Ans.$$

(c)   The specific gas constant is found by

$$R = \frac{R_0}{M} = \frac{1,545}{29.5} = 52.37 \frac{\text{lb}_f \text{ ft}}{\text{lb}_m \text{ °R}} \qquad Ans.$$

(d)   The volume of the mixture is also the volume of each constituent,

$$V = \frac{n R_0 T}{p} = \frac{n_O R_0 T}{p_O} = \frac{n_N R_0 T}{p_N} = \frac{m R T}{p} = \frac{0.8475(1,545)(520)}{50(144)} = 94.5 \text{ ft}^3 \qquad Ans.$$

The density of the mixture is

$$\rho = \frac{1}{v} = \frac{p}{R_0 T} = \frac{50(144)}{1,545(520)} = 0.00896 \text{ mole/ft}^3 \qquad Ans.$$

and since $M = 29.5$,

$$\rho = (29.5 \text{ lb}_m/\text{mole})(0.00896 \text{ mole/ft}^3) = 0.264 \text{ lb}_m/\text{ft}^3 \qquad Ans.$$

(e) The partial pressures are found by Eqs. (11-22b):

$$p_i = x_i p$$
$$p_O = 0.369(50) = 18.45 \text{ psia} \qquad Ans.$$
$$p_N = 0.631(50) = 31.55 \text{ psia} \qquad Ans.$$
$$\overline{\phantom{p_N = 0.631(50) =} 50.00 \text{ psia}}$$

and the partial volumes by Eqs. (11-23b),

$$V_i = x_i V$$
$$V_O = 0.369(94.5) = 34.85 \text{ ft}^3 \qquad Ans.$$
$$V_N = 0.631(94.5) = 59.65 \text{ ft}^3 \qquad Ans.$$
$$\overline{\phantom{V_N = 0.631(94.5) =} 94.5 \text{ ft}^3}$$

## 11-11. Volumetric and Gravimetric Analyses.

The constituents of a mixture are reported as volume or mass fractions of the entire mixture. The analysis based upon the measurement of volumes is called the *volumetric analysis*, while the analysis based upon measurements of mass is called the *gravimetric analysis*.

Experimentally, the volumetric analysis is usually made on a sample of the mixture at atmospheric pressure and temperature, and the volumes of the constituents are also measured at these conditions. Since the pressure is low, the gases should behave ideally and the volume analysis should be directly proportional to the partial volumes,

$$V = V_a + V_b + V_c \Big]_{p,T}$$

and, by Eqs. (11-23b),

$$x_i = \frac{n_i}{n} = \frac{V_i}{V} = \text{volume fraction of constituent } i$$

The mass, or gravimetric, analysis reports each constituent on a mass basis,

$$m = m_a + m_b + m_c$$
$$\frac{m_i}{m} = \text{mass fraction of constituent } i \qquad (11\text{-}24)$$

The mass analysis may also be an *ultimate analysis;* here the mass fractions are for the basic chemical elements of the mixture.

**Example 6.**   A gas analyzes by volume 12 per cent $CO_2$, 4 per cent $O_2$, and the remainder $N_2$.   Determine the gravimetric analyses.

*Solution*

$$\text{Volumetric } \% \div 100 = \text{mole fraction}$$
$$(\text{Mole fraction}) \times M = \text{relative mass}$$

| Component | Volumetric % | Mole fraction | $M$ | Relative mass $\left(\dfrac{\text{lb}_m \text{ constituent}}{\text{mole mixture}}\right)$ |
|---|---|---|---|---|
| $CO_2$........ | 12.0 | 0.12 | 44.0 | 5.28 |
| $O_2$......... | 4.0 | 0.04 | 32.0 | 1.28 |
| $N_2$......... | 84.0 | 0.84 | 28.02 | 23.52 |
| | 100.0 | 1.00 | ..... | 30.08 $= M$ for mixture |

Note that the average molecular weight of the mixture is obtained by adding the figures in the last column of the table.

The gravimetric analysis for the constituents is

$$\frac{m_{CO_2}}{m} = \frac{5.28}{30.08} = 0.175, \text{ or } 17.5\% \qquad Ans.$$

$$\frac{m_{O_2}}{m} = \frac{1.28}{30.08} = 0.043, \text{ or } 4.3\% \qquad Ans.$$

$$\frac{m_{N_2}}{m} = \frac{23.52}{30.08} = 0.782, \text{ or } 78.2\% \qquad Ans.$$

$$\overline{1.000} \qquad \overline{100.0\%}$$

The ultimate analysis is

Carbon in carbon dioxide:

$$1\tfrac{2}{44}(5.28) = 1.44, \text{ and } \frac{1.44}{30.08} \text{ is } 0.0478, \text{ or } 4.78\% \text{ C} \qquad Ans.$$

Oxygen in carbon dioxide:

$$3\tfrac{2}{44}(5.28) = 3.84$$
$$\text{Oxygen in gas} = \underline{1.28}$$
$$\text{Total oxygen} = 5.12, \text{ and } \frac{5.12}{30.08} \text{ is } 0.1702, \text{ or } 17.02\% \text{ O} \qquad Ans.$$

$$\text{Nitrogen in gas} = 23.52, \text{ and } \frac{23.52}{30.08} \text{ is } 0.782, \text{ or } 78.2\% \text{ N} \qquad Ans.$$

$$\text{Total} = \overline{100.0\%}$$

**11-12. The Properties of the Ideal-gas Mixture.**   A mixture of ideal gases is characterized by the complete indifference of each constituent to the presence of other gases.   For example, the pressure of the mixture is but the sum of the partial pressures,

$$p = p_a + p_b + p_c + \cdots \Big]_{T,V} \qquad (11\text{-}22a)$$

In similar manner, the internal energy of the mixture is equal to the sum of the internal energies of the constituents; each constituent is considered to exist alone in the mixture volume, and the value of the whole is equal to the sum of the values for the parts. It then follows that the same additive rule applies to other properties:

$$nu = n_a u_a + n_b u_b + n_c u_c + \cdots \Big]_{T,V} \tag{11-25a}$$

$$nh = n_a h_a + n_b h_b + n_c h_c + \cdots \Big]_{T,V} \tag{11-25b}$$

$$nc = n_a c_a + n_b c_b + n_c c_c + \cdots \Big]_{T,V} \tag{11-25c}$$

$$ns = n_a s_a + n_b s_b + n_c s_c + \cdots \Big]_{T,V} \tag{11-25d}$$

In the above equations, substitution of mass $m$ can be made for moles $n$, and the properties of the mixture will be based on a mass unit such as the pound.

These equations specify that the properties of the constituents should be evaluated at the temperature and volume of the mixture, for each constituent occupies the entire volume. But for ideal gases the properties of internal energy, enthalpy, and heat capacity are dependent only upon temperature, and therefore the volume need be considered only when Eqs. (11-25a) to (11-25c) are arbitrarily extended to evaluate mixtures of real gases. (It should be remembered, however, that the entropy of either an ideal or a real gas is not dependent on temperature alone.)

The changes during a process of the properties of internal energy, enthalpy, and entropy of the mixture are of importance. Upon differentiating Eqs. (11-25) and dividing by $n$, the moles of mixture (and the division restricts the following equations to mole units),

$$du = x_a \, du_a + x_b \, du_b + x_c \, du_c + \cdots$$
$$dh = x_a \, dh_a + x_b \, dh_b + x_c \, dh_c + \cdots$$
$$ds = x_a \, ds_a + x_b \, ds_b + x_c \, ds_c + \cdots \tag{11-25e}$$

Introducing the ideal-gas relationships, Eqs. (3-8b) and (3-8d),

$$du = (x_a c_{va} + x_b c_{vb} + x_c c_{vc}) \, dT = c_{vm} \, dT \tag{11-26a}$$
$$dh = (x_a c_{pa} + x_b c_{pb} + x_c c_{pc}) \, dT = c_{pm} \, dT \tag{11-26b}$$

Therefore,     $c_{vm} = \Sigma x_i c_{vi}$      $c_{pm} = \Sigma x_i c_{pi}$

Equations for the entropy of a mixture can be found from Eqs. (11-3). For example, when Eq. (11-3b) is substituted in Eq. (11-25e) (considering a mixture of only two constituents),

$$ds = x_a \left( c_{pa} \frac{dT}{T} - R_0 \frac{dp_a}{p_a} \right) + x_b \left( c_{pb} \frac{dT}{T} - R_0 \frac{dp_b}{p_b} \right) \tag{a}$$

but the mole fraction $x$ is a constant; hence, from Eqs. (11-22b),

$$p_i = x_i p \quad \text{and} \quad \frac{dp_i}{dp} = x_i = \frac{p_i}{p} \quad \text{therefore,} \quad \frac{dp_i}{p_i} = \frac{dp}{p} \quad (b)$$

Substituting Eq. (b) in Eq. (a) gives

$$ds = (x_a c_{pa} + x_b c_{pb}) \frac{dT}{T} - R_0 \frac{dp}{p} \quad (11\text{-}27a)$$

Equations (11-3a) and (11-3c) can be treated in similar manner, it being noted that

$$\frac{dv_i}{v_i} = \frac{dv}{v} \quad \text{because} \quad n_i v_i = V = nv$$

to yield

$$ds = (x_a c_{va} + x_b c_{vb}) \frac{dT}{T} + R_0 \frac{dv}{v} \quad (11\text{-}27b)$$

$$ds = (x_a c_{pa} + x_b c_{pb}) \frac{dv}{v} + (x_a c_{va} + x_b c_{vb}) \frac{dp}{p} \quad (11\text{-}27c)$$

Of course, these results could have been anticipated. Equations (11-27) merely show that, after the heat capacities of the mixture have been determined, the mixture can be treated as if it were a single-constituent system. Such a procedure has already been followed whenever air, which is a gas mixture, was the main substance studied.

**Example 7.** A mixture of 1 mole of oxygen and 2 moles of nitrogen is in a tank at a temperature of 86.6°F and a pressure of 12.7 psia. Determine the entropy of the mixture if each constituent is assigned a datum-state value of zero entropy at 1 atm and 0°F.

*Solution.* The mole fractions are

$$x_O = \tfrac{1}{3} \qquad x_N = \tfrac{2}{3}$$

and the partial pressures are

$$p_O = \tfrac{1}{3}(12.7) = 4.23 \text{ psia} \qquad p_N = \tfrac{2}{3}(12.7) = 8.46 \text{ psi}$$

From Table B-6,

$$c_{pO} = 7.01 \text{ Btu/(mole)}(°R) \qquad c_{pN} = 6.96 \text{ Btu/(mole)}(°R)$$

Equation (11-3b) can be used for each constituent,

$$ds_i = c_{pi} \frac{dT}{T} - R_0 \frac{dp_i}{p_i}$$

$$s_O = \Delta s_O = 7.01 \ln \frac{546.6}{460} - 1.986 \ln \frac{4.23}{14.7}$$
$$= 1.21 + 2.48 = 3.69 \text{ Btu/(mole)}(°R)$$

$$s_N = \Delta s_N = 6.96 \ln \frac{546.6}{460} - 1.986 \ln \frac{8.46}{14.7}$$
$$= 1.202 + 0.318 = 1.52 \text{ Btu/(mole)}(°R)$$

With these values substituted in Eq. (11-25d),

$$S = ns = n_O s_O + n_N s_N = 1(3.69) + 2(1.52) = 6.73 \text{ Btu}/°\text{R} \qquad Ans.$$

**Example 8.** The mixture in Example 7 is cooled at constant volume to a temperature of 50°F. Determine the change in internal energy and in entropy.
    *Solution.* From Table B-6,

$$c_{vN} = 4.97 \text{ Btu}/(\text{mole})(°\text{R}) \qquad c_{vO} = 5.02 \text{ Btu}/(\text{mole})(°\text{R})$$

and by Eq. (11-26a) written in the form

$$n \, \Delta u = (n_N c_{vN} + n_O c_{vO})(T_2 - T_1)$$
$$\Delta U = [2(4.97) + 1(5.02)](-36.6) = -547 \text{ Btu} \qquad Ans.$$

The change in entropy for the mixture is found by Eq. (11-27b) integrated for constant-valued heat capacities:

$$n \, \Delta s = (n_O c_{vO} + n_N c_{vN}) \ln \frac{T_2}{T_1} + 0$$
$$\Delta S = [1(5.02) + 2(4.97)] \ln \frac{510}{546.6} = -1.012 \text{ Btu}/°\text{R}$$

## 11-13. The Irreversible Mixing Process.†

Suppose that several gases, all at the same temperature and pressure, are to be mixed together. For example, in an isolated system each gas could occupy a separate partitioned space in a large insulated tank; when the partitions are removed, the gases will individually expand to fill the entire volume. For real gases, such an adiabatic mixing process is not necessarily either an isothermal or a constant-pressure process. But when ideal gases are mixed, each gas undergoes a free expansion to the total volume and to its partial pressure in the mixture; the total pressure remains unchanged at the common initial value for the unmixed constituents because a free expansion of an ideal gas does not change the temperature (Art. 3-5). For this reason, mixing does not affect the values of internal energy, enthalpy, or heat capacity for the constituent gases because these properties are functions only of temperature. However, Eq. (11-4a) shows that the entropy of each constituent will increase during the free expansion, and thus the entropy change of mixing is positive. This increase is to be expected because mixing is an irreversible process: whenever an irreversible process occurs, the entropy of the isolated system will increase (Art. 7-6). Moreover, if the temperatures of the gases before mixing are unequal, the irreversibility of mixing is increased, as shown by the loss in availability (Art. 7-9).

**Example 9.** A mole of oxygen at 30 psia and 60°F is in a container that is connected through a valve to a second container filled with 2 moles of nitrogen at 10 psia and 100°F. The valve is opened, and adiabatic mixing occurs. Determine the equilibrium temperature and pressure of the mixture.

---

† The reversible mixing process is introduced (indirectly) in Art. 16-7.

*Solution.* For this isolated system, the First Law shows that the change in internal energy is zero. Values of $Mc_v$ from Table B-6 are substituted in Eq. (11-26a), and the equilibrium temperature is computed,

$$n \, \Delta u = n_O c_{vO}(\Delta T)_O + n_N c_{vN}(\Delta T)_N = 0$$
$$1(5.02)(t - 60) + 2(4.97)(t - 100) = 0 \qquad t = 86.6°F \qquad Ans.$$

The pressure is found from the ideal-gas equation of state after the volumes of the tanks have been computed,

$$pV = nR_0T$$
$$V_O = \frac{1(1,545)(520)}{30(144)} = 186.2 \text{ ft}^3$$
$$V_N = \frac{2(1,545)(560)}{10(144)} = 1,203.0 \text{ ft}^3$$
$$\text{Total} = \overline{1,389.2 \text{ ft}^3}$$

The mixture pressure is

$$p = \frac{nR_0T}{V} = \frac{3(1,545)(546.6)}{1,389.2(144)} = 12.7 \text{ psia} \qquad Ans.$$

**Example 10.** Compute the change in entropy for each gas and for the mixing process in Example 9.

*Solution.* The change in entropy of each constituent is found by

$$\Delta s = c_v \ln \frac{T_2}{T_1} + R_0 \ln \frac{v_2}{v_1} \qquad (11\text{-}4a)$$

$$\Delta s = (\Delta s \text{ from change in temperature}) + (\Delta s \text{ from change in volume})$$

For the oxygen:

$$n_O \, \Delta s_O = 1(5.02) \ln \frac{546.6}{520} + 1.986 \ln \frac{1389.2}{186.2}$$
$$= 0.246 + 3.985 = +4.231 \text{ Btu/°R} \qquad Ans.$$

For the nitrogen:

$$n_N \, \Delta s_N = 2(4.97) \ln \frac{546.6}{560} + 2(1.986) \ln \frac{1389.2}{1,203}$$
$$= -0.236 + 0.556 = +0.320 \text{ Btu/°R} \qquad Ans.$$

The increase in entropy for the process is

$$\Delta S = \Sigma n \, \Delta s = 4.231 + 0.320 = +4.551 \text{ Btu/°R} \qquad Ans.$$

This can be divided into two parts: that part caused by the temperature equalization and that part caused by the expansion (mixing).

$$\Sigma n \, \Delta s = (+0.246 - 0.236) + (3.985 + 0.556)$$
$$\Delta S = +0.01 + 4.541 = +4.551 \text{ Btu/°R}$$

**11-14. The Isentropic Process.** After the properties of the gas mixture have been determined, analysis of state changes during a process can be made in the same manner as for a single-constituent system. The isentropic process is of special interest. Note that if constant values of the constituent heat capacities are substituted in Eq. (11-27c) the equation will integrate into the familiar form $pv^k = C$. If greater accuracy is desired, mean values of the heat capacities can be used in determining $k$, as illustrated in Example 2, or, better, Eqs. (11-27$a$) and (11-27$b$) can be integrated in the manner illustrated in Art. 11-8.

The isentropic process *for the mixture* does not necessarily dictate an isentropic process *for each constituent* but, rather, that

$$\Delta S_{\text{mixture}} = 0 \qquad (a)$$

and therefore, for a two-constituent mixture,

$$\Delta S_a + \Delta S_b = 0 \qquad (b)$$

Consider that the derivation of Eq. (11-27$a$) shows that

$$ds_{\text{mix}} = c_{p,\text{mix}} \frac{dT}{T} - R_0 \frac{dp}{p} \qquad (c)$$

$$ds_a = c_{pa} \frac{dT}{T} - R_0 \frac{dp}{p} \qquad (d)$$

$$ds_b = c_{pb} \frac{dT}{T} - R_0 \frac{dp}{p} \qquad (e)$$

If it is premised that the constituents, as well as the mixture, are to be isentropically compressed or expanded, inspection of the above equations decrees that this is possible only if

Therefore,
$$c_{p,\text{mix}} = c_{pa} = c_{pb}$$
$$k_{\text{mix}} = k_a = k_b$$

The other alternative for satisfying Eqs. ($a$) and ($b$) would be that the entropy of one constituent increased while the entropy of the second constituent decreased. In a compression process (that is, $T_2 > T_1$, $p_2 > p_1$), suppose that $\Delta S_a$ is positive ($+$); then, $\Delta S_b$ must be negative ($-$), and Eqs. ($c$) to ($e$) show for this case that

$$c_{pa} > c_{pb} \qquad \text{then } k_a < k_b$$

because $c_p - c_v = R_0$ (a constant).

**Example 11.** A mixture of 1 lb$_m$ air and 0.94 lb$_m$ steam has a pressure of 50 psia and a temperature of 250°F. If this mixture is isentropically compressed to 100 psia, find ($a$) the final temperature and ($b$) the change in entropy for each constituent.

*Solution.* ($a$) The mole fraction of each constituent is

$$x_a = \frac{\frac{1}{29}}{\frac{1}{29} + 0.94/18} = 0.398 \qquad x_w = 1.0 - x_a = 0.602$$

Equation (11-27$a$), for the isentropic process, is

$$ds = (x_a c_{pa} + x_w c_{pw}) \frac{dT}{T} - R_0 \frac{dp}{p} = 0$$

The heat capacities will be assumed to be constants, for simplicity, with values selected from Table B-6,

$$[0.398(6.96) + 0.602(8.02)] \ln \frac{T_2}{T_1} = R_0 \ln \frac{p_2}{p_1}$$

$$7.59 \frac{\text{Btu}}{\text{mole }°\text{R}} \ln \frac{T_2}{T_1} = 1.986 \frac{\text{Btu}}{\text{mole }°\text{R}} \ln \frac{p_2}{p_1} \qquad (f)$$

$$\frac{T_2}{T_1} = \left(\frac{p_2}{p_1}\right)^{0.262}$$

$$T_2 = 710(2)^{0.262} = 851°\text{R} \qquad Ans.$$

Note that Eq. (f) is equivalent to

$$c_{p,\text{mix}} \ln \frac{T_2}{T_1} = R_0 \ln \frac{p_2}{p_1}$$

$$\frac{T_2}{T_1} = \left(\frac{p_2}{p_1}\right)^{R_0/c_p} = \left(\frac{p_2}{p_1}\right)^{(k-1)/k}$$

Hence, the average $k$ for this mixture is equal to

$$\frac{k-1}{k} = 0.262 \qquad \text{or} \qquad k = 1.355$$

(b) The change in specific entropy of the air equals

$$\Delta s_a = c_{pa} \ln \frac{T_2}{T_1} - R_0 \ln \frac{p_2}{p_1}$$
$$= 6.96 \ln {}^{851}\!/_{710} - 1.986 \ln 2 = -0.105 \text{ Btu/(mole)}(°\text{R}) \qquad Ans.$$

and the total change in entropy of the air is

$$\Delta S_a = n_a \Delta s_a = -0.00362 \text{ Btu/}°\text{R} \qquad Ans.$$

The change in entropy of the water vapor is equal to the change in entropy of the air but of opposite sign,

$$\Delta S_w = +0.00362 \text{ Btu/}°\text{R} \qquad Ans.$$

and therefore the change in specific entropy of the water vapor is

$$\Delta s_w = +0.0693 \text{ Btu/(mole)}(°\text{R}) \qquad Ans.$$

**11-15. Gas and Vapor Mixtures.** When a liquid is placed in a greater space than its volume, it will evaporate as a vapor into the space above the liquid. This is because the molecules within the liquid move at various velocities; the higher-velocity molecules have sufficient energy to overcome the surface restraint and leave the liquid. When the liquid is confined in a closed volume, vapor molecules will strike the liquid surface and be condensed or held by the potential attractive forces of the liquid. Thus, the pressure exerted by the vapor will assume an equilibrium and maximum value when the rate of vaporization of liquid is balanced by the rate of condensation of vapor. For this equilibrium between *saturated* liquid and *saturated* vapor, the pressure is called the *vapor pressure*, and it is a function of the temperature (Fig. 9-2). (As commonly stated, the boiling temperature of the liquid is controlled by the pressure.) The vapor pressure is independent of the relative amounts of vapor and liquid present because both the rate of vaporization and the rate of condensation are proportional to the liquid surface; changing the area of the liquid surface will change the time for reaching equilibrium, but such a change will not affect the equilibrium pressure.

At low pressures, other gases than the vapor from the liquid can be present in the confined space without seriously interfering with the molecular activities of vaporization and condensation. Because of this fact, the vapor-pressure–temperature relationship for the pure substance can be applied to a low-pressure system of one liquid constituent and several gaseous constituents to predict the maximum partial pressure of the vapor. In this instance, the mixture of gases is commonly said to be *saturated* with the vapor from the liquid constituent, although, to be precise, it is the vapor and the liquid which are saturated. If the temperature of this mixture is raised, the liquid may disappear and, for any additional increase in temperature, the vapor will be *superheated*. Here, no liquid component is present—only gases; hence, the partial pressure of the superheated vapor is proportional to its mole fraction.

**Example 12.** A pound of air saturated with water vapor is in a tank at a temperature of 250°F and a pressure of 50 psia. Determine the volume of the tank and the amount of water vapor present in the mixture.

*Solution.* The vapor pressure and specific volume of the saturated vapor are found from the Steam Tables for a temperature of 250°F. (Here the subscript $s$ will be used instead of $w$ because the water vapor is saturated.)

$$t = 250°F \qquad p_s = 29.825 \text{ psia} \qquad v_s = v_g \text{ of the Steam Tables} = 13.821 \text{ ft}^3/\text{lb}_m$$

The pressure of the air is found by Dalton's law,

$$p = p_a + p_s \qquad p_a = 50 - 29.825 = 20.175 \text{ psia}$$

The volume of the tank is the volume occupied by the air,

$$pV = nRT \qquad V = \frac{\frac{1}{2}9(1,545)(710)}{20.175(144)} = 13.02 \text{ ft}^3 \qquad Ans.$$

This volume is also the volume occupied by the water vapor:

$$\frac{V}{v_s} = m_s \qquad m_s = \frac{13.02}{13.821} = 0.94 \text{ lb}_m \text{ steam} \qquad Ans.$$

(Note that these data are the same as in Example 11.) The answer can be closely checked by Eqs. (11-22b),

$$x_s = \frac{p_s}{p} = \frac{29.825}{50} = 0.597$$

and by Eqs. (11-20a),

$$x_s = \frac{n_s}{n_s + n_a} = \frac{m_s/18}{m_s/18 + \frac{1}{2}9} = 0.597$$
$$m_s = 0.922 \text{ lb}_m \qquad Ans.$$

**Example 13.** Ethyl alcohol is to be used in an automobile carburetor that is adjusted to give a 9:1 air-fuel ratio (a mass ratio). If the temperature in the manifold is 60°F and the pressure is atmospheric, what percentage of the alcohol will be evaporated, assuming that equilibrium between vapor and liquid is reached? (Vapor pressure of the alcohol at 60°F is 0.64 psia, and its molecular weight is 46.)

*Solution*

$$x_s = \frac{p_s}{p} = \frac{0.64}{14.7} = 0.0435 \tag{11-22b}$$

$$x_s = \frac{n_s}{n_s + n_a} = \frac{m_s/46}{m_s/46 + \frac{9}{29}} = 0.0435 \tag{11-20a}$$

$$m_s = 0.65 \text{ lb}_m$$

Hence, for every pound of liquid fuel metered by the carburetor, 0.65 lb$_m$ should be evaporated in the manifold, or 65 per cent vaporization under ideal conditions (*Ans.*).

In the actual engine, a lesser amount is vaporized because sufficient time is not available to allow phase equilibrium to be reached.

**Example 14.** A mixture contains 1 lb$_m$ of dry air and 0.01 lb$_m$ of water at a pressure of 20 psia and a temperature of 80°F. Determine the partial pressure of the water vapor.

*Solution.* Here whether or not the air is saturated with water vapor is unknown. If it were saturated, the pressure of the vapor would be (Steam Tables, 80°F)

$$p_s = 0.5069 \text{ psia}$$

and this is the maximum pressure that water vapor can exert at 80°F. On the other hand, if the 0.01 lb$_m$ of water were superheated vapor, its partial pressure would be proportional to the mole fraction,

$$p_w = x_w p = \frac{0.01/18}{0.01/18 + \frac{1}{29}} \, 20 = 0.318 \text{ psia} \qquad Ans.$$

Since this pressure is less than the saturated pressure, the vapor must be superheated. Thus, for a temperature of 80°F, a greater amount of water vapor than 0.01 lb$_m$ is required to saturate 1 lb$_m$ of air (and so raise the pressure to the limiting value of 0.5069 psia).

## PROBLEMS

The following points are to be borne in mind in solving the problems:

1. A diagrammatic sketch of the system must be drawn showing the boundaries and:

(a) Initial and final properties
(b) All transfers of heat, work, and mass
(c) All other given data (*n* values, etc.)

2. Note especially whether or not the initial and final states can be identified (recall that two properties usually fix the state). In most problems, only one property is apparently given for either the initial or the final state. Here the path equation ($pv^n = C$) or the wording (isentropic or $s = C$, etc.) will yield the missing data.

3. Determine from the data whether or not the process is reversible, and use this knowledge in selecting equations. For example, the First Law energy-balance equations [Eqs. (4-4d) and (5-7d)] are valid for reversible or irreversible processes and are simple to use. Note that a solution from Eq. (4-4d) or (5-7d) will agree with that from, for example, Eq. (11-6) or (11-7a) if the process is reversible.

4. Beware of involved equations such as Eq. (11-7b). Use, instead, Eq. (11-7a) in the manner of Example 1.

5. Illustrating the process on the $Ts$ and $pv$ diagrams may make the solution obvious.

6. Gas-constant values can be found in Table B-6.

7. Answers to be Btu units, and for work answers show also units of foot pound force. Mass unit is to be the pound mole or pound mass as implied by the statement of the problem. If in doubt, answers are to be expressed in both units.

8. Heat-capacity values are assumed constant unless otherwise stated.

9. All gases are assumed to behave perfectly unless otherwise stated.

10. Atmospheric pressure is 14.7 psia.

**1.** A tank contains 5 ft$^3$ of nitrogen at 200 psia and 200°F. The tank is cooled until the temperature of the nitrogen is 60°F. Determine the amount of heat transferred from the nitrogen, the final pressure, the change in entropy, and the change in enthalpy.

**2.** A constant volume of air is cooled by transfer of 40 Btu of heat until the pressure is 30 psia. If the initial temperature is 300°F and the initial pressure is 100 psia, determine the final temperature, change in entropy, change in enthalpy, and the work done.

**3.** Three pounds of air confined in a closed tank is heated until the enthalpy is increased 40 Btu/lb$_m$. If the initial pressure is 100 psia and the final temperature is 400°F, determine the change in internal energy.

**4.** Oxygen is confined in a tank under a pressure of 100 psia and at a temperature of 300°F. If this gas is cooled until the entropy decreases by 0.264 Btu/(lb$_m$)(°R), what will be the final pressure?

**5.** A pound of air at 100°F and 20 psia is expanded at constant pressure until the entropy has changed by 0.3 Btu/(lb$_m$)(°R). Determine the final temperature, heat transferred, and change in enthalpy and internal energy.

**6.** A mole of air at 14.7 psia and 60°F is heated at constant pressure until the volume is tripled. Determine the work, change of internal energy, entropy, and amount of heat transferred.

**7.** A gas is compressed at constant temperature from 15 psia and 25 ft$^3$ to 105 psia. Compute the work and heat transferred and the change in specific entropy, internal energy, and enthalpy. Repeat, assuming a flow process with negligible kinetic energy.

**8.** Air at 60°F is isothermally compressed to 200 psia while the volume is halved and the entropy decreases by −0.5 Btu/°R. Compute the initial volume, mass, heat, and work transferred.

**9.** Two pounds of methane is isothermally compressed at 60°F while 200 Btu of heat is transferred. (a) Find the pressure and volume ratios of compression. (b) If the final pressure is 80 psia, find the change in entropy and the original pressure.

**10.** A pound of air is expanded in a polytropic nonflow process with $n = 1.2$ from $p_1 = 100$ psia and $t_1 = 260°F$ to $t_2 = 60°F$. Find the heat and work transferred and the change in entropy, enthalpy, and internal energy.

**11.** A pound of air is compressed in a polytropic nonflow process from $p_1 = 20$ psia to $p_2 = 100$ psia while the entropy increases by 0.046 Btu/(lb$_m$)(°R). Determine the value for $n$.

**12.** A gas is expanded from a pressure of 100 to 10 psia with corresponding volume of 2.3 ft$^3$ while work of amount 10 Btu is done.   Find the value of $n$ for this nonflow process.   (The final equation can be solved by trial.)

**13.** Hydrogen is compressed along a polytropic path for which $n = 1.2$.   The initial pressure is 20 psia, temperature is 80°F, and volume is 5 ft$^3$.   The work added is 10 Btu (closed system).   Compute the heat transferred and change in entropy, enthalpy, and internal energy.

**14.** One mole of nitrogen expands isentropically in a nonflow process from $p_1 = 100$ psia to $p_2 = 10$ psia.   If the initial temperature is 100°F, find the heat and work transferred and change in internal energy, enthalpy, and entropy.

**15.** Air at 15 psia with volume of 10 ft$^3$ is isentropically compressed in a nonflow process by transfer of 60 Btu of work until the temperature is 375°F.   Find the mass of air, change in enthalpy and internal energy, and initial temperature.

**16.** A pound of air at a pressure of 25 psia and a temperature of 60°F is isentropically compressed until its enthalpy has increased by 50 Btu.   It is then isothermally expanded to the original pressure and, finally, cooled at constant pressure to the original state.   Find the work and heat transferred and the change in entropy for each nonflow process and for the cycle.   Find the cycle efficiency.

**17.** A mole of air at 60°F is compressed along a polytropic path for which process $n = 1.5$, expanded isentropically to the initial temperature, and, finally, isothermally compressed back to the initial state with abstraction of 700 Btu of heat.   Compute the heat and work transferred for each nonflow process and for the cycle.

**18.** A pound of nitrogen at 60°F is heated at constant volume until the pressure is doubled, expanded along a polytropic path ($n = 1.2$) to the initial pressure, and, finally, cooled at constant pressure back to the initial state.   Compute the heat, work, and change in entropy for each nonflow process and for the cycle.

**19.** A pound of methane at 80°F is heated at constant pressure until the volume is doubled, compressed isothermally to the original entropy, expanded along a polytropic path ($n = 1.2$) to the initial volume, and finally cooled to the initial state. Compute the heat and work transferred for each nonflow process and for the cycle.

**20.** A steady flow of air enters a system at a pressure of 15 psia and 60°F and leaves the system with pressure of 15 psia but temperature of 600°F.   Find the heat and work transferred and change in entropy and enthalpy ($\Delta KE \approx 0$).

**21.** Nitrogen is isothermally compressed in a steady-flow system from 35 psia and 60°F to a state with entropy $-0.2$ Btu/(lb$_m$)(°R) less than the initial value. Determine the heat and work transferred if the mass-flow rate is 2 lb$_m$/sec and $\Delta KE \approx 0$.

**22.** Methane enters a system at low velocity and with pressure of 100 psia and temperature of 400°F and leaves the system at a pressure of 10 psia.   (a) What is the maximum amount of kinetic energy that can be realized from this expansion if no heat or work is transferred?   (b) Compute the maximum amount of work that can be obtained under the same conditions.   (c) If the kinetic energy realized in (a) is dissipated within the flow system (throttling), what will be the temperature of the exit fluid (at pressure $p_2 = 10$ psia)?

**23.** Air enters a compressor at $p_1 = 14.7$ psia and $t_1 = 60$°F and leaves at $p_2 = 40$ psia and $t_2 = 252$°F.   Determine the work required if no heat is transferred.   Can you prove that this process is irreversible?

**24.** A mole of carbon dioxide at 500 psia undergoes a constant-pressure nonflow process with the temperature changing from 1000 to 1100°R.   Calculate the reversible work by the van der Waals equation and by the perfect-gas laws.

**25.** A mole of carbon dioxide at 500 psia undergoes an isothermal and reversible nonflow process at 1000°R until the volume is doubled. Calculate the work by the van der Waals equation and by the ideal-gas laws.

**26.** A pound of air is heated at constant pressure from 60.3 to 1050.3°F. Determine the change in internal energy, enthalpy, and entropy and the values for heat and work transfers for this process.†

**27.** Air at 14.7 psia and 60.3°F is reversibly and adiabatically compressed to 147 psia. Determine the final temperature.†

**28.** Air at 14.7 psia and 60.3°F is reversibly and adiabatically compressed to 700.3°F. Determine the final pressure.†

**29.** Air at 14.7 psia and 60.3°F is irreversibly and adiabatically compressed to 147 psia and a temperature of 540.3°F. Compute the change in entropy.†

**30.** Compressed air at 50 psia and 100°F is throttled through a globe valve to a pressure of 30 psia. The initial and final velocities of the air are negligible. For this adiabatic process, determine the work and changes in entropy, enthalpy, and internal energy.†

**31.** Is Eq. (11-11) valid for irreversible isothermal processes? Discuss.

**32.** For a nonflow compression process between definite pressure limits, determine whether isothermal or isentropic compression will require the least work, or whether a decision can be made. What conclusions can be made for the inverse expansion process?

**33.** A mixture of 5 lb$_m$ of argon, 10 lb$_m$ of nitrogen, and 10 lb$_m$ of methane has a pressure of 20 psia and a temperature of 100°F. Determine for the mixture (a) the molecular weight, (b) the partial pressure and partial volumes, and (c) the molar volume (volume of 1 mole of mixture).

**34.** Repeat Prob. 33, but assume that the mixture consists of 2 moles of oxygen and 3 moles of hydrogen.

**35.** A mixture, containing 26 lb$_m$ of nitrogen and the remainder oxygen, occupies a volume of 100 ft³ at a pressure of 75 psia and temperature of 60°F. Determine the molecular weight of the mixture and the partial pressure of the nitrogen.

**36.** Repeat Prob. 35, but assume that the mixture contains 0.3 mole fraction of nitrogen and 0.7 mole fraction of oxygen.

**37.** A mixture of 0.1 mole of oxygen and the remainder carbon dioxide occupies a volume of 100 ft³ at a pressure of 200 psig. If the partial pressure of the carbon dioxide is 150 psia, what is the temperature of the gas?

**38.** Derive the relationship $mR = m_aR_a + m_bR_b + m_cR_c$.

**39.** Determine the volumetric analysis for the data of Probs. 33 to 37.

**40.** Determine the gravimetric and ultimate analysis for the data of Probs. 33 to 37.

**41.** A mixture of 30 per cent nitrogen and 70 per cent carbon dioxide (by volume) has a temperature of 200°F. Calculate the internal energy and enthalpy of this mixture above a datum of 0°F for the internal energy (use variable-heat-capacity relationships) on a mole and also a pound mass basis.

**42.** Repeat Prob. 41, but for a temperature of 600°F.

**43.** Ten pounds mass oxygen at 100 psia and 200°F is in a container that is connected through a valve to a second container filled with 20 lb$_m$ carbon dioxide at 50 psia and 100°F. The valve is opened, and adiabatic mixing occurs. Determine the final pressure and temperature.

**44.** Repeat Prob. 43, but assume that the second tank contained 0.5 mole of nitrogen.

† Use Table B-9.

**45.** Repeat Prob. 43, assuming that the process was not adiabatic and that 10 Btu of heat was transferred to the surroundings during (and not after) the mixing process. (Use constant values for the heat capacities.)

**46.** Determine the probable volumetric analysis for the data of Prob. 43.

**47.** Compute the change in entropy for the data of Prob. 43.

**48.** Compute the change in entropy for the data of Prob. 44.

**49.** Compute the change in entropy and enthalpy for the data of Prob. 45.

**50.** A mixture containing 0.3 mole of air and 0.7 mole of methane is isentropically compressed from $p_1 = 14.7$ psia and $t_1 = 60°F$ to $p_2 = 120$ psia. Compute the final temperature of the mixture and the change in entropy for each component. (Use constant values for the heat capacities.)

**51.** Repeat Prob. 50, using variable-heat-capacity relationships.

**52.** Repeat Prob. 50, assuming that the final state has a volume one-sixth of the original volume (and that $p_2$ is unknown).

**53.** A mole of air saturated with water vapor is in a tank at a temperature of 300°F and a pressure of 100 psia. Determine the amount of water vapor in the mixture.

**54.** If the mixture in Prob. 53 is cooled to 200°F, what will be the pressure? How much heat must be transferred in this process? (Use constant value of heat capacity for the air and Steam Table values for the water.)

**55.** Compute the change in entropy for the data of Prob. 53. (Use Steam Table data for the water.)

**56.** Repeat Example 13, assuming that the temperature in the manifold is 50°F and that the vapor pressure is 0.45 psia.

**57.** What must be the pressure in the manifold of the engine in Example 13 if the air mixture is saturated? For complete evaporation? (Temperature is 60°F.)

**58.** A mixture of 0.2 $lb_m$ steam and 0.2 $lb_m$ air is in a tank at a temperature of 200°F. Determine the volume and the pressure in the tank, if possible.

**59.** Determine the heat that must be transferred to raise the temperature of the mixture of Prob. 58 to 300°F.

## REVIEW PROBLEMS

**1.** For a constant-pressure process, the value of $n$ in $pv^n = C$ is 0, $\infty$, 1.0, $k$, not given.

**2.** For a reversible adiabatic process, the heat capacity is 0, $\infty$, $c_p$, $c_v$, $k$, not given.

**3.** For a reversible constant-temperature nonflow process, the heat capacity is 0, $c_p$, $c_v$, not given.

**4.** For a reversible constant-temperature process from small to large volume, the heat transfer is $+$, $-$, 0, $\infty$, not given.

**5.** When heat is reversibly added to a gas during a process, then work is $+$, $-$, 0, not given.

**6.** When heat is reversibly added to a gas while it expands and while its temperature drops, then $\Delta S$ is $+$, $-$, 0, $\infty$, not given.

**7.** For a polytropic expansion to a larger volume with falling temperature, the heat exchange is $+$, $-$, 0, $\infty$, not given.

**8.** In a reversible nonflow process the pressure (can) (cannot) rise if the volume is increasing.

**9.** The entropy change of an isolated system can be $+$, $-$, 0, $+$ or 0, $+$ or $-$ or 0, not given.

**10.** The entropy change of an adiabatic system can be $+$, $-$, 0, $+$ or 0, $+$ or $-$ or 0, not given.

**11.** The equation $T\ ds - p\ dv = du$ holds for (all substances studied) (ideal gases) (ideal or real gases).

## SELECTED REFERENCES

1. Obert, E. F.: "Concepts of Thermodynamics," McGraw-Hill Book Company, Inc., New York, 1960.
2. Keenan, J. H., and J. Kaye: "Gas Tables," John Wiley & Sons, Inc., New York, 1948.

# CHAPTER 12

# NONSTEADY FLOW, FRICTION, AND AVAILABILITY

Give me the money that has been spent in war . . . and I will
build a schoolhouse in every valley over the entire earth.

*Charles Sumner*

The subjects of nonsteady flow, friction, and availability were introduced previously, but development necessarily had to be delayed until
the Second Law and process equations were studied.

**12-1. The Emptying Process.** The inverse problem to the filling
process (discussed in Art. 5-10) occurs when the fluid flows *from* the
system *to* the surroundings. If the leaving stream has constant values of
$e_f$, then, since the algebraic sign of $m_2 - m_1$ compensates for the flow
direction, the same equation as before is applicable:

$$\Delta Q - \Delta W = m_2 u_2 - m_1 u_1 - (m_2 - m_1)e_f \Big]_{e_f = C} \qquad (5\text{-}4f)$$

In most expansion processes, the velocity $V_f$ will change considerably
with time, and $e_f$ is not constant. An example of this type of flow is the
reversible adiabatic expansion process with fluid flowing to surroundings
at constant pressure (Fig. 12-1). Since all elements of fluid leaving the
system have identical values of pressure and, also, of entropy, all have
the same final thermostatic state (but each element has a different velocity). The first element of fluid to escape will expand isentropically in
the nozzle from $T_1$, $p_1$ to $p_2$; it will acquire a relatively high velocity
while its temperature is reduced to $T_2$. Succeeding elements of fluid will
expand from lower system pressures and temperatures to the same exit
pressure $p_2$. Throughout the process, the fluid within the system is also
expanding isentropically (and, in so doing, pushing fluid from the system).
Thus, each element escaping from the system expands from the same
initial pressure to the same final pressure, temperature, entropy, enthalpy,
etc. But the velocity of each element leaving the system is progressively
less as time passes. The first element to leave utilized the entire pressure
drop to create velocity, succeeding elements experienced a part of the
expansion within the system and the remainder within the nozzle, and
consequently the last element to leave the system had negligible velocity.

Suppose that a reversible impulse turbine were to be attached to the exit of the system of Fig. 12-1. In this manner, the leaving kinetic energy (of amount indicated by $\Sigma KE$) could have been converted completely into work without changing the enthalpy of the fluid. For this process, Eq. (5-4f) becomes

$$-\Delta W_{rev} = -\Sigma KE = m_2 u_2 - m_1 u_1 - (m_2 - m_1)h_f \Big]_{s=C} \qquad (12\text{-}1)$$

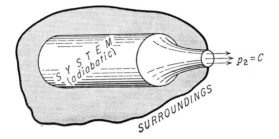

FIG. 12-1. The emptying process into surroundings of infinite extent.

**Example 1.** A tank with volume of 50 ft$^3$ is filled with air at $p_1 = 100$ psia and $t_1 = 240°F$. If this air is used to drive a turbine, calculate the maximum adiabatic work for expansion to atmospheric pressure.

*Solution.* The final temperature of the air in the tank is also that of the high-velocity stream,

$$\frac{T_1}{T_f} = \frac{T_1}{T_2} = \left(\frac{p_1}{p_2}\right)^{(k-1)/k} = \left(\frac{100}{14.7}\right)^{0.286} = 1.73 \qquad \text{or} \qquad T_f = T_2 = 405°R$$

Hence,

$$m_1 = \frac{p_1 V}{R T_1} = \frac{100(144)(50)}{(1,545/29)(700)} = 19.3 \text{ lb}_m \qquad m_2 = \frac{14.7(144)(50)}{(1,545/29)(405)} = 4.9 \text{ lb}_m$$

Equation (12-1) for ideal gases becomes

$$-\Delta W_{rev} = c_v(m_2 T_2 - m_1 T_1) - (m_2 - m_1)c_p T_f$$
$$= 0.171[19.3(700) - 4.90(405)] - (19.3 - 4.9)(0.24)(405)$$
$$\Delta W_{rev} = 572 \text{ Btu} \qquad Ans.$$

**12-2. Availability and the Steady-flow System.** Consider a system in steady flow. The *availability* of the fluid at a coordinate $L$ (at an initial state) is the maximum work that can be done while the fluid passes in steady flow to the *dead state*—the state which is in equilibrium with the surroundings (Art. 6-1). The dead state can be achieved by:

1. Reversible steady-flow processes which convert all of the potential

energy and essentially all of the kinetic energy into work ($V_{final} \to 0$, but cannot be zero since flow must continue):

$$\Delta w_{rev \atop sf} = -\Delta KE - \Delta PE \approx \frac{V_1^2}{2g_c} + \frac{gZ_1}{g_c} \qquad (a)$$

2. An isentropic (reversible adiabatic) process to the temperature $T_0$ (isentropic, so that availability is neither added nor taken away by transfer of heat):

$$\Delta w_{rev \atop s=C} = -\Delta h_{12} = h_1 - h_2 \qquad (b)$$

3. A reversible isothermal process at $T_0$ to the dead-state pressure $p_0$ (and heat transfer at $T_0$ has zero availability):

$$\Delta w_{rev \atop T_0=C} = -\Delta h_{20} + \int T \, ds = h_2 - h_0 + T_0(s_0 - s_2) \qquad (c)$$

All of the work of the steady-flow system, unlike that of the closed system, is useful since the boundaries of the system are fixed. The total useful work is found by adding together Eqs. ($a$), ($b$), and ($c$):

$$\Sigma \, \Delta w_{rev \atop sf} = \left( h_1 + \frac{V_1^2}{2g_c} + \frac{gZ_1}{g_c} \right) - h_0 + T_0(s_0 - s_1) \qquad (d)$$

This work is, by definition, the availability in steady flow:

$$\mathcal{Q}_{sf} = e_f - T_0 s - (h_0 - T_0 s_0) \qquad (12\text{-}2a)$$

(and $\mathcal{Q}_{sf}$ can be positive or negative). The change in available energy between two coordinates of the flow system (between two states) is

$$\Delta \mathcal{Q}_{sf} = \Delta e_f - T_0 \, \Delta s \qquad (12\text{-}2b)$$

Equation (12-2$b$) evaluates the change in available energy for either irreversible or reversible processes since $\mathcal{Q}_{sf}$ is a property. When the process between two states is reversible, the change in available energy of the flow stream is reflected by an exact compensating change in the surroundings; when the process between the same two states is irreversible, the surroundings receive a lesser amount of available energy than before (or give a greater amount).

The *effectiveness* of a process is defined by

$$\varepsilon = \left| \frac{\text{increase in available energy}}{\text{decrease in available energy}} \right| \qquad (12\text{-}3)$$

The irreversibility because of the process is the algebraic sum of numerator and denominator in Eq. (12-3):

$$\Delta \mathcal{I} = \Sigma(\text{increases and decreases in available energy}) \qquad (12\text{-}4)$$

When *all* of the surroundings are included in the analysis, Eq. (12-4) reduces to

$$\Delta \mathcal{G} = -T_0 \Sigma \, \Delta S \qquad (7\text{-}12)$$

In most commercial work, the compression and expansion efficiencies are used:

$$\eta_c = \frac{\text{isentropic work}}{\text{actual work}} \qquad \eta_e = \frac{\text{actual work}}{\text{isentropic work}} \qquad (12\text{-}5)$$

**Example 2.** Superheated steam at 400 psia and 600°F enters a turbine and emerges at a pressure of 50 psia and quality of 99.5 per cent. Compute the engine efficiency, effectiveness, and loss of available energy for the adiabatic process ($T_0$ is 60°F).
*Solution.* From the Steam Tables (with $b \equiv h - T_0 s$ for convenience),

$$p_1 = 400 \text{ psia} \qquad t_1 = 600°F \qquad h_1 = 1{,}306.9 \qquad s_1 = 1.5894 \qquad b_1 = 480.9$$

With isentropic expansion to 50 psia (Mollier chart),

$$p_2 = 50 \text{ psia} \qquad x = 0.945 \qquad h_2 = 1{,}122 \qquad s_2 = 1.5894$$

At the end state for the irreversible process,

$$p_2 = 50 \text{ psia} \qquad x = 0.995 \qquad h_3 = 1{,}169.5 \qquad s_3 = 1.6522 \qquad b_3 = 310.9$$

With these values,

$$\Delta w_{13} = h_1 - h_3 = 1306.9 - 1169.5 = 137.4 \text{ Btu/lb}_m$$
$$\Delta w_{12} = h_1 - h_2 = 1306.9 - 1122 = 184.9 \text{ Btu/lb}_m$$
$$\Delta \mathcal{A}_{sf} = \Delta b_{13} = b_3 - b_1 = 310.9 - 480.9 = -170.0 \text{ Btu/lb}_m$$

Hence,

$$\eta_e = \frac{\Delta w_{13}}{\Delta w_{12}} = \frac{137.4}{184.9} = 0.745, \text{ or } 74.5\% \qquad\qquad Ans.$$

$$\varepsilon = \left| \frac{\Delta w_{13}}{\Delta b_{13}} \right| = \left| \frac{137.4}{-170.0} \right| = 0.81, \text{ or } 81\% \qquad\qquad Ans.$$

$$\Delta \mathcal{G} = \Delta w_{13} + \Delta b_{13} = 137.4 - 170.0 = -32.6 \text{ Btu/lb}_m \qquad Ans.$$

**12-3. Availability and the Closed System.** For the closed system, the dead state can be achieved by:

1. Reversible processes which entirely convert the kinetic and potential energies of the system into work:

$$\Delta W_{\text{rev}} = -\Delta KE - \Delta PE = \frac{mV_1^2}{2g_c} + \frac{mgZ_1}{g_c} \qquad (a)$$

2. An isentropic (reversible adiabatic) process to the temperature $T_0$:

$$\Delta W_{\substack{\text{rev} \\ s=C}} = -\Delta U_{12} = U_1 - U_2 \qquad (b)$$

3. A reversible isothermal process at $T_0$ to the dead-state pressure $p_0$:

$$\Delta W_{\substack{\text{rev} \\ T_0 = C}} = -\Delta U_{20} + \int T \, dS = U_2 - U_0 + T_0(S_0 - S_2) \qquad (c)$$

But a part of the work of the closed system, of amount $p_0 \, \Delta V$, must be used to push aside the atmosphere. Therefore, the net useful work is obtained by adding together Eqs. $(a)$, $(b)$, and $(c)$ and subtracting $p_0 \, \Delta V_{10}$ (and $S_2 = S_1$):

$$\Sigma \, \Delta W_{\substack{rev \\ net}} = \left( U_1 + \frac{mV_1{}^2}{2g_c} + \frac{mgZ}{g_c} \right) - U_0 + T_0(S_0 - S_1) - p_0(V_0 - V_1)$$

$$(d)$$

This work is, by definition, the availability in the closed system:

$$\mathfrak{A} \equiv E + p_0 V - T_0 S - (U_0 + p_0 V_0 - T_0 S_0) \qquad (12\text{-}6a)$$

(and $\mathfrak{A}$ is never negative). The change in available energy between two states is

$$\Delta \mathfrak{A} = \Delta E + p_0 \, \Delta V - T_0 \, \Delta S \qquad (12\text{-}6b)$$

As before, $\Delta \mathfrak{A}$ evaluates the change in available energy between two states for either reversible or irreversible processes, but only in the reversible process do the surroundings experience an exact compensation. For the irreversible process, the loss of available energy is evaluated in the same manner as for the steady-flow system [Eqs. (12-4) and (7-12)].

**12-4. Mechanical Friction and the Closed System.** Consider an irreversible closed-system process that can be approximated on the $pv$ diagram:

$$dQ - dW = dU \Big]_{1 \text{ to } 2, \text{ irreversibly}} \qquad (4\text{-}4d)$$

Suppose that the path of the irreversible process is followed by a reversible process:

$$dQ_{rev} - dW_{rev} = dU \Big]_{1 \text{ to } 2, \text{ reversibly}}$$

And, upon subtracting,

$$d\mathfrak{F} \equiv dQ_{rev} - dQ = dW_{rev} - dW \qquad (12\text{-}7a)$$

Equation (12-7a) defines friction as before (Art. 4-3b):

**Friction is the dissipation of work or available energy into a heating effect.**

Friction can be measured in two different manners:

**Friction is evaluated as the difference in work or the difference in heat between the reversible and the actual process, both of which traverse the same series of states.**

Thus, by definition of the reversible work and Eq. (12-7a),

$$d\mathfrak{F} + dW = p\, dV \qquad \text{(closed system)} \tag{12-8a}$$

Or, by definition of the reversible heat and Eq. (12-7a),

$$d\mathfrak{F} + dQ = T\, dS \qquad \text{(closed system)} \tag{12-9a}$$

**Example 3.** Air at 100 psia and 800°R expands adiabatically in a cylinder to 10 psia along a polytropic path such that $n = 1.3$. Compute, per pound of air, the friction of the process (for ideal gases).

*Solution.* The temperature at the end of the expansion is found by Eq. (3-7b),

$$T_2 = T_1 \left(\frac{p_2}{p_1}\right)^{\frac{n-1}{n}} = 800 \left(\frac{1}{10}\right)^{0.230} = 470°\text{R}$$

The irreversible adiabatic work equals [Eq. (4-4d)]

$$\Delta w_{\text{irrev}}\Big]_{\Delta q = 0} = u_1 - u_2 = c_v(T_1 - T_2) = 0.172(330) = 56.6 \text{ Btu/lb}_m$$

The reversible work is found by Eq. (11-6),

$$\Delta w_{\text{rev}} = \frac{p_2 v_2 - p_1 v_1}{1 - n} = \frac{R(T_2 - T_1)}{1 - n} = \frac{1.986(330)}{29(0.3)} = 75.4 \text{ Btu/lb}_m$$

Therefore, by Eq. (12-7a),

$$\Delta \mathfrak{F} = \Delta w_{\text{rev}} - \Delta w = 75.4 - 56.6 = 18.8 \text{ Btu/lb}_m \qquad Ans.$$

**12-5. Mechanical Friction and the Steady-flow System.** The friction within the closed system arose because of a pressure difference which created a mass velocity, and the velocity became an aimless turbulence which was dissipated by the viscosity of the fluid—by internal shear forces. The same type of dissipation is apparent in the turbulent flow of fluids (and in the internal shear of a laminar flow); in addition, the over-all mass velocity of the fluid gives rise to shearing forces at the boundaries. However, there is an orderliness to the friction in the steady-flow process (which was not present in the closed system) since the properties of the flowing fluid change with position but not with time. Because of this, the measurement of the approximate irreversible path of the steady-flow system is a much more *practical* undertaking than that for the closed system. (And therefore the measurement of friction in the closed system is rarely mentioned, since a path for the usual irreversible process is a difficult, if not impossible, concept.)

The friction equations for the steady-flow system are obtained from the same reasoning as for the closed system: The energy equation is applied to a reversible and to an irreversible process *that follow the same sequence of states,*

$$dq_{rev} - dw_{rev} = dh + \frac{V}{g_c} dV + \frac{g}{g_c} dZ$$

$$dq - dw = dh + \frac{V}{g_c} dV + \frac{g}{g_c} dZ$$

By subtracting one equation from another, the same concept for friction is obtained as for the closed system,

$$d\mathcal{F}_{sf} = dq_{rev} - dq = dw_{rev} - dw \qquad (12\text{-}7b)$$

[which differs in independent variable from Eq. (12-7a)]. By definition of the reversible work [Eq. (6-3c)] and with Eq. (12-7b),

$$d\mathcal{F}_{sf} + dw = -v\,dp - \frac{V}{g_c}dV - \frac{g}{g_c}dZ \qquad (12\text{-}8b)$$

Or, by definition of the reversible heat and Eq. (12-7b),

$$d\mathcal{F}_{sf} + dq = T\,ds \qquad (12\text{-}9b)$$

**Example 4.**   Water under a pressure of 100 psia enters a 2-in. pipe with mass-flow rate of 10 $lb_m$/sec.   The exit is through a 1-in. pipe at an elevation of 30 ft above the inlet, and the water leaves with pressure of 50 psia.   If no work is transferred, determine the friction.   (Assume a constant density of 62.3 $lb_m$/ft$^3$.)

*Solution.*   At entrance and exit,

$$V_1 = \frac{\dot{m}_f}{A\rho} = \frac{10}{0.0218(62.3)} = 7.36 \text{ ft/sec}$$

$$V_2 = \frac{\dot{m}_f}{A\rho} = \frac{10}{0.00545(62.3)} = 29.44 \text{ ft/sec}$$

By Eq. (12-8b) with $\Delta w = 0$,

$$\Delta\mathcal{F}_{sf} = -\int_{p_1}^{p_2} v\,dp - \Delta KE - \Delta PE$$

$$= -\frac{144}{62.3}(50 - 100) - \frac{29.44^2 - 7.36^2}{2g_c} - \frac{g}{g_c}(30 - 0)$$

$$= 115.6 - 12.65 - 30 = 73 \text{ ft-lb}_f/lb_m \qquad Ans.$$

To *predict* values of the function $\mathcal{F}$ for a complex system is a hopeless task because of the wide varieties of devices that can be a part of a steady-flow process.   For processes not involving work, a *friction factor* can be defined:

$$\frac{d\mathcal{F}_{sf}}{dL} \equiv 4f \frac{V^2}{2g_c D} \qquad (12\text{-}10a)$$

wherein $D$ is defined as the *hydraulic diameter,*

$$D \equiv 4 \frac{\text{cross-sectional area}}{\text{wetted perimeter}} \qquad (12\text{-}10b)$$

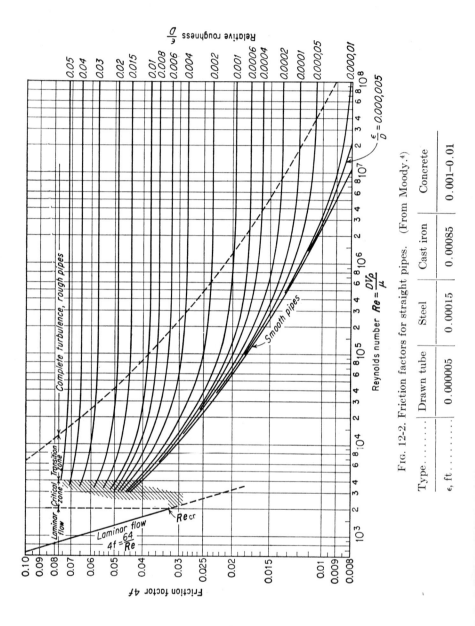

FIG. 12-2. Friction factors for straight pipes. (From Moody.[4])

| Type | Drawn tube | Steel | Cast iron | Concrete |
|---|---|---|---|---|
| ε, ft | 0.000005 | 0.00015 | 0.00085 | 0.001–0.01 |

216

It is found experimentally that, for turbulent subsonic flow in straight conduits of constant cross section (usually round),

$$f = f\left(\frac{DV\rho}{\mu}, \frac{\epsilon}{D}\right) \tag{12-10c}$$

where $\epsilon/D$ is the *roughness factor*. Values of $\epsilon$ for straight round pipes are shown in Fig. 12-2.

Equation (12-10a) can be substituted into Eq. (12-8b),

$$v\,dp + \frac{V}{g_c}\,dV + \frac{g}{g_c}\,dZ + 4f\frac{V^2}{2g_cD}\,dL = 0\bigg]_{\Delta w = 0} \tag{12-11a}$$

In general, the first and fourth terms of this equation must be solved by graphical integration. By dividing by $v^2$, noting that

$$G \equiv \frac{\dot{m}_f}{A} = \frac{V}{v} \tag{12-12}$$

then, for cases where $\Delta Z = 0$, $G = C$,

$$\frac{dp}{v} + \frac{G^2}{g_c}\frac{dv}{v} + 4f\frac{G^2}{2g_cD}\,dL = 0\bigg]_{\substack{\Delta Z, w = 0 \\ A = C}} \tag{12-11b}$$

Integrating, with $f_m$ representing an average value,

$$\int_{L_1}^{L_2} \frac{dp}{v} + \frac{G^2}{g_c}\ln\frac{v_2}{v_1} + 4f_m\frac{G^2}{2g_cD}(L_2 - L_1) = 0\bigg]_{\substack{\Delta Z, w = 0 \\ G = C}} \tag{12-11c}$$

[and this form agrees† with Keenan's equation (181)]. By assuming ideal gases, so that $v = RT/p$, Eq. (12-11c) can be shown as

$$\frac{p_2{}^2 - p_1{}^2}{2RT_m} + \frac{G^2}{g_c}\ln\frac{v_2}{v_1} + 4f_m\frac{G^2}{2g_cD}(L_2 - L_1) = 0\bigg]_{\substack{\Delta Z, w = 0 \\ G = C}} \tag{12-11d}$$

wherein $T_m$ (and $f_m$) are average values [and this form agrees‡ with McAdams's equation (6-9d)].

## PROBLEMS

$$T_0 = 60°F, \quad p_0 = 14.7 \text{ psia}$$

**1.** A diesel engine with 4-in. bore and $4\frac{1}{2}$-in. stroke is to be started at $-30°F$, using a special valve arrangement which opens the intake valve at bottom dead center. The compression ratio is 16:1. Assume an isentropic expansion, and find the temperature of the mixture (air, ideal gas) after the valve is opened (adiabatic process).

**2.** A tank filled with air at 50 psia and 100°F is to be used to drive a turbogenerator. The system is the emergency power for a radio transmitter that requires 5 watts for

† When the $4f_m$ is replaced by $8f_m$.
‡ Except for a velocity-distribution factor.

10 min of operation. Assume an over-all efficiency of 50 per cent and ideal gases. What should be the size of the tank?

**3.** A pound of saturated steam at 100 psia is contained in a rigid tank. The steam expands adiabatically and reversibly to the atmosphere through a reversible turbine. Calculate the work and the mass of steam remaining in the tank.

**4.** An insulated and rigid tank contains 1 mole of air and 1 mole of helium at 100 psia and 300°F. The two gases are separated by an elastic and adiabatic wall. The ideal gases reversibly expand, and $\frac{1}{4}$ mole of air escapes through a reversible turbine to the atmosphere. Calculate the maximum work.

**5.** The state of the gas in the cylinder of an engine when the exhaust valve opens is 91 psia, 3720°R. The gas blows down to atmospheric pressure, with consequent waste of available energy. Calculate the ideal work that could be obtained from this blowdown energy (basis of 1 lb$_m$ air originally in the cylinder, ideal gas).

**6.** A blowdown wind tunnel has a tank capacity of 300 ft$^3$, with air stored at a pressure of 3,000 psia and 80°F. The tanks are honeycombed with metal so that isothermal expansion in the tank is approached within 10°F from initial to final state of 225 psia [heat capacity of metal is 0.2 Btu/(lb$_m$)(°R)]. From the tank, the air passes through a pressure regulator and heated maze (to dissipate the kinetic energy and to hold a constant temperature), then into a calming section, which is maintained at 200 psia, and finally into the test section. (a) Determine the approximate blowdown time for a constant flow rate of 100 lb$_m$/sec. (b) The tank is pumped up by a compressor, with the air entering the tank at about 80°F since the line from the compressor is long. What will be the approximate final temperature of air and tank, assuming that the filling process is adiabatic?

**7.** A steel cylinder is being rapidly filled with oxygen from a large reservoir at room temperature (70°F). Oxygen initially in the cylinder is at 300 psia, 70°F. Mass of cylinder is 50 lb$_m$, $c_p = 0.122$ Btu/(lb$_m$)(°F), and volume is 1.3 ft$^3$. To what pressure must the cylinder be overfilled so that a final pressure of 1,800 psia at room temperature is achieved? (Use compressibility chart; assume heat transfer between cylinder and ambient air is negligible while instantaneous between cylinder and oxygen.)

**8.** The turbine of a Rankine cycle receives saturated steam at 400 psia and exhausts into a condenser at 1 in. Hg pressure. Calculate the loss of available energy. Where is this loss located?

**9.** For the conditions of Prob. 8, the real turbine and pump had expansion efficiencies of 0.75. Calculate $\varepsilon$ and $\Delta s$ for the cycle, turbine, and pump.

**10.** Saturated steam at 400 psia enters a turbine and exhausts at 1 in. Hg, with adiabatic work of 306 Btu/lb$_m$. Compute $\eta_e$, $\varepsilon$, and $\Delta s$ for the process.

**11.** A turbine receives steam at 400 psia, 600°F and exhausts at 50 psia. The loss of available energy for the adiabatic process is 34 Btu/lb$_m$. Compute $\varepsilon$ and $\eta_e$.

**12.** Steam at 94 psia with enthalpy of 1204.96 Btu/lb$_m$ enters an open heater and mixes with water at 94 psia, enthalpy of 47.33 Btu/lb$_m$, and entropy of 0.0914 Btu/(lb$_m$)(°R). The water leaving the heater is saturated at 94 psia. Compute $\varepsilon$ and $\Delta s$ for the heater, based upon 1 lb$_m$ at exit.

**13.** Saturated steam at 100 psia is adiabatically throttled to 20 psia ($\Delta KE \approx 0$). Compute $\Delta s$.

**14.** Repeat Prob. 13, but for air (ideal gas) at 600°R.

**15.** Calculate the availability of steam in steady flow at 1,200 psia, 800°F, $V = 800$ ft/sec, and $Z = 1,000$ ft.

**16.** Calculate the availability of air at $p = 100$ psia, $T = T_0$; at $p = p_0$ and $T = T_0$; at $p = p_0$ and $T = 300$°R (negligible velocity and height, closed system).

**17.** Coal, considered to be pure carbon, is burned with excess air so that the maximum combustion temperature is 2800°F [mean $c_p$ of gases is 0.25 Btu/(lb$_m$)(°R)].

(a) Calculate the loss of available energy, per pound of carbon, because of the combustion process.

(b) Actually, the combustion gases supply heat to a Rankine cycle that operates with saturated steam at 200 psia to the turbine, and with atmospheric pressure in the condenser. The stack temperature is 600°F. Calculate the loss of available energy for the stack and condenser and for the temperature difference between furnace and cycle; calculate, too, the useful work of the cycle (all per pound of carbon) and the effectiveness for the entire process.

**18.** Heat is supplied to a Carnot cycle by cooling the furnace gases from 2600°F $[c_p = 0.25$ Btu/$(lb_m)(°R)]$ to within 100°F of the minimum temperature. The cycle operates between 2,000 psia (saturated liquid to saturated vapor) and 1 in. Hg in the condenser. (a) Calculate the stack loss (in Btu per pound mass gas). (b) Calculate the condenser loss. (c) Calculate the loss arising from the temperature difference between furnace and cycle. (d) Calculate the effectiveness of the cycle for this heat reservoir.

**19.** A desuperheater and pressure reducer receives steam at 1,000 psia, 1000°F and water at 1,500 psia, 100°F. Vaporization of the water reduces the superheat to 800°F while the pressure is also reduced to 800 psia. Calculate the loss of available energy (in Btu per pound water through boiler).

**20.** Compute the availability of air in steady flow at 300 psia, 1000°F, $V = 1,000$ ft/sec.

**21.** Air enters the regenerator of a gas turbine at 850°R, 60 psia and leaves at 1000°R, 50 psia. The hot exhaust gas (air) enters at 1300°R, 16 psia and leaves at 14.7 psia. Calculate $\varepsilon$ and $\Delta \vartheta$, assuming ideal gases and equal mass-flow rates.

**22.** Air flows isothermally at 60°F through a horizontal 3-in. (diameter) pipe at the rate of 10 $lb_m$/sec. Initial pressure is 100 psia, and final pressure is 80 psia. Determine the friction (ideal gas).

**23.** Air flows isothermally at 60°F through a horizontal 2-in. (diameter) pipe at the rate of 5 $lb_m$/sec. Initial pressure is 100 psia, and final pressure is 90 psia. Determine the friction (ideal gas).

**24.** Determine the pressure drop in 1,000 ft of horizontal 3-in. (diameter) steel pipe if the velocity is 5 ft/sec. The fluid is crude oil, with specific gravity of 0.92, viscosity of 0.04 $lb_m$/(ft)(sec), and temperature of 70°F.

**25.** Repeat Prob. 24, but for a velocity of 25 ft/sec.

**26.** Air at an initial pressure of 150 psia and temperature of 60°F is to be carried in a 6-in. (diameter) steel pipe at a rate of 300 $lb_m$/min. Determine the pressure drop in 1 mile. $[\mu_m = 0.0000151$ $lb_m$/(ft)(sec), isothermal, ideal gas.]

**27.** The exhaust of a gas turbine is through a 1-ft (diameter) steel tube, 10 ft in length, with $V_1 = 2,000$ ft/sec, $t_1 = 1200°F$, $p_2 = p_0$. Determine the pressure drop. $[\mu = 2.62(10^{-5})$ $lb_m$/(ft)(sec).]

**28.** A steel gas line is to carry methane at the rate of 300,000 ft³/hr at 15 psia and 60°F. How large should be the pipe if the pressure drop is to be about 20 in. $H_2O$ per 1,000 ft? Let $\mu = 22(10^{-8})$ $lb_f$ sec/ft².

**29.** Air is adiabatically compressed from 14.7 psia, 520°R to 147 psia and 1120°R along a path such that $n$ is constant. Calculate the friction and the work required for this ideal-gas process of a closed system.

**30.** Check Example 3 by a different method of solution.

**31.** Compute the change in availability and the loss in available energy for the air of Example 3, and compare with the friction.

**32.** Compute the change in available energy and the loss of availability for the air of Prob. 29 and compare with the friction.

**33.** For the data of Prob. 23, compute the loss of available energy and compare with the friction.

## SELECTED REFERENCES

1. Obert, E. F.: "Concepts of Thermodynamics," chap. 15, McGraw-Hill Book Company, Inc., New York, 1960.
2. Keenan, J. H.: "Thermodynamics," John Wiley & Sons, Inc., New York, 1941.
3. McAdams, W. H.: "Heat Transmission," 3d ed., McGraw-Hill Book Company, Inc., New York, 1954.
4. Moody, L. F.: Friction Factors for Pipe Flow, *Trans. ASME*, **66**: 671 (November, 1944).

CHAPTER 13

# FLUID FLOW

There is a tide in the affairs of men, which,
taken at the flood, leads on to fame and fortune.
*Shakespeare* (Julius Caesar)

In this chapter one-dimensional flow of both real and ideal fluids will be studied. Changes in elevation, in general, will not be considered.

**13-1. Reference States.** A convenient concept is the *stagnation state:*

**The stagnation state is the limit that could be approached by a stream in steady flow if it were decelerated reversibly and adiabatically to essentially zero velocity without doing work.**

The stagnation properties are designated by the subscript 0.

Another reference is the *sonic state* (Art. A-4):

**The sonic state is that where the stream velocity equals the acoustic velocity (M = 1).**

The sonic properties are designated by the superscript *.

**13-2. Energy and Continuity Equations.** A stream in steady flow must satisfy the *continuity equation*

$$\frac{dA}{A} + \frac{dV}{V} + \frac{d\rho}{\rho} = 0 \qquad (5\text{-}6b)$$

or
$$\dot{m}_f = AV\rho = C \qquad (5\text{-}6a)$$

and the *energy equation:*

$$dq - dw = dh + \frac{V\,dV}{g_c}\Bigg]_{\text{rev or irrev}} \qquad (5\text{-}7c)$$

By applying the integrated energy equation between the flow and the stagnation state ($\Delta q$, $\Delta w$, zero; $V_0$, closely zero), the *stagnation enthalpy* is defined:

$$h_0 = h + \frac{V^2}{2g_c} \qquad (13\text{-}1a)$$

$$s_0 = s \qquad (13\text{-}1b)$$

221

The stagnation enthalpy has the same *numerical* value whether or not the restriction $s_0 = s$ is considered. Therefore, Eq. (5-7$c$) can be shown as

$$dq - dw = dh_0 \bigg]_{\text{rev or irrev}} \tag{13-2a}$$

For processes that involve neither heat nor work, the stagnation enthalpy is constant:

$$h_{01} = h_{02} = \text{constant} \bigg]_{\substack{\Delta q, \Delta w = 0 \\ \text{rev or irrev}}} \tag{13-2b}$$

For ideal gases, enthalpy is a function of temperature alone; therefore, the stagnation temperature is also constant:

$$T_{01} = T_{02} = \text{constant} \bigg]_{\substack{\Delta q, \Delta w = 0 \\ \text{rev or irrev} \\ \text{ideal gas}}} \tag{13-2c}$$

The change in entropy of the flow stream is equal to that of the corresponding stagnation states (since these states were each attained at constant entropy). By Eqs. (11-4$b$) and (13-2$c$),

$$\Delta s_{1-2} = \Delta s_{01-02} = c_p \ln \frac{T_{02}}{T_{01}} - R \ln \frac{p_{02}}{p_{01}} = R \ln \frac{p_{01}}{p_{02}} \bigg]_{\substack{\Delta q, \Delta w = 0 \\ \text{ideal gas}}} \tag{13-2d}$$

Thus the irreversibility of the process is reflected by a loss of stagnation pressure (although the stagnation temperature is not affected). Conversely, *for the reversible adiabatic process (without work) the stagnation pressure is constant throughout the flow region.*

The time required for a fluid to pass through a nozzle (or an orifice) is short, since the velocity is high and the length traveled is small. Therefore, the real process is essentially adiabatic. The energy equation can be solved for $V_2$:

$$V_2 \bigg]_{\substack{\Delta q, \Delta w = 0 \\ \text{rev or irrev}}} = \sqrt{2g_c(h_1 - h_2) + V_1^2} \tag{13-3a}$$

wherein $V_1$ is called the *velocity of approach*. Alternatively, with Eq. (13-1$a$),

$$V_2 \bigg]_{\substack{\Delta q, \Delta w = 0 \\ \text{rev or irrev}}} = \sqrt{2g_c(h_0 - h_2)} \tag{13-3b}$$

**13-3. Momentum Equations.** A fluid in flow obeys the laws and principles of thermodynamics and mechanics. Thus a flow problem must satisfy the equations of *state, continuity, energy, momentum,* and *entropy.* (The entropy restriction is usually tacit: the flow is allowed

only one direction, friction is assumed to be always a degradation, etc.)
By Newton's second law,

**The resultant of the forces acting upon a stream in steady flow
equals the change in the flow of momentum (change in the
momentum flux) through the region under observation.**

The forces acting on an element of a stream consist of pressure forces
($p\mathbf{A}$) on the ends of the element (positive in the direction of flow);
body forces, such as weight (which will be neglected); and wall forces
on the sides of the element. The wall forces ($d\mathbf{F}_{wall}$) are the reactions
of the wall to the pressure and viscous forces of the fluid, and thus are
influenced by changes in area or direction, or from internal struts or from
side-wall shear. (The direction of the wall forces is, in general, not
obvious.) By equating the forces acting upon the element with the
change in momentum flux,

$$d\mathbf{F}_{\substack{wall \\ on\ fluid}} - d(p\mathbf{A}) = \frac{\dot{m}_f}{g_c}\, d\mathbf{V} \qquad (13\text{-}4a)$$

The momentum equation is vector for the general coordinate $L$.
Equation (13-4a) can be integrated for fixed directions $x$, $y$, and $z$:

$$F_{x,wall} + p_1 A_{1x} - p_2 A_{2x} = \frac{\dot{m}_f}{g_c}\,(V_{2x} - V_{1x})$$

$$F_{y,wall} + p_1 A_{1y} - p_2 A_{2y} = \frac{\dot{m}_f}{g_c}\,(V_{2y} - V_{1y}) \qquad (13\text{-}4b)$$

$$F_{z,wall} + p_1 A_{1z} - p_2 A_{2z} = \frac{\dot{m}_f}{g_c}\,(V_{2z} - V_{1z})$$

In these equations, forces and velocities in the direction of flow are
considered positive in sign, and therefore subscript 1 identifies the inlet
and subscript 2 the outlet of the flow stream.

**13-4. Design of a Nozzle.** The ideal velocity will be achieved when
the expansion is reversible; therefore, since the process is also adiabatic,
the entropy remains constant. Thus two properties are available to
fix the state at each stage of the process: a selected pressure and the
initial entropy. The area at each selected state can be found from continuity, Eq. (5-6a).

**Example 1.** Determine the variation in area throughout a nozzle that is to expand
steam reversibly and adiabatically from the stagnation state of $p_0 = 100$ psia and
$t_0 = 600°F$ to $p_2 = 20$ psia.
*Solution.* From Table B-4,

$$v_0 = 6.218 \text{ ft}^3/\text{lb}_m \qquad h_0 = 1329.1 \text{ Btu}/\text{lb}_m \qquad s_0 = 1.7581 \text{ Btu}/\text{lb}_m$$

Selection of pressure at any stage of the expansion is arbitrary, and therefore pressures of 100, 80, 60, 54.6, 40, and 20 psia are selected. Each of these pressures and the common entropy, $s = 1.7581$, enable values of $h$ and $v$ to be found in Steam Table 3. With these values, the velocity is found from Eq. (13-3$b$) and the area for unit mass flow from Eq. (5-6$a$):

| $p$ psia | $h$ Btu/lb$_m$ | $v$ ft$^3$/lb$_m$ | $V$ ft/sec | $A$ ft$^2$ |
|---|---|---|---|---|
| 100 | 1329.1 | 6.218 | 0 | $\infty$ |
| 80 | 1304.1 | 7.384 | 1,119 | 0.0066 |
| 60 | 1273.8 | 9.208 | 1,660 | 0.00556 |
| 54.6 | 1265.0 | 9.844 | 1,790 | 0.00550 |
| 40 | 1234.2 | 12.554 | 2,175 | 0.00577 |
| 20 | 1174.8 | 21.279 | 2,775 | 0.00768 |

The velocities, specific volumes, and areas found in Example 1 are plotted in Fig. 13-1 against the corresponding pressures. Study of Fig. 13-1 reveals an odd result. The area of the nozzle decreases as the pressure decreases until a minimum area, called the *throat*, is reached; from here on the area increases. The explanation, however, is indicated

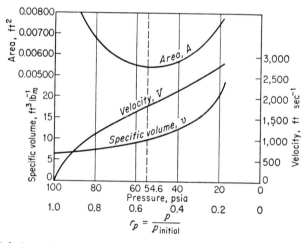

FIG. 13-1. Relations of area, velocity, and specific volume in a convergent-divergent nozzle.

by the continuity equation. Since $\dot{m}_f$ is constant, the area demanded by the flow is proportional inversely to the velocity and directly to the specific volume. (Thus a constant-density fluid would require only a converging passageway.) In the early stages of the expansion in Fig. 13-1, the velocity demands are stronger than those of specific volume and the area decreases; at the throat both demands compensate; after the throat, larger areas must be supplied to accommodate the greatly increased specific volume of the flow.

If the expansion does not proceed over a wide pressure range, only the convergent section may be necessary. In Example 1, this would be true for expansions from 100 psia down to the throat value of approximately 55 psia. A nozzle of this type is called a *convergent nozzle*. However, when the pressure at the exit of the nozzle is less than 55 psia (as it is in Example 1), a divergent section must also be used to attain the maximum velocity. Nozzles of this type are called *convergent-divergent nozzles*.

The calculations in Example 1 determined the area of the nozzle at several different states of the expansion. At each state there will be a definite pressure and a definite velocity. Since the length of the nozzle does not enter the calculations for either

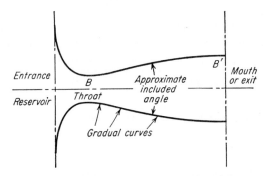

Fig. 13-2. Convergent-divergent nozzle.

pressure or velocity, the nozzle can have any length. If the nozzle is very short, the transition from the high- to the low-pressure region is quickly accomplished. If the nozzle is made longer, the transition will occur over a proportionately longer period of time, although obviously the same sequence of areas is visited in each case. The nozzle should be constructed in the manner illustrated in Fig. 13-2. At the start of the expansion the velocity is zero, and therefore the area should be infinite. To approach this condition the walls of the nozzle could be made tangent to the walls of the reservoir, as illustrated in Fig. 13-2, although practically the shape of the nozzle at this location is not critical since the velocity is low. The converging section of the nozzle need follow no particular type of curve, and any shape is acceptable that allows the area to decrease continually. It has been found from experience that the convergent part can be made quite short in length with no ill effects, and this is done to reduce friction. For the same reason, the walls of the nozzle should be well polished. However, the divergent part of the nozzle must be more carefully treated, because here the velocities are quite high. The throat of the nozzle should merge into the divergent section, as shown at $B$ in Fig. 13-2. The curve must have no corner or discontinuity that would cause an abrupt change in the direction of the fluid and set up a disturbance in the high-velocity stream. Similarly, if the fluid is to leave the nozzle without disturbances created by an abrupt change in direction, the diverging walls should continue to curve gradually and become parallel at exit $B'$, for here the velocity of the fluid is extremely high. While the diverging section should be short if friction losses are to be avoided, the divergence cannot be too great or the fluid will break away from the walls and complete the expansion in a wildly turbulent manner. If the maximum included angle of the nozzle is small, the design will be less critical, and any smooth curve between throat and exit will yield good results. Thus small industrial nozzles are made by simply taper-reaming the divergent section. The maximum included angle of most nozzles is less than 20°, and smaller values of the order of 6 to 12° are preferred by some designers.

With the foregoing thoughts in mind, it can be seen that the basic dimensions for a convergent-divergent nozzle reduce to those of the exit and throat areas, and for a convergent nozzle, to those of the exit area alone. Since the pressure at exit is presumably known, the area and reversible velocity at this section can be directly computed. However, the throat area cannot be computed until the throat pressure is predicted (Art. 13-5).

**13-5. The Throat Pressure Ratio.** When flow is adiabatic and without work, Eqs. (13-1a) and (13-2a) show that the stagnation enthalpy is constant:

$$dh_0 = dh + \frac{V\,dV}{g_c} = 0$$

If the flow is also reversible, Eq. (7-4b) yields

$$dh = v\,dp$$

By combining, the *Euler equation* is obtained:

$$v\,dp + \frac{V\,dV}{g_c} = 0 \bigg]_{\substack{rev \\ dq,dw\,=\,0}} \qquad (13\text{-}5)$$

The Euler and continuity equations yield, with Eq. (A-2b) (Prob. 12),

$$\frac{dA}{dp} = \frac{A}{\rho V^2/g_c}\left(1 - \frac{V^2}{a^2}\right) = \frac{A}{\rho V^2/g_c}(1 - \mathbf{M}^2) \qquad (13\text{-}6a)$$

Note in Fig. 13-1 that the slope $dA/dp$ is positive from the entrance to the throat, zero at the throat, and negative from the throat to the exit. From Eq. (13-6a),

$$\frac{dA}{dp} > 0 \quad \mathbf{M} < 1.0 \qquad \text{(subsonic velocities, entrance to throat)}$$

$$\frac{dA}{dp} = 0 \quad \mathbf{M} = 1.0 \qquad \text{(sonic velocity at throat)}$$

$$\frac{dA}{dp} < 0 \quad \mathbf{M} > 1.0 \qquad \text{(supersonic velocities, throat to exit)}$$

Equation (13-5) shows that pressure and velocity changes oppose each other ($dp/dV$ is negative). By combining Eqs. (13-5) and (13-6$a$),

$$\frac{dA}{dV} = -\frac{A}{V}(1 - \mathbf{M}^2) \tag{13-6$b$}$$

Hence

$$\frac{dA}{dV} < 0 \qquad \text{when } \mathbf{M} < 1 \text{ (subsonic flow)}$$

$$\frac{dA}{dV} > 0 \qquad \text{when } \mathbf{M} > 1 \text{ (supersonic flow)}$$

Figure 13-3 illustrates the possible flow configurations for accelerating (nozzle) and decelerating (diffuser) flow in the subsonic and supersonic regions.

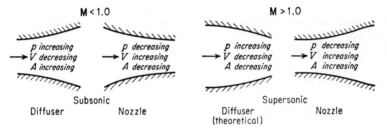

FIG. 13-3. Flow configurations for subsonic and supersonic flow.

When the sonic velocity is reached, both Eqs. (13-3$b$) and (A-2$b$) are valid:

$$V_{\text{rev}}^* = \sqrt{2g_c(h_0 - h^*)_{s=c}} = v\sqrt{-g_c\left(\frac{\partial p}{\partial v}\right)_s} \tag{$a$}$$

Unfortunately, an analytical relationship between $p$ and $v$ may not be available for fluids such as steam. However, the ideal-gas laws are usually adequate and ($a$) reduces (Prob. 13) to

$$r_p^* = \frac{p^*}{p_0} = \left(\frac{2}{k+1}\right)^{\frac{k}{k-1}} \tag{13-7}$$

Equation (13-7) is the means for predicting the throat pressure $p^*$ in a reversible adiabatic nozzle. The ratio $r_p^*$ is the *minimum throat pressure ratio*, often called the *critical pressure ratio* (a misleading name; Art. 10-3).

Equation (13-7) can be used as an approximation for real fluids. For the data of Example 1, values of $k$ can be computed to fit the relationship

$pv^k = C$ (or $k$ values can be selected from Fig. 8 of the Keenan and Keyes Steam Tables). If $k = 1.30$ is selected as the probable value,

$$r_p^* = \frac{p^*}{p_0} = \left(\frac{2}{k+1}\right)^{\frac{k}{k-1}} = 0.546$$
$$p^* = 100(0.546) = 54.6 \text{ psia}$$

And this value is confirmed by Fig. 13-1.

**Example 2.** Find the necessary dimensions for an ideal nozzle that will allow a mass-flow rate of 1 $\text{lb}_m$/sec of air from a large tank at 50 psia and 60°F to a discharge region at 10 psia.

*Solution.* Since the over-all pressure ratio is 0.1, it seems probable that a convergent-divergent nozzle is required. By Eq. (13-7),

$$r_p^* = \frac{p^*}{p_0} = \left(\frac{2}{k+1}\right)^{\frac{k}{k-1}} = \left(\frac{2}{2.4}\right)^{\frac{1.4}{0.4}} = 0.528$$
$$p^* = 50(0.528) = 26.4 \text{ psia}$$

The value of 26.4 psia is greater than the discharge pressure; therefore, a minimum area is necessary for reversible flow. The velocity at this throat is found by transforming Eq. (13-3b) with ideal-gas relationships:

$$V^* = \sqrt{2Jg_c(h_0 - h^*)} = \sqrt{2Jg_c c_p T_0\left(1 - \frac{T^*}{T_0}\right)} = \sqrt{2Jg_c c_p T_0[1 - (r_p^*)^{\frac{k-1}{k}}]}$$

Selecting $c_p$ from Fig. B-1 and noting that $\sqrt{2g_c J} = 223.8$ will give

$$V_{rev}^* = 223.8 \sqrt{0.24(520)[1 - (0.528)^{0.286}]} = 1,017 \text{ ft/sec}$$

The initial specific volume can be found by the equation of state:

$$v_1 = \frac{RT_1}{p_1} = \frac{1,545(520)}{29(50)(144)} = 3.86 \text{ ft}^3/\text{lb}_m$$

The specific volume at the throat is equal to

$$v^* = v_1\left(\frac{p_1}{p^*}\right)^{1/k} = 3.86(1.893)^{0.175} = 6.09 \text{ ft}^3/\text{lb}_m$$

By the continuity equation,

$$A^* = \frac{\dot{m}_f v}{V} = \frac{1.0(6.09)}{1,017} = 0.00598 \text{ ft}^2 \qquad Ans.$$

The area at the exit is found in a similar manner.

### 13-6. Mass-flow Rate of a Nozzle.

Consider *any* section in the converging part of a nozzle. As the velocity at this section is increased from zero (by decreasing the downstream pressure), the mass-flow rate is also increased. But when the downstream pressure is reduced to a value such that the sonic velocity is reached at the throat, no further increase in velocity can be obtained at any section in the converging part of the nozzle. Thus the mass-flow rate reaches its maximum value when the throat† section attains its maximum velocity—the sonic velocity.

In Fig. 13-4 the mass-flow rate is shown plotted against the pressure ratio at the throat of the nozzle. If it were to be assumed that $r_p$ could vary from 1 to zero, curve $ABC$ would result. But since $r_p$ at the throat could never fall below the critical ratio $r_p^*$, the curve from $B$ to $C$ is imaginary and the actual flow rate is shown by curve $ABD$.

A convergent nozzle operating in the region $A$ to $B$ presents no particular problem since the pressure at the throat is also the receiver pressure. But if the receiver pressure is reduced below the critical value, the pressure at the throat remains unchanged at the critical value! The fluid expands to this limiting pressure, then completes its expansion to the receiver pressure outside the nozzle. (And in so doing, supersonic

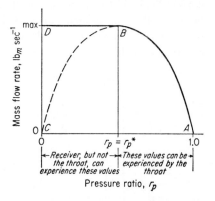

FIG. 13-4. Variation of mass-flow rate with throat-pressure ratio for nozzle or venturi.

velocities may be attained; but these velocities are downstream from the nozzle exit, and therefore the flow rate remains unchanged. In the convergent-divergent nozzle, the divergent section guides and controls the expansion in a much closer approach to reversibility, and therefore, much higher supersonic velocities are induced.)

**13-7. Phenomena in Converging-Diverging Passageways.** The converging-diverging passageway is characteristic of both a nozzle and a venturi (Fig. 6-2). Consider the flow conditions‡ in Fig. 13-5. Tests $A$ and $B$ illustrate *venturi flow:* a pressure at the exit slightly lower than that at the entrance causes a greatly decreased pressure at the throat. In the convergent section, expansion proceeds with the attainment of high velocity and low pressure at the throat. In the divergent section, compression occurs because the velocity is progressively reduced while the pressure is increased. This *diffuser action* continues as long as the throat pressure is greater than, or equal to, the critical value.

In test $A$ (and probably test $B$), the mass-flow rate will be less than

---

† Up to this point, the name *throat* has been applied only to the minimum area in the nozzle, where the velocity is sonic. From here on, a more general usage of the word throat will be followed, to designate the minimum area in any flow device. Thus the exit of a converging nozzle and the minimum area of a venturi are called *throats,* even though the velocities at these sections are not sonic.

‡ Construction drawings for this laboratory device can be found in Ref. 6.

that of the others, for here the minimum throat pressure has not been reached and the velocity at the throat has therefore not attained its maximum value.

When the pressure at the exit of the nozzle is less than the throat pressure and the exit area has been proportioned in accordance with the correct values for pressure and velocity, the expansion line for the nozzle will be a smooth curve such as $E$ (Fig. 13-5). Here the critical pressure ratio occurs at the throat, but lower pressures and higher velocities occur in the divergent section.

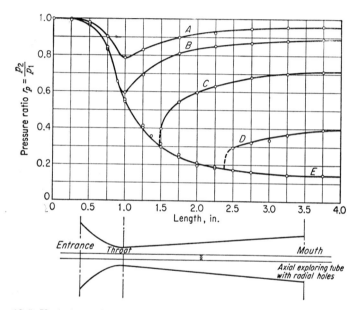

Fig. 13-5. Variations of pressure conditions in a convergent-divergent nozzle.

If the pressure at the exit is gradually raised, the fluid will persist in expanding almost to the former exit pressure. However, at some stage, a *compression shock* will occur when the supersonic stream strikes the higher-density fluid at or near the exit. When compression shock occurs, the velocity becomes subsonic and the fluid is irreversibly compressed. The pressure increases, but never to a degree that could be attained by an isentropic compression. Because of this irreversibility, the pressure after the shock may be lower than the pressure at the throat (curve $D$). Note that by the continuity equation the velocity can vary widely at any area because of the compensating effect of the specific volume. After the shock the diverging section of the nozzle acts as a diffuser, and the subsonic velocity is reduced to a still lower value while the pressure is increased to the discharge pressure.

Since the throat is not affected by the shock, which occurs downstream, the mass-flow rates are equal for tests $C$, $D$, and $E$ (and the exit pressures are all different).

Thus a convergent-divergent passageway may act either as a nozzle or as a venturi, and because of this, the throat pressure and the mass-flow rate cannot readily be predicted. For the convergent nozzle, either the pressure at the throat is that dictated by the critical ratio or else it is the exhaust pressure. For the convergent-divergent nozzle, the pressure at the throat is either the pressure dictated by the critical ratio or else (venturi) a higher pressure that is less than the exhaust pressure.

**13-8. Nozzle and Diffuser Coefficients.** The *efficiency* of the nozzle is defined as the ratio of the actual kinetic energy to the maximum value that could be realized by an isentropic expansion from the initial state to the final state. This ideal kinetic energy can be found from

$$\frac{V_{2\,rev}^2}{2Jg_c} = \frac{V_1^2}{2Jg_c} + (h_1 - h_2)_{s=C}$$

And by definition

$$\eta_n = \frac{\text{actual } KE}{\text{ideal } KE} = \frac{V_2^2}{V_{2\,rev}^2} = \frac{V_2^2/2Jg_c}{V_1^2/2Jg_c + (h_1 - h_2)_{s=C}} \qquad (13\text{-}8)$$

The *velocity coefficient* is defined as the ratio of the actual velocity to the ideal velocity:

$$C_V = \frac{\text{actual velocity}}{\text{ideal velocity}} = \frac{V_2}{V_{2\,rev}} = \sqrt{\eta_n} \qquad (13\text{-}9)$$

The nozzle is a remarkably efficient device, with the efficiency varying in general with the length and roughness of the inner surface. For convergent nozzles, efficiencies of 98 per cent and velocity coefficients of 99 per cent are quite usual values. Convergent-divergent nozzles must be quite long as compared with convergent nozzles if the cone angle of the divergent section is to be held to small values. The velocity in the divergent section is high, being greater than the acoustic velocity, and friction lose is increased. For this reason, the over-all efficiency of a convergent-divergent nozzsl is usually lower than that for a convergent nozzle, although values above 90 per cent are quite usual. If venturi action occurs and the velocity at the throat is not high, the compression process in the divergent section can be accomplished without seriously interfering with the efficiency values given above for convergent nozzles. However, if the throat velocity approaches the acoustic value, the compression is not readily performed without turbulence that causes lower efficiencies.

**Example 3.** Determine the dimensions for a nozzle that is to expand 1 $lb_m$/sec of steam from $p_0 = 100$ psia and $t_0 = 600°F$ to $p_2 = 20$ psia if the velocity coefficient from entrance to throat is assumed to be 0.99, and from entrance to exit 0.96. (Data are basically the same as in Example 1.)

*Solution.* The same method of solution should be used as in Example 1. Assume a pressure and find the velocity and specific volume at that pressure. With this information, the area corresponding to each pressure can be found. Suppose the first pressure selected is 54.6 psia, which was the throat pressure in Example 1:

$$h_0 = 1329.1 \text{ Btu/lb}_m \qquad h_2 = 1265.0 \text{ Btu/lb}_m \qquad V_{2\,\text{rev}} = 1{,}790 \text{ ft/sec}$$

With these data

$$V_{2'} = C_V V_{2\,\text{isen}} = 0.99(1{,}790) = 1{,}773 \text{ ft/sec}$$

But the actual velocity also equals

$$V_{2'} = \sqrt{2Jg_c(h_0 - h_{2'})} = C_V \sqrt{2Jg_c(h_0 - h_2)_{\text{isen}}} = \sqrt{2Jg_c\eta_n(h_0 - h_2)_{\text{isen}}}$$

Therefore,

$$h_0 - h_{2'} = C_V{}^2(h_0 - h_2)_{\text{isen}} = \eta_n(h_0 - h_2)_{\text{isen}} = (0.99)^2(64.1)$$
$$= 62.8 \text{ Btu/lb}_m$$
$$h_{2'} = h_0 - 62.8 = 1329.1 - 62.8 = 1266.3 \text{ Btu/lb}_m$$

Since enthalpy and pressure are known to be

$$p_2 = 54.6 \text{ psia} \qquad h_{2'} = 1266.3 \text{ Btu/lb}_m$$

the Steam Tables yield

$$v_{2'} = 9.946 \text{ ft}^3/\text{lb}_m$$

With these values the area is found from the continuity equation

$$A_{2'} = \frac{\dot{m}_f v}{V} = \frac{1.0(9.946)}{1{,}773} = 0.0056 \text{ ft}^2 \qquad Ans.$$

However, this area is not necessarily the throat area since the derivation of Art. 13-5 presupposed reversibility. But it is not unusual to assume that the answer found above is the corrected throat area if only because the equations in this chapter rest upon the assumption that the velocity distribution is uniform at any section of the nozzle (and, too, the amount of irreversibility is small). Actually, some nonuniformity must be present, and this nonuniformity of velocity will be increased by friction since friction will probably be initiated at the walls. These elementary equations, then, are not precisely true, and therefore it can be assumed, for simplicity, that the answer given above is the throat area. With this reasoning, it is not necessary to chart out the complete area-pressure relationships since a higher degree of precision is not warranted.

In the same manner the conditions at the mouth of the nozzle can be found, although here the pressure is fixed by the receiver.

A *coefficient* of *discharge* is defined as the ratio of the actual mass rate of flow to the ideal rate of flow that theoretically could be attained by a reversible expansion from the initial state to the final state:

$$C = \frac{\text{actual mass rate of flow}}{\text{ideal mass rate of flow}} = \frac{\dot{m}_{f\,\text{actual}}}{(VA/v)_{\text{isen}}} \qquad (13\text{-}10)$$

The actual flow rate is found by test while the ideal flow rate is calculated. In this manner the coefficient of discharge and the variations in the coefficient that will accompany any change in operating conditions are determined. If sufficient data of this kind are available, then it is possible to predict the value of the coefficient of discharge for a test and therefore find, quite simply, the mass-flow rate.

**13-9. Metastable Flow.** It has been assumed that the fluid in flow passes from one equilibrium state to another without lag from molecular readjustment (zero *relaxation time*). Actually, this assumption is not true, since time is required to redistribute the energy among the molecules. For example, the heat capacity (and $k$) values vary quite markedly with temperature (Fig. B-1). If the fluid passes quickly from one temperature to another, the relaxation time may be greater than the time of the process, and the heat capacities cannot adjust to the equilibrium values.

When the fluid is all gas or all liquid, the effects of relaxation are usually relatively insignificant. But when the flowing fluid reaches the state where a phase change is demanded, relatively great changes must be accomplished, and the lag in reaching equilibrium may be quite apparent.

When a fluid is expanded through a nozzle from the vapor region to the two-phase region, at some stage of the expansion the saturation state is reached and the vapor should begin to condense. However, tests show that the condensation may be delayed until a lower temperature and pressure are attained, at which time sudden condensation occurs in the form of a vast number of extremely small droplets. In the interval between the reaching of the saturation state and the beginning of condensation, the steam is in a *metastable state* and is said to be *supersaturated*.

Supersaturation of steam has been studied by several investigators. In the normal course of events condensation of steam begins when the kinetic energy of the molecules is not sufficiently great to overcome the attractive forces between molecules; here and there throughout the gas, molecules are attracted to each other and group together to form a nucleus for further growth. As these molecular groups get larger, they become tiny droplets of condensate. Such a molecular process does not require a great amount of time, but expansion in a nozzle happens in an extremely small interval of time, and the temperature may be falling at a faster rate than the speed of agglomeration. When this delay in condensation occurs, the driving force for the molecular grouping is greatly increased because all through the gas will be molecules with insufficient energy to resist the molecular interattraction. Condensation begins at a rapid rate because, for the pressure experienced, the temperature is lower than that demanded for phase equilibrium.

From tests[3] on convergent-divergent nozzles, it can be concluded that, for initial pressures up to 300 psia, the condensation pressure is below the critical pressure ratio computed for superheated steam; for this reason condensation occurs in the divergent section of the nozzle. Thus, supersaturation prevails beyond the throat of a nozzle, and the flow should be calculated on the basis of the metastable equilibrium instead of on the basis of thermal equilibrium between vapor and liquid. To calculate the state of supersaturated steam, the properties of the vapor can be extrapolated into the metastable region by use of the equation of state for the vapor (and the ideal-gas equations are the simplest approach). In other words, although a phase boundary is crossed in the expansion, condensation is delayed, and the ideal-gas equations are closer approximations to the supersaturated state than the equilibrium data of the Steam Tables.

The flow of the metastable fluid from entrance to throat of the nozzle is closely reversible since, conceivably, a diffuser would allow the initial state to be essentially regained. With the onset of condensation the process becomes highly irreversible since the same series of states could not be retraced, and the change of state into dry vapor would occur at a temperature higher than the condensation temperature experienced in the expansion process.

**Example 4.** Saturated steam at 50 psia flows through a round nozzle, with throat diameter of 1 in., to a discharge pressure of 30 psia. Calculate the ideal velocity and the ideal flow rate if velocity of approach is zero and phase equilibrium is assumed at each stage of the expansion.

*Solution.* If phase equilibrium is maintained, condensation will begin at the very start of the expansion, and the $pv$ relationship will follow, approximately, the empirical equation $pv^{1.13} = C$. The critical ratio is closely 0.58 [Eq. (13-7)], and the corresponding pressure is $p = 50(0.58) = 29$ psia. In this case, the nozzle can be made convergent because the exit pressure of 30 psia is above the value for the critical ratio. From the Steam Tables,

| *At 50 psia* | *At 30 psia* | |
|---|---|---|
| $s_0 = 1.6585$ Btu/$(lb_m)(°F)$ | $s_f = 0.3680$ | $s_{fg} = 1.3313$ |
| $h_0 = 1174.1$ Btu/$lb_m$ | $v_g = 13.746$ | $h_{fg} = 945.3$ |
| $v_0 = 8.515$ ft$^3$/$lb_m$ | | $h_f = 218.82$ |

Solving for the state at $p = 30$ psia, $s = 1.6585$,

$$x = \frac{1.6585 - 0.3680}{1.3313} = 0.97$$
$$v_2 \approx 13.746(0.97) = 13.3 \text{ ft}^3/lb_m$$
$$h_2 = 218.82 + 0.97(945.3) = 1134.8 \text{ Btu}/lb_m$$

Substituting these values in Eq. (13-3b) (and $\sqrt{2g_c J} = 223.8$),

$$V_{2\text{ rev}} = 223.8 \sqrt{1,174.1 - 1,134.8} = 1,398 \text{ ft/sec}$$

The ideal mass-flow rate for the equilibrium expansion is

$$\dot{m}_f = \frac{AV}{v} = \frac{(\pi/4)(1/144)1,398}{13.3} = 0.574 \text{ lb}_m/\text{sec} \qquad Ans.$$

**Example 5.** Repeat Example 4, but assume that the flow is supersaturated and that the perfect-gas equations will adequately portray the true conditions.

*Solution.* Equation (13-3$b$) is expanded into the form

$$V_{2\,\text{rev}} = \sqrt{\frac{2g_c k}{k-1}\, p_0 v_0 \left[1 - \left(\frac{p_2}{p_0}\right)^{\frac{k-1}{k}}\right]}$$

Substitute

$$p_0 = 50(144)\ \text{lb}_f/\text{ft}^2 \qquad v_0 = 8.515\ \text{ft}^3/\text{lb}_m \qquad k = 1.31$$

(where $k$ is found from Fig. 8, Keenan and Keyes Steam Tables)

$$V_{2\,\text{rev}} = 8.02\,\sqrt{\frac{1.31}{0.31}\,7200(8.515)[1 - (0.6)^{0.237}]} = 1{,}372\ \text{ft/sec}$$

Hence, the velocity attained in the supersaturated state is less than the velocity for phase-equilibrium expansion.

The specific volume of the gas at the throat is

$$v_2 = v_0 \left(\frac{p_0}{p_2}\right)^{\frac{1}{k}} = 8.515(1.667)^{0.765} = 12.55\ \text{ft}^3/\text{lb}$$

Note that the specific volume is less than that found in Example 4; therefore, the density is greater.

The mass-flow rate is

$$\dot{m}_f = \frac{AV}{v} = \frac{(\pi/4)(1/144)1{,}372}{12.55} = 0.597\ \text{lb}_m/\text{sec} \qquad Ans.$$

The mass-flow rate is greater than that found in Example 4. In actual test the value of 0.597 lb$_m$/sec would be approached and not 0.574 lb$_m$/sec; hence if the test had been based on the latter value as the ideal value, the discharge coefficient would approach

$$C = \frac{0.597}{0.574} = 1.04$$

Coefficient of discharge over 1.0, such as this, led to the study of supersaturation.

The temperature at the throat can be calculated from the polytropic relationship

$$\frac{T_2}{T_1} = \left(\frac{p_2}{p_1}\right)^{\frac{k-1}{k}} \qquad \text{or} \qquad T_2 = 741(0.887) = 658°R\ \text{or}\ 198°F$$

The phase-equilibrium temperature corresponding to the throat pressure of 30 psia can be found from the Steam Tables:

$$t = 250°F$$

Hence, the metastable undercooling is $250 - 198 = 52°F$.

**13-10. Compressible and Incompressible Flow.** Let ideal-gas relationships be applied to Eq. (13-3$b$):

$$V_{2\,\text{rev}}^2 = 2g_c(h_0 - h_2) = 2g_c c_p T_0 \left(1 - \frac{T_2}{T_0}\right) = 2g_c c_p T_0 \left[1 - \left(\frac{p_2}{p_0}\right)^{\frac{k-1}{k}}\right]$$

And with Eq. (11-1$a$),

$$V_{2\,\text{rev}} = \sqrt{\frac{2g_c k R T_0}{k-1}\left[1 - \left(\frac{p_2}{p_0}\right)^{\frac{k-1}{k}}\right]} \qquad (13\text{-}11a)$$

When Eq. (13-11a) is used for pressure ratios near unity, the difference term approaches zero. Any slight error in calculating this difference leads to relatively large errors in the computed velocity. To remedy this difficulty, the term can be expanded into a power series that is rapidly convergent (Prob. 20):

$$V_{2\,rev} = \sqrt{2g_cRT_0\left[\frac{p_0 - p_2}{p_0} + \frac{1}{2k}\left(\frac{p_0 - p_2}{p_0}\right)^2 + \cdots\right]} \qquad (13\text{-}11b)$$

For small pressure differences, only the first term within the brackets may be significant (and therefore $p_0$ is essentially $p_1$). Hence

$$V_{2\,rev} \approx \sqrt{2g_cv_1(p_1 - p_2)} \qquad (13\text{-}11c)$$

The pressure difference could be balanced by a column of the flowing fluid H ft in height and of constant density $\rho$:

$$p_1 - p_2 = \rho H \frac{g}{g_c}$$

This expression can be substituted into Eq. (13-11c) (which is an approximate formula for compressible fluids) to obtain the familiar hydraulic formula (which is an exact formula for incompressible fluids):

$$V_{2\,rev} = \sqrt{2gH} \qquad (13\text{-}12a)$$

where H is the *head* of flow in feet of the fluid flowing.

The *head* is indirectly found by measuring the pressure with a manometer that uses a fluid other than that flowing. By equating the system head to that of the manometer,

$$\rho H \frac{g}{g_c} = Z_m(\rho_m - \rho)\frac{g}{g_c}$$

$$H = \frac{Z_m(\rho_m - \rho)}{\rho} \qquad (13\text{-}12b)$$

where $Z_m$ = manometer reading, ft
$\rho_m$ = density of fluid in manometer, $lb_m/ft^3$
$\rho$ = density of fluid flowing, $lb_m/ft^3$

**13-11. Standards for the Venturi and the Flow Nozzle.** In many applications, the measuring device is a permanent part of an industrial system; therefore, the device should not introduce a loss, for a loss will cause an additional expense in pumping costs. For this reason, the *venturi* is usually preferred as a means for measuring flow and is the standard meter for water or steam flows in many industrial applications. In Fig. 13-6 a venturi and manometer are illustrated. The venturi is made of cast iron or steel, with bronze or monel linings to minimize corrosion. Most styles use a piezometer ring or annular space at both the inlet and throat communicating with the flow chamber by a series of radially drilled holes. This construction ensures that the

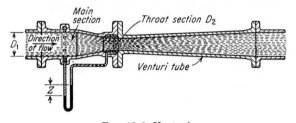

FIG. 13-6. Venturi.

average pressure is being measured, and it also acts as a safety factor against the possibility of one hole becoming plugged. The approach to the throat is a frustrum of a cone with angle of 25 to 30°, the diameter of the throat being one-fourth to one-half of the pipe diameter, and the length being one-half of its own diameter. The downstream section or diffuser is a cone with angle of 7° or less. The over-all loss in pressure can be estimated to be roughly 10 per cent of the differential pressure. The coefficient of discharge is affected mainly by the Reynolds number, and for high **Re**, that is, highly turbulent flow conditions, a value of 0.98 can be used as an approximate value.

A *flow nozzle* consists of a short cylinder with one end flared to form the entrance to the nozzle (Fig. 13-7). The purpose of the cylindrical throat is to ensure that the fluid will leave the nozzle without contraction in order that the area of the throat and the jet may be considered equal. In effect, the flow nozzle is a venturi without a diffuser; hence, the loss of head from failure to recompress the high-velocity flow may vary from 30 to 90 per cent of the differential. However, the coefficient of discharge is essentially that for the venturi, say 0.98.

**13-12. Standards for the Orifice.** Probably the oldest device for measuring the mass-flow rate is the thin-plate orifice constructed in the manner of Fig. 13-8. The orifice is installed with the sharp edge (or, better, *square edge*) on

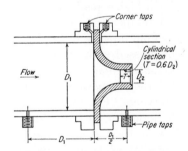

FIG. 13-7. Long-radius flow nozzle.

the upstream side and beveling, if any, on the downstream side. The pressure in front of the orifice may be slightly higher than the true pressure because of impact. In passing through the orifice the pressure drops abruptly, reaching a minimum value at the *vena contracta* or location of smallest jet diameter. At this point all filaments of the flow become parallel. Beyond this point the pressure increases as the fluid is decelerated, and this action is accompanied by considerable turbulence from lack of a diffuser section. Hence the downstream pressure is considerably lower than for a venturi and of the same order as for a flow nozzle.

In flowing through the orifice plate, the jet contracts to an area of about 0.6 that of the orifice, and therefore the discharge coefficient, which is based upon the orifice area, is much less than unity. The orifice must be carefully constructed if predetermined discharge coefficients are to be used without a calibration test because, for example, rounding of the inlet corner will decrease the jet contraction, thus increasing the real flow area and increasing the quantity of flow. The phenomenon of the critical ratio does not appear for sharp-edged orifices, because experiments show that the mass-flow rate continues to increase as the pressure ratio is decreased below the critical (Ref. 5).

A coefficient of discharge of 0.60 can be used for routine calculations.

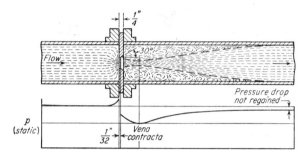

FIG. 13-8. Flow through thin-plate sharp-edge orifice.

## 13-13. Temperature of a Moving Stream.

The true temperature of a moving fluid is the temperature that would be shown by a measuring instrument moving with the same velocity as the fluid. However, the practical measurement of temperature is accomplished by instruments that are at rest, and the mass-flow impact on the measuring device will give a higher reading of temperature than the true reading.

A temperature-measuring device in a high-velocity stream will cause the fluid directly in front of the obstacle to be more or less isentropically compressed with attainment of a temperature that would approach the stagnation temperature. However, the thermal meter may not measure the true stagnation temperature because a number of other factors will intervene. For example, there will be radiation loss from the hotter thermometer to the colder gas and walls; frictional effects of the fluid flow on the indicator will be affected by properties of the fluid, such as viscosity, while the shape and relative position of the thermometer in the fluid stream will also enter the problem. For these reasons the measured temperature may be greater or less than the stagnation temperature. The recovery factor of the measuring device is defined as

$$r_{\text{test}} = \frac{T_{\text{measured}} - T}{T_{\text{stagnation}} - T} \tag{13-13}$$

For air and most gases the recovery factor is less than 1. Twisted wire thermocouples in air have recovery factors[†] of 0.73 to 0.84. The higher

---

[†] H. C. Hottel and A. Kalitinsky, Temperature Measurements in High Velocity Air Streams, *J. Appl. Mech.*, **12**(1): A25–A32 (March, 1945).

values are obtained by axial flow over the couple, and the lower values are for transverse flow (see, also, Art. 19-8*e*).

For high velocities, probes are used to assist in the deceleration of the fluid and to guard against radiation errors.    Figure 13-9 shows a probe for high-velocity stream measurements under conditions where the stream temperature and wall temperature are of the same order.    In this probe the high-velocity fluid is brought essentially to rest and the temperature measured by the thermocouple. (Bleed holes allow a small but continuous flow past the thermocouple.)    Radiation losses

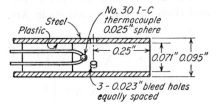

FIG. 13-9. Stagnation temperature probe (Pratt and Whitney design).

are negligible because of the shielding effect of the enclosed tube, which also acts as a diffuser.    Recovery factors of 0.98 are reported for this probe for velocities under and slightly over the acoustic velocity.

The problem is more complicated if the temperature of the fluid is high compared with that of the wall, for radiation losses will increase considerably.    To reduce such losses, multiple shields must be used (Chap. 19, Prob. 45).

**13-14. Pressure of a Moving Stream.**    The true static pressure of a fluid in motion is relatively easy to measure compared with the true static temperature.    Either a piezometer ring or a small radial hole in the pipeline will enable the pressure to be measured.    However, the hole must be so located that no opportunity exists for impact, or a higher pressure will be encountered.    If the flow is pulsating, standing waves of pressure variations may exist, and erratic static pressures will be experienced along the length of the pipe.

It would appear that a diffuser would need to be constructed as a means for measuring the stagnation pressure, but experience shows that better results are obtained by allowing the fluid stream to form its own diffuser. This is accomplished by inserting a small blunt body into the flow so that the fluid stream is divided without undue turbulence.    By so doing, fluid approaching the center line of the body undergoes a deceleration which is almost reversible (at subsonic velocities), and the impact pressure at this stagnation point may approach the ideal stagnation pressure. The impact pressure is measured by locating a small opening near the stagnation point and connecting this opening by a tube to a pressure gauge.

An instrument which creates a stagnation point and also communicates the pressure at this point to a pressure gauge is called a *pitot tube*.    One type, Fig. 13-10*a*, is designed only for impact pressures, while another type, Fig. 13-10b, allows both static and impact pressures to be measured. (The arrangement in Fig. 13-10a is preferable, however, since the static pressure is measured upstream from the pitot tube and therefore disturbances arising from the instrument in the flow stream are avoided.)

The decelerating process is the direct opposite of nozzle expansion, and therefore the equations in Art. 13-10 appear as

$$V_{1\,\text{rev}} = \sqrt{\frac{2g_c k R T_0}{k-1}\left[1 - \left(\frac{p_1}{p_0}\right)^{\frac{k-1}{k}}\right]} \tag{13-14a}$$

$$V_{1\,\text{rev}} = \sqrt{2g_c R T_0\left[\frac{p_0 - p_1}{p_0} + \frac{1}{2k}\left(\frac{p_0 - p_1}{p_0}\right)^2 + \cdots\right]} \tag{13-14b}$$

$$V_{1\,\text{rev}} = \sqrt{2gH} \tag{13-14c}$$

where $V_1$ = velocity approaching pitot tube, ft/sec
  $p_1$ = static pressure
  $p_0$ = stagnation pressure
  $T_0$ = stagnation temperature

In the case of a circular duct, it is customary to divide the cross section into one central area and four concentric rings each of equal area. By placing the pitot tube at the mean radius of each of these areas, the velocity pressure for that area is found. The locations† of these points in a duct of radius $R$ are $0.316R$, $0.548R$, $0.707R$, $0.837R$, and $0.949R$. It is better practice to take these readings on both sides of the center line of the duct; thus, there are 10 readings in all, giving rise to the descriptive name *ten-point method*. For some work two traverses are made at right angles to each other, and thus there are 20 readings.

**Example 6.** A 4-ft-diameter duct carrying air of density 0.0736 lb$_m$/ft$^3$ is traversed by a pitot tube using the ten-point method. The readings in inches of water at 72°F from one side of the duct to the other are, respectively, 0.210, 0.216, 0.220, 0.219, 0.220, 0.220, 0.218, 0.219, 0.220, and 0.216. Find the average velocity and the mass flow (local $g = g_0$).

† Consider that 10 equal areas are desired to be established in a circular duct of radius $R$. Then,

$$\pi r_1^2 = \pi(r_2^2 - r_1^2) = \pi(r_3^2 - r_2^2) \cdots \pi(R^2 - r_9^2)$$

If these areas are equal,

$$r_2^2 = 2r_1^2 \qquad r_3^2 = 3r_1^2 \cdots R^2 = 10r_1^2$$

Therefore,

$$r_1 = \frac{1}{\sqrt{10}}R \qquad r_2 = \frac{\sqrt{2}}{\sqrt{10}}R \qquad r_3 = \frac{\sqrt{3}}{\sqrt{10}}R \qquad r_4 = \frac{\sqrt{4}}{\sqrt{10}}R \qquad \text{etc.}$$

And now radii $r_1$, $r_3$, $r_5$, etc., are the mean values for *five* equal areas:

$$r = 0.316R, \ 0.548R, \ 0.707R, \ \text{etc.}$$

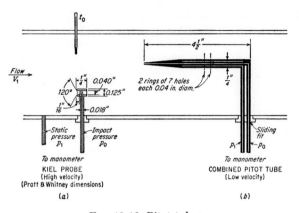

Fɪɢ. 13-10. Pitot tubes.

*Solution.* The velocity at each point can be calculated and then averaged, but it is quicker to average the square roots of the readings:

$$\sqrt{Z_m} = \frac{\sqrt{0.210} + \sqrt{0.216} + \cdots + \sqrt{0.216}}{10} = 0.4668$$

$$\text{or} \qquad Z_m = 0.216 \text{ in. } H_2O$$

The densities of water and air are

$$\rho_m = 62.3 \text{ lb}_m/\text{ft}^3 \qquad \rho = 0.0736 \text{ lb}_m/\text{ft}^3$$

By Eq. (13-12b),

$$H = \frac{Z_m(\rho_m - \rho)}{\rho} = \frac{0.216(62.3 - 0.074)}{(12)0.074} = 15.1 \text{ ft of fluid flowing}$$

By Eq. (13-14c),

$$V = \sqrt{2gH} = \sqrt{64.4(15.1)} = 31.2 \text{ ft/sec} \qquad Ans.$$

while, by the continuity equation,

$$\dot{m}_f = \frac{AV}{v} = \frac{\pi}{4}(4^2)(31.2)(0.0736) = 29 \text{ lb}_m/\text{sec} \qquad Ans.$$

**13-15. Jet Pumps.** Ejectors are commonly used as vacuum pumps because of their low cost, simplicity, and dependability. The principle of the ejector depends upon the entrainment of a gas by a high-velocity fluid jet, the resulting mixture being compressed in a diffuser and discharged at a pressure higher than that in the gas chamber. Figure 13-11 is a diagrammatic sketch of a single-stage ejector that can be used to produce a vacuum of 26 in. of mercury. Steam enters the ejector at high pressure and expands in a convergent-divergent nozzle. The high-velocity fluid leaving the nozzle passes through the air chamber and entrains part of the air to be evacuated. The mixture of air and fluid then enters a convergent-divergent diffuser that recompresses the mixture to the discharge pressure. The convergent-divergent diffuser is necessary because in normal operation the pressure in the air chamber will be low and the velocity of the steam jet will be supersonic. To recompress such a stream the diffuser must first converge and then diverge as the velocity falls below the acoustic.

For creating a low vacuum or discharging to a higher back pressure, several ejectors can be connected in series and called, respectively, first, second, etc., stages with the last stage discharging directly or indirectly into the atmosphere. A small condenser called the *inter-condenser* can be used between stages to condense the steam of each stage, thus relieving subsequent stages that need only compress those gases that are noncondensable.

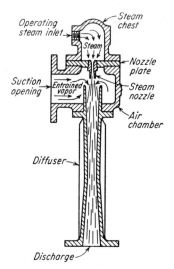

A jet pump used to pump water is called an *injector*. In Fig. 13-12 steam from the boiler passes to the injector and expands in the nozzle to a low pressure while acquiring a high velocity. Water is drawn into the chamber around the nozzle and accelerated in the combining tube by the steam jet, which is itself condensed. The mixture of water and steam is compressed in the delivery tube while being slowed down, and a pressure is finally achieved that can be greater than the boiler pressure. At first glance such action may seem improbable, but note that the expansion of the steam transforms available energy of amount $-\int v\,dp$ *for a vapor* into kinetic energy. This kinetic energy is used to lift and accelerate a water column, and the action is most inefficiently done because the process is highly irreversible. But the kinetic energy necessary for the compression *of the liquid* ($-\int v\,dp$ for the condensed steam and pumped water) back to the initial pressure is but a small fraction of the energy that was available, and no trouble is experienced in achieving pressures much higher than the initial pressure. This is but another example of Art. 5-8d; the work which must be done to compress a vapor or which is realized from expanding a vapor is much greater than the work required to compress a liquid.

FIG. 13-11. Steam ejector (El-liott Co.).

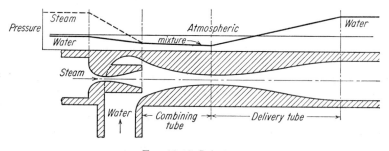

FIG. 13-12. Injector.

When the injector is started, condensation of the steam will not occur until water enters the combining tube and the discharge escapes through an overflow pipe (not illustrated). After condensation begins, the pressure will build up and delivery can be made to the boiler. If hot water is led to the injector, vaporization of the water occurs as the pressure is decreased and satisfactory operation is difficult to achieve.

It should be realized that mixing hot and cold fluids results in a loss in available energy. A boiler installation not using feedwater heaters must pump cold water into the boiler. An injector combines in part this irreversible operation with the pumping, but the inefficiency of the injector is no greater loss than the inefficiency of the mixing process. For this reason injectors are used where cold water is pumped into steam boilers such as the locomotive boiler. The efficiency of the injector as a pump is of the order of 1 per cent, but this is immaterial because the water is introduced into the boiler without supplying work from an external source. However, in modern plants, to reduce the loss in availability that is caused by mixing hot and cold fluids, the feedwater is passed through numerous feedwater heaters before the water enters the boiler. Since the injector needs cold water for best operation, in most plants it is more convenient to use centrifugal pumps.

**13-16. Isentropic Flow Tables.** For steady flow through a nozzle or diffuser, the process is completely specified by:
1. The continuity equation
2. The energy equation
3. The restrictions of reversible flow without heat or work
It then follows that all stagnation ($h_0$, $T_0$, $p_0$, $\rho_0$, $s = s_0$, etc.) and sonic ($p^*$, $\rho^*$, $V^*$, $A^*$, etc.) properties are constants since there is but one stagnation state and but one sonic state for a specified flow state.

For simplicity in solving problems, other restrictions are added:
4. One-dimensional flow
5. Ideal gas (with fixed $k$)
The properties of the flow stream can be expressed as functions of a reference state (such as the stagnation or the sonic state) and the Mach number. For example, the temperature and the stagnation temperature are related by Eq. (13-1a):

$$h_0 = h + \frac{V^2}{2g_c}$$

Then for ideal gases with constant heat capacities,

$$T_0 = T + \frac{V^2}{2g_c c_p}$$

The sonic velocity is substituted, Eq. (A-2c), and with Eq. (11-1a),

$$\frac{T_0}{T} = 1 + \frac{k-1}{2} \mathbf{M}^2 \tag{13-15}$$

The advantage of Eq. (13-15) is that the ratio $T_0/T$ can be computed and tabled with Mach number as the single independent variable ($k$ is a constant for a specified table). Then to find the ratio of two temperatures $T_1$ and $T_2$ at different sections of the flow stream from Eq. (13-15), it is merely necessary to divide one tabled value by the other, since $T_0$ is constant for a particular problem:

$$\frac{T_2}{T_1} = \frac{T_2}{T_0}\frac{T_0}{T_1} = \frac{1 + [(k-1)/2]M_1^2}{1 + [(k-1)/2]M_2^2} \tag{13-16}$$

For example, let $T_1 = 520°R$ at a section where $M_1 = 0.3$. What will $T_2$ be when $M_2 = 0.5$? From Table B-17 (for $k = 1.4$),

$$M = 0.3 \qquad \frac{T}{T_0} = 0.982 \qquad\qquad M = 0.5 \qquad \frac{T}{T_0} = 0.952$$

Hence
$$\frac{T_2}{T_1} = \frac{T_2}{T_0}\frac{T_0}{T_1} = \frac{0.952}{0.982} = 0.97 \qquad \text{or} \qquad T_2 = 504°R$$

and calculation of the quantities $1 + [(k-1)/2]M^2$ is avoided.

Table B-17 also uses the sonic state as a reference. For example, to find the ratios of two areas at two sections of the flow stream,

$$\frac{A_2}{A_1} = \frac{A_2}{A^*}\frac{A^*}{A_1} = \frac{M_1}{M_2}\left\{\frac{1 + [(k-1)/2]M_2^2}{1 + [(k-1)/2]M_1^2}\right\}^{\frac{k+1}{2(k-1)}} \tag{13-17}$$

Hence calculation of the $M$, $k$ fraction on the right side of Eq. (13-17) can be avoided.

**Example 7.** Find the dimensions for an ideal nozzle that will allow a mass-flow rate of 1 $lb_m/sec$ of air from a large tank at 50 psia, 60°F, to a discharge region at 10 psia (same data as in Example 2).

*Solution.* These data and $pv = RT$ yield

$$p_0 = 50 \text{ psia} \qquad T_0 = 520°R \qquad \rho_0 = 0.259 \text{ lb}_m/\text{ft}^3 \qquad p_2 = 10 \text{ psia}$$

At the throat $M = 1$, and from Table B-17 for $k = 1.4$,

$$\frac{p}{p_0} = \frac{p^*}{p_0} = 0.528 \qquad \frac{T}{T_0} = \frac{T^*}{T_0} = 0.833 \qquad \frac{\rho}{\rho_0} = \frac{\rho^*}{\rho_0} = 0.634$$

Hence
$$p^* = 26.4 \text{ psia} \qquad T^* = 433°R \qquad \rho^* = 0.164 \text{ lb}_m/\text{ft}^3$$

The velocity at the throat is found by Eq. (A-2d):

$$V^* = a = 49.1 \sqrt{T^*} = 1,017 \text{ ft/sec}$$

And by the continuity equation,

$$A^* = \frac{\dot{m}_f}{V^*\rho^*} = \frac{1}{1,017(0.164)} = 0.00598 \text{ ft}^2 \qquad Ans.$$

At exit,

$$\frac{p_2}{p_0} = \frac{10}{50} = 0.20$$

Interpolating in Table B-17, $\qquad M_2 = 1.71$

and $\qquad \frac{V}{V^*} = 1.487 \qquad \frac{T}{T_0} = 0.631 \qquad \frac{A}{A^*} = 1.35 \qquad \frac{\rho}{\rho_0} = 0.317$

Hence

$$V_2 = 1,513 \text{ ft/sec} \qquad T_2 = 328°R \qquad A_2 = 0.00808 \text{ ft}^2 \qquad \rho_2 = 0.082 \text{ lb}_m/\text{ft}^3$$

The answers and related data agree with the values found in Example 2.

**13-17. Fanno Flow.** When flow is steady, without heat or work or change in elevation, the *energy equation* reduces to

$$h_0 = h + \frac{V^2}{2g_c}\Bigg]_{\Delta q, w, Z=0} \tag{13-1a}$$

With the *continuity equation* in the form for constant-area flow,

$$G \equiv \frac{\dot{m}_f}{A} = \frac{V}{v} \tag{12-12}$$

Eq. (13-1a) is converted to

$$h_0 = h + \frac{G^2 v^2}{2g_c}\Bigg]_{\substack{\Delta q, w, Z=0 \\ A=C}} \tag{13-18}$$

Equation (13-18), called the *Fanno equation*, defines the states achieved by adiabatic flow at constant area with fluid (mechanical) friction (Art. 12-5).

Suppose that Eq. (13-18) is solved for three specified values of $G$ (other than zero), with fixed $h_0$, for a fluid such as steam (a laborious task), as illustrated in Fig. 13-13. Each path, since it has a different mass-flow rate per unit area, has a different amount of friction, inherent within the value of $G$, arbitrarily assigned. (The length of pipe, however, and

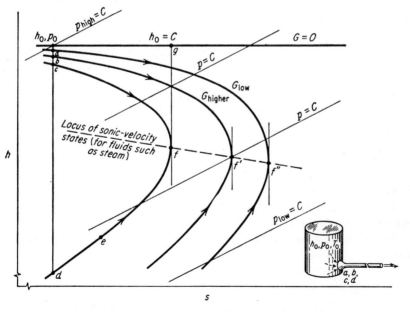

FIG. 13-13. Fanno lines on $hs$ diagram (schematic).

therefore the law of friction imposed, is not specified by the diagram.) Since the effect of friction is to *increase* the entropy, the flow directions in an adiabatic process must be as shown. Also, observe from Eq. (13-1a) that lines of constant enthalpy are also lines of constant velocity; it follows that fluid flow in an upper branch (*cf*, for example) must be at a lower velocity than that in the corresponding lower branch (*def*, for example). A few trial calculations will show that branches such as *cf* have subsonic velocities, branches such as *def* have supersonic velocities, and the common points $f$, $f'$, $f''$ at maximum entropy have the sonic velocities.

It follows from Fig. 13-13 that *in subsonic flow the pressure decreases in the direction of flow with consequent acceleration; in supersonic flow the pressure increases in the direction of flow with consequent deceleration. The limiting velocity for either type of Fanno flow is the sonic velocity.*

Note, in Fig. 13-13, that all paths approach as a limit the horizontal locus of $h_0 = C$, which is the Fanno line of zero flow and, therefore, the locus of stagnation pressures (which decrease continually in the direction of flow). The locus of sonic-velocity states for fluids such as steam is not at constant enthalpy (as it is for ideal gases) since the sonic velocity is a function of both temperature and pressure. Note, too, that adding pipe adds restriction, thus decreasing $G$, and therefore the acoustic velocity will occur at a lower exit pressure.

To achieve the $G$ values specified for states $a$, $b$, $c$, $d$, a nozzle expansion must take place between the pipe and the reservoir at $h_0$ and $p_0$; a convergent nozzle is sufficient to pass from $h_0$, $p_0$ to $a$, $b$, $c$ and a convergent-divergent nozzle from $h_0$, $p_0$ to $d$. This process, represented by the reversible adiabatic $h_0p_0abcd$, is not at constant area and, therefore, it is not at constant $G$.

In summary, Fanno flow is specified by:

1. The continuity equation
2. The energy equation
3. The restrictions of constant area without work or heat

It follows that:

*a.* All sonic properties ($p^*$, $\rho^*$, $V^*$, $A^*$, etc.) are constants since there is but one sonic state for a specified process (Fig. 13-13).

*b.* The stagnation properties *at the sonic state* are also constants for a specified process ($g$, for example, in Fig. 13-13).

For simplicity in solving problems, other restrictions are added as before:

4. One-dimensional flow
5. Ideal gas (with fixed $k$)

By following the same procedures as in Art. 13-16, the properties of the flow stream, relative to sonic-state properties, can be expressed as functions of the Mach number.

For example, the pressure at any state $x$ on a Fanno line can be related to the sonic pressure by

$$\frac{p_x}{p^*} = \frac{1}{M_x} \sqrt{\frac{k+1}{2\left(1 + \frac{k-1}{2}M_x^2\right)}} \qquad (13\text{-}19a)$$

Values for the Mach number relationships are found in Table B-18. Then, as before,

$$\frac{p_y}{p_x} = \frac{p_y}{p^*}\frac{p^*}{p_x} = \frac{M_x \sqrt{1 + \frac{k-1}{2}M_x^2}}{M_y \sqrt{1 + \frac{k-1}{2}M_y^2}} \qquad (13\text{-}19b)$$

and the labor of solving the right side of Eqs. (13-19) is avoided.

In particular, Eq. (12-11a) can be rearranged and integrated (a messy job, at best) to yield

$$4f\frac{L^* - L}{D} = \frac{1 - M^2}{kM^2} + \frac{k+1}{2k}\ln\frac{(k+1)M^2}{2\left(1 + \frac{k-1}{2}M^2\right)}$$

The right side of this equation is also evaluated in Table B-18. It then follows that between sections $L_1$ and $L_2$ (and $f$ is a mean value)

$$\frac{4f}{D}(L_2 - L_1) = 4f\frac{L^* - L_1}{D} - 4f\frac{L^* - L_2}{D} \qquad (13\text{-}20)$$

With Eq. (13-20) and Fig. 12-2, problems involving friction are simplified.

**Example 8.** Determine the length of 6-in.-ID commercial steel pipe required to change the flow of air from $M = 0.2$ to $M = 0.4$ in Fanno flow.

*Solution.* Figure 12-2 shows that

$$\epsilon = 0.00015 \qquad \text{or} \qquad \frac{\epsilon}{D} = 0.0003 \qquad \text{hence } 4f \approx 0.015$$

From Table B-18,

$$M = 0.2 \qquad \frac{4fL}{D} = 14.5 \qquad\qquad M = 0.4 \qquad \frac{4fL}{D} = 2.31$$

or

$$\frac{4f}{D}(L_2 - L_1) = \frac{0.015}{0.5}(L_2 - L_1) = 14.5 - 2.31$$

$$L_2 - L_1 = 407 \text{ ft} \qquad Ans.$$

**13-18. One-dimensional Shock.** It is well known that a stationary shock can be set up in a high-velocity stream (Figs. 13-5 and A-2) that is, closely, a plane discontinuity in the flow. Since the shock occurs at essentially constant area, the state of the fluid before and after the change must lie on a Fanno line of Fig. 13-13. Now a fluid in flow must obey *all* the laws of science; therefore, the momentum equation will be selected to give added information:

$$F_{\text{wall}} + p_1A_1 - p_2A_2 = \frac{\dot{m}_f}{g_c}(V_2 - V_1) \qquad (13\text{-}4b)$$

Equation (13-4b) holds, of course, for the entire Fanno flow. Since area is constant, $F_{\text{wall}}$ arises only from fluid friction. However, at those particular sections where a shock is permissible, the discontinuity occurs in a very short length of duct, and friction forces are not only negligible but become zero when the discontinuity is assumed to be plane. Hence $F_{\text{wall}}$ is set equal to zero and the equation manipulated to give (with the help of the continuity equation)

$$\frac{p_2}{p_1} = \frac{1 + k\mathbf{M}_1{}^2}{1 + k\mathbf{M}_2{}^2}\bigg]_{\text{plane section}} \tag{a}$$

Equation (a) is the new restriction placed upon Fanno flow for the sections where shock is possible. By combining (a), a particular momentum restriction, with Eq. (13-19b), the general Fanno pressure relationship, the allowable flow states for shock are found:

$$\mathbf{M}_y{}^2 = \frac{\mathbf{M}_x{}^2 + 2/(k-1)}{[2k/(k-1)]\mathbf{M}_x{}^2 - 1} \tag{13-21}$$

Solution of Eq. (13-21) bears out experimental observations: *Shock can occur only with a supersonic stream, and the velocity after the shock is always subsonic.* Thus, in Fig. 13-14, the permissible flow pattern for shock is supersonic flow a, with the shock changing the state to b at higher entropy; shock flow from c to d at higher entropy, however, is not permissible from momentum considerations (the fluid flow does not have sufficient momentum to cause the irreversible compression process called *shock*).

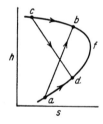

FIG. 13-14. Shock path on Fanno line.

The procedure in solving problems could be to solve Eq. (13-21) for $\mathbf{M}_y$, for a specified $\mathbf{M}_x$, and then use the Fanno table (Table B-18), since the Fanno reference constants (Art. 13-17) are unaffected by shock. However, this labor is avoided by using Table B-19, which combines Eq. (13-21) with Table B-18.

**Example 9.** An impact tube is in a supersonic air stream where the static pressure is 3.55 psia. A *normal shock* (a plane shock) occurs before the tip of the tube with the velocity then isentropically reduced to the stagnation state, where the pressure is 20 psia (the impact tube reading). What was the initial Mach number of the flow and the entropy change in the shock?

*Solution.* The stagnation pressure after the shock is divided by the initial static pressure:

$$\frac{p_{0y}}{p_x} = \frac{20}{3.55} = 5.64$$

And by Table B-19,

$$\mathbf{M}_x = 2.00 \quad Ans. \qquad \mathbf{M}_y = 0.577 \qquad \frac{p_{0y}}{p_{0x}} = 0.721$$

By Eq. (13-2d),

$$\Delta s = R \ln \frac{p_{0x}}{p_{0y}} = \frac{1.986}{29} \ln \frac{1}{0.721} = 0.0226 \text{ Btu/(lb}_m)(°R) \qquad Ans.$$

**13-19. Thrust.**† The forces acting on a fluid stream can be found from the momentum equation. For flow in a fixed direction,

$$F_{\text{wall}} = \frac{\dot{m}_f}{g_c}(V_2 - V_1) + p_2 A_2 - p_1 A_1 \qquad (13\text{-}4b)$$

Here $F_{\text{wall}}$ is the force of the *walls on the fluid* (Fig. 17-21). The *reaction* to this force is called the *internal thrust—the force of the fluid on the walls.* (Rather than set up a new sign convention, the thrust of propulsive devices will always be considered a positive number.)

Since the system is usually surrounded by an atmosphere at a fixed pressure $p_0$, a portion of $F_{\text{wall}}$ can be furnished by the atmosphere. Hence the net force is defined by

$$F_{\text{net}} \equiv F_{\text{wall}} - p_0(A_2 - A_1) = \frac{\dot{m}_f}{g_c}(V_2 - V_1) + (p_2 - p_0)A_2$$
$$- (p_1 - p_0)A_1 \quad (13\text{-}22a)$$

The *reaction* to $F_{\text{net}}$ is called the *net thrust* (and this is the useful thrust for a propulsive device).

A *force function* is defined as

$$F \equiv pA + \frac{\dot{m}_f}{g_c}V \qquad (13\text{-}23a)$$

which is equivalent to

$$F \equiv pA(1 + kM^2) \qquad (13\text{-}23b)$$

Comparison of Eqs. (13-4b), (13-22a), and (13-23a) shows, for a horizontal straight duct,

$$F_{\text{wall}} = F_2 - F_1 \qquad \qquad |\text{internal thrust}| \quad (13\text{-}23c)$$
$$F_{\text{net}} = (F_2 - F_1) - p_0(A_2 - A_1) \qquad |\text{net thrust}| \quad (13\text{-}23d)$$

A Mach relationship can also be derived:

$$\frac{F}{F^*} = \frac{1 + kM^2}{M\sqrt{2(k + 1)\left(1 + \dfrac{k-1}{2}M^2\right)}} \qquad (13\text{-}23e)$$

Values of the right side of Eq. (13-23e) can be found in either Table B-17 or B-18.

**Example 10.** A turbojet engine has an inlet area of $\frac{1}{2}$ ft² and an exit area of 1 ft². At the inlet, $M_1 = 0.5$ and $p_1 = 1$ atm, while at exit, $M_2 = 1.0$ and $p_2 = 1$ atm. Determine the internal and net thrusts.

† See also Art. 17-11.

*Solution.* The magnitude (but not the direction) of the internal thrust is given by either Eq. (13-4b) or (13-23c):

$$|\text{Internal thrust}| = F_{\text{wall}} = p_2 A_2 (1 + k M_2{}^2) - p_1 A_1 (1 + k M_1{}^2)$$

$$= 14.7(144)1(1 + 1.4) - 14.7(144)\tfrac{1}{2}\left(1 + \frac{1.4}{4}\right)$$

$$= 5{,}075 - 1{,}430 = 3{,}645 \text{ lb}_f \qquad Ans.$$

And by either Eq. (13-22a) or (13-23d),

$$|\text{Net thrust}| = (F_2 - F_1) - p_0(A_2 - A_1) = 3{,}645 - 1{,}060 = 2{,}585 \text{ lb}_f \qquad \textbf{\textit{Ans.}}$$

When the fluid changes direction, turning forces must be supplied, and either the vector momentum equation [Eq. (13-4a)] or the scalar momentum equations [Eqs. (13-4b)] become necessary. The procedure is illustrated in Fig. 13-15 for a two-dimensional system at constant elevation. First, the $x$ and $y$ axes are oriented to facilitate the solution; in this instance the $x$ axis is aligned in the direction of the leaving flow. Then, from Eq. (13-22a) and Fig. 13-15,

$$F_x = \frac{\dot{m}_f}{g_c}(V_2 - V_1 \cos \alpha) + (p_2 - p_0)A_2 - (p_1 - p_0)A_1 \cos \alpha$$

$$\tag{13-22b}$$

$$F_y = \frac{\dot{m}_f}{g_c}(-V_1 \sin \alpha) - (p_1 - p_0)A_1 \sin \alpha \tag{13-22c}$$

A *rocket* (Fig. 13-16) carries its entire working substance, and therefore only one flow stream crosses the boundary of the system. The propellant

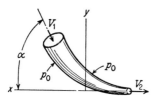

FIG. 13-15. Forces on two-dimensional system.

is usually a liquid or a solid which decomposes or explodes to create a high pressure within the combustion chamber. If the combustion chamber were completely closed, the net force from the combustion pressures would be zero. But the rear of the rocket is open (the exit of the nozzle), and therefore a greater force is exerted by the gas pressure on the forward walls of the combustion chamber, and on the diverging walls of the nozzle, than that arising from the gas pressure on the converging walls of the nozzle. Thus the net force of the gases on the walls is to the left in Fig. 13-16. The pressure is maintained by allowing the products of combustion to escape through the nozzle while new fuel oxidant is fed from the reservoir. Since the rocket is independent of the atmosphere as a source of oxygen, it is the device which makes interplanetary travel an interesting and engaging topic of conversation. More important, it is the engine which has allowed man to produce artificial satellites (and international headaches).

The rocket is analyzed by applying Eqs. (13-22) to the system (dashed line) in Fig. 13-16. Steady flow of the exit gases can be obtained for a relatively long time by regulating the flow of propellant. It follows from Eq. (13-22c) that the net force in the $y$ direction is zero since the feed tubes are symmetrical. In the $x$ direction, since $\alpha$ is 90°, Eq. (13-22b) reduces to

$$F_x = \frac{\dot{m}_f}{g_c} V_2 + (p_2 - p_0) A_2 \qquad \text{(net thrust)} \qquad (13\text{-}24a)$$

By reducing $p_2$, $V_2$ increases, and the thrust becomes a maximum when the exit pressure is $p_0$ (Prob. 44):

$$F_{\max} = \frac{\dot{m}_f}{g_c} V_2 \qquad \text{(maximum net thrust)} \qquad (13\text{-}24b)$$

Thus the thrust from the pressure of the gases acting on the net pro-

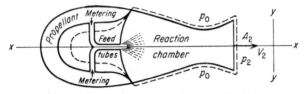

FIG. 13-16. Rocket.

jected walls of the rocket can be evaluated by measuring the exhaust-gas velocity and the mass-flow rate.

## PROBLEMS

**1.** Steam expands adiabatically and reversibly from $p = 500$ psia, $t = 700°F$, to 300 psia. Find the velocity attained (a) if the velocity of approach is zero, (b) if the velocity of approach is 300 ft/sec.

**2.** Repeat Prob. 1 for an exhaust pressure of 150 psia.

**3.** For the data in Prob. 1, find the highest approach velocity that can be neglected if the calculated final velocity is to be within 2 per cent of the correct answer.

**4.** Steam expands in an ideal nozzle from $p_0 = 500$ psia, $t_0 = 700°F$, to 300 psia. Find the exit area to pass 1 $lb_m$/sec.

**5.** Steam expands in an ideal nozzle from $p_0 = 500$ psia, $t_0 = 700°F$, to 150 psia. Find the throat and exit areas to pass 1 $lb_m$/sec.

**6.** Air expands reversibly and adiabatically from $p_0 = 100$ psia, $t_0 = 400°F$, to atmospheric pressure. Calculate the final velocity (ideal gas).

**7.** Find the throat and exit areas for the data of Prob. 6 to pass 1 $lb_m$/sec.

**8.** Carbon dioxide flows through an ideal nozzle from $p_0 = 500$ psia, $t_0 = 300°F$, to 25 psia. Find throat and exit areas for a mass-flow rate of 0.25 $lb_m$/sec (ideal gas).

**9.** Repeat Prob. 8, but use Fig. B-3.

**10.** An ideal nozzle expands air from $p_0 = 50$ atm, $t_0 = 400°R$, to 5 atm. Find the area at the throat for a mass-flow rate of 0.25 $lb_m$/sec (ideal gas).

**11.** An ideal nozzle expands nitrogen from $p_0 = 100$ psia, $t_0 = 300°F$, to 10 psia. Find the throat and exit areas (ideal gas).

**12.** Derive Eq. (13-6a).

**13.** Derive Eq. (13-7).

**14.** For flow through an ideal nozzle, derive the equation $\dot{m}_f = (\text{constant})A^*p_0/\sqrt{T_0}$.

**15.** Steam expands from $p_0 = 500$ psia, $t_0 = 700°F$, to 150 psia; nozzle efficiency is 0.95. Find the velocity, specific volume, and probable discharge coefficient.

**16.** Nitrogen at $p_0 = 100$ psia, $t_0 = 300°F$, passes through a nozzle to a discharge region at 10 psia. If the velocity coefficient is 0.98 from entrance to throat and 0.96 over all, find the throat and exit areas.

**17.** Saturated steam at $p_0 = 100$ psia flows through an ideal nozzle to a discharge region at 60 psia. Calculate the mass-flow rate if $A_2 = 1$ in.$^2$ Assume phase equilibrium.

**18.** Repeat Prob. 17, but assume supersaturated flow and compare answers.

**19.** Superheated steam at $p_0 = 100$ psia, $t_0 = 360°F$, passes through an ideal nozzle to $p_2 = 40$ psia. Calculate the mass-flow rate if $C = 0.98$ and $A^* = 0.5$ in.$^2$

**20.** Derive Eq. (13-11b). Use MacLaurin's series,

$$f(x) = f(0) + f'(0)x + \frac{f''(0)}{2!}x^2 + \cdots$$

and let
$$\frac{p_1 - p_2}{p_1} = x \qquad m = \frac{k-1}{k}$$

so that
$$\left(\frac{p_2}{p_1}\right)^{\frac{k-1}{k}} = (1 - x)^m$$

**21.** Air is flowing through an 8-in. pipe and through a 2-in. thin-plate orifice. The "adiabatic" coefficient of discharge is 0.65, and the differential manometer reading is 30 in. water. The upstream temperature of the air is 80°F, and upstream pressure is 10 psig. Compute the flow rate.

**22.** A 4- by 2-in. venturi has a coefficient of discharge of 0.98. Air is flowing to the venturi under a pressure of 50 psig and temperature of 60°F. The pressure drop to the throat of the venturi is 12 in. water. Determine the mass-flow rate.

**23.** A 1-in. (diameter) flow nozzle is installed in a 2-in. pipe that carries air at 12 in. water gauge pressure. The temperature of the air is 60°F, the discharge coefficient for the nozzle is 0.97, and the pressure drop across the nozzle is 6 in. water. Compute the mass-flow rate.

**24.** Air is flowing through a 4-in. pipe under a static pressure of 3 in. water. A thermometer indicates a temperature of 100°F, while the mean velocity head calculated from pitot-tube readings is 2.08 in. water. Determine the average velocity of the air and the mass-flow rate. (Barometer shows 29.90 in. Hg.)

**25.** Repeat Prob. 24, but assume that the static pressure is 10.25 psig and the mean pitot-tube pressure is 30.06 psia.

**26.** Derive Eq. (13-17).

**27.** Find the throat and exit areas for an ideal nozzle to pass 1 $\text{lb}_m$/sec of air from $p_0 = 100$ psia, $t_0 = 200°F$, to 10 psia. (Use Table B-17.)

**28.** Air at $M_1 = 0.5$ enters a diffuser with area ratio $A_1/A_2$ of 0.5. Find $M_2$ for isentropic flow. (Use Table B-17.)

**29.** A gas ($k = 1.4$) enters an ideal nozzle with $M_1 = 0.1$, $t = 200°F$, $p = 100$ psia and leaves with $M_2 = 2.0$. Find the throat area for a flow rate of 1 $\text{lb}_m$/sec.

**30.** Derive Eq. (13-19a).

**31.** Deduce from Fig. 13-13 what will happen if additional pipe is added to a system which has sonic velocity at exit (exit at atmospheric pressure).

**32.** In Fanno flow of air, $M_1 = 0.1$, $p_1 = 50$ psia, $M_2 = 0.8$. Find $p_2$ and $\Delta s$.

**33.** For the data of Prob. 32, $T_1 = 500°R$. Find $T_2$ and $\rho_2$.

**34.** Find the length of steel tubing (2 in. ID) wherein the flow of air is from $\mathbf{M} = 3.0$ to $\mathbf{M} = 1.5$ in Fanno flow. $(4f = 0.003.)$

**35.** Find the $L/D$ values $(4f = 0.003)$ to get $\mathbf{M} = 1$ in Fanno flow if $\mathbf{M}_1 = 0$, $\frac{1}{2}$, 1, $1\frac{1}{2}$, 2, $\infty$.

**36.** A constant-area steel pipe, 1 ft ID, carries air adiabatically from $\mathbf{M} = 0.1$ to $\mathbf{M} = 0.3$. Calculate the length of pipe and the final pressure if the initial pressure is 100 psia and $4f = 0.015$.

**37.** If the static pressure of the supersonic flow in Example 9 is 5 psia, with $\mathbf{M} = 2.0$, find the pitot-tube reading and the entropy increase.

**38.** Find the Mach number and the per cent of rise in pressure for a shock at $\mathbf{M} = 2.0$.

**39.** An ideal nozzle, with an exit-to-throat-area ratio of 2.5, is connected to a large tank of air at 100 psia, 40°F. Determine the exit pressure so that a shock is located at the exit.

**40.** Air expands from $p_0 = 160$ psia to 40 psia through an ideal nozzle and then enters a 6-in.-ID commercial steel pipe. What is the longest length of pipe that can be used without disturbing the mass-flow rate of the nozzle? (Fanno flow.)

**41.** A 12-in.-ID pipe carries 5,000 cfm of air at $p_1 = 14.7$ psia, $t_1 = 100°F$. At exit from the pipe, $p_2 = 11.7$ psia and $t_2 = 100°F$. Calculate the force required to hold the pipe in place. Show or explain the direction of the force. Specifically, what made this force necessary?

**42.** Water flows through a 90° elbow in a 4-in.-ID steel pipeline at the rate of 1 ft³/sec with the initial pressure at 1.5 psig. Calculate the force on the elbow from the flow. [The elbow is equivalent to 12 ft of pipe; assume that $\mu = 1.88(10^{-5})$ lb$_f$ sec/ft².]

**43.** A rocket has an exit-to-throat-area ratio of 4 and a throat of 6 in.² The reaction chamber has a pressure of 200 atm. Find internal and net thrusts. (Use Table B-17; atmospheric pressure is 14.7 psia.)

**44.** A jet engine is tested on a thrust stand in surroundings of $p_0 = 14.7$ psia. Air enters the inlet of 2 ft² at 400 ft/sec, 14.7 psia, and 60°F, and liquid fuel enters at 50 psia, 60°F. The products of combustion leave at 16 psia, 1,400 ft/sec, and 2500°F. (a) Neglecting the fuel, calculate the net and internal thrusts. (b) Consider that the liquid fuel enters with negligible velocity but increases the mass flow of leaving gases, and recompute the net and internal thrusts. (c) Repeat the calculations with $p_2 = 14.7$ psia.

**45.** Try to redraw Fig. 13-16 so that the propellant does not enter radially yet the rocket mass remains symmetric about its axis.

## SELECTED REFERENCES

1. American Society of Mechanical Engineers: "Report of ASME Special Research Committee on Fluid Meters," New York, 1937.
2. Stodola, A.: "Steam and Gas Turbines," vols. I, II, McGraw-Hill Book Company, Inc., New York, 1927.
3. Yellott, J. I.: Supersaturated Steam, *Trans. ASME*, **56**(6): 411–430 (June, 1934).
4. American Society of Mechanical Engineers: Flow Measurements by Means of Standardized Nozzles and Orifice Plates, *Power Test Codes*, part **5**, chap. 4, New York, 1940.
5. Cunningham, R. G.: Super-critical Compressible Flow through Square-edged Orifices, *Trans. ASME*, **73**: 625–638 (July, 1951).

6. Obert, E. F., and D. A. Gyorog: The Mechanical Engineering Laboratory, Design and Equipment, *Univ. Wisconsin Eng. Expt. Sta. Rept.* 16, September, 1960.
7. Obert, E. F.: "Concepts of Thermodynamics," McGraw-Hill Book Company, Inc., New York, 1960.
8. Spink, L. K.: "Principles and Practice of Flow Meter Engineering," The Foxboro Company, Foxboro, Mass., 1958.
9. Hall, N. A.: "Thermodynamics of Fluid Flow," Prentice-Hall, Inc., Englewood Cliffs, N.J., 1951.

# CHAPTER 14

# PSYCHROMETRICS

Foul weather comes, not in the heavens,
but in the ways of men.

*John Buchan*

The conditioning of air is one of the important tasks of the practicing engineer.

**14-1. Psychrometric Principles.**† Dry air is a mixture of gases that has a representative volumetric analysis in per cent as follows: oxygen, 20.99; nitrogen, 78.03; argon, 0.94, including traces of the rare gases neon, helium, and krypton; carbon dioxide, 0.03; and hydrogen, 0.01. For most calculations, it is sufficiently accurate to consider dry air as consisting of 21 per cent of oxygen and 79 per cent of inert gases taken as nitrogen (3.76 parts $N_2$ to 1 part $O_2$ by volume).

TABLE 14-1. MASS ANALYSIS OF DRY AIR

| Gas | Volumetric analysis, % | Mole fraction | Molecular weight | Relative weight $\left(\dfrac{\text{lb}_m \text{ constituent}}{\text{mole mixture}}\right)$ |
|---|---|---|---|---|
| $O_2$..... | 20.99 | 0.2099 | 32.00 | 6.717 |
| $N_2$.... | 78.03 | 0.7803 | 28.016 | 21.861 |
| A..... | 0.94 | 0.0094 | 39.944 | 0.376 |
| $CO_2$... | 0.03 | 0.0003 | 44.003 | 0.013 |
| $H_2$.... | 0.01 | 0.0001 | 2.016 | |
| | 100.00 | 1.0000 | ...... | 28.967 = $M$ for air |

*Dew Point.* An important constituent of the usual air mixture is water vapor existing either as saturated or as superheated steam. The mixture can be cooled at constant pressure, and if the water vapor is initially superheated, each constituent will be cooled at constant partial pressure because the composition of the gaseous mixture remains constant. With continued cooling, the water vapor will reach the saturated state, and any further decrease in temperature will cause condensation and thus a change in composition of the gaseous phase. The temperature that

† In this article, the terminology of Ref. 2 has been adopted wherever feasible; similarly, all empirical constants have been obtained from this source.

255

marks the appearance of liquid water is called the *dew point*. In Fig. 14-1, the path of the cooling process for the water-vapor constituent is shown on the *Ts* diagram. At state 1, the vapor is superheated at a temperature called the *dry-bulb* temperature, which can be measured with the usual thermometer. Cooling at constant partial pressure occurs from state 1 to state 2. At state 2, the *dew point*, the water vapor is saturated. If the temperature of the air is lowered beneath the dew point, the air remains saturated, although the partial pressure of the water vapor progressively decreases because of condensation while the state changes from 2 to 3. The dew-point temperature allows the vapor pressure of the superheated vapor to be found from the Steam Tables; the partial pressure of the saturated vapor at the dew point is also the partial pressure of the superheated vapor at state 1.

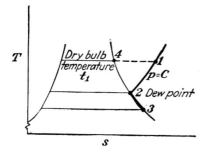

FIG. 14-1. Relationship of dry-bulb and dew-point temperatures.

*Relative Humidity.* The *relative humidity* $\phi$ is best defined as the ratio of the actual partial pressure of the vapor to the saturated partial pressure of the vapor at the same temperature,

$$\phi \equiv \frac{p_w}{p_s}\bigg]_T \qquad (14\text{-}1a)$$

where $w$ = water vapor

$s$ = saturated water vapor

In Fig. 14-1, the relative humidity of the superheated vapor in the mixture at $t_1$ is the ratio of the partial pressure at state 1 (or 2) to the partial pressure at state 4. Since the perfect-gas laws are quite accurate at the low pressures encountered with atmospheric air, $\phi$ can be expressed as

$$\phi = \frac{p_w}{p_s} = \frac{R_w T/v_w}{R_w T/v_s} = \frac{v_s}{v_w} = \frac{\rho_w}{\rho_s} \qquad (14\text{-}1b)$$

it being noted that the temperature is the same at states 1 and 4. As shown by Eq. (14-1b), relative humidity (at low pressures) is the ratio of the density of the steam present in the mixture to the saturation density of steam at the same temperature. When the relative humidity of the mixture is 1.0, the mixture is at the dew point.

*Humidity Ratio, or Specific Humidity.* The *humidity ratio*, or *specific humidity*, $W$ is defined as the ratio of the mass of water vapor to the mass of dry air in the mixture,

$$W \equiv \frac{m_w}{m_a}\bigg]_{V,T} \qquad (14\text{-}2a)$$

Thus, a mixture containing 1 $lb_m$ of dry air would contain $W$ $lb_m$ of steam, and, at low pressure,

$$W = \frac{m_w}{m_a} = \frac{\rho_w}{\rho_a} = \frac{p_w/R_w T}{p_a/R_a T} = \frac{p_w}{p_a} \frac{R_a}{R_w} = 0.622 \frac{p_w}{p - p_w} \quad (14\text{-}2b)$$

($R_w$ and $R_a$ are specific gas constants for water vapor and air.) The specific and the relative humidities are related; by Eq. (14-1b),

$$p_w = \phi p_s$$

and, upon substitution in Eq. (14-2b),

$$W = \phi \frac{p_s}{p_a} \frac{R_a}{R_w} = \phi \frac{p_s}{\rho_a} = \phi \frac{v_a}{v_s} \quad (14\text{-}2c)$$

*Degree of Saturation.* The degree of saturation $\mu$ is defined as the ratio of actual specific humidity (humidity ratio) to the specific humidity of saturated air at the dry-bulb temperature,

$$\mu \equiv \frac{W}{W_{sat}} \Big]_T \quad (14\text{-}3a)$$

It can be related to relative humidity as follows:

$$\mu = \frac{W}{W_{sat}} = \frac{0.622 p_w/(p - p_w)}{0.622 p_s/(p - p_s)} = \frac{p_w(p - p_s)}{p_s(p - p_w)} = \phi \frac{p - p_s}{p - p_w} \quad (14\text{-}3b)$$

Values of both $p_s$ and $p_w$ are small relative to $p$; thus, the degree of saturation is approximately equal to the relative humidity. The latter is commonly used in psychrometric work, while the degree of saturation is applied mainly to gas-vapor mixtures of other than air and water.

**Example 1.** The temperature in a room is 80°F, and the relative humidity is 30 per cent. Determine (a) the partial pressure of the steam and the dew point, (b) the density of each constituent, (c) the specific humidity, (d) degree of saturation. (Barometer reads 29.92 in. Hg.)

*Solution.* (a) The vapor pressure of saturated steam at 80°F is obtained from the Steam Tables,

$$p_s = 0.5069 \text{ psia} \qquad \text{and} \qquad v_s = 633.1 \text{ ft}^3/\text{lb}_m$$

Since in this problem the relative humidity is 30 per cent,

$$p_w = \phi p_s = 0.30(0.5069) = 0.15207 \text{ psia} \qquad Ans.$$

This partial pressure of the superheated vapor is also the saturation pressure that defined the dew point. Thus, from the Steam Tables,

$$t_{\text{dew}} \approx 46°F \qquad Ans.$$

(b) The density of the saturated steam at 80°F is

$$\rho_s = \frac{1}{v} = \frac{1}{633.1} = 0.00158 \text{ lb}_m/\text{ft}^3$$

and since the relative humidity is 30 per cent, by Eq. (14-1b)

$$\rho_w = \phi \rho_s = 0.30(0.00158) = 0.000474 \text{ lb}_m/\text{ft}^3 \qquad Ans.$$

This answer can be checked by the ideal-gas equation using 85.6 ft-lb$_f$/(lb$_m$)(°R) for the specific gas constant,

$$\rho_w = \frac{p}{RT} = \frac{0.15207(144)}{85.6(540)} = 0.000474 \text{ lb}_m/\text{ft}^3 \qquad Ans$$

The density of the dry air is similarly found,

$$p_a = p - p_w = 14.696 - 0.1521 = 14.544 \text{ psia}$$
$$\rho_a = \frac{p_a}{R_a T} = \frac{14.544(144)}{53.3(540)} = 0.0729 \text{ lb}_m/\text{ft}^3 \qquad Ans.$$

(c) The specific humidity is found by Eq. (14-2b),

$$W = \frac{\rho_w}{\rho_a} = \frac{0.000474}{0.0729} = 0.0065 \qquad Ans.$$

or, more expressively,

$$W = 0.0065 \frac{\text{lb}_m \text{ steam}}{\text{lb}_m \text{ dry air}} \qquad Ans.$$

(d) The degree of saturation is found by Eq. (14-3b),

$$\mu = \phi \frac{p - p_s}{p - p_w} = 0.30 \left( \frac{14.696 - 0.507}{14.696 - 0.152} \right) = 0.292 \qquad Ans.$$

or, more directly,

$$W_s = 0.622 \frac{p_s}{p - p_s} = \frac{0.622(0.507)}{14.696 - 0.507} = 0.0223$$

and

$$\mu = \frac{W}{W_s} \bigg]_T = \frac{0.0065}{0.0223} = 0.291 \qquad Ans.$$

**Example 2.** Atmospheric air at 30°F and 60 per cent relative humidity is conditioned to 80°F and 50 per cent relative humidity. Determine the amount of water added to the air. (Barometer reads 29.92 in. Hg.)

*Solution.* The pressure of saturated vapor at 30°F is found in the Steam Tables,

$$p_s = 0.0808 \text{ psia}$$

The partial pressure of the water vapor initially in the atmosphere is

$$p_w = \phi p_s = 0.60(0.0808) = 0.0485 \text{ psia}$$

and by Eq. (14-2b) the initial humidity ratio is

$$W_1 = 0.622 \frac{p_w}{p - p_w} = 0.622 \left( \frac{0.0485}{14.696 - 0.0485} \right) = 0.00206 \frac{\text{lb}_m \text{ steam}}{\text{lb}_m \text{ dry air}}$$

Each of the above steps is repeated for the air after conditioning to 80°F,

$$p_s = 0.5069 \text{ psia}$$
$$p_w = \phi p_s = 0.50(0.5069) = 0.2533 \text{ psia}$$
$$W_2 = 0.622 \frac{p_w}{p - p_w} = 0.622 \left( \frac{0.2533}{14.443} \right) = 0.01090 \frac{\text{lb}_m \text{ steam}}{\text{lb}_m \text{ dry air}}$$

The change in moisture content of the air during the process equals

$$W_2 - W_1 = 0.01090 - 0.00206 = 0.00884 \frac{\text{lb}_m \text{ water}}{\text{lb}_m \text{ dry air}} \qquad Ans.$$

Expressed in *grains*, where 7,000 grains = 1 $\text{lb}_m$,

$$W_2 - W_1 = 0.00884(7,000) = 61.88 \frac{\text{grains water}}{\text{lb}_m \text{ dry air}} \qquad Ans.$$

**14-2. Multiple-stream Steady-flow Processes.** The conditioning of large quantities of air or water invariably demands a steady-flow process, and in most instances, more than one flow path will be present. In Fig. 14-2, air with its contained moisture enters a system and leaves with a greater or lesser amount of vapor. Because of this humidification or dehumidification, water must also enter or leave if the process is to be

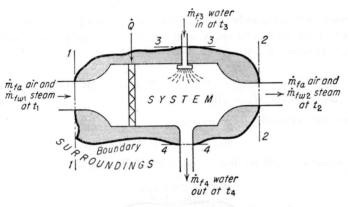

FIG. 14-2. General steady-flow air-conditioning system.

continuous, while the system may or may not be adiabatic. An energy balance can be made for this general system from Eq. (5-4c) to yield

$$\dot{Q} = \Sigma \dot{m}_f h_{\text{out}} - \Sigma \dot{m}_f h_{\text{in}}$$

and therefore

$$\dot{Q} = \underset{\text{Dry air}}{\dot{m}_{fa}(h_{a2} - h_{a1})} + \underset{\text{Water vapor}}{(\dot{m}_{fw2}h_{w2} - \dot{m}_{fw1}h_{w1})} + \underset{\text{Water}}{(\dot{m}_{f4}h_4 - \dot{m}_{f3}h_3)} \qquad (14\text{-}4a)$$

Upon dividing by $\dot{m}_{fa}$ the units for heat are changed:

$$\Delta_L q' = (h_{a2} - h_{a1}) + (W_2 h_{w2} - W_1 h_{w1}) + \frac{\dot{m}_{f4}h_4 - \dot{m}_{f3}h_3}{\dot{m}_{fa}} \qquad (14\text{-}4b)$$

where $\Delta_L q'$ = heat transferred from $L_1$ to $L_2$, Btu/lb$_m$ dry air

　　　　1 = entrance conditions for air-steam mixture

　　　　2 = exit conditions for air-steam mixture

　　　　3 = entrance conditions for water

　　　　4 = exit conditions for water

The amounts of water entering or leaving the system at 3 and 4 are not equal because a part of the water is either added or taken away from the air mixture. For continuity of mass flow of the water,

$$W_2 - W_1 = \frac{\dot{m}_{f3} - \dot{m}_{f4}}{\dot{m}_{fa}} \qquad (14\text{-}4c)$$

The equations can be simplified by defining the enthalpy of the mixture of 1 $lb_m$ of dry air plus $W$ $lb_m$ of water vapor to be

$$H \frac{\text{Btu}}{lb_m \text{ dry air}} \equiv h_a + W h_w$$

When this identity is substituted in Eq. (14-4b),

$$\Delta q' = H_2 - H_1 + \frac{\dot{m}_{f4} h_4 - \dot{m}_{f3} h_3}{\dot{m}_{fa}} \qquad (14\text{-}4d)$$

In some instances, only one flow stream of water is associated with the system; for example, the spray water could be entirely vaporized by the air, and therefore $\dot{m}_{f4}$ would be zero. For this case, Eq. (14-4b) reduces to

$$\Delta q' = (h_{a2} - h_{a1}) + (W_2 h_{w2} - W_1 h_{w1}) - (W_2 - W_1)h_{f3} \qquad (14\text{-}5a)$$
$$\Delta q' = H_2 - H_1 - (W_2 - W_1)h_{f3} \qquad (14\text{-}5b)$$

because, for continuity of mass flow,

$$\dot{m}_{fa}(W_2 - W_1) = \dot{m}_{f3} \qquad (14\text{-}5c)$$

These equations will be illustrated in the following articles.

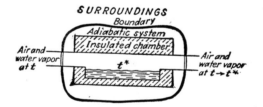

FIG. 14-3. Process of adiabatic saturation.

**14-3. Temperature of Adiabatic Saturation.** Consider a process wherein the humidity of air is increased by passing it through an insulated chamber that contains a large surface of water (Fig. 14-3). Here the air is cooled in passing over the water, and the water is cooled as vaporization occurs. For this adiabatic process, the water will reach a steady-state temperature when the thermal energy transferred from the air to the water is equal to the thermal energy required to vaporize the water. This equilibrium temperature $t^*$ is called† the *temperature of adiabatic*

† The asterisk (*) designates that the property is to be evaluated at the state of adiabatic saturation.

*saturation;* it is the lowest temperature reached *by the water* in the process of Fig. 14-3.

Suppose that the insulated chamber of Fig. 14-3 is infinitely long; then, the air leaving the chamber will be at the same temperature as the water, and, moreover, the air will be saturated with water vapor. Thus, as the air passes through the chamber, the dry-bulb temperature of the air progressively decreases until the limiting temperature of adiabatic saturation is reached. The adiabatic saturation temperature of the air, however, is a concept that has but one value at all sections of Fig. 14-3. The condition of the water vapor as it passes through the chamber is illustrated in Fig. 14-4 by path 1-2. (Note that the dew point and the adiabatic saturation temperatures have quite different values except for the one case of saturated air.)

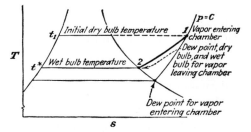

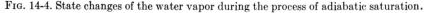

FIG. 14-4. State changes of the water vapor during the process of adiabatic saturation.

The system of Fig. 14-3 can be converted into a steady-flow system by supplying make-up water at the adiabatic saturation temperature. Assume that the air leaves the system at the temperature of adiabatic saturation. For this process, when it is noted that

$$\Delta q' = 0$$
$$W_2(h_{w2} - h_3) = W^*(h_g^* - h_f^*) = W^*h_{fg}^*$$

then Eq. (14-5a) can be arranged into the form

$$W = \frac{(h_a^* - h_a) + W^*h_{fg}^*}{h_w - h_f^*} \tag{14-6a}$$

Equation (14-6a) could be used to find the specific humidity of air since all other terms in the equation can be computed or measured with an apparatus like that in Fig. 14-3.

An important generalization can be seen by applying Eq. (14-5b) to the process of adiabatic saturation,

$$H_1 + (W^* - W_1)h_f^* = H_2^* \tag{14-7a}$$

Since the middle term is relatively insignificant,†

$$H_1 \approx H_2^* \tag{14-7b}$$

Equation (14-7b) shows that the total enthalpy of the mixture remains essentially constant during the process of adiabatic saturation. Since the adiabatic saturation temperature also remains constant during the process, it can be concluded that the total enthalpy of the air-vapor mixture is a function of the adiabatic saturation temperature. This approximation is sufficiently accurate for most meteorological and air-conditioning work.

*Enthalpy of Moist Air.* The enthalpy of moist air is usually expressed in terms of the enthalpy of a mixture of 1 $lb_m$ of dry air plus $W$ $lb_m$ of water vapor, or

$$H = h_a + W h_w \tag{14-8a}$$

The dry-air enthalpy $h_a$ for pressures near atmospheric can be accurately evaluated from the ideal-gas relationship,

$$h_a = 0.24t \tag{14-9}$$

from an arbitrary datum of 0°F.

The enthalpy of the water vapor, $h_w$, is equal to

$$h_w = 1.0(t^* - 32) + h_{fg}^* + 0.45(t - t^*) \tag{14-10a}$$

where the asterisked properties properly should refer to the dew-point temperature. However, no serious error is introduced by the use of any other temperature intermediate between the dew-point and dry-bulb, such as the temperature of adiabatic saturation. If the air is saturated (that is, the water vapor is saturated, not superheated), the last term, representing the superheat, becomes zero. The datum for the enthalpy of the water vapor is that of the Steam Tables: $h_w = 0$ for saturated liquid at 32°F.

With the foregoing relationships, the enthalpy of the moist air becomes

$$H = 0.24t + W[(t^* - 32) + h_{fg}^* + 0.45(t - t^*)] \tag{14-8b}$$

which can be shown as

$$\begin{aligned} H &= (0.24 + 0.45W)t + W[(t^* - 32) + h_{fg}^* - 0.45t^*] \\ &= c_m t + W(0.55t^* + h_{fg}^* - 32) \end{aligned} \tag{14-8c}$$

Here $c_m$ is called the *humid heat capacity*. It is the heat capacity of $(1 + W)$ $lb_m$ of mixture, or the heat capacity of the mixture per pound

---

† For this reason, the make-up water added to the adiabatic saturation process can be at a quite different temperature from that of the wet-bulb temperature without perceptibly changing the final condition of the air.

of dry air. Although the datum states for the dry air and for the water vapor are different, enthalpy differences are normally required and the datum state is of no significance in the calculations.

Usually, the enthalpy of the water vapor is found from the empirical formula

$$h_w = 1,060 + 0.45t \tag{14-10b}$$

Therefore,

$$H = 0.24t + W(1,060 + 0.45t) \tag{14-8d}$$

$$H = c_m t + W(1,060) \tag{14-8e}$$

A property similar to enthalpy, called the *sigma function*, is defined as

$$\Sigma \equiv H - Wh_f^* \approx H - W(t^* - 32) \tag{14-11}$$

This function is exactly constant during the process of adiabatic saturation [note Eq. (14-7a)]. In the past, it has been used in place of the enthalpy function on some psychrometric charts.

**14-4. The Wet-bulb Temperature.** Figure 14-5 illustrates a *sling psychrometer*, which consists of a *wet-bulb* and a *dry-bulb thermometer*. The wet-bulb is so-called because the sensing element is covered with a wick moistened with water (and a thermocouple can be used as well as a glass thermometer). The temperature attained by causing air to flow over the wetted wick is known as the *wet-bulb temperature*. The process involved is similar to that of Fig. 14-3 except that the ratio of air flow over the relatively small quantity of water is sufficient to cause only negligible change in the temperature and humidity of the air. If the air is unsaturated, the water temperature decreases because of evaporation and approaches an equilibrium temperature between the dew-point and dry-bulb temperatures of the air. *The equivalence of wet-bulb and adiabatic saturation temperatures is not a general relationship, and for liquids other than water, the wet-bulb temperature is significantly different from the adiabatic saturation temperature.*

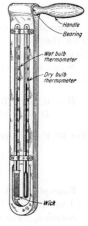

FIG. 14-5. Sling psychrometer.

Minimum errors in measuring the wet-bulb temperature are obtained with an air velocity of approximately 1,000 ft/min over the wetted wick and a supply water temperature slightly higher than the wet-bulb temperature being measured. Too high a supply water temperature will result in complete evaporation before the steady-state condition is reached, while too low a water temperature will result in a rising temperature with only a short and difficult-to-recognize stage at the true wet-bulb temperature.

Equation (14-6a) for the adiabatic saturation process can be written as

$$W = \frac{(h_a^* - h_a) + W^* h_{fg}^*}{h_w - h_f^*} = \frac{0.24(t^* - t) + W^* h_{fg}^*}{h_{fg}^* + 0.45(t - t^*)} \tag{14-6b}$$

**Example 3.** Determine (a) the humidity ratio, (b) the relative humidity, (c) the dew-point temperature for air at 80°F dry-bulb and 60°F wet-bulb. (Barometer is 14.696 psia.)

*Solution.* (a) At 60 and 80°F, from the Steam Tables,

$$p^* = 0.256 \text{ psia} \qquad p_s = 0.507 \text{ psia}$$

$$W^* = 0.622 \left( \frac{0.256}{14.696 - 0.256} \right) = 0.0111 \text{ lb}_m/\text{lb}_m \text{ dry air}$$

By Eq. (14-6b),

$$W = \frac{0.24(60 - 80) + 0.0111(1,060)}{1,060 + 0.45(80 - 60)} = \frac{6.96}{1,069} = 0.0065 \text{ lb}_m/\text{lb}_m \text{ dry air} \qquad Ans.$$

(b) From Eqs. (14-2b) and (14-1a),

$$p_w = \frac{Wp}{0.622 + W} = \frac{0.0065(14.696)}{0.6285} = 0.152 \text{ psia}$$

$$\phi = \frac{p_w}{p_s} = \frac{0.152}{0.507} = 0.30, \text{ or } 30\% \qquad Ans.$$

(c) At $p_w = 0.152$ psia, from the Steam Tables,

$$\text{Dew-point temperature} = 46°F \qquad Ans.$$

**Example 4.** Compare the enthalpy of air at 80°F dry-bulb and 60°F wet-bulb with that of air saturated at the same wet-bulb temperature. Repeat, but for the sigma function.

*Solution.* From Example 3, for 80°F dry-bulb and 60°F wet-bulb, $W = 0.0065$ $\text{lb}_m/\text{lb}_m$.

$$H = 0.24t + W(1,060 + 0.45t) = 19.2 + 0.0065(1,096) = 26.32 \text{ Btu/lb}_m \text{ dry air}$$
$$\Sigma = H - W(t^* - 32) = 26.32 - 0.0065(28) = 26.14 \text{ Btu/lb}_m \text{ dry air}$$

For 60°F dry- and wet-bulb, $W^* = 0.0111$ (Example 3).

$$H = 14.4 + 0.0111(1,087) = 26.46 \text{ Btu/lb}_m \text{ dry air}$$
$$\Sigma = 26.46 - 0.0111(28) = 26.15 \text{ Btu/lb}_m \text{ dry air}$$

**Example 5.** Compute the heat added for the heating and humidifying process of Example 2. Assume that make-up water is supplied at 60°F.

*Solution.* For the initial state of 30°F and $W = 0.00206$,

$$c_{m1} = 0.24 + 0.45W = 0.24 + 0.45(0.00206) = 0.2409 \text{ Btu/(lb}_m \text{ dry air)(°F)}$$
$$H_1 = c_{m}t + W(1,060) = 0.2409(30) + 0.00206(1,060) = 9.41 \text{ Btu/lb}_m \text{ dry air}$$

For the final, or leaving, state at 80°F and $W = 0.01090$,

$$c_{m2} = 0.24 + 0.45(0.01090) = 0.2449 \text{ Btu/(lb}_m \text{ dry air)(°F)}$$
$$H_2 = 0.2449(80) + 0.01090(1,060) = 31.14 \text{ Btu/lb}_m \text{ dry air}$$

The enthalpy of the supply water is

$$h_f = 60 - 32 = 28 \text{ Btu/lb}_m \text{ water}$$

By Eq. (14-5b),

$$\Delta q' = H_2 - H_1 - (W_2 - W_1)h_3$$
$$= (31.14 - 9.41) - (0.01090 - 0.00206)(28) = 21.73 - 0.25$$
$$= 21.48 \text{ Btu/lb}_m \text{ dry air}$$

**14-5. The Psychrometric Chart.** Although the properties of the air and water-vapor mixture can be readily calculated, it is generally more convenient to use a psychrometric chart (Fig. B-5). The typical psychrometric chart is plotted in the form of specific humidity versus dry-bulb temperature with superimposed lines of constant relative humidity, wet-bulb temperature, and specific volume. The chart is constructed for a given total pressure, usually 29.92 in. Hg, but can be used with engineering accuracy for most calculations when the real barometric pressure differs slightly from standard, say 29.0 to 31.0 in. For large deviations from standard barometric pressure, as encountered at altitudes of about 2,000 ft and over, or for precise solutions, values should be calculated or corrections applied to the chart values.

Particular values of specific humidity and total pressure set the partial pressure of water vapor and, in turn, the dew-point temperature. A given specific humidity and dry-bulb temperature give the wet-bulb, or adiabatic saturation, temperature; relative humidity; specific volume; and enthalpy. Thus, all the psychrometric relationships can be presented graphically to give a chart on which the state point is located from any two of the above properties, with all remaining properties determined directly without calculation.

For convenience, lines of constant wet-bulb (adiabatic saturation) temperature are shown as lines of constant enthalpy, although it has been demonstrated that the enthalpy increases slightly along a constant wet-bulb line as saturation is approached [see Eq. (14-7a)]. The magnitude of this deviation is small and can be neglected in many engineering calculations. Where needed, it is available directly from lines of enthalpy deviation plotted on the chart. This enthalpy deviation is equal to $(W - W^*)h^*$, where $h^*$ is the enthalpy of the water at the wet-bulb temperature [or, for temperatures above 32°F, $(W - W^*)(t - 32)$, thus allowing for the relatively insignificant middle term of Eq. (14-7a)].

To make an accurate enthalpy balance for a process of varying specific humidity, the enthalpy of added or rejected water must also be considered [Eq. (14-5b)]. This is determined from the product of the amount of water added or rejected multiplied by the enthalpy of the water at the temperature of addition or rejection, or $(W_1 - W_2)(t - 32)$ for liquid water (see Example 7).

**Example 6.** Repeat Example 1, using the psychrometric chart. Determine (a) the partial pressure of the water vapor and the dew point, (b) the density of each constituent, (c) the humidity ratio.

*Solution.* Given air at 80°F and $\phi = 0.30$.

(a) On the chart, locate 80°F on the abscissa; then ascend vertically to the curved line marked "30%" ($\phi$). Proceed horizontally from $t = 80$°F, $\phi = 30$ per cent, to the saturation curve (which is identified by the equality of the wet- and dry-bulb temperatures), and read the dew point, 46°F (*Ans.*), which agrees with Example 1. The partial pressure of the water vapor is obtained, as before, from the Steam Tables for the temperature of 46°F.

(*b*) Note the diagonal lines sloping downward to the right marked "cu ft/$lb_m$ dry air." At $t = 80°F$, $\phi = 30$ per cent, read (interpolate) $v_a = 13.74$ ft³/$lb_m$ dry air; therefore,

$$\rho_a = \frac{1}{v_a} = 0.0728 \text{ lb}_m/\text{ft}^3 \qquad Ans.$$

compared with $\rho_a = 0.0729$ $lb_m$/ft³ of Example 1. In this same volume is contained 46 grains of water vapor (scale on right ordinate) and

$$\rho_w = \frac{m_w}{V} = \frac{46/7,000}{13.74} = 0.000478 \text{ lb}_m/\text{ft}^3 \qquad Ans.$$

compared with 0.000474 $lb_m$/ft³ of Example 1.

(*c*) The specific humidity is directly the mass of water vapor associated with 1 $lb_m$ dry air, and as found in (*b*),

$$W = 46 \frac{\text{grains water}}{\text{lb}_m \text{ dry air}} = 0.00657 \frac{\text{lb}_m \text{ water}}{\text{lb}_m \text{ dry air}} \qquad Ans.$$

compared with 0.0065 of Example 1.

**Example 7.** Repeat Examples 2 and 5, using the psychrometric chart.

*Solution.* Enter chart at 30°F and 60 per cent relative humidity, and read, on right-hand ordinate, $W_1 = 14.4$ grains/$lb_m$, or 0.00206 $lb_m$/$lb_m$ dry air.

Enter chart at 80°F and 50 per cent relative humidity, and read $W_2 = 76.3$ grains/$lb_m$, or 0.01090 $lb_m$/$lb_m$ dry air. Hence,

$$W_2 - W_1 = 0.00884 \text{ lb}_m \qquad \text{or } 61.9 \text{ grains/lb}_m \text{ dry air}$$

At 30°F and $\phi = 0.60$, the chart shows that

$$H_1 = 9.30 + 0.11 = 9.41 \text{ Btu/lb}_m \text{ dry air}$$

The chart enthalpy deviation could have been calculated from $(W - W^*)h^*$, where $h^*$ is the enthalpy of ice at the wet-bulb temperature, or

$$(W - W^*)h^* = \frac{14.4 - 20}{7,000}(-144,4) = +0.115 \text{ Btu/lb}_m \text{ dry air}$$

At 80°F and $\phi = 0.50$,

$$H_2 = 31.30 - 0.11 = 31.19 \text{ Btu/lb}_m \text{ dry air}$$

Here $\quad (W - W^*)h^* = \frac{76.3 - 98.5}{7,000}(66.8 - 32) = -0.110 \text{ Btu/lb}_m \text{ dry air}$

Correction for the make-up water equals

$$(W_2 - W_1)(t - 32) = 0.00884(28) = 0.25 \text{ Btu/lb}_m \text{ dry air}$$

Hence, $\quad \Delta q' = H_2 - H_1 - (W_2 - W_1)h_3 = 31.19 - 9.41 - 0.25$
$$= 21.53 \text{ Btu/lb}_m \text{ dry air}$$

**14-6. Air-conditioning Applications.** *Adiabatic Mixing.* A steady-flow process frequently encountered in air conditioning is the adiabatic mixing of several streams of air to form a conditioned mixture. Suppose that two flow streams enter a system, adiabatically mix, and one flow stream emerges (Fig. 14-6). For this system,

$$\Sigma \dot{m}_f h_{\text{out}} = \Sigma \dot{m}_f h_{\text{in}}$$

Therefore,
$$\dot{m}_{fa1} H_1 + \dot{m}_{fa2} H_2 = (\dot{m}_{fa1} + \dot{m}_{fa2}) H_3$$
$$\dot{m}_{fa1} W_1 + \dot{m}_{fa2} W_2 = (\dot{m}_{fa1} + \dot{m}_{fa2}) W_3$$

Upon rearranging,
$$\frac{\dot{m}_{fa1}}{\dot{m}_{fa2}} = \frac{H_3 - H_2}{H_1 - H_3} = \frac{W_3 - W_2}{W_1 - W_3} \qquad (14\text{-}12a)$$

On the psychrometric chart, the final state 3 of the mixture lies on a straight line connecting the initial states of the two streams before mixing.

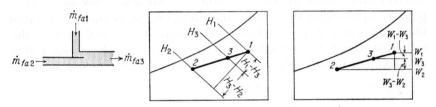

FIG. 14-6. Adiabatic mixing process on the psychrometric chart.

This is proved by Fig. 14-6 and Eq. (14-12a) (since the $H$ and $W$ scales are linear),

$$\frac{\text{Line 2-3}}{\text{Line 3-1}} = \frac{H_3 - H_2}{H_1 - H_3} = \frac{W_3 - W_2}{W_1 - W_3} = \frac{\dot{m}_{fa1}}{\dot{m}_{fa2}} \qquad (14\text{-}12b)$$

Moreover, as shown by Eq. (14-12b), the final state 3 divides the line into two parts that are in the same ratio as were the two mass-flow rates of the *dry air* before mixing.

**Example 8.** A stream of 2,000 cfm of saturated air at 50°F is mixed with 1,500 cfm of air at 80°F and $\phi = 60$ per cent. Determine the final condition of the mixture.

*Solution.* From the chart at 50°F, saturated, $v_a = 13.0$ ft³/lb$_m$ dry air; at 80°F, $\phi = 60$ per cent, $v_a = 13.88$ ft³/lb$_m$ dry air. Hence,

$$\dot{m}_{fa2} = \frac{2,000}{13.0} = 154 \ \frac{\text{lb}_m \ \text{dry air}}{\text{min}}$$

$$\dot{m}_{fa1} = \frac{1,500}{13.88} = 108 \ \frac{\text{lb}_m \ \text{dry air}}{\text{min}}$$

The distance between the two initial states can be multiplied by $108\!/\!262$ and distance 2-3 is located,

$$t = 62.5°F \qquad \phi = 0.83 \qquad Ans.$$

*Heating and Cooling without Change in Specific Humidity.* If air is heated or cooled without change in moisture content, the path for the process must lie on a horizontal line on the psychrometric chart, such as line 1-2 (Fig. 14-8). The heat transferred can be found from Eq. (14-4d) for the condition that

$$\dot{m}_{f3} = \dot{m}_{f4} = 0$$

*Cooling and Dehumidifying.* Suppose that a cooling coil is placed in the air stream, as illustrated in Fig. 14-7. One portion of the air will contact the coil surface and be cooled along path $ab$ to the dew point, and condensation will then take place as the temperature is reduced from $b$ to $c'$ or $c''$. A second portion of air, called the *bypass air*, may never strike the coil but instead may be cooled by mixing with the first and colder portion. Now, if the coil is infinitely long, the air will mix together and assume an average saturated end point such as $c$. Here point $c$, representing the *mean* surface temperature of the coil, is called the *apparatus dew point* (ADP). However, in the real system this saturated state is never reached, and the final state of the stream will be at a point such as $d$. Since the entire process involves mixing as well as

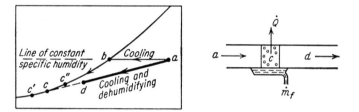

Fig. 14-7. Cooling and dehumidifying process on the psychrometric chart.

cooling, point $d$ will lie on a line between the initial state $a$ and the apparatus-dew-point state $c$. The heat transfer from this system is evaluated by modifying Eq. (14-4d).

The degree to which the actual process approaches the theoretical is described by the *bypass factor* (BF), defined as the ratio of line length $cd$ to line length $ca$. The bypass factor equals the amount of air bypassed, expressed as a decimal,

$$\text{BF} = \frac{t_d - t_c}{t_a - t_c} = 1 - \frac{t_a - t_d}{t_a - t_c}$$

Its usefulness arises from the fact that a given coil has essentially a fixed value of bypass factor for each value of air flow over it.

The relationship between the sensible and latent heat removals from air during a process of cooling and dehumidifying is expressed in terms of the *sensible heat factor* (SHF), defined as the ratio of sensible heat removal to the total heat removal, or

$$\text{SHF} = \frac{0.24(t_a - t_d)}{H_a - H_d} = \frac{0.24(t_a - t_c)}{H_a - H_c}$$

All cooling and dehumidifying lines of given slope on the psychrometric chart have the same value of sensible heat factor. Horizontal lines (sensible cooling only) have an SHF of 1.0, and the SHF decreases as the slope increases. The sensible heat factor is of importance in the design of air-conditioning systems because of the need to maintain desired conditions of both temperature and humidity. There is a definite relationship between the latent and sensible parts of the load which can be expressed as a *load sensible heat factor*. The *load* SHF must bear a definite relationship to the supply-air SHF in order to obtain the desired room condition. If the *load* SHF is very low (high percentage of latent load), it may be impractical to operate with a sufficiently low apparatus dew point to obtain the needed supply-air SHF. In this case, it may become necessary to cool and dehumidify to a practical apparatus dew point and then reheat the air (on chart move horizontally to right from the condition leaving the coil) enough to establish a net slope (that is, SHF) to satisfy the load.

Air can be cooled and dehumidified by passing the air stream through a spray or curtain of cold water that is chilled by an external refrigerator to a temperature lower than the dew point of the initial air. The air will be humidified and cooled if the water is supplied at a higher temperature than the dew point of the air stream and, of course, at a lower temperature than the air temperature. Such processes can be evaluated by Eqs. (14-4).

*Evaporative Cooling.* If the quantity of water sprayed into an air stream were reduced until no leaving stream of water left the system, it would be found that the air would be cooled and humidified. The air would be cooled even though the temperature of the entering water was higher than that of the air. In this instance, the heat capacity of the

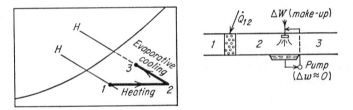

FIG. 14-8. The processes of heating without change in specific humidity and evaporative cooling on the psychrometric chart.

small amount of water that enters the system cannot markedly increase the temperature of the much greater amount of air. But when evaporation occurs and the high latent-heat demands of water are satisfied, a pronounced drop in temperature of the air takes place to supply the thermal energy. In a similar manner, if a small amount of very cold water is introduced and entirely evaporated, the chilling effect of the cold water on the hot air is negligibly small in comparison with the thermal demands of vaporization.

In evaporative cooling, a large amount of water is pumped into a spray chamber and constantly recirculated without transfer of heat. Thus, the temperature of the water will gradually decrease and approach as a limit the temperature of adiabatic saturation. To this circulating water must be added make-up water to replace the water vapor that has been formed and carried away by the air. The amount of make-up water will be small; thus, the effect of its temperature, being other than the wet-bulb temperature, on the equilibrium temperature is negligible.

This type of air washer, with recirculating water that is neither heated nor cooled from an external source, is a common means for cooling and humidifying air. The path followed by the air during the cooling is essentially at constant wet-bulb temperature (path 2-3, Fig. 14-8), and the energy balance for the process is given by Eq. (14-7a). (Note that process 1-2 in Fig. 14-8 is quite independent of process 2-3.)

**Example 9.** Air at a temperature of 90°F and 30 per cent relative humidity is cooled in an adiabatic spray humidifier that uses recirculated water. If the process reduces the air temperature by 80 per cent of the original wet-bulb depression, what will be the final condition of the air?

*Solution.* At $t = 90°F$, $\phi = 0.30$; then, from the chart $t^* = 67.2°F$.

$$\text{Original wet-bulb depression} = 90 - 67.2 = 22.8°F$$

Then, the temperature reduction is

$$\Delta t = 22.8(0.80) = 18.24°F$$

and the dry-bulb temperature equals

$$t = 90 - 18.24 = 71.76°F \qquad Ans.$$

(The final wet-bulb temperature is the same as the initial value: 67.2°F.) From the chart,

$$\phi = 0.80 \qquad Ans.$$

*Cooling Tower.* A cooling tower is often used to cool water by evaporation. At the top of the tower, hot water is sprayed downward through baffles that help to break up the water into small streams or drops to present large areas for evaporation. Atmospheric air enters the tower at the base and flows upward against the liquid particles; the hot water is cooled, mainly by vaporization of a part of the water, while the air is raised in temperature and also saturated (essentially) with water vapor. Because of the evaporation, the water may be cooled below the dry-bulb temperature of the entering air (but it cannot be cooled below the wet-bulb temperature). Equation (14-4d) can be directly applied to this adiabatic process.

Cooling towers are commonly rated in terms of *approach* of the cooled water to the atmospheric- (entering-) air wet-bulb temperature for a given *cooling range*, the decrease in water temperature. Although values vary considerably according to the economics of the selection of the components of the entire system, typical values are 7°F approach and 10°F cooling range for refrigeration-condenser water cooling and 10°F approach and 25°F cooling range for process water cooling.

**Example 10.** Water is cooled in a cooling tower from a temperature of 100 to 75°F. Air enters the tower at a temperature of 82°F with a relative humidity of 40 per cent and leaves at a temperature of 95°F with relative humidity of 98 per cent. Determine (a) the cooling range and approach, (b) the amount of water cooled per pound of dry air, and (c) the percentage of water lost by evaporation.

*Solution.* (a) Cooling range = $100 - 75 = 25°F$ (*Ans.*). Entering wet-bulb temperature at 82°F dry-bulb and 40 per cent relative humidity is 65°F, and approach = $75 - 65 = 10°F$ (*Ans.*).

(b) From the psychrometric chart,

$$t = 82°F \qquad \phi = 40\% \qquad H_1 = 30.0 \text{ Btu} \qquad W_1 = 65 \text{ grains}$$
$$t = 95°F \qquad \phi = 98\% \qquad H_2 = 62.2 \text{ Btu} \qquad W_2 = 254 \text{ grains}$$

The enthalpy of the water is found from Eq. (14-10a),

$$t = 100°F \qquad h_{f3} = 68 \text{ Btu/lb}_m \qquad t = 75°F \qquad h_{f4} = 43 \text{ Btu/lb}_m$$

With these values substituted into Eq. (14-4d), noting Eq. (14-4c),

$$62.2 - 30 + \frac{\dot{m}_{f4}(43) - (0.0270\dot{m}_{fa} + \dot{m}_{f4})68}{\dot{m}_{fa}} = 0$$

Hence,

$$\frac{\dot{m}_{f4}}{\dot{m}_{fa}} = 1.214 \frac{\text{lb}_m \text{ water}}{\text{lb}_m \text{ dry air}} \qquad Ans.$$

(c) The water lost by evaporation equals $\dot{m}_{f3} - \dot{m}_{f4} = \dot{m}_{fa}(W_2 - W_1) = 0.0270$ lb$_m$ water/lb$_m$ dry air, or

$$\frac{0.0270}{1.214 + 0.0270} = 0.0218, \text{ or } 2.18\% \qquad Ans.$$

Thus, for every pound of hot water entering the tower, 0.9782 lb$_m$ of cold water leaves.

## PROBLEMS

**1.** The temperature in a room is 68°F, and the specific humidity is 0.006. Calculate the relative humidity and the density of each constituent, and find the dew point.

**2.** Determine how much moisture must be added to or removed from 1,000 cfm of the air in Prob. 1 to change the temperature to 80°F and the relative humidity to 50 per cent.

**3.** Calculate the heat that must be transferred during the process of Prob. 2.

**4.** Air has a dry-bulb temperature of 60°F and a wet-bulb temperature of 50°F. Calculate the relative humidity, the humidity ratio, the dew point, and the density of the mixture.

**5.** Air has a dry-bulb temperature of 75°F and a relative humidity of 50 per cent. Determine the temperature of adiabatic saturation, the humidity ratio, the dew point, and the enthalpy of the mixture.

**6.** Check Probs. 4 and 5 by means of the psychrometric chart.

**7.** A stream of 1,000 cfm of saturated air at 60°F is mixed with 1,500 cfm of recirculated air at 75°F and 60 per cent. Determine the final condition of the mixture.

**8.** Air at 60°F and 30 per cent relative humidity is to be added to 1,000 cfm of air at 80°F and 60 per cent relative humidity to achieve a resultant mixture at 70°F. Determine the amount of air that must be added and the final conditions of the mixture.

**9.** A stream of 1,000 cfm of air at 90°F and 90 grains specific humidity is to be cooled to 75°F using a cooling coil with surface temperature above the dew point of the air. How much heat must be transferred?

**10.** The air in Prob. 9 is to be cooled and dehumidified by using a coil with a mean surface temperature of 50°F. If the resultant air stream after mixing and cooling has a relative humidity of 90 per cent, compute the amount of heat that was removed.

**11.** Air at 80°F dry-bulb and 50 per cent relative humidity passes over a cooling and dehumidifying coil and leaves at 60°F dry-bulb and 10 per cent relative humidity. From the psychrometric chart, find the apparatus dew point, the coil bypass factor, the heat removed per pound of dry air, and the sensible heat factor for the process.

**12.** Air at 75°F dry-bulb and 60 per cent relative humidity passes over a cooling and dehumidifying coil at a velocity such that the coil bypass factor is 0.10. The apparatus dew point is 50°F. Determine from the psychrometric chart the leaving-air dry-bulb temperature and relative humidity; the latent, sensible, and total heat removed; and the sensible heat factor.

**13.** The air leaving the coil of Prob. 12 is reheated to 60°F dry-bulb. Determine the net sensible heat factor of the combined processes.

**14.** Air at 50°F, $\phi = 40$ per cent, is to be conditioned to 72°F and $\phi = 50$ per cent, in the following processes: (a) heating at constant humidity ratio; (b) humidified at constant wet-bulb temperature (to saturation); (c) heating at constant humidity ratio. Determine the condition of the air after each process and the heat transferred during each process.

**15.** Repeat Prob. 14, assuming that the air temperature in the evaporative cooling process is reduced by 80 per cent of the original wet-bulb depression.

**16.** Water at 120°F enters a cooling tower and leaves at 77°F. Air enters the tower at 80°F and 30 per cent relative humidity and leaves at 110°F. Determine the air flow necessary to furnish 100,000 $lb_m$/hr of cooled water and the water make-up rate.

## SELECTED REFERENCES

1. Carrier, W. H., R. E. Cherne, W. A. Grant, and W. H. Roberts: "Modern Air-conditioning, Heating and Ventilating," Pitman Publishing Corporation, New York, 1959.
2. Heating, Ventilating, Air Conditioning Guide of the American Society of Heating and Air-conditioning Engineers, Inc., New York.

# CHAPTER 15

# VAPOR CYCLES AND PROCESSES

Let us, while waiting for new monuments,
preserve the ancient monuments.

*Victor Hugo*

In the past, the vapor cycle has had heat supplied by a combustion process; now, and in the future, the nuclear reactor has replaced, and will continue to replace, the furnace.

**15-1. The Cycle Work Ratio.** A theoretical power cycle may have a high thermal efficiency; yet the real prototype may have an extremely low thermal efficiency. A warning of this possibility is given by the *work ratio* $r_W$,

$$r_W = \frac{\text{net work from cycle}}{\text{plus work of cycle}} = \frac{\Sigma \, \Delta W}{\Sigma + \Delta W} \tag{15-1}$$

For example, the work quantities and work ratio for a Carnot cycle are found to be (Example 1)

$$\Delta W_{\text{turb}} = 452.7 \text{ Btu} \qquad \Delta W_{\text{comp}} = -123.1 \text{ Btu} \qquad r_W = \frac{329.6}{452.7} = 0.729$$

Suppose that the expansion and compression efficiencies of the real prototype cycle are 0.50; then

$$\Delta W_{\text{turb}} = 226.35 \text{ Btu} \quad \Delta W_{\text{comp}} = -246.2 \text{ Btu} \quad r_W = \frac{-19.85}{226.35} = -0.0825$$

And the real cycle cannot operate as a power cycle because work must be supplied! Moreover, the *fixed costs* of the installation are high when the work ratio is low. For a cycle with work ratio of 0.1, it would be necessary to buy, install, and maintain a turbine of 10,000 hp and a compressor of 9,000 hp to produce 1,000 hp!

**15-2. The Carnot Power Cycle.** Suppose that a Carnot cycle operates within the two-phase region of a substance (Fig. 15-1a). Here processes $bc$ and $da$ are at constant pressure as well as at constant temperature. If the processes are nonflow, the substance must be confined in a cylinder by a piston. An essentially adiabatic compression or expansion process could be readily accomplished, but the isothermal processes ($bc$ and $da$)

would involve both heat and work.   In practice, it is easier to construct machines that involve only work or only heat because the combination of work and heat in one process requires the real machine to be operated at very slow speeds.    Also, the changes in volume are large, and therefore the piston-cylinder size tends to be excessive for any process.    For these reasons, a series of flow processes is best used for executing the vapor cycle.    Thus, the isothermal processes $bc$ and $da$ of Fig. 15-1$a$ (which are difficult processes in a nonflow cycle) can be easily approached by using flow processes: a *boiler* for process $bc$ and a *condenser* for process $da$.

In the flow system of Fig. 15-1$b$, saturated vapor $c$ leaves the boiler, enters the turbine, and expands to state $d$.   The two-phase mixture at $d$ enters the condenser, where it is cooled to state $a$, and then it enters a

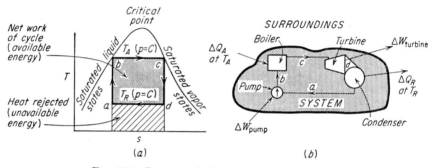

Fig. 15-1. Carnot cycle for a vapor such as steam.

compressor, where it is compressed to the saturated-liquid state $b$.    The cycle is completed by vaporizing the liquid in the boiler at constant pressure and temperature.   Since each process involves only heat or only work, high-speed operation can be accomplished and therefore the Carnot cycle could serve as the model for a power plant.

The heat passing to the power cycle is obtained, usually, from a combustion process and, for high furnace efficiency (Art. 8-4), the gases of combustion should be cooled as close as practicable† to the surrounding temperature $T_0$.    But consider Fig. 15-2.    If a Carnot cycle were to be used with this heat reservoir, the combustion gases could be cooled, at the most, to $T_A$ (and the practical limit would be $T_A$ plus several hundred degrees in order to have speed of heat transfer).   It follows from Fig. 15-2 that the furnace for a Carnot cycle would have an extremely low efficiency.

---

† Actually, a temperature of about 300°F is desired to avoid condensation of water, which leads to corrosion of the chimney of the furnace (and sulfur dioxide formed by combustion of sulfur in the fuel unites with water to form sulfurous acid).

It can be concluded that, although the Carnot cycle could serve as the model for a power cycle, a better selection can undoubtedly be made because of the following objections:

1. The low work ratio indicates that the real power cycle will probably have a low thermal efficiency.

2. The low work ratio predicts high fixed costs for a given output.

3. The furnace efficiency will be low since low-temperature heat is not required by the cycle.

Because of objections 1 and 3, it can be surmised that the power plant which uses a Carnot cycle as the model will have a low over-all efficiency (Examples 2 and 4).

**15-3. The Rankine Power Cycle.** A cycle with a high work ratio can be obtained by substituting a liquid-compression process for the work-consuming vapor-compression process of the Carnot cycle. Thus, in Fig. 15-3, the Carnot cycle is modified into a *Rankine cycle* by completing the condensation process to the saturated-liquid state in the condenser and then reversibly pumping the saturated liquid to the boiler pressure, state $b$. By so doing, two objectives are attained:

1. The work ratio is raised considerably since relatively little work is required to pump a liquid (Art. 5-8$d$ and Chap. 9, Example 7).

2. The furnace efficiency is raised since heat is now required to raise the temperature of the liquid from state $b$ to state $c$.

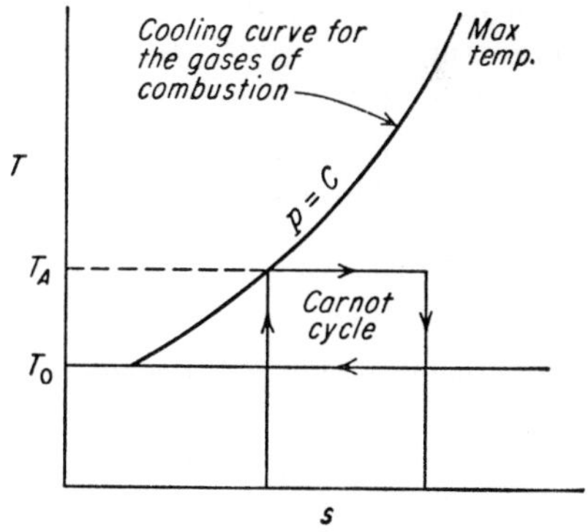

FIG. 15-2. Carnot cycle can cool the combustion gases only to $T_A$.

Thus, heat is added at constant pressure and with increasing temperature from $b$ to $c$, and at constant temperature and pressure from $c$ to $d$: processes that take place in the boiler. The remaining processes of the cycle are reversible and adiabatic expansion $de$ in a turbine, constant-temperature and constant-pressure condensation $ea$ in a condenser, and reversible adiabatic compression $ab$ in a liquid pump.

In an idle steam power plant, the condenser contains air at atmospheric pressure. When the turbine is started, steam can expand only to the condenser pressure, and this will be atmospheric pressure. An air vent on the condenser is connected to an air pump, and the air (and its contained steam vapor) is pumped from the condenser. The exhaust steam from the turbine, on condensing upon the cold condenser tubes, suffers a tremendous contraction of volume, and therefore the pressure falls. The pressure continues to fall until all the air is removed; then, the pressure in the condenser will be that dictated by the vapor pressure of the condensate. This pressure is controlled by the temperature of the cooling-water supply. For example, when steam condenses, a high vacuum exists at the normal temperature of the coolant; at 60°F the saturation pressure of steam is 0.2563 psia. Thus, the air pump must be kept in continuous operation to remove air that leaks into the high vacuum system, and this pump work is an added cost of operation. Note, especially, that the steam from the turbine can expand only to the *mixture pressure* in the condenser; hence, the presence of air prevents the attainment of the maximum work.

Although this argument is based upon pressure, it should be remembered that pressure is only the incidental factor in obtaining the maximum work. The prime essential is to operate the cycle between the highest and the lowest possible *tempera-tures*. But the attainment of the lowest temperature is also reflected by the attainment of a definite pressure, which is dictated by the pressure-temperature characteristics of the fluid. Therefore, fluids, such as steam, which have low vapor pressures at atmospheric temperatures are undesirable because leakage of air into the system will prevent expansion to the pressure dictated by the *lowest possible temperature*. (But water is inexpensive.)

**Example 1.** Determine the thermal efficiency of the cycle and the over-all efficiency for a power plant using the Carnot cycle of Fig. 15-1 if the furnace efficiency is 75 per cent, the pressure at $bc$ is 600 psia, and the temperature at $ad$ is 60°F.

*Solution.* The work of each process will be found in view of Example 2,

$$p = 600 \text{ psia} \qquad h_g = 1{,}203.2 \qquad h_f = 471.6 \qquad h_{fg} = 731.6$$
$$t = 486.21°F \qquad s_g = 1.4454 \qquad s_f = 0.6720 \qquad s_{fg} = 0.7734$$

$$p = 0.2563 \text{ psia} \qquad h_g = 1{,}088.0 \qquad h_f = 28.06 \qquad h_{fg} = 1{,}059.9$$
$$t = 60°F \qquad s_g = 2.0948 \qquad s_f = 0.0555 \qquad s_{fg} = 2.0393$$

The enthalpy at state $d$ can be found from the Mollier chart or by calculation,

$$s_c = s_d = s_f + xs_{fg}$$
$$1.4454 = 0.0555 + x(2.0393) \qquad x_d = 0.6816$$
$$h_d = h_f + xh_{fg}$$
$$= 28.06 + 0.6816(1{,}059.9) \qquad h_d = 750.5$$

and, in the same manner,

$$x_a = 0.3023 \qquad h_a = 348.5$$

The work for each process is

$$\Delta w_{bc} = \Delta w_{da} = 0 \qquad \text{(constant-pressure flow process)}$$
$$\Delta w_{cd} = -\Delta h = 1203.2 - 750.5 = 452.7 \text{ Btu/lb}_m$$
$$\Delta w_{ab} = -\Delta h = 348.5 - 471.6 = -123.1$$
$$\Sigma \Delta w = 329.6 \text{ Btu/lb}_m$$

This answer can be checked:

$$\Sigma \Delta w_{rev} = \Delta h_{bc} - T_R \Delta s_{bc} = 731.6 - 519.7(0.7734) = 329.7 \text{ Btu/lb}_m$$

With these data,

$$\eta_t = \frac{T_A - T_R}{T_A} = \frac{486.21 - 60}{486.21 + 459.7} = 0.45, \text{ or } 45\% \qquad Ans.$$

The over-all efficiency is the product of furnace and cycle efficiencies,

$$\eta = \eta_t \eta_f = 0.45(0.75) = 33.7\% \qquad Ans.$$

**Example 2.** Upon test of the cycle of Example 1, the expansion efficiency was found to be 0.85 and the compression efficiency 0.50, although both processes were essentially adiabatic. The furnace efficiency was not changed by the irreversibilities in the cycle. Determine the thermal efficiency of the cycle and the over-all efficiency of the plant.

*Solution.* The cycle for this problem would resemble that in Fig. *A*. For the work of the compressor and the state at $b'$,

$$\eta_c = \frac{\Delta w_{ab}}{\Delta w_{ab'}} = \frac{-\Delta h_{ab}}{-\Delta h_{ab'}} = \frac{-123.1}{-\Delta h_{ab}} = 0.50$$

$$\Delta w_{ab'} = \frac{-123.1}{0.50} = -246.2 \text{ Btu/lb}_m$$

$$h_{b'} = h_a + \Delta h_{ab} = 348.5 + 246.2 = 594.7 \text{ Btu/lb}_m$$

For the work of the turbine,

$$\eta_e = \frac{\Delta w_{cd'}}{\Delta w_{cd}} = \frac{-\Delta h_{cd'}}{-\Delta h_{cd}} = \frac{-\Delta h_{cd'}}{452.7} = 0.85$$

$$\Delta w_{cd'} = 452.7(0.85) = 384.8 \text{ Btu/lb}_m$$

The heat transferred to the cycle is

$$\Delta q_A = h_c - h_{b'} = 1203.2 - 594.7 = 608.5 \text{ Btu/lb}_m$$

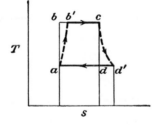

FIG. *A*

Note that less heat need be added than in Example 2 because of the dissipation of work in fluid friction. With these data,

$$\eta_t = \frac{\Sigma \Delta w}{\Delta q_A} = \frac{384.8 - 246.2}{608.5} = 0.228, \text{ or } 22.8\% \qquad Ans.$$

$$\eta = \eta_t \eta_f = 0.228(0.75) = 0.171, \text{ or } 17.1\% \qquad Ans.$$

**Example 3.** Repeat Example 1, but for a Rankine cycle (Fig. 15-3) and therefore with an improved furnace efficiency, say 85 per cent.

*Solution.* The pump work for the liquid is closely $-v \Delta p$ (Chap. 9, Example 7), and at $t = 60°F$, $v_f = 0.01604 \text{ ft}^3/\text{lb}_m$; hence,

$$\Delta w = -\frac{0.01604(600 - 0.2563)(144)}{778.16} = -1.78 \text{ Btu/lb}_m$$

The properties at each state are (Steam Tables and Example 1)

$$h_a = 28.06 \qquad h_b = 29.84 \qquad h_d = 1203.2 \qquad h_e = 750.5$$
$$s_a = 0.0555 \qquad s_b = 0.0555 \qquad s_d = 1.4454 \qquad s_e = 1.4454$$

Then,
$$\Delta q_A = h_d - h_b = \quad 1173.4 \text{ Btu/lb}_m$$
$$\Delta q_R = h_a - h_e = - \ 722.4$$
$$\Sigma \ \Delta w = \quad 451.0 \text{ Btu/lb}_m$$

$$\Delta w_{\text{turb}} = h_d - h_e = 452.7 \text{ Btu/lb}_m$$
$$\Delta w_{\text{pump}} = h_a - h_b = -1.78 \text{ Btu/lb}_m$$

With these data,

$$\eta_t = \frac{\Sigma \ \Delta w}{\Delta q_A} = \frac{451.0}{1173.4} = 0.384, \text{ or } 38.4\% \qquad Ans.$$

$$\eta = \eta_t \eta_f = 0.384(0.85) = 0.326, \text{ or } 32.6\% \qquad Ans.$$

Note that the thermal efficiency of the Rankine cycle is *less* than that of the Carnot cycle but that the over-all efficiency of the Rankine power plant is about the same as that of the Carnot power plant (Example 1; see also Example 4).

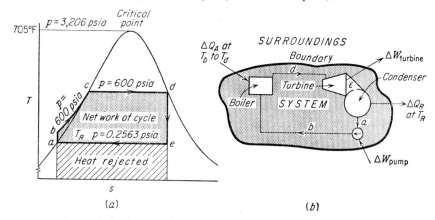

FIG. 15-3. Rankine cycle (temperature rise *ab* greatly exaggerated).

**Example 4.**   Repeat Example 2, but for the Rankine cycle and the data of Example 3.

*Solution.*   The cycle for this problem would resemble that in Fig. *B*.   For the work of the pump and the state at *b'*,

$$\eta_c = \frac{\Delta w_{ab}}{\Delta w_{ab'}} = \frac{-\Delta h_{ab}}{-\Delta h_{ab'}} = \frac{-1.78}{-\Delta h_{ab'}} = 0.50$$

$$\Delta w_{ab'} = \frac{-1.78}{0.50} = -3.56 \text{ Btu/lb}_m$$

$$h_{b'} = h_a + \Delta h_{ab'} = 28.06 + 3.56 = 31.62 \text{ Btu/lb}_m$$

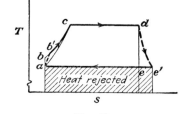

FIG. *B*

The work of the turbine will be the same as in Example 2, 384.8 Btu/lb$_m$.  The heat transferred to the cycle is

$$\Delta q_A = h_d - h_{b'} = 1203.2 - 31.6 = 1171.6 \text{ Btu/lb}_m$$

Therefore, $\quad \eta_t = \dfrac{\Sigma \Delta w}{\Delta q_A} = \dfrac{384.8 - 3.56}{1171.6} = 0.325$, or 32.5%  *Ans.*

$$\eta = \eta_t \eta_f = 0.325(0.85) = 0.276, \text{ or } 27.6\% \qquad Ans.$$

Hence, the Rankine cycle as a model for the power plant is far more desirable than a Carnot model since the furnace efficiency can be higher and the cycle itself is less sensitive to irreversibilities: the over-all efficiency is *greater* than that of the Carnot model.

The data from Examples 1, 2, 3, and 4 can be compared:

|  | Example 1 | Example 2 | Example 3 | Example 4 |
|---|---|---|---|---|
|  | Theoretical Carnot | Carnot model | Theoretical Rankine | Rankine model |
| Thermal efficiency, %.......... | 45 | 22.8 | 38.4 | 32.5 |
| Over-all efficiency, %.......... | 33.7 | 17.1 | 32.6 | 27.6 |
| Work ratio................. | 0.729 | 0.360 | 0.995 | 0.990 |

The foregoing examples indicate that, although the Carnot cycle is theoretically most desirable, practically, the Rankine cycle is less sensitive to irreversibilities that are always present in real systems.  Since the net work of the Carnot cycle is low and, therefore, equipment size and cost high, it becomes evident that improvements in the Rankine cycle are the most logical fields to explore.

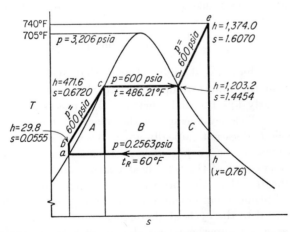

FIG. 15-4. Rankine cycle with superheated steam (temperature rise *ab* greatly exaggerated).

The thermal efficiency of any cycle can be increased by increasing the temperature of the heat-addition process. But for water, as is evident in Fig. 15-4, the increase in temperature is accompanied by a considerable increase in pressure. (In fact, for a temperature of 705°F—the critical temperature—the critical pressure is 3,206 psia.) Because of this, steam must be superheated if the highest possible temperature is to be attained in the cycle. In Fig. 15-4 is shown a Rankine cycle that uses steam superheated to 740°F (the name *Rankine cycle* does not limit the state of the steam). This cycle can be analyzed in the same manner as illustrated in Examples 1 and 3. However, it is instructive to consider the cycle to be made up of three independent cycles, A, B, and C. Cycle A is the cycle added to the Carnot cycle B to eliminate the vapor-compression process. Cycle C can be called the superheat cycle. Then for these cycles:

|  | 1 | | 2 | 3 | 4 |
|---|---|---|---|---|---|
|  | Heat added | | Heat rejected | Work (1) + (2) | Thermal efficiency, % (3) ÷ (1) |
|  | Btu | % of total | | | |
| Cycle A | $h_c - h_b$ | .... | $-T_R(s_c - s_b)$ | | |
|  | 441.8 | 32.8 | $-320.4$ | 121.4 | 27.5 |
| Cycle B | $h_d - h_c$ | .... | $-T_R(s_d - s_c)$ | | |
|  | 731.6 | 54.5 | $-401.9$ | 329.7 | 45.1 |
| Cycle C | $h_e - h_d$ | .... | $-T_R(s_e - s_d)$ | | |
|  | 170.8 | 12.7 | $-84.0$ | 86.8 | 50.8 |
|  | $\Sigma \Delta Q_A = 1344.2$ | 100.0 | $\Sigma \Delta Q_R = -806.3$ | $\Sigma \Delta W = 537.9$ | |

$$\eta_t = \frac{\Sigma \Delta W}{\Sigma \Delta Q_A} = 0.400, \text{ or } 40.0 \text{ per cent}$$

The tabled values reveal, quite clearly, that cycle A, wherein water is heated from the lowest temperature to the saturation temperature, has a particularly low thermal efficiency that exerts a strong influence on the over-all efficiency because of the relatively large amount of heat added. Cycle C does not radically affect the over-all thermal efficiency (or the efficiency of the Carnot cycle B) because the superheat accounts for only 12.7 per cent of the total heat transferred to the system. This analysis indicates that the properties of water are not particularly suited to the prime requirement for maximum thermal efficiency: heat to be transferred to the cycle at the highest possible temperature.

An advantage of superheat, not shown by a theoretical analysis, is the elimination, in part, of moisture from the steam in the last stages

of the expansion in the turbine. Note that increasing the pressure†
of the Rankine cycle and so increasing the temperature of heat addition also cause the final condition of the steam to become increasingly
wet (Fig. 15-4). The presence of over 10 per cent moisture in the
steam will cause quite serious erosion of the turbine blades and, also, a
decreased engine efficiency; these conditions can be corrected by super-
heating the steam.

**15-4. The Ideal Fluid for the Rankine Cycle.** In view of the dis-
closures in Art. 15-3, a digression will be made here to consider the
properties of an ideal fluid. The ideal fluid for the Rankine cycle, and

therefore the ideal fluid for the industrial power plant, should have the
following properties:

1. The latent heat of vaporization
should be large and the heat capacity
of the liquid should be small [at least
relative to each other, for then the
effect of cycle $A$ (Fig. 15-4) on cycle
$B$ would be negligible and the efficiency
of the Rankine cycle (cycles $A + B$)
would be essentially that of the
Carnot cycle $B$].

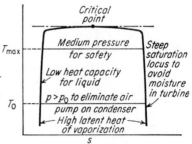

FIG. 15-5. Characteristics of an ideal
fluid for the Rankine cycle.

2. The critical point should be above the highest operating temperature. [This would allow the temperature of heat addition to cycle $B$
(Fig. 15-4) to be increased to the highest possible value that the material
can withstand.]

3. The vapor pressure, at the highest operating temperature, should
not be high. (High pressures not only are dangerous but also increase
design costs and maintenance problems.)

4. The vapor pressure, at the lowest operating temperature, should
be higher than atmospheric pressure. (This would prevent air from
leaking into the condenser, thus raising the mixture pressure, which is
the pressure of exhaust for the turbine and therefore controls the work
output to a great extent.)

5. The entropy of the saturated vapor should not change markedly
with change of pressure; this would require the locus of saturated-vapor
states on the temperature-entropy diagram to be steep. [However, it
should not be at such an angle that the steam is in the superheat region
as it enters the condenser, for then the steam has not expanded to the

† Most modern large central stations are designed for 800 to 900 psia, 900 to 925°F
operation, although pressures of 1,500 psia are not uncommon and several *super-
critical* (4,500 psi) plants are in operation. Smaller industrial plants use pressures
of 300 to 600 psia and temperatures of 450 to 750°F.

lowest possible temperature, the condensation temperature at the exhaust pressure. Too, the rate of heat transfer is higher for wet steam than for superheated steam. But neither should the steam, in passing through the turbine, become too wet (more than 10 per cent moisture); else erosion of the turbine blades and reduced blading efficiency will become problems.]

6. The properties of the fluid should be conducive to high rates of heat transfer in order that both surface areas and temperature differences can be small in the heat exchanger.

7. The fluid should be cheap in cost, stable, nonexplosive, and non-corrosive under all conditions of operation, and nonpoisonous for safety of personnel.

No fluid is known that satisfies all these desirable characteristics, which are illustrated, in part, by Fig. 15-5. Water, of course, is the usual fluid for the power cycle because it is cheap and plentiful. The problem, then, is either to discover a perfect fluid or else to adapt the Rankine cycle to the properties of water in order to obtain the highest possible thermal efficiency.

**15-5. The Reheat Cycle.** It should now be evident that raising the average temperature of heat additions to the water in a Rankine cycle requires not only superheat but also a high pressure, which causes an excessive increase in the moisture content of the steam in the expansion process through the turbine. This fault can be corrected by the *reheat cycle* (Fig. 15-6), which is a modification of the Rankine cycle. Here steam is expanded in a turbine to an intermediate pressure ($h$) where the moisture content is not excessive. The steam is piped back to the furnace, reheated to the initial temperature (process $hi$), and then expanded in a second turbine to the condenser temperature (process $ij$). Inspection of Fig. 15-6 shows that expansion within the two-phase region has been decreased by the reheat.

Whether or not the reheat cycle will have a greater efficiency than the Rankine cycle can be visualized by means of Fig. 15-6. Note that the reheat cycle can be considered to be made up of cycles $A$, $B$, $C$, and $D$. Hence, reheating the steam will increase the thermal efficiency only if the efficiency of cycle $D$ is greater than the efficiency of the Rankine cycle ($A + B + C$). The thermal efficiency of cycle $D$ depends directly upon the temperature of state $h$. If the temperature of $h$ is close to the temperature of $d$, then reheating will raise the thermal efficiency of cycle $ABCD$ above that of cycle $ABC$. If reheating is delayed until the expansion has proceeded to the neighborhood of point $k$, then cycle $D$ may have a low thermal efficiency, and the combined efficiency of cycle $ABCD$ may be less than the efficiency of the Rankine cycle $ABC$.

Although reheat can be employed as a means for raising the thermal efficiency of a cycle, the thermodynamic gain is small, and therefore the main purpose of the reheat is to eliminate erosion caused by moisture. Because of this, reheating is used when high operating pressures cause a high percentage of moisture to appear in the expansion process. Not only is erosion relieved, but also the engine efficiency of the turbine is increased by the decrease in moisture. Accompanying these advantages, however, are the disadvantages caused by the increased cost and complexity of the system. Since the steam to be reheated is at low pressure, the volume is large and the friction losses are high in transporting the steam to the reheater in the furnace and then back again to the low-pressure turbine. For these reasons, increasing the superheat, without reheating, may be an easier but less effective means of approaching the same end point.

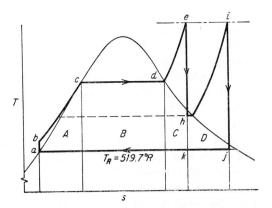

Fig. 15-6. Reheat cycle (temperature rise $ab$ greatly exaggerated).

**15-6. The Regenerative Cycle.** The thermal efficiency of the Rankine cycle is less than that of the Carnot cycle because heat is added at other than the highest temperature. But this deficiency can be eliminated by *regenerative heating*, although the method to be first discussed is highly impractical. In Fig. 15-7, the condensate is pumped to boiler pressure and then led through heat-transfer coils installed in the turbine. By this method, the fluid can be conceived to be reversibly raised in temperature from $b$ to $c$, while the expanding fluid can be conceived to be reversibly cooled from $d$ to $e$. The thermal efficiency of this *regenerative cycle* is equal to the Carnot efficiency. The proof of this statement is contained in the following three conditions: (1) Heat is added to the cycle at one constant temperature $T_A$. (2) Heat is rejected from the cycle at another constant temperature $T_R$. (3) All processes are (or can be conceived to be) reversible. Then, by the reasoning in Art. 7-1, the

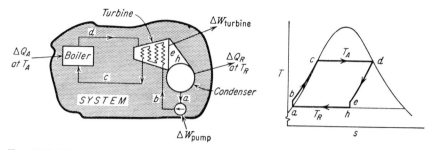

FIG. 15-7. Keenan's theoretical regenerative cycle (temperature change *ab* and, therefore, *eh* greatly exaggerated).

efficiency of this reversible cycle is equal to that of a Carnot cycle, which employs the same two heat reservoirs.

However, even if such a regenerative turbine could be constructed, it would be undesirable for fluids such as water, because the moisture content of the expanding steam in the turbine would be increased by the cooling. An alternative method can be proposed. In Fig. 15-8a is shown a regenerative system wherein a small fraction $y$ of the steam is extracted from the turbine before it has completely expanded to the final temperature. The extracted steam is mixed with the feedwater in an "open" heater.† In this manner the temperature of the water can be increased (from $b$ to $c$) by condensing the extracted steam (from $i$ to $c$) without changing the quality of the steam within the turbine. With an infinity of extraction points, each at a different temperature of the expansion process, and an infinity of heaters, the temperature difference between the extracted steam and the feedwater will be made infinitely small and the irreversibility of mixing will be similarly decreased.

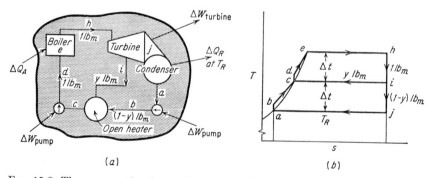

FIG. 15-8. The regenerative (extraction) system (temperature rise in pumps greatly exaggerated).

† An *open* heater is a *mixing* or *direct-contact* heater. In a *closed* heater, the hot and cold fluids are separated, heat being transferred through metal coils (Fig. 15-9).

For this hypothetical system, heat will be transferred only at the highest and lowest temperatures that are experienced by the system. With a finite number of extraction points, the irreversibility of each mixing process is a loss in available energy. Despite this loss, the thermal efficiency of the irreversible regenerative "cycle" can be greater than the thermal efficiency of the reversible Rankine cycle. This is so because heat is added at a higher average temperature in the regenerative system, and therefore a greater percentage of this heat can be converted into work. The proof of this statement is easily shown. Let 1 lb$_m$ of steam enter the condenser of Fig. 15-8 for either an extraction cycle or a Rankine cycle so that $\Delta q_R$ is a fixed number:

$$\eta_t = \frac{\Sigma \, \Delta W}{\Sigma \, \Delta W + \Delta q_R} = \frac{1}{1 + \Delta q_R/(\Sigma \, \Delta W)}$$

Since $\Sigma \, \Delta W$ will be greater for the extraction cycle of Fig. 15-8 [(1 + y) lb$_m$ of steam must enter the turbine at $h$], its thermal efficiency must also be greater than that of the nonextraction cycle.

The temperature-entropy diagram for the fluid in a one-heater regenerative system is constructed, more usually, for 1 lb$_m$ and $(1 - y)$ lb$_m$ of fluid, and the quantity of fluid present in each process is shown on the diagram (Fig. 15-8b). Note that the quantities of steam $y$ and $(1 - y)$ undergo different cycles. The path *abcdehija* is followed by the fraction $(1 - y)$; the path *cdehic* is followed by the fraction $y$. Thus, the name regenerative cycle is not precisely correct because a combination of cycles is present; usage, however, justifies the name. In other words, the irreversible regenerative system, itself, cannot be shown on the *Ts* diagram, but the paths followed by each portion of the fluid in the real system can be accurately portrayed. However, because of the irreversibilities, proper interpretation must be made as to which of the areas on the diagram represent transfers of heat to the system. The heat added to the system (determined from Fig. 15-8a) is

$$\Delta Q_A = h_h - h_d \frac{\text{Btu}}{\text{lb}_m \text{ water through boiler}}$$

The heat rejected from the system involves only the fraction $(1 - y)$:

$$\Delta Q_R = (1 - y)(h_a - h_j) \frac{\text{Btu}}{\text{lb}_m \text{ water through boiler}}$$

The work done in processes *hi* and *cd* is

$$\Delta W_{hi} = (h_h - h_i) \frac{\text{Btu}}{\text{lb}_m \text{ water through boiler}}$$

$$\Delta W_{cd} = (h_c - h_d) \frac{\text{Btu}}{\text{lb}_m \text{ water through boiler}}$$

Processes $ij$ and $ab$ involve only the fraction $(1 - y)$, and the work is

$$\Delta W_{ij} = (1 - y)(h_i - h_j) \frac{\text{Btu}}{\text{lb}_m \text{ water through boiler}}$$

$$\Delta W_{ab} = (1 - y)(h_a - h_b) \frac{\text{Btu}}{\text{lb}_m \text{ water through boiler}}$$

For a finite number of heaters, the irreversibility of mixing can be held to a minimum by dividing the temperature rise equally among the heaters. Thus, in Fig. 15-8 the cold water at $b$ could be pumped into a boiler and the temperature difference $be$ would exist between the cold water and the boiler water. This irreversibility is cut in half by using one extraction heater to raise the temperature from $b$ to $c$ before the water enters the boiler. (In effect the boiler acts as a heater to raise the temperature from $d$ to $e$.) If two heaters are used, the temperature rise from $b$ to $e$ is divided into three parts, one part for each heater and the third part for the boiler. (Even if the cycle is a superheat cycle, the same reasoning is valid: The temperature rise from condenser to boiler is divided into equal increments without regard for the temperature of superheat.)

The amount of steam that must be extracted for each heater is readily found from the First Law. The heater in Fig. 15-8 can be arbitrarily

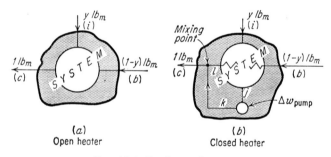

(a)
Open heater

(b)
Closed heater

FIG. 15-9. Feedwater heaters.

enclosed by a boundary to form a new system (Fig. 15-9a). For this flow system, which is essentially adiabatic,

$$\Sigma \, \Delta Q - \Sigma \, \Delta W = \Sigma m \, \Delta h$$

Therefore,

$$\Sigma m \, \Delta h = 0$$
$$h_c = y h_i + (1 - y) h_b$$
$$y = \frac{h_c - h_b}{h_i - h_b}$$

If two or more heaters are used, each heater can be considered to be an independent system and analyzed in the same manner as shown here. The heater closest to the boiler must be first analyzed, or else there will be more than one unknown.

In commercial applications, *closed* heaters, such as that illustrated in Fig. 15-9b, are more usually encountered. Here the feedwater is passed through coils in the heater and not directly mixed with the extracted steam, although the condensate from the extracted steam is pumped into the feedwater line, as shown in Fig. 15-9b. For the system of Fig. 15-9b, the First Law can be applied:

$$\Sigma \, \Delta Q - \Sigma \, \Delta W = \Sigma m \, \Delta h \qquad \text{and} \qquad \Sigma \, \Delta Q = 0$$

and
$$-y \, \Delta w = h_c - y h_i - (1 - y) h_b$$

$$y = \frac{h_c - h_b}{h_i - h_b - \Delta w}$$

(In this equation, the value substituted for $\Delta W$ will be negative, by convention.) In the actual installation the saturation temperature $t_i$ of the extracted steam is hotter than the temperature $t_t$ of the leaving feedwater, and this *terminal difference* may range in value from 5 to 20°.

**Example 5.** A Rankine cycle operates with steam between the limits of 600 psia, 740 and 60°F. (The same data were used in Art. 15-3 for the superheated Rankine cycle.) Determine the thermal efficiency if a heater is installed to heat the feedwater in the manner illustrated in Fig. 15-8.

*Solution.* The optimum extraction pressure for the heater is first determined:

$$p = 600 \text{ psia} \qquad t_e = 486.21°F$$
$$t_b = \phantom{0}60°F$$
$$\overline{546.21} \div 2 = 273.1°F$$

The pressure corresponding to this temperature is 44 psia. The properties at different states throughout the cycle are shown in Fig. *C* (Mollier chart and Steam Table values).

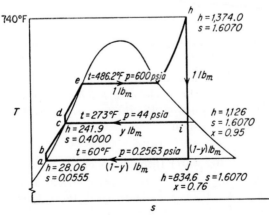

Fig. *C*

The enthalpies at states $b$ and $d$ are calculated:

$$h_d = h_c + v\,\Delta p = 241.9 + \frac{0.0172(600 - 44)(144)}{778.2} = 243.72 \text{ Btu/lb}_m$$

$$h_b = h_a + v\,\Delta p = 28.06 + \frac{0.0160(44 - 0.25)144}{778.2} = 28.19 \text{ Btu/lb}_m$$

The fraction of extracted steam is calculated:

$$y = \frac{h_c - h_b}{h_i - h_b} = \frac{241.95 - 28.19}{1128 - 28.19} = 0.195$$

The heat added is equal to

$$\Delta Q_A = h_h - h_d = 1374 - 243.7 = 1130.3 \text{ Btu}$$

The heat rejected is equal to

$$\Delta Q_R = (h_a - h_j)(1 - y) = (28.06 - 834.6)(0.805) = -649.3 \text{ Btu}$$

and the thermal efficiency is

$$\eta_t = \frac{\Delta Q_A + \Delta Q_R}{\Delta Q_A} = \frac{481.0}{1130.3} = 0.426, \text{ or } 42.6 \text{ per cent} \qquad Ans.$$

Compare this with the superheated Rankine cycle efficiency, 40.0 per cent.

Figure 15-10 illustrates a two-heater system which does not require auxiliary pumps because the condensate of the extracted steam is "flashed"

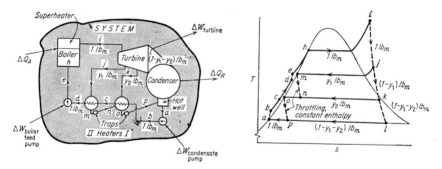

Fig. 15-10. A two-heater regenerative system (terminal differences and temperature rise in pumps greatly exaggerated).

or "cascaded" back to a low-pressure region. Here the fraction $(1 - y_1 - y_2)$ passes through the cycle $abcdehijkla$; the fraction $y_1$ passes through the cycle $abcdehijmnopa$; the fraction $y_2$ passes through the cycle $abcdehijkopa$. For this system

$$\Delta Q_A = h_i - h_e \frac{\text{Btu}}{\text{lb}_m \text{ water through boiler}}$$
$$\Delta Q_R = (h_a - h_l)(1 - y_1 - y_2) + (h_a - h_p)(y_1 + y_2)$$
$$\Delta W_{\text{turbine}} = (h_i - h_j) + (h_j - h_k)(1 - y_1) + (h_k - h_l)(1 - y_1 - y_2)$$
$$\Delta W_{\text{turbine}} = (h_i - h_l)(1 - y_1 - y_2) + (h_i - h_j)y_1 + (h_i - h_k)y_2$$

Note that the terminal difference for heater 1 is $t_o - t_c$; for heater 2 it is $t_m - t_d$.

In some instances the extracted steam is used for process work other than to warm the feedwater. A system of this type can be called an *extraction system* but not a *regenerative system*.

**15-7. The Binary-vapor System.** The discussion in previous articles has indicated that an ideal fluid and a Rankine cycle would offer

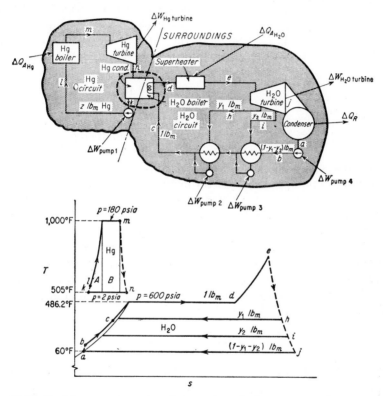

Fig. 15-11. The binary-vapor system (temperature rise in pumps greatly exaggerated; complete detail not shown for $H_2O$ $Ts$ diagram).

the best practical solution to the problem of converting heat into work. In the absence of an ideal fluid, the Rankine cycle can be modified into a regenerative cycle, but here the low critical temperature of water mitigates against adding all the heat at the highest possible temperature. This disadvantage of water for high-temperature operation can be overcome, in part, by "topping" the regenerative cycle with a cycle that uses a fluid with a higher critical temperature than that of water and, also, a lower heat capacity for the liquid phase. Thus, in Fig. 15-11, *a binary-vapor* system is illustrated, consisting of a regenerative cycle, which uses water as the fluid, and a Rankine-model cycle, which uses mercury as the fluid.

The properties of mercury (given in Table B-5) are such that the pressure is only 180 psia for a vaporization temperature of 1000°F.

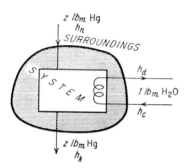

FIG. 15-12. Flow system of mercury condenser and steam boiler.

Since the heat capacity of the liquid is small, a regenerative cycle is not demanded,† and the Rankine-model cycle can be used. Heat is transferred to the mercury cycle to heat the liquid at state $l$ to the saturated vapor state $m$. The mercury vapor then expands in a turbine to the temperature $n$ with corresponding pressure of 2 psia. (Expansion cannot be made to 60°F because the vacuum in the mercury condenser would be extremely high and therefore impractical to maintain.) Condensation of the mercury takes place in a condenser that is also a boiler for the water cycle. The mercury is pumped back into the mercury boiler to complete the cycle.

The bottom fluid of the system passes through the processes dictated by the extraction cycle. Here heat is added to superheat the steam and so decrease the moisture content in the final stages of expansion.

However, the mercury cycle must maintain a greater mass-flow rate than the steam cycle since the latent heat of vaporization of steam is considerably greater than that of mercury. The relative flow rates can be determined by an energy balance on the adiabatic system of Fig. 15-12 (which is also shown in Fig. 15-11):

$$z(h_n - h_k) = h_d - h_c$$

The work and heat transfers for the binary-vapor system will be computed on the basis of 1 lb$_m$ flow of water through the boiler:

$$\Delta Q_A = z(h_m - h_l) + (h_e - h_d) \frac{\text{Btu}}{\text{lb}_m \text{ H}_2\text{O through boiler}}$$

$$\Delta Q_R = (h_a - h_j)(1 - y_1 - y_2) \frac{\text{Btu}}{\text{lb}_m \text{ H}_2\text{O through boiler}}$$

$$\Delta W_{\text{turbines}} = z(h_m - h_n) + (h_e - h_h) + (h_h - h_i)(1 - y_1)$$
$$+ (h_i - h_j)(1 - y_1 - y_2) \frac{\text{Btu}}{\text{lb}_m \text{ H}_2\text{O through boiler}}$$

† A regenerative cycle would be undesirable because the efficiency of the furnace would be decreased.

The pump work is computed on the same basis noting that $\Delta W_{p1}$ is for $z$ lb$_m$ Hg, $\Delta W_{p2}$ is for $y_1$ lb$_m$ H$_2$O, $\Delta W_{p3}$ is for $y_2$ lb$_m$ H$_2$O, $\Delta W_{p4}$ is for $(1 - y_1 - y_2)$ lb$_m$ H$_2$O.

Thus, to achieve the objective of adding heat at the highest possible temperature, a binary-vapor cycle can be used since no one fluid is known that possesses all of the desirable characteristics of an ideal fluid. Mercury has certain advantages as a topping fluid (items 1, 2, 3, and 5 of Art. 15-4). It also has disadvantages:

1. High cost: The Kearny Station of the Public Service Electric and Gas Company of New Jersey produces 50,000 kw from a mercury-steam binary cycle. The binary system requires about 400,000 lb$_m$ Hg or approximately 8 lb$_m$ per net kilowatt output. Assuming a market value of $1.00 lb$_m$ Hg, the mercury in the system represents an investment of $400,000.

2. Not only are leaks from the system expensive, but also mercury vapor is quite toxic. Maintenance costs are increased to guard against leakage.

3. Mercury does not "wet" steel surfaces, and therefore heat transfer is hindered. The addition of small amounts of magnesium and titanium has largely eliminated this difficulty.

4. Iron is soluble in mercury at high temperatures, and therefore special steels must be used.

The mercury cycle at the Kearny Station operates between limits of 140 psig, 975°F, and 1 in. Hg pressure while developing 20,000 kw. The steam cycle operates between limits of 365 psig, 750°F, and 1 in. Hg pressure while developing 30,000 kw. The thermal efficiency of the system is about 37 per cent. This efficiency has been exceeded only by the diesel engine, although the reciprocating engine is built in much smaller sizes (the largest engine in this country is about 8,000 hp). Moreover, internal-combustion engines require a more expensive fuel than external-combustion engines.

Other fluids than mercury and water have been proposed for binary cycles: ammonia, methyl chloride, ethyl chloride, sodium dioxide, ethyl bromide, aluminum bromide, diphenyl, and diphenyl oxide. Too, "topping" gas turbines are now being used to supply heat to the steam cycle.

**15-8. Other Performance Factors.** The *heat rate* is a logical factor for the cycle or for the complete station of furnace and cycle:

$$\text{Heat rate} = \frac{\text{heat transferred over a period of time}}{\text{work output during the same period of time}} \quad (15\text{-}2)$$

The *cycle heat rate* is defined as the heat received by the *cycle* per unit output of the *cycle*. The *station heat rate* is the ratio of the heating value of the fuel (the limiting amount of heat that can be transferred from the furnace) to the work output of the cycle. Both of these ratios usually include the efficiency of the electrical equipment.

A heat rate for the turbine cannot be logically justified because the turbine represents only one process in a cycle of processes. But it is quite usual to define a *turbine heat rate* which is based upon definitions and interpretations proposed by the ASME. Recourse must be made to this source for detailed methods when an official performance test on a steam turbine is to be made. The salient points, however, of the Code† are as follows:

*Complete Expansion Turbine.* The turbine is charged with the enthalpy $h_1$ of the steam supplied and credited with the enthalpy $h_{f2}$ of the saturated liquid corresponding to the exhaust pressure. Then by the logic of Eq. (15-2),

$$\text{Turbine heat rate} = \frac{h_1 - h_{f2}}{\Delta w} \left(\frac{\text{Btu}}{\text{kwhr}}\right) \qquad (15\text{-}3)$$

Here
$$\Delta w = \frac{\Delta w_{\text{turbine}} \ (\text{Btu/lb}_m)}{3413 \ (\text{Btu/kwhr})} = \frac{\text{kwhr}}{\text{lb}_m}$$

*Reheating-cycle Turbine.* The turbine is charged with the enthalpy $h_1$ of the steam supplied plus the increase in enthalpy from reheating, and credited with the enthalpy $h_{f2}$ of the saturated liquid corresponding to the exhaust pressure:

$$\text{Turbine heat rate} = \frac{h_1 + \Delta h_{\text{reheat}} - h_{f2}}{\Delta w} \left(\frac{\text{Btu}}{\text{kwhr}}\right) \qquad (15\text{-}4)$$

*Regenerative-cycle Turbine.* The turbine is charged with the enthalpy $h_1$ of the steam supplied plus the enthalpy increase from pumps located *between* condenser and final heater and credited with the enthalpy $h_8$ of the water leaving the final heater:

$$\text{Turbine heat rate} = \frac{\dot{m}_{f1}(h_1 - h_8) + \Sigma \dot{m}_f (\Delta h)_{\text{pumps}}}{\dot{m}_{f1} \, \Delta w} \left(\frac{\text{Btu}}{\text{kwhr}}\right) \qquad (15\text{-}5)$$

and
$$\dot{m}_{f1} = \text{mass-flow rate to throttle of turbine}$$

Allowances are also made for make-up water, steam used for sealing glands, and subcooling of the condensate in the condenser.

The *steam rate* is applied only to straight-expansion turbines, and it is the mass flow into the turbine per specified unit of output, pounds

---

† "Power Test Codes; Steam Turbines," American Society of Mechanical Engineers, New York, 1941.

of steam per kilowatthour (or horsepower-hour) output,

$$\text{Steam rate} = \frac{3413 \ (\text{Btu/kwhr})}{\Delta w_{\text{turbine}} \ (\text{Btu/lb}_m)} = \left(\frac{\text{lb}_m \ \text{steam}}{\text{kwhr}}\right) \qquad (15\text{-}6a)$$

$$= \frac{2545 \ (\text{Btu/hp-hr})}{\Delta w_{\text{turbine}} \ (\text{Btu/lb}_m)} = \left(\frac{\text{lb}_m \ \text{steam}}{\text{hp-hr}}\right) \qquad (15\text{-}6b)$$

In terms of the steam rate, the turbine heat rate equals

$$\text{Turbine heat rate} = (h_1 - h_{f2}) \frac{\text{Btu}}{\text{lb}_m} \left(\text{steam rate} \ \frac{\text{lb}_m}{\text{kwhr}}\right) = \frac{\text{Btu}}{\text{kwhr}}$$

$$= (h_1 - h_{f2}) \left(\frac{3413}{\Delta w_{\text{turbine}}}\right) \qquad (15\text{-}7a)$$

And since the work ratio of the steam cycle is close to unity,

$$\text{Turbine heat rate} \approx \frac{3413}{\eta_t} = \text{cycle heat rate} \qquad (15\text{-}7b)$$

**15-9. The Heat Reservoir and the Cycle.** It has been implied, in previous articles, that transfer of heat to the cycle was reversibly accomplished. This implication was adopted because the cycle was the center of attention and the presence of irreversibilities between cycle and heat reservoir would only confuse and not aid in the selection of the most efficient cycle. Now, however, let attention be directed to the heat reservoir.

The Rankine cycle is able to abstract a greater amount of heat from the practical heat reservoir than the Carnot cycle is able to obtain. When the Rankine cycle is regenerated by using extraction heaters, the thermal efficiency of the cycle increases, but the efficiency of the furnace may decrease because the flue gases leave at higher temperatures than before. *The remedy is to find means for maintaining a constant efficiency of the furnace, no matter what changes are made in the cycle, for then the cycle of highest efficiency will also result in the highest over-all station efficiency.* The practical remedy is to install air preheaters near the exit of the furnace to cool the gases by transferring energy to the air supplied for fuel combustion. In this manner, the efficiency of the furnace is held constant, while the temperature of combustion is raised and the transfer of heat is facilitated.

The type of firing limits the amount of air preheat. With stoker operation, the air temperature is limited to 300 or 400°F; with higher temperatures, stoker maintenance problems arise. Pulverized fuel, oil, or gas allows higher air temperatures of 500 to 600°F. For a simple cycle such as the Rankine, air heaters are unnecessary, of course, because the flue-gas temperature can be held to a low and desirable value by an *economizer*. An economizer is a heat exchanger, placed near the exit of the furnace, wherein the cold feedwater is heated by cooling the hot furnace gases.

## PROBLEMS

The following items may also be specified by the teacher in the solution of the problems:

*a.* Thermal efficiency
*b.* Heat added
*c.* Heat rejected
*d.* Turbine work
*e.* Pump work
*f.* Work ratio
*g.* Error arising by neglecting pump work
*h.* $Ts$ diagram to illustrate heat transfers
*i.* $pv$ diagram

*j.* $hs$ diagram
*k.* Heat rate for cycle (kwhr)
*l.* Heat rate for cycle (hp-hr)
*m.* Station heat rate (kwhr) (Assume 10,000 Btu/$lb_m$ coal and furnace efficiency of 85 per cent)
*n.* Steam rate (kwhr)
*o.* Steam rate (hp-hr)
*p.* Mollier chart not permitted for solution
*q.* Neglect pump work

Unless other information is given in the problem, the temperature ($T_0$) and pressure ($p_0$) of the atmosphere can be assumed to be 60°F and 14.7 psia.

**1.** Determine the thermal efficiency, the work of each process, and the work ratio for a Carnot cycle that uses steam as the working fluid in a series of flow processes. The condition of the steam at the start of the isothermal addition of heat is 600 psia and 90 per cent moisture; at the end of the heating process the state of the steam is 600 psia and 5 per cent moisture. The heat-rejection process is at a pressure of 1 in. Hg. Repeat the problem, assuming that each process is nonflow with the fluid being confined in a cylinder by a piston.

**2.** Assume that upon test of the cycle of Prob. 1 the compression and expansion efficiences are each found to be 75 per cent and both processes are adiabatic. Compute the thermal efficiency and the work ratio.

**3.** A Rankine cycle operates with saturated steam at 400 psia, and the pressure in the condenser is 1 in. Hg. Determine the thermal efficiency, work of each flow process, heat rejected, and the work ratio. Repeat the problem, assuming that the processes are all nonflow.

**4.** Assume that, upon test of the cycle of Prob. 3, the compression and expansion efficiencies are each found to be 75 per cent and both processes are adiabatic. Compute the thermal efficiency and the work ratio.

**5.** A Rankine cycle operates with superheated steam at 400 psia and 700°F, and the pressure in the condenser is 1 in. Hg. Determine the thermal efficiency, heat rejected, and the work ratio.

**6.** Upon test of the cycle of Prob. 5, the compression and expansion efficiencies are found to be 75 per cent, and both processes are adiabatic. Compute the thermal efficiency and the work ratio.

**7.** Repeat Prob. 5, using the method of analysis demonstrated in Art. 15-3.

**8.** The nozzles in the turbine of Prob. 5 pass 20,000 $lb_m$/hr of steam. Compute the horsepower of the turbine and the quantity of cooling water demanded by the condenser if the temperature rise of the coolant is 10°F.

**9.** Upon test of the cycle in Prob. 5, the turbine steam rate is found to be 9.5 $lb_m$/kwhr. Determine, for the actual cycle, the thermal efficiency and the expansion efficiency if the compression efficiency is 60 per cent. (The work processes are adiabatic.)

**10.** The Rankine cycle of Prob. 5 is converted into a reheat cycle. The steam is reheated at constant pressure from an initial state of 1 per cent moisture to 700°F. Determine the thermal efficiency and the heat rejected for the cycle.

**11.** Repeat Prob. 10 but for the data of Prob. 6 and reheat pressure of 54 psia.

**12.** Steam at 1,200 psia and 800°F enters a high-pressure turbine and adiabatically expands with expansion efficiency of 85 per cent to a state of 5 per cent moisture. The steam is then reheated to 600°F but with pressure drop of 10 psia before it enters the low-pressure turbine where the expansion efficiency is 82 per cent for the adiabatic expansion. Determine the thermal efficiency of the cycle if the compression efficiency of the adiabatic water pump is 60 per cent and the condenser vacuum is 28.93 in. Hg.

**13.** The cycle of Prob. 6 is governed by throttling, and at part load the steam enters the turbine nozzles at a pressure of 300 psia. Compute the thermal efficiency, and draw the $Ts$ diagram for this cycle.

**14.** Repeat Prob. 9, assuming that the steam rate includes a generator efficiency of 85 per cent.

**15.** A regenerative cycle operates between limits of 1,200 psia, 1000°F, and 1 in. Hg. Compute the thermal efficiency if one heater is used, of the type shown in Fig. 15-9a, at the optimum extraction pressure. Engine efficiency of turbine and compression efficiency of pumps are each 100 per cent.

**16.** For the data of Prob. 15 vary the extraction pressure between limits of 1 in. Hg and 1,200 psia, and construct a curve showing the relationship between thermal efficiency and extraction pressure.

**17.** Repeat Prob. 15 for two heaters.

**18.** Repeat Prob. 15 for three heaters.

**19.** Repeat Prob. 15, assuming that the expansion efficiency is 85 per cent up to the extraction point and 82 per cent for the entire expansion. (Pump efficiency is unchanged.)

**20.** Repeat Prob. 15, assuming that the expansion efficiency of the regenerative turbine is 82 per cent.

**21.** Repeat Prob. 15, but also assume the steam is reheated to the initial temperature at the extraction pressure. (Extracted steam is not to be reheated.)

**22.** A reheat-regenerative cycle has two extraction heaters similar to Fig. 15-9b. For the real system terminal differences and throttling exist. Draw a $Ts$ diagram showing throttling and temperature drop from superheater to high-pressure turbine, irreversible expansion in high-pressure turbine, pressure drop in reheater, pressure and temperature drop from reheater to low-pressure turbine, and irreversible expansion in low-pressure turbine. All pumps are irreversible. Label reasons (causes) on the diagram for all irreversibilities.

**23.** Top the regenerative cycle of Example 5 with a reversible mercury cycle that operates between limits of 180 psia (saturated) and 2 psia. Compute the thermal efficiency for the binary cycle and for each cycle considered alone.

**24.** Top the cycle of Prob. 6 with a mercury cycle that operates between limits of 180 psia (saturated) and 1 psia. The compression and expansion efficiencies are 75 per cent in both cycles, and all work processes are adiabatic. Compute the thermal efficiencies for the binary cycle and for each cycle considered alone.

**25.** Calculate the effectiveness and loss of available energy for the heater of Example 5.

**26.** For the data of Prob. 12, calculate the loss of available energy for each process (reasonable assumptions are allowed).

**27.** Compute the loss of available energy for the adiabatic process of Prob. 13.

**28.** Compute the loss of available energy and the effectiveness for each process in Examples 2 and 4.

# CHAPTER 16

# COMBUSTION

Human history becomes more and more a
race between education and catastrophe.

*H. G. Wells*

The chemistry of the combustion process is an engineering problem of practical and, also, theoretical significance. The practicing engineer must be able to calculate an energy balance for the commercial power system, and he should be aware of the limitations to the process of combustion that theory is able to predict.

**16-1. Complete-reaction Equations.** Consider the reaction taking place when carbon unites with oxygen to form carbon dioxide,

$$C + O_2 \rightarrow CO_2 \qquad (a)$$

This equation implies that 1 molecule of carbon unites with 1 molecule of oxygen to form 1 molecule of carbon dioxide,

$$1 \text{ molecule C} + 1 \text{ molecule O}_2 \rightarrow 1 \text{ molecule CO}_2 \qquad (b)$$

The relative masses of these molecules are shown by the molecular weights,

$$C\ 12 \qquad O_2\ 32 \qquad CO_2\ 44$$

and, therefore,

$$12 \text{ units mass C} + 32 \text{ units mass O}_2 = 44 \text{ units mass CO}_2$$
$$12 \text{ lb}_m \text{ C} + 32 \text{ lb}_m \text{ O}_2 = 44 \text{ lb}_m \text{ CO}_2 \qquad (c)$$

In this manner, the chemical-reaction equation is converted into a mass equation. Each mass shown in Eq. (*c*) is, by definition, a mole (Art. 1-2), and therefore Eq. (*a*) is equivalent to

$$1 \text{ mole C} + 1 \text{ mole O}_2 \rightarrow 1 \text{ mole CO}_2 \qquad (d)$$

All perfect gases, under fixed conditions of temperature and pressure, have identical mole volumes, and this condition is approximated by real gases at low pressures. Thus, Eq. (*d*) can be written

$$1 \text{ volume gaseous C} + 1 \text{ volume O}_2 \rightarrow 1 \text{ volume CO}_2 \Big]_{p,T} \qquad (e)$$

Comparison of Eqs. (a), (b), (d), and (e) shows that Eq. (a) can be interpreted to be a molecular, a molar, or, with some approximation, a volumetric equation. This basic form of the chemical equation can always be converted into a mass equation by multiplying each term by the appropriate molecular weight [Eq. (c)].

By the same line of reasoning, the equation for the reaction of hydrogen and oxygen can be written in any of the following forms:

$$H_2 + \tfrac{1}{2}O_2 \rightarrow H_2O$$
$$1 \text{ mole } H_2 + \tfrac{1}{2} \text{ mole } O_2 \rightarrow 1 \text{ mole } H_2O$$
$$1 \text{ volume } H_2 + \tfrac{1}{2} \text{ volume } O_2 \rightarrow 1 \text{ volume } H_2O]_{p,T}$$
$$2.016 \text{ lb}_m \text{ } H_2 + 16 \text{ lb}_m \text{ } O_2 = 18.016 \text{ lb}_m \text{ } H_2O$$
$$18.016 \text{ lb}_m \text{ mixture} = 18.016 \text{ lb}_m \text{ products}$$
$$1\tfrac{1}{2} \text{ mole mixture} \neq 1 \text{ mole products}$$

These equations show that the mass of mixture must equal the mass of products, although the number of moles (and volumes) of mixture and products are not necessarily equal.

In most instances, the combustion process is with atmospheric air. By Table 14-1,
$$M_{air} = 28.967 \approx 29$$

The molecular weight of the *apparent nitrogen* is found by including the inert gases in Table 14-1 with the nitrogen,

$$M_{\substack{apparent \\ nitrogen}} = 28.161$$

Thus, the value 28.161 will be assigned to the apparent nitrogen (rather than the value 28.016 for pure nitrogen). For every mole of oxygen supplied by the air,

$$\frac{79.01}{20.99} = 3.764 \, \frac{\text{moles apparent nitrogen}}{\text{mole oxygen}}$$

Or, on a mass basis,

$$\frac{22.25}{6.717} = 3.313 \, \frac{\text{lb}_m \text{ apparent nitrogen}}{\text{lb}_m \text{ oxygen}}$$

For example, the reaction of carbon and pure oxygen is

$$C + O_2 \rightarrow CO_2$$

and when the oxygen is supplied by dry air, 3.76 moles of nitrogen and other inert gases accompany each mole of oxygen,

$$C + O_2 + 3.76N_2 \rightarrow CO_2 + 3.76N_2$$

Upon multiplying each term by the appropriate molecular weight, the mass equation is determined,

$$12 \text{ lb}_m \text{ C} + 32 \text{ lb}_m \text{ O}_2 + 106 \text{ lb}_m \text{ N}_2 = 44 \text{ lb}_m \text{ CO}_2 + 106 \text{ lb}_m \text{ N}_2$$

The steps in balancing the chemical equation can be illustrated by the complete combustion of octane ($C_8H_{18}$) with the *theoretical* amount of dry air,

$$C_8H_{18} + \quad O_2 + \quad N_2 \rightarrow \quad CO_2 + \quad H_2O + \quad N_2$$
$$\text{(unbalanced equation)}$$

First, a *carbon balance* is made ($C_{\text{mixture}} = C_{\text{products}}$),

$$C_8 \qquad\qquad\qquad \rightarrow 8CO_2$$

then *a hydrogen balance* ($H_{\text{mixture}} = H_{\text{products}}$),

$$H_{18} \qquad\qquad\qquad \rightarrow \qquad 9H_2O$$

followed by an *oxygen balance* ($O_{\text{products}} = O_{\text{mixture}}$),

$$12\frac{1}{2}O_2 \qquad\qquad \leftarrow 8CO_2 + 9H_2O$$

and, finally, a *nitrogen balance* ($N_2 = 3.76O_2$),

$$12\frac{1}{2}(3.76)N_2 \rightarrow \qquad\qquad 47N_2$$

The complete-combustion equation is

$$C_8H_{18} + 12\frac{1}{2}O_2 + 47N_2 \rightarrow 8CO_2 + 9H_2O + 47N_2$$

The relative amount of air and fuel taking part in the reaction is called the *air-fuel ratio* and the inverse, the *fuel-air ratio*,

$$AF = \frac{\text{mass air}}{\text{mass fuel}} = \frac{(12\frac{1}{2} + 47)(29)}{8(12) + 18} = 15.1 \frac{\text{lb}_m \text{ air}}{\text{lb}_m \text{ fuel}}$$

$$FA = \frac{\text{mass fuel}}{\text{mass air}} = \frac{1}{15.1} = 0.0662 \frac{\text{lb}_m \text{ fuel}}{\text{lb}_m \text{ air}}$$

The foregoing values (and the related chemical equation) are said to be *chemically* or *theoretically correct*, or to be values for the *complete-combustion equation*. The criterion is that neither of the reactants could be increased in amount without a surplus of some form of reactant appearing also as a product. The usual combustion problem has either insufficient air or insufficient fuel, relative to the theoretical amount; hence, the chemically correct equation and the chemically correct air-fuel ratio serve to show the amount of the deficiency (Art. 16-3).

**16-2. Combustible Elements in Fuel.** The combustible elements in solid fuels, such as coal, consist of carbon, hydrogen, and oxygen, with other elements appearing in small amounts. A representative ultimate analysis of a dry bituminous coal, exclusive of ash and moisture, might appear as C, 88; H, 6; O, 4; N, 1; and S, 1. Liquid fuels are mixtures of complex hydrocarbons, although for combustion calculations gasoline or fuel oil can be assumed to average† the molecular formula $C_8H_{17}$. The same procedure cannot be used for coals because analyses of different coals vary widely.

**Example 1.** What would be an equivalent formula for a hydrocarbon fuel that analyzes 85 per cent C and 15 per cent H?

*Solution.* The formula will be of the form $C_aH_b$, and by the analysis and molecular weights

$$12a = 85 \quad \text{or} \quad a = 7.08$$
$$1b = 15 \quad \text{or} \quad b = 15$$

The result is

$$C_{7.08}H_{15} \quad Ans.$$

If desired, this answer can be multiplied by 1.13 to obtain whole numbers,

$$C_8H_{17} \quad Ans.$$

**Example 2.** Determine the complete-combustion equation for the representative coal analysis listed in this article.

*Solution.* The formula for the coal can be represented by

$$C_aH_bS_cO_dN_e$$

The ultimate analysis (mass) is:

| C | H | O | N | S |
|---|---|---|---|---|
| 88% | 6% | 4% | 1% | 1% |

The formula is equivalent to

$$C_{88/12}H_{6/1}O_{4/16}N_{1/14}S_{1/32}$$

and a "mole" of this coal will have a mass of 100 $lb_m$. The reaction equation (unbalanced) for this artificial mole unit is

$$C_{7.333}H_6O_{0.25}N_{0.071}S_{0.0312} + O_2 + (3.76O_2)N_2 \rightarrow CO_2 + H_2O + SO_2 + \text{(in air and fuel)} N_2$$

As before, a *carbon balance,*

$$7.333C \qquad\qquad \rightarrow 7.333CO_2$$

A *hydrogen balance,*

$$6H \qquad\qquad \rightarrow \qquad 3H_2O$$

A *sulfur balance,*

$$0.0312S \qquad\qquad \rightarrow \qquad 0.0312SO_2$$

† This value is probably more exact for mid-continent gasolines than a value obtained from a *single* ultimate analysis.

An *oxygen balance,*

$$\frac{0.25}{2} \qquad + O_2 \qquad\qquad \to 7.333 + \frac{3}{2} + 0.0312$$

or
$$O_2 = 8.74 \text{ (oxygen supplied by air)}$$

Finally, a *nitrogen balance,*

$$N_2 = \frac{0.071}{2} \text{ (fuel)} + 3.76(8.74)\text{(air)}$$

$$= 0.035 + 32.85 = 32.88\text{(nitrogen in products)}$$

The complete-combustion equation for 1 "mole" (100 $lb_m$) of dry and ash-free coal is

$$(C_{7.333}H_6S_{0.0312}, \text{ etc.}) + 8.74O_2 + 32.85N_2 \to 7.333CO_2 + 3H_2O$$
$$+ 0.0312SO_2 + 32.88N_2$$

Note that the oxygen, hydrogen, and sulfur in the fuel could have been neglected for simplicity without significantly affecting the accuracy of the solution. Note, too, that ash, water, etc., could have been included in the "formula" for the coal to obtain a more general solution for the real conditions in the plant.

**16-3. Products of Combustion.** The products of combustion for the real process cannot be assumed in the manner of Art. 16-1 since the extent of reaction is unknown (Art. 16-6) and partial products, such as CO, $H_2$, and $O_2$ (not to mention C, H, OH, etc.), may be present in significant amounts (especially true at high temperatures). For example,[†]

$$C_aH_b + cO_2 + 3.76cN_2 \to mCO_2 + nCO + xH_2O + yH_2 + zO_2$$
$$+ 3.76cN_2$$

The reaction equation has five variables ($m$, $n$, $x$, $y$, $z$) once the fuel and air quantities ($a$, $b$, $c$) are specified. Only two of these variables are unknowns, however, since the constants $a$, $b$, $c$ allow three variables to be expressed in terms of the other two. The remaining two variables can be found, at a specified temperature, by the relationships decreed by the equilibrium constants for the $CO_2$ and $H_2O$ reactions (Art. 16-7).

If the objective is to find the air-fuel ratio, the products of combustion can be cooled and chemically analyzed. One form of laboratory equipment for analyzing gas mixtures is the *Orsat* apparatus, which measures the percentages of $CO_2$, $O_2$, and CO by volume (the usual products from burning fuels such as coal). In the case of hydrocarbon fuels, when precise measurements are not available, the amounts of $H_2$ and $CH_4$ can be approximated by the empirical relationships (for rich or slightly lean mixtures):

$$H_2 = \tfrac{1}{2}CO\% \qquad CH_4 = 0.3\% \text{ (constant)} \qquad (16\text{-}1)$$

[†] And methane is usually present in the products at low temperatures and rich (in fuel) mixtures.

The Orsat analysis is a volumetric analysis and appears on a dry basis, although the original gas sample is saturated with water. That the Orsat will ignore the water vapor and report the analysis for a hypothetical dry mixture is shown in Example 3.

**Example 3.** A mixture of carbon dioxide and nitrogen at constant temperature and atmospheric pressure is contained over water in an Orsat. Show that the Orsat will measure the dry percentage of carbon dioxide.

*Solution.* Let

$n_1$ = moles of water vapor in original mixture
$n_2$ = moles of water vapor after absorption of carbon dioxide
$n_C$ = moles of carbon dioxide
$n_N$ = moles of nitrogen

The partial pressure of the saturated water vapor is constant since temperature is constant throughout the test [Eqs. (11-22*b*)],

$$p_{H_2O} = \frac{\text{moles of water vapor}}{\text{moles of mixture}} \text{ (14.7 psia)} = \text{constant} \qquad (a)$$

Inspection of Eq. (*a*) shows that, as the moles of mixture decrease from absorption of a component, the moles of water vapor must also decrease if the partial pressure is to remain constant. Thus, whenever a gas is absorbed in one of the pipettes, a proportional amount of water vapor is also condensed. The original mixture contains

$$n_1 + n_C + n_N \text{ moles}$$

After removal of the $CO_2$, the mixture contains

$$n_2 + n_N \text{ moles}$$

Then
$$p_{H_2O} = \frac{n_1}{n_1 + n_C + n_N} (14.7) = \frac{n_2}{n_2 + n_N} (14.7) \qquad (b)$$

Equation (*b*) can be reduced to

$$\frac{n_1}{n_2} = \frac{n_C + n_N}{n_N} \quad \text{and} \quad \frac{n_1 - n_2}{n_2} = \frac{n_C}{n_N} \qquad (c)$$

The Orsat will absorb both the $CO_2$ and $n_1 - n_2$ moles of water vapor in the first pipette. The percentage absorption will equal

$$\text{Orsat } \% \ CO_2 = \frac{n_C + (n_1 - n_2)}{n_1 + n_C + n_N}$$

Substituting from Eqs. (*c*) for $n_1$ and $n_1 - n_2$ and reducing,

$$\text{Orsat } \% \ CO_2 = \frac{n_C}{n_C + n_N} \qquad Ans.$$

But this percentage is the *dry percentage* of $CO_2$ in the original mixture. Thus, the Orsat measures the percentage of gas in the dry mixture and not the actual per cent in the real mixture.

Several methods are available for calculating the air-fuel ratio of the mixture from a gas analysis of the products of combustion:

a. Carbon balance
b. Hydrogen balance
c. Carbon-hydrogen balance
    1. With known fuel
    2. With unknown fuel
d. Oxidized-products method

Methods a, b, c1, and d require that the chemical composition of the fuel be known; method c2 does not require this knowledge.   Methods b and c are impractical for fuels with little hydrogen (coal).

Since the Orsat analysis is volumetric, the percentage of each constituent can be considered to be the moles of that constituent in the products.   The methods are best illustrated by examples.

a. *Carbon Balance.*   This is probably the least involved method, and it is quite accurate for mixtures with excess air.   Here the percentage of $CO_2$ is relatively high, and therefore slight errors in the analysis may not be critical.   The method assumes that free (solid) carbon is not formed and that the nitrogen in the products (and therefore in the mixture) is the gas remaining in the Orsat at the conclusion of the analysis (found by difference).   The oxygen in the mixture is then found from the known composition of air ($O_2 = N_2/3.76$).

**Example 4.**   The dry components of the exhaust gas from a spark-ignition engine using mid-continent gasoline as fuel were reported to be:

| $CO_2$ | $O_2$ | CO | $H_2$ | $CH_4$ | Total | $N_2 = 100 - 21.9$ |
|--------|-------|-----|-------|--------|-------|--------------------|
| 8.7% | 0.3% | 8.9% | 3.7% | 0.3% | 21.9% | 78.1% |

Determine the air-fuel ratio, and compare with the measured air-fuel value of 11:1.
*Solution.*   The composition of the fuel, with good precision, can be assumed to be $C_8H_{17}$.   The unbalanced reaction equation is

$$ZC_8H_{17} + \frac{78.1}{3.76} O_2 + 78.1N_2 \rightarrow$$
$$8.7CO_2 + 8.9CO + 0.3CH_4 + 0.3O_2 + 3.7H_2 + 78.1N_2 + \text{(condensed)} \ H_2O$$

A carbon balance is made by summing the combined carbon,

$$8Z = 8.7 + 8.9 + 0.3 = 17.9 \qquad Z = 2.238$$

Thus, the mixture is

$$2.238C_8H_{17} + 20.8O_2 + 78.1N_2$$

The air-fuel ratio, as usual by mass, is

$$AF = \frac{\text{mass of air}}{\text{mass of fuel}} = \frac{98.9(29)}{2.238(113)} = 11.3 \qquad Ans.$$

The computed air-fuel ratio is 11.3, and the measured value is 11. The error is $(0.3/11)(100)$, or 2.7 per cent.

*b. Hydrogen Balance.* In a few instances, a hydrogen-balance method may be desirable, for example, in the case of a compression-ignition engine under heavy load with free carbon appearing in the exhaust gas. Since the amounts of gases that contain hydrogen are small, extreme care must be exercised in the gas analysis.

**Example 5.** Repeat Example 4, but use the hydrogen-balance method.
*Solution.* The unbalanced reaction equation would appear (Example 4) as

$$ZC_8H_{17} + 20.8O_2 + 78.1N_2 \rightarrow 8.7CO_2 + 8.9CO + 0.3CH_4 + 0.3O_2 + 3.7H_2$$
$$+ 78.1N_2 + X(\text{condensed})H_2O + (\text{solid})C$$

*Oxygen Balance.* The amount of condensed $H_2O$ is computed by balancing the oxygen on both sides of the equation,

$$20.8 = 8.7 + \frac{8.9}{2} + 0.3 + \frac{X}{2} \quad \text{or} \quad X = 14.7$$

*Hydrogen Balance.* The hydrogen is balanced on both sides of the equation,

$$17Z = 0.3(4) + 3.7(2) + 14.7(2) \quad \text{or} \quad Z = 2.235$$

Thus, the mixture is
$$2.235C_8H_{17} + 20.8O_2 + 78.1N_2$$
And the air-fuel ratio is

$$AF = \frac{\text{mass of air}}{\text{mass of fuel}} = \frac{98.9(29)}{2.235(113)} = 11.3 \quad Ans.$$

*c. Carbon-Hydrogen Balance.* When the composition of the fuel is unknown and cannot be closely estimated, the carbon- and hydrogen-balance methods can be combined to find the solution.

**Example 6.** For the same data as in Examples 4 and 5, compute the air-fuel ratios by a carbon-hydrogen balance.
*Solution.* The unbalanced reaction equation is written for the unknown fuel,

$$C_xH_y + 20.8O_2 + 78.1N_2 \rightarrow$$
$$8.7CO_2 + 8.9CO + 0.3CH_4 + 0.3O_2 + 3.7H_2 + X(\text{condensed})H_2O + 78.1N_2$$

Carbon balance: $x = 17.9$
Oxygen balance: $X = 14.7$
Hydrogen balance: $y = 38$

Accordingly, the mixture is

$$C_{17.9}H_{38} + 20.8O_2 + 78.1N_2$$
with
$$\frac{H}{C} = \frac{38}{17.9(12)} = 0.177$$

The air-fuel ratio is

$$AF = \frac{98.9(29)}{17.9(12) + 38(1)} = 11.3 \quad Ans.$$

When the accuracy of the Orsat analysis is questionable and the analysis of the fuel is known, a carbon-hydrogen balance, but without a nitrogen balance, may prove to be the best solution. The method is illustrated in Example 7.

**Example 7.** For the same data as in Examples 4 to 6, compute the air-fuel ratio by a carbon-hydrogen balance for the fuel $C_8H_{17}$.

*Solution.* The unbalanced reaction equation is written for the known fuel,

$$ZC_8H_{17} + aO_2 + bN_2 \rightarrow$$
$$8.7CO_2 + 8.9CO + 0.3CH_4 + 0.3O_2 + 3.7H_2 + X(\text{condensed})H_2O + bN_2$$

*Carbon Balance.* $Z = 2.238$ (Example 4).

*Hydrogen Balance.* The amount of condensed $H_2O$ is computed by balancing the hydrogen on both sides of the equation,

$$2.238(17) = 0.3(4) + 3.7(2) + X(2) \qquad \text{or} \qquad X = 14.7$$

(which agrees with Example 5).

*Oxygen Balance.* The oxygen in the mixture is evaluated from the known products,

$$a = 8.7 + \frac{8.9}{2} + 0.3 + \frac{14.7}{2} = 20.8$$

(which agrees with Examples 4 and 5).

*Nitrogen.* The nitrogen in the air is found from the $N_2/O_2$ relationship,

$$b = 3.764a = 3.764(20.8) = 78.1$$

Accordingly, the mixture is

$$2.238C_8H_{17} + 20.8O_2 + 78.1N_2$$

(which agrees with Example 4), and the ratio is

$$AF = \frac{98.9(29)}{2.238(113)} = 11.3 \qquad Ans.$$

In this instance, the answer agrees with those in Examples 4 to 6. In other, more usual cases, a difference will be found because the values from the Orsat analysis were in error (because of either faulty technique or faulty sampling of the gases). The method illustrated in this example is usually assumed to be a better solution than those in Examples 5 and 6. Of course, with a faulty Orsat analysis, no method can be precise, and the best solution is derived more from judgment than from method.

*d. Oxidized Products.* It is desirable in analyzing gases to have a high percentage of each constituent in order that slight errors in volumetric measurements will be insignificant. It is also desirable to have components, like $CO_2$, that are susceptible to easy analysis. The products may contain slight amounts of hydrogen, methane, and other hydrocarbon gases that are especially difficult to recognize or to evaluate accurately. For these reasons, the *oxidized-products method* is especially valuable. In practice, the oxidizer is made of 1-in. stainless-steel tubing and filled with cupric oxide wire of 0.020 in. diameter. The tube is placed in a furnace and held at about 1200°F. The gases pass through the tube, and the products CO, $H_2$, $CH_4$, etc., are oxidized to $CO_2$ and $H_2O$, while $O_2$ in the products is removed by absorption. The gas leaving the oxidizer is passed into a conventional Orsat for measurement of the $CO_2$.

**Example 8.** Determine the air-fuel ratio for the data of Examples 4 to 7, assuming that the exhaust gas were to be completely oxidized.

*Solution.* The exhaust gas entering the oxidizer consists of (Example 6)

$$8.7CO_2 + 0.3O_2 + 8.9CO + 3.7H_2 + 0.3CH_4 + 78.1N_2 + XH_2O$$

In the oxidizer, the following reactions would occur:

$$8.9CO + 8.9O \rightarrow 8.9CO_2$$
$$3.7H_2 + 3.7O \rightarrow 3.7H_2O$$
$$0.3CH_4 + 1.2O \rightarrow 0.3CO_2 + 0.6H_2O$$

The gas leaving the oxidizer would be a mixture of

$$(8.7 + 8.9 + 0.3)CO_2 + (3.7 + 0.6 + X)H_2O + 78.1N_2$$

This gas when tested in the Orsat would indicate

$$\% \ CO_2 = \frac{17.9(100)}{17.9 + 78.1} = 18.6\%$$

Therefore, $N_2 = 100 - 18.6 = 81.4$, and

$$aC_8H_{17} + \frac{81.4}{3.76}O_2 + 81.4N_2 \rightarrow 18.6CO_2 + 81.4N_2$$

By a carbon balance,
$$a = 2.325$$
and the air-fuel ratio is

$$AF = \frac{103(29)}{2.325(113)} = 11.35 \qquad Ans.$$

This answer agrees with those found in Examples 4 to 7. (The excellent agreement of all methods shows that the original analysis was precise.)

**16-4. Heat of Combustion.†** The heat liberated by a chemical reaction is called the *heat of combustion*‡ when oxygen is one of the reactants. The heat of combustion at constant volume is measured by a *bomb* or *constant-volume calorimeter*. The procedure is to place a measured mass of fuel into the calorimeter and then to charge the bomb with, relatively, a great amount of oxygen to a pressure of 20 or 30 atm (to ensure essentially complete conversion of the fuel into products). The bomb is then placed into a water bath so that, after ignition (spark), the final temperature of the bomb and bath will be essentially equal to the

† This article summarizes and extends Arts. 4-3f, 8-3, and 8-4.
‡ The heat of combustion is usually considered to be a positive number.

initial temperature (a fraction of a degree higher; Chap. 4, Prob. 16). For this process at constant volume, by Eq. (4-4d),

$$\Delta Q_V = \Delta U = U_{\text{products}} - U_{\text{mixture}} \Big]_T \qquad (16\text{-}2a)$$

Although initial and final temperatures are closely equal, the internal energy of the mixture does not equal the internal energy of the products because chemical internal energy was liberated in the process.

The heat of combustion of gaseous fuels can be measured by a *flow*, or *constant-pressure, calorimeter* (which is simply a water-cooled furnace). Air and fuel flow into the calorimeter and are burned, with the products being cooled to the initial temperature of the mixture. For this steady-flow process, Eq. (8-8) shows that

$$\Delta Q_p = \Delta H = H_{\text{products}} - H_{\text{mixture}} \Big]_{T,p} \qquad (16\text{-}2b)$$

wherein $\Delta Q_p$ is called the *heat of combustion at constant pressure*.

The heat of combustion at constant pressure differs from that at constant volume if a volume change accompanies the reaction. By subtracting Eq. (16-2a) from (16-2b) (note comments in Art. 4-3f on the approximations),

$$\Delta Q_p - \Delta Q_V = \Delta H - \Delta U \Big]_T = \Delta p V \Big]_T = \Delta n \, R_0 T \qquad (16\text{-}3)$$

If the volume of the products is greater than that of the mixture, work will be done *by* the system during the constant-pressure process and therefore $\Delta Q_p$ will be *less* than $\Delta Q_V$ (ignoring the algebraic sign).

After the heat of reaction has been measured at one temperature, it can be calculated for any other temperature. Consider Eq. (16-2a),

$$\Delta Q_{V.T_1} = (U_{\text{products}} - U_{\text{mixture}})_{T_1}$$

and at a temperature $T_2$,

$$\Delta Q_{V.T_2} = (U_{\text{products}} - U_{\text{mixture}})_{T_2}$$

By subtracting one equation from the other,

$$\Delta Q_{V.T_2} - \Delta Q_{V.T_1} = \Delta U_{\text{products}} \Big]_{T_1}^{T_2} - \Delta U_{\text{mixture}} \Big]_{T_1}^{T_2} \qquad (16\text{-}4a)$$

Equation (16-2b) can be similarly treated,

$$\Delta Q_{p.T_2} - \Delta Q_{p.T_1} = \Delta H_{\text{products}} \Big]_{T_1}^{T_2} - \Delta H_{\text{mixture}} \Big]_{T_1}^{T_2} \qquad (16\text{-}4b)$$

Equation (16-4a) is illustrated in Fig. 16-1. Note that the change in internal energy from reactants to products at constant temperature is the heating value, by definition, and is measured by $ab$ at $T_1$ and $a'b'$ at $T°$. If the heat capacities of reactants and products are the same in value, then the heating value will not change with temperature.

Whenever a fuel contains hydrogen, one of the products of combustion will be water, which will exist as a liquid, a gas, or a two-phase mixture. If the water formed by combustion of the hydrogen in the fuel can be condensed, a greater amount of heat can be obtained from the calorimeter

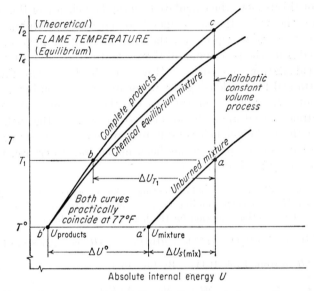

FIG. 16-1. Pictorial representation of the relationships between the energies of the unburned mixture and the burned products.

test than if the water existed in the vapor state. Because of this fact, two heating values can be recognized: the *higher heating value (gross)* is obtained when water formed by combustion is entirely condensed in the calorimeter test; the *lower heating value (net)* of the fuel is obtained when water formed by combustion exists entirely in the vapor state. The difference between these two heating values is equal to the latent energy of vaporization of the water at the test temperature.

In the constant-volume calorimeter, a few drops of water can be placed in the bomb to saturate the oxygen atmosphere before the test is begun. In this manner, a higher heating value of the fuel is obtained because (practically) all the water formed by combustion must condense and the latent internal energy of vaporization of the water is transferred to the coolant.

In the constant-pressure calorimeter, no water vapor can condense if the gases leaving the calorimeter are at a sufficiently high temperature. The constant-pressure calorimeter will give directly the lower heating value if the exit temperature is above the condensation (dew-point) temperature of the combustion products and the incoming air and fuel are warmed to that temperature. However, most calorimeters are operated to give the higher heating value by cooling the products down to the initial conditions. Correcting for any additional (or less) moisture remaining in the products over that supplied in the fuel and air, by adding (or subtracting) the latent heat of vaporization not delivered to the calorimeter from these sources, gives the higher heating value at the given test temperature.

The heat of combustion of a fuel at constant pressure is, in many instances, a measure of the energy transferred in a thermodynamic system. For this reason, the calculated thermal efficiency of the system will depend on the value that is assigned to the heat of combustion. The higher heating value represents the true amount of heat that can be transferred from the reaction, and therefore it is the value that should be used in calculating efficiencies. However, the heat that theoretically can be attained by condensing the water formed by combustion is, practically, not attainable. For this reason, thermal-efficiency calculations are sometimes based upon the lower heating value of the fuel. In comparing the thermal efficiencies of different systems, it is well to ensure that the same basis of comparison has been made, for, otherwise, misleading conclusions may be drawn.

**Example 9.** Calculate the higher heating value of gaseous octane at constant volume and 77°F if the corresponding value at constant pressure is known.

*Solution.* The reaction equation for the higher heating value is

$$C_8H_{18}(g) + 12\tfrac{1}{2}O_2 \rightarrow 8CO_2 + 9H_2O(l) \qquad \Delta H° = -2,369,859 \text{ Btu}$$

(value for $\Delta H°$ is from Table B-11). By Eq. (16-3),

$$\Delta Q_{vh} = \Delta Q_{ph} - \Delta n\, R_0 T$$

The value of $\Delta n$ for the reaction is $-5.5$ (or $8 - 13.5$) since the volume of liquid is negligible. Hence,

$$\Delta Q_{vh} = -2,369,859 + 5.5(1.986)(536.7) = -2,369,859 + 5,863$$
$$= -2,363,996 \text{ Btu} \qquad Ans.$$

**Example 10.** Calculate the lower heating value of gaseous octane at constant volume and 77°F.

*Solution.* From the data in Example 9,

$$C_8H_{18}(g) + 12\tfrac{1}{2}O_2 \rightarrow 8CO_2 + 9H_2O(l) \qquad \Delta U°_{77°F} = -2,363,996 \text{ Btu} \qquad (a)$$

At 77°F, $h_{fg}$ is 1050.4 Btu/lb$_m$ (Steam Tables), or

$$u_{fg} = h_{fg} - pv_{fg} = 1050.4 - \frac{0.4593(144)(694.9)}{778.2} = 991.4 \text{ Btu/lb}_m$$

Hence, for 9(18) lb$_m$ water,

$$162 \text{ lb}_m \text{ H}_2O(l) \rightarrow 162 \text{ lb}_m \text{ H}_2O(g) \qquad \Delta U_{77°F} = +160,607 \text{ Btu} \qquad (b)$$

Equations $(a)$ and $(b)$ could be added together if each constituent were in the same standard state (Art. 8-3). This requirement is not fulfilled because the data of $(b)$ are for a pressure of 0.4593 psia, while the $H_2O(l)$ in $(a)$ is at 1 atm pressure. This discrepancy is not serious because the internal energy is influenced more by temperature than by pressure. Hence, $(a)$ and $(b)$ are added, and approximately (but within experimental accuracy),

$$\Delta Q_{vl} = -2,203,389 \text{ Btu} \qquad Ans.$$

**Example 11.** Compute the lower heat of combustion of gaseous octane at constant volume and 1000°R.

*Solution.* The reaction equation is

$$C_8H_{18} + 12\tfrac{1}{2}O_2 \to 8CO_2 + 9H_2O$$

From Table B-10,

$$\Delta U_{\text{products}} = 8(U_{1000°} - U_{537°})_{CO_2} + 9(U_{1000°} - U_{537°})_{H_2O}$$
$$= 8(3852 - 115) + 9(3009 - 101) = 56,068 \text{ Btu}$$
$$\Delta U_{\text{mixture}} = 1(U_{1000°} - U_{537°})_{C_8H_{18}} + 12\tfrac{1}{2}(U_{1000°} - U_{537°})_{O_2}$$
$$= 24,773 - 640 + 12\tfrac{1}{2}(2539 - 83) = 54,833 \text{ Btu}$$

With these values and $\Delta Q_{V.537°}$ (Example 10) substituted into Eq. (16-4a),

$$\Delta Q_{V,1000°} = -2,203,389 + 56,068 - 54,833$$
$$= -2,202,154 \text{ Btu} \qquad Ans.$$

## 16-5. Theoretical Flame Temperature.

Consider a constant-volume and adiabatic combustion of a fuel-air mixture,

$$\Delta Q - \Delta W = \Delta U$$

Then, $\qquad \Delta Q = 0 \qquad \Delta W = 0 \qquad \Delta U = 0$

Therefore, $\qquad U_{\text{products},T_2} = U_{\text{mixture},T_1}$ $\qquad\qquad (a)$

The heat of reaction at constant volume is defined as

$$\Delta Q_{V,T_1} = U_{\text{products},T_1} - U_{\text{mixture},T_1} \qquad\qquad (16\text{-}2a)$$

and, upon substituting $(a)$ in Eq. (16-2a),

$$\Delta Q_{V,T_1} = U_{\text{products},T_1} - U_{\text{products},T_2} \qquad\qquad (16\text{-}5a)$$

while, for adiabatic combustion at constant pressure,

$$\Delta Q_{p,T_1} = H_{\text{products},T_1} - H_{\text{products},T_2} \qquad\qquad (16\text{-}5b)$$

Equations (16-5) show that the combustion process can be visualized as burning the mixture completely into products at the initial temperature $T_1$ and with the heat of reaction so obtained, the products can be raised to the final temperature $T_2$. This visualization is permissible since internal energy (or enthalpy) is a property. Thus, irrespective of the actual series of states in the real combustion process, a hypothetical series of states can be assumed between the two end states. In Fig. 16-1, the real adiabatic process at constant volume must follow path $ac$ (constant internal energy; the hypothetical path is $abc$).

Equations (16-5) can be converted to different forms by Eqs. (16-4):

$$\Delta Q_{V,T_2} = U_{\text{mixture},T_1} - U_{\text{mixture},T_2} \qquad (16\text{-}6a)$$

$$\Delta Q_{p,T_2} = H_{\text{mixture},T_1} - H_{\text{mixture},T_2} \qquad (16\text{-}6b)$$

Equations (16-6) imply that the mixture is raised in temperature from $T_1$ to $T_2$ and then converted into products. The heat of reaction available to perform this task is that evaluated at the temperature $T_2$. Since this value cannot be computed until after the unknown temperature $T_2$ has been determined, Eqs. (16-6) are not so convenient as Eqs. (16-5).

**Example 12.** Compute the theoretical flame temperature for complete combustion of gaseous octane with the theoretical amount of air at constant volume from an initial temperature of 537°R.

*Solution.* The reaction equation is

$$C_8H_{18}(g) + 12\tfrac{1}{2}O_2 + 47N_2 \rightarrow 8CO_2 + 9H_2O(g) + 47N_2$$
$$\Delta U_{77°F} = -2{,}203{,}389 \text{ Btu} \qquad \text{(Example 10)}$$

The lower heating value is selected because then all the products are gases. From Table B-10,

$$U_{\text{products},T_1} = 8(115) + 9(101) + 47(81) = 5636 \text{ Btu}$$

And, by Eq. (16-5a),

$$U_{\text{products},T_2} = U_{\text{products},T_1} - \Delta Q_{V,T_1}$$
$$= 5636 + 2{,}203{,}389 = 2{,}209{,}025 \text{ Btu}$$

Assuming that $T_2 = 5300°R$, with Table B-10,

$$U_{\text{products},T_2} = 8(55{,}265) + 9(43{,}187) + 47(29{,}648) = 2{,}224{,}259 \text{ Btu}$$

Assuming that $T_2 = 5200°R$,

$$U_{\text{products},T_2} = 2{,}171{,}267$$

Upon interpolating,

$$T_2 = 5271°R \qquad Ans.$$

**16-6. Equilibrium Flame Temperature.** The theoretical flame temperatures of Art. 16-5 cannot be attained because reaction is never complete. For example, the constant-volume combustion process in Fig. 16-1 cannot reach the temperature $T_2$ but has, as a limit, the chemical equilibrium temperature $T_\epsilon$.

A new problem arises when equilibrium flame temperatures are to be computed. Consider the complete conversion of CO into $CO_2$,

$$CO + \tfrac{1}{2}O_2 + 1.89N_2 \rightarrow CO_2 + 1.89N_2$$

Here the moles of products are fixed, and the temperature of combustion is the *single* unknown; thus, an energy equation is the only requirement for solution (Example 12),

$$-\Delta Q_{V,T_1} = n_{CO_2}(U_{T_2} - U_{T_1}) + n_{N_2}(U_{T_2} - U_{T_1})$$

But the equilibrium conversion equation is

$$CO + \tfrac{1}{2}O_2 + 1.89N_2 \rightarrow \epsilon CO_2 + (1 - \epsilon)CO + \frac{1 - \epsilon}{2} O_2 + 1.89N_2$$

and *two* unknowns are present: the *extent of reaction* $\epsilon$ and the *temperature* of combustion.   To solve a problem with two unknowns, two equations of restraint are required: an energy equation as before and an equilibrium conversion equation to relate temperature and $\epsilon$.   The solution of the two simultaneous equations then yields the required information.

**16-7. The van't Hoff Equilibrium Box.**   The concept of the equilibrium constant can be illustrated by studying a reversible chemical reaction in a van't Hoff reaction chamber (Fig. 16-2).   Here a mixture of con-

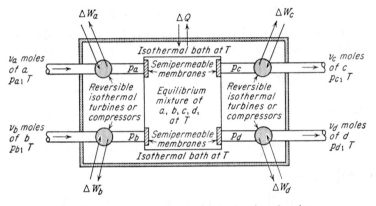

Fig. 16-2. A van't Hoff (modified) reaction chamber.

stituents $a$, $b$, $c$, $d$ is in equilibrium within the reaction chamber, and each constituent is also in equilibrium with pure constituent through one of the four semipermeable membranes.   In the operation of the reversible system, reactants $a$ and $b$ pass from their initial pressures $p_{a1}$ and $p_{b1}$ to their equilibrium pressures $p_a$ and $p_b$ by means of the isothermal turbomachines.   Work and heat are transferred between system and surroundings.   As reactants $a$ and $b$ enter† the equilibrium chamber, the equilibrium adjusts to form products $c$ and $d$ (a catalyst can be premised present to speed or to allow the reaction), and these products are displaced through their respective membranes.   The products are then changed in pressure from the equilibrium values ($p_c$ and $p_d$) to the reservoir values $p_{c1}$ and $p_{d1}$ by means of the reversible isothermal turbo-

† For the real (irreversible) process, the pressures of $a$ and $b$ before their semipermeable membranes must be greater than those of the equilibrium partial pressures in the mixture to effect the transfer.   But for a reversible process no pressure difference is necessary since the operation is hypothetical.

machines. Too, the reactants are converted into products in proportions dictated by the stoichiometric equation

$$\nu_a a + \nu_b b = \nu_c c + \nu_d d \qquad (a)$$

no matter what equilibrium composition or equilibrium pressure is contained within the reaction chamber. [And a similar argument can be made for the reversible mixing process (Prob. 28).]

The work of the system arises from the isothermal expansion or compression of each constituent as dictated by its equilibrium pressure and by its pressure at the flow boundary; no work is obtained from the equilibrium conversion process in the reaction chamber. The work from the four turbomachines equals

$$\Delta W_{sf} = \Delta W_a + \Delta W_b + \Delta W_c + \Delta W_d \qquad (b)$$

The reversible isothermal work for compressing or expanding an ideal gas without change in kinetic or potential energy is [Eq. (11-8)]

$$\Delta W_{rev} \Big]_{sf} \Big]_T = -n \int_{p_1}^{p_2} v \, dp = n R_0 T \ln \frac{p_1}{p_2}$$

For the entire system of Fig. 16-2 and Eq. (a), the work equals

$$\Delta W_{rev} \Big]_{sf} \Big]_T = R_0 T \left( \nu_a \ln \frac{p_{a1}}{p_a} + \nu_b \ln \frac{p_{b1}}{p_b} + \nu_c \ln \frac{p_c}{p_{c1}} + \nu_d \ln \frac{p_d}{p_{d1}} \right)$$

which reduces to

$$\Delta W_{rev} \Big]_{sf} \Big]_T = R_0 T \left( \ln \frac{p_c^{\nu_c} p_d^{\nu_d}}{p_a^{\nu_a} p_b^{\nu_b}} - \ln \frac{p_{c1}^{\nu_c} p_{d1}^{\nu_d}}{p_{a1}^{\nu_a} p_{b1}^{\nu_b}} \right) \qquad (c)$$

Let the pressure of each flow stream at the boundary be 1 atm (a *standard state* designated by the degree superscript; Art. 8-3) so that the last term in Eq. (c) is zero,

$$\Delta W_{rev}^{\circ} \Big]_{sf} \Big]_{\substack{T = C \\ p_1 = 1 \text{ atm}}} = R_0 T \ln \frac{p_c^{\nu_c} p_d^{\nu_d}}{p_a^{\nu_a} p_b^{\nu_b}} \qquad (d)$$

Equation (d) evaluates the work done by the steady-flow system as $\nu_a$ and $\nu_b$ moles of reactants (each at 1 atm) are reversibly converted into $\nu_c$ and $\nu_d$ moles of products (each at 1 atm), with transfer of heat restricted to reversible exchanges with the surroundings at a constant temperature $T$. Under these conditions, the work evaluated by Eq. (d) can have but one value at the temperature $T$, no matter what equilibrium conditions are in the reaction chamber. If this were not true, a perpetual-motion machine of the second kind could be devised from two van't Hoff systems. It follows that the total pressure, the presence of inert gases, and the particular composition of the equilibrium mixture are matters of indifference.

It also follows that the single-valuedness of the work function between two states dictates an equivalent change n some property of the system. The reversible work of a steady-flow system for the conditions of Fig. 16-2 (and, also, the reversible net work of a closed system in temperature and pressure equilibria with the surroundings) equals $-\Delta G$ (Art. 8-5). With Eq. (d) and Eq. (8-10c),

$$-\Delta G^\circ \Big]_T = R_0 T \ln \frac{p_c{}^{\nu_c} p_d{}^{\nu_d}}{p_a{}^{\nu_a} p_b{}^{\nu_b}} = R_0 T \ln K_p \qquad (16\text{-}7)$$

Equation (16-7) holds for *any* system of ideal gases since all variables are properties. Since $K_p$ is defined by Eq. (16-7) in terms of intensive properties, it need not be associated with any particular process—it can be used to find the proportions of the constituents at an equilibrium state of a closed or open system.

**Example 13.**  Calculate the degree of dissociation of 1 mole of $CO_2$ at 5170°R for total pressures of 1 and 10 atm.

*Solution.*  Figure B-7 shows, at 5170°R,

$$CO + \tfrac{1}{2}O_2 \rightarrow CO_2 \qquad K_p = 5 = \frac{p_{CO_2}}{p_{CO}(p_{O_2})^{\frac{1}{2}}}$$

The equilibrium mixture is

$$(1 - \epsilon)CO + \tfrac{1}{2}(1 - \epsilon)O_2 + \epsilon CO_2 = \frac{3 - \epsilon}{2} \text{ moles} = n$$

The partial pressures of the constituents are

$$p_{CO_2} = \left(\frac{\text{moles } CO_2}{\text{moles mixture}}\right)(\text{total pressure}) = \frac{\epsilon}{n}\, p$$

$$p_{CO} = \frac{1 - \epsilon}{n}\, p$$

$$p_{O_2} = \frac{1 - \epsilon}{2n}\, p$$

and, for $p = 1$ atm,

$$K_p = \frac{(\epsilon/n)p}{\dfrac{1 - \epsilon}{n}\, p \left(\dfrac{1 - \epsilon}{2n}\right)^{\frac{1}{2}} p^{\frac{1}{2}}} = \frac{\epsilon(3 - \epsilon)^{\frac{1}{2}}}{(1 - \epsilon)^{\frac{3}{2}}} = 5.00$$

Solving, by trial,

$$\epsilon = 0.656, \text{ or } 34.4\% \text{ dissociation of } CO_2 \qquad Ans.$$

At a pressure of 10 atm,

$$K_p = \frac{\epsilon(3 - \epsilon)^{\frac{1}{2}}}{(1 - \epsilon)^{\frac{3}{2}}(10)^{\frac{1}{2}}} = 5.00$$

Upon solving,

$$\epsilon = 0.82, \text{ or } 18\% \text{ dissociation of } CO_2 \qquad Ans.$$

Note that the increase in pressure tended to shift the equilibrium to a smaller volume. In this reaction, the volume of $CO_2$ formed is less than the volume of the reactants CO and $O_2$; therefore, the extent of reaction was increased by the increased pressure. The opposite effect, of course, is encountered when the products have a greater volume than that of the mixture.

What will happen when the concentration of one (or more) of the constituents is increased? Here the excess constituent tends to drive the reaction in a direction to relieve the excess (Example 14).

**Example 14.** If three times the stoichiometric correct amount of oxygen is supplied in burning CO, what will be the extent of reaction at 5170°R and 1 atm?
*Solution.* The combining equation is

$$CO + \tfrac{3}{2}O_2 \rightarrow CO_2 + O_2 \qquad\qquad (a)$$

The mixture at equilibrium would be

$$(1 - \epsilon)CO + \tfrac{1}{2}(1 - \epsilon)O_2 + \epsilon CO_2 + O_2 = \frac{5 - \epsilon}{2} = n$$

The partial pressures of the constituents are

$$p_{CO_2} = \frac{\epsilon}{n}\,p \qquad p_{CO} = \frac{1 - \epsilon}{n}\,p \qquad p_{O_2} = \frac{3 - \epsilon}{2n}\,p$$

The excess $O_2$ in $(a)$ cancels from the equilibrium constant equation,

$$K_p = \frac{p_{CO_2}p_{O_2}}{p_{CO}(p_{O_2})^{3/2}} = \frac{p_{CO_2}}{p_{CO}(p_{O_2})^{1/2}} = \frac{\epsilon(5 - \epsilon)^{1/2}}{(1 - \epsilon)(3 - \epsilon)^{1/2}} = 5.00$$

Solving by trial,

$$\epsilon = 0.78, \text{ or } 78\% \text{ extent of reaction} \qquad Ans.$$

Compare with Example 13. The presence of excess oxygen drives the reaction further toward completion. Also, the excess oxygen would lower the flame temperature and so decrease the tendency to dissociate the $CO_2$.

## PROBLEMS

**1.** Write the combustion equation with the theoretical amount of air for acetylene ($C_2H_2$) and for decane ($C_{10}H_{22}$).

**2.** Repeat Prob. 1, but for methyl alcohol ($CH_3OH$).

**3.** Repeat Prob. 1, but assume that only 85 per cent of the theoretical amount of air is used and that the moles of hydrogen in the products are only one-half the moles of carbon monoxide.

**4.** A fuel oil analyzes 87 per cent carbon and 13 per cent hydrogen (gravimetric). What would be a representative molecular formula?

**5.** (a) A coal analysis (gravimetric) is (dry basis):

| C | H | O | N | S | Ash | Total |
|---|---|---|---|---|-----|-------|
| 71 | 4 | 9 | 1 | 3 | 12 | 100 |

Determine the minimum mass of air necessary to obtain complete combustion.

(b) The coal, as fired, had 10 per cent moisture.   Change the dry analysis in (a) to include the moisture.

(c) Determine the minimum mass of air necessary to obtain complete combustion of the wet coal in (b).

(d) Compute the sensible heat loss for the products of (b) if the gases leave the furnace at 500°F.   (Note that the moisture in the coal will increase the amount of water vapor in the products; 77°F datum.)

**6.** Repeat Prob. 5, but for a coal that analyzes:

| C | H | O | N | $H_2O$ | Ash | Total |
|---|---|---|---|--------|-----|-------|
| 70 | 3 | 2 | 2 | 10 | 13 | 100 |

**7.** The products of combustion of the coal in Prob. 6 are analyzed by an Orsat:

| $CO_2$ | CO | $O_2$ | $N_2$ | Total |
|--------|-----|-------|-------|-------|
| 14.6 | 0.2 | 5.5 | 79.7 | 100 |

(a) Determine the mass of air burned per pound of coal.

(b) Determine the mass of dry products per pound of coal.

(c) Assume that the humidity is 0.01 $lb_m$ of water vapor for each pound of dry air, and determine the amount of water vapor in the products per pound of coal.

**8.** Determine the condensation temperature of the water vapor when dodecane ($C_{12}H_{26}$) is burned with the theoretical amount of air at atmospheric pressure.   The initial temperature of the air is 60°F, and relative humidity is 50 per cent.

**9.** Octane ($C_8H_{18}$) and air are burned, and the Orsat products of combustion are:

| $CO_2$ | CO | $H_2$ | $CH_4$ | $N_2$ |
|--------|-----|-------|--------|-------|
| 9.9 | 7.2 | 3.3 | 0.3 | Remainder |

Compute the air-fuel ratio by a carbon balance, and compare with the answer found by making a hydrogen balance.

**10.** The Orsat products of combustion of a hydrocarbon fuel are:

| $CO_2$ | $O_2$ | CO | $H_2$ | $CH_4$ |
|--------|-------|-----|-------|--------|
| 13.0 | 2.2 | 0.2 | 0.0 | 0.1 |

Determine the air-fuel ratio by means of a carbon-hydrogen balance (fuel, $C_aH_b$).

**11.** The Orsat products of combustion of octane and air are:

| $CO_2$ | $O_2$ | CO | Apparent $N_2$ |
|--------|-------|-----|----------------|
| 8.7 | 0.2 | 8.6 | Remainder |

The measured air-fuel ratio is 11.3:1. Compute the air-fuel ratio in the best possible manner.

**12.** A gas analyzes by volume:

| CO₂ | CO | H₂ | CH₄ | C₂H₄ | N₂ | Total |
|-----|-----|-----|-----|------|-----|-------|
| 3 | 19 | 41 | 25 | 9 | 3 | 100 |

Determine the air-fuel ratio for complete combustion and the temperature of condensation of the water vapor in the products (atmospheric pressure).

**13.** An exhaust analysis (volumetric) of a spark-ignition engine (fuel, $C_8H_{17}$) showed the following:

| CO₂ | CO | H₂ | CH₄ | O₂ |
|-----|-----|-----|-----|-----|
| 9.9% | 7.2% | 3.3% | 0.3% | 0% |

Compute the air-fuel ratio, and compare with the measured value of 11.3:

(a) By the carbon-balance method
(b) By the hydrogen-balance method
(c) By the carbon-hydrogen-balance method and unknown fuel
(d) By the carbon-hydrogen-balance method and known fuel
(e) By the oxidized-products method

**14.** Given the higher heat value of methane at constant pressure and 77°F (Table B-11), compute the lower heating value and, also, the higher and lower values at constant volume.

**15.** Compute the higher heat of reaction of methane at constant pressure and 1000°R (value at 537°R is in Table B-11).

**16.** Determine the dew point for the products of combustion of methane with the chemically correct amount of air at atmospheric pressure.

**17.** Given the higher heating value of gaseous decane at constant pressure (Table B-11), compute seven other heating values for gaseous and liquid fuel. Find the maximum per cent difference.

**18.** Determine the theoretical flame temperature by burning methane with 100 per cent excess air at constant volume and 537°R.

**19.** Repeat Prob. 18, but at constant pressure.

**20.** Determine the theoretical flame temperature when burning liquid $C_8H_{18}$ at 537°R with double the amount of air required for complete combustion and at constant pressure.

**21.** Compute the theoretical flame temperature when burning CO at constant pressure with the chemically correct amount of air at 537°R.

**22.** Repeat Prob. 21, but for constant-volume combustion.

**23.** Compute the theoretical flame temperature when burning H₂ with 300 per cent of the chemically correct air at constant pressure and 537°R.

**24.** Repeat Prob. 23, but for combustion at constant volume.

**25.** Calculate the degree of dissociation of 1 mole of $CO_2$ at 5000°R under total pressures of 10 and 100 atm.

**26.** Repeat Example 14, but for total pressures of 10 and 100 atm.

**27.** Calculate the degree of dissociation of $CO_2$ when CO is burned with the theoretical amount of air if the equilibrium temperature is 4000°R and the total pressure is 20 atm.

**28.** Determine the minimum work (per pound of air) that is required to separate atmospheric air at 77°F into nitrogen and oxygen each at 1 atm pressure and at 77°F. (Use the reaction box of Art. 16-7 for this separation process.) Explain why the usual process of mixing is an irreversible process. Show that the minimum work is also given directly by Eq. (8-10c).

**29.** Calculate the ideal-gas equilibrium constant at 77°F for the reaction of carbon monoxide with oxygen to form carbon dioxide.

**30.** Include, now, the equilibrium constant $K_p$ in the discussion of Eq. (a), Art. 8-6.

### SELECTED REFERENCES

1. Lewis, B., and G. von Elbe: "Combustion, Flames, and Explosions of Gases," Academic Press, Inc., New York, 1951.
2. Jost, W.: "Explosion and Combustion Processes in Gases," McGraw-Hill Book Company, Inc., New York, 1946.
3. Lichty, L. C.: "Thermodynamics," 2d ed., McGraw-Hill Book Company, Inc., New York, 1948.
4. Obert, E. F.: "Concepts of Thermodynamics," McGraw-Hill Book Company, Inc., New York, 1960.

# CHAPTER 17

# GAS CYCLES AND PROCESSES

Duty is like a man's shadow—he can flee
from it—but it will always pursue him.
*Marcus Aurelius*

For the production of power in amounts of 50,000 and 100,000 kw, the large central station with a vapor (steam) cycle is the universal system. When the power requirements are small (less than 10,000 kw) or when light weight is demanded or variable-speed operation is required, the internal-combustion engine is the preferred source of power. However, development work now in progress on gas turbines and fuel cells forecasts that these forms of prime movers may well prove to be active competitors of both steam and the internal-combustion engine.

**17-1. Gas Cycles.** The changes in state of a gas during the processes of the Carnot cycle are illustrated in Fig. 17-1. Comparison of Figs. 15-1

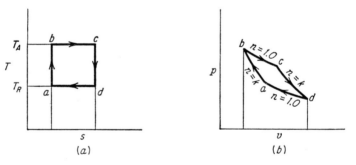

FIG. 17-1. Carnot cycle for a gas.

and 17-1 shows that the cycle on the $Ts$ diagrams has the same appearance for either a gas or a wet vapor (but not on the $pv$ diagrams). The difference is that the isothermal processes for the gas are not also at constant pressure as they are for the wet vapor; hence, the gas cycle cannot be simplified by using flow processes. Recall that the Carnot cycle with a wet vapor and with flow processes is a practical cycle for power generation (although other cycles are more desirable; Art. 15-3). But the Carnot cycle with a gas and with either flow or nonflow processes is quite impractical since the heat-addition and -rejection processes also require work transfers. If such processes were to be attempted, the cycle would be restricted to extremely slow operation.

The practical method for simulating processes that involve both heat and work is to burn a fuel internally within an engine and so dispense with the necessity for heat-transfer surfaces. The temperature attained in a nonflow combination process can be extremely high, far above the melting temperature of the enclosing metal, because the high temperature is only momentarily experienced, and succeeding processes are at relatively low temperatures.†

The efficiency of the combustion engine (also, the over-all efficiency of a thermodynamic cycle with an accompanying combustion process) is defined as (loosely called the thermal efficiency)

$$\eta_{\substack{\text{combustion} \\ \text{engine}}} \equiv \frac{\Delta W_{\text{output}}}{-\Delta H_{p_0, T_0}} \qquad (17\text{-}1)$$

wherein $-\Delta H$ is the change in enthalpy from reactants to products (the heat of combustion of the fuel) at the temperature and pressure of the surroundings. Note that Eq. (17-1) has nothing to do with the Second Law restrictions on thermal efficiency since it applies to a process, not to a cycle (Art. 8-5).

As a matter of historical interest, Rudolf Diesel in 1893 tried to simulate the Carnot cycle by gradually injecting a fuel during the expansion stroke of a piston engine (*bc* in Fig. 17-1). He reasoned that fuel introduced at this stage should spontaneously ignite since the air within the cylinder had been compressed to a high temperature. Then, as the piston descended on the working stroke and the temperature tended to fall, fuel could be continuously injected and burned at a rate such that the process would be isothermal. The other processes of the Carnot cycle would then follow. Diesel soon found that isothermal expansion and isothermal compression could not be accomplished‡ and that the extremely high pressures arising from the double compression (isothermal compression *da* followed by isentropic compression *ab* in Fig. 17-1) were undesirable.

It should be emphasized that *the attainment of the Carnot processes is neither necessary nor desirable when a combustion process (or a chemical reaction) occurs.* Consider that air and fuel enter the combustion system while the leaving products are of different chemical composition. Thus, the system involves *not* a thermodynamic cycle but, rather, a *process.* The maximum work of a process, such as this, was analyzed in Art. 8-5 and found to be equal to the change in the Gibbs free energy of the flow

---

† A stagnant layer of gas protects the metal because a large temperature gradient exists between the metal and the main body of the gas; also, the metal walls are externally cooled by water or air (Art. 19-6).

‡ It is interesting to note that isothermal compression was to be simulated by injecting water.

stream. The combustion engine would deliver this maximum amount of work, without regard for the Carnot processes, if the processes selected were all reversible. No practical means are known for conducting the combustion process in a manner approaching reversibility.†

A theoretical means can be devised as suggested by Prof. J. Keenan. Consider that the combustion process at equilibrium contains both mixture‡ and products in definite proportions and in amounts predicted by the equilibrium constant. When the temperature of the equilibrium mixture is raised, the equilibrium shifts and the degree of completion of the reaction diminishes; therefore, a state will exist at an *extremely high* temperature where the mixture can exist in equilibrium with an infinitesimally small proportion of products. A theoretical manner for approaching a reversible combustion process is now apparent. The mixture can be isentropically compressed to the high temperature where essentially no reaction occurs. (It is assumed that the speed of the compression process is extremely swift, and therefore reaction does not begin during the compression, or else a negative catalyst is present.) Then, the mixture can be slowly and isentropically expanded; as the temperature falls, reaction reversibly occurs; and at each temperature level of the expansion a greater and greater amount of products will be formed. In this sequence of processes (at constant§ entropy) chemical reaction can be conceived to be reversibly executed although, of course, not at constant temperature.

Although the processes of a reversible internal-combustion engine bear no resemblance to the processes of the Carnot cycle, the same conclusions are evident for either system: *For the highest efficiency, the combustion engine and the thermodynamic cycle must operate between the highest and lowest temperature that can be attained.* In a cycle, heat should be added at the highest possible temperature; in an internal-combustion engine, combustion should begin at the highest possible temperature, for then the irreversibility of the chemical reaction is reduced. Moreover, in both the cycle and the combustion engine, expansion should proceed to the lowest possible temperature in order to obtain the maximum amount of work. Because of these similarities, the combustion engine can be analyzed as if it were a cycle by assuming that the combustion process is equivalent to a transfer of heat and that no change in composition is undergone by the working substance. Such an analysis, of course, will not be exact, but it does indicate design considerations that improve the efficiency of the real engine. (Recall that the predictions of Art. 8-5 gave no indications as to how the efficiency of a chemical-reaction process could be improved.)

† But in a galvanic cell a chemical reaction can approach reversible operation (Art. 6-8).

‡ The fuel must be a simple substance, such as gaseous carbon, and not a complex compound. Also, the mixture must be reversibly obtained (see, for example, Fig. 16-2).

§ The entropy is constant because both processes are reversible and adiabatic.

To illustrate the procedure, consider that the *Otto cycle* has four processes (Fig. 17-2) that are similar to those of a spark-ignition engine:

*ab*  Isentropic compression
*bc*  Constant-volume addition of heat
*cd*  Isentropic expansion
*da*  Constant-volume rejection of heat

The thermal efficiency of this thermodynamic cycle is defined by Eq. (4-6), and, with Eq. (3-8*b*),

$$\eta_t = \frac{\Delta Q_A + \Delta Q_R}{\Delta Q_A} = 1 - \frac{T_d - T_a}{T_c - T_b} \qquad (a)$$

The temperatures $T_d$ and $T_a$ can be expressed in terms of $T_c$ and $T_b$, respectively, by Eq. (3-7*c*). Hence, (*a*) reduces to [note Eq. (11-12*b*)]

$$\eta_t = 1 - \frac{1}{r_v{}^{k-1}} \qquad (17\text{-}2a)$$

Equation (17-2*a*) shows that the thermal efficiency of the Otto cycle (and the efficiency of the Otto spark-ignition engine) can be increased by:

1. Raising the expansion ratio $r_v$ (which is also the compression ratio)
2. Using a gas with a high $k$ value

Conclusion 1 was not widely known at the start of the development of the spark-ignition engine, and engines with extremely low expansion ratios were marketed (1900). Through the years, the efficiency of the engine has been increased by raising the expansion ratio considerably (about 10:1 in modern spark-ignition automotive engines versus about 4 or 5:1 in 1920).

The second conclusion is of pedagogical interest. Thus, the Otto cycle with helium ($k = 1.6$) as the working substance has a higher thermal efficiency than when air ($k = 1.4$) is used. Also, the efficiency of the theoretical cycle is independent of the load (which is varied by varying the quantity of heat added to the cycle).

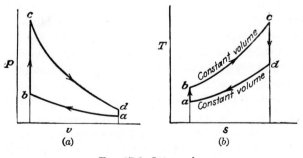

FIG. 17-2. Otto cycle.

If the same reasoning were applied to the Carnot cycle of Fig. 17-1, the same equation and the same conclusions would be obtained. But the thermal efficiency of a Carnot cycle is independent of the working substance and depends only upon the temperature of source and sink (Art. 7-1). The answer to this apparent contradiction is that changing $k$ or $r_v$ automatically changes the temperature limits [Eq. (11-12c)].

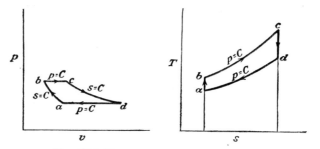

FIG. 17-3. Brayton (Joule) cycle for a gas.

The gas-turbine process for analysis can be considered to be a cycle as represented by Fig. 17-3—the *Brayton* or *Joule cycle:*

ab   Isentropic compression
bc   Constant-pressure addition of heat
cd   Isentropic expansion
da   Constant-pressure rejection of heat

The thermal efficiency of this cycle is also given by Eq. (17-2a). Since, however, the Brayton cycle operates between definite pressure limits, it is preferable to convert Eq. (17-2a) by Eq. (11-12c) into the equivalent form

$$\eta_t = 1 - \frac{1}{r_p^{(k-1)/k}} \qquad (17\text{-}2b)$$

It is sometimes helpful to compare two cycles by superimposing both cycles on the $Ts$ diagram. In Fig. 17-4, the Otto cycle $abcd$ is compared with a cycle $abc'd'$, which has a constant-pressure addition of heat (the *Diesel cycle*). The conditions of similarity shown by Fig. 17-4 are that both cycles (1) start from the same initial state $a$, (2) have the same compression ratio, and (3) have the same heat input. To fulfill the premise of equal compression ratios, the compression process $ab$ must be the same for both cycles; to fulfill the premise of equal heat supplied, the $Ts$ areas $fbce$ and $fbc'e'$ must be equal. Since constant-volume processes are steeper on the $Ts$ diagram than constant-pressure processes, construction of equal areas for the heat supplied shows state $c'$ to have a greater entropy than that of state $c$. It then follows that the Diesel cycle must reject more heat than the Otto cycle (area $fad'e'$ is greater than area $fade$).   Hence, since

$$\eta_t = 1 - \frac{|\Delta Q_R|}{\Delta Q_A} = 1 - \frac{|\Delta Q_R|}{\text{constant}}$$

the efficiency of the Diesel cycle is *less* than that of the Otto cycle for the given restrictions.

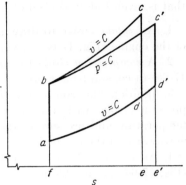

The thermal efficiency of some cycles can be raised by *regeneration*.   Consider the *Stirling cycle* of Fig. 17-5. Here heat is added to the gas in the constant-volume process $bc$ and also in the isothermal process $cd$.   Heat is rejected in the constant-volume process $da$ and in the isothermal process $ab$.   Note that the temperature of initial heat rejection is $T_d$ and the

Fig. 17-4. Comparison of Otto and Diesel cycles at the same compression ratio and with the same heat input.

temperature of initial heat addition is $T_b$.   Since $T_b$ is less than $T_d$, it should be possible to find means for transferring a part of the rejected heat to the heat-addition process.   When such means are found (a heat exchanger), less heat need be supplied by the surroundings, although the work of the cycle is unchanged, and therefore the thermal efficiency will be increased.   If the Stirling cycle could be perfectly regenerated, no heat would be necessary for process $bc$ since the energy

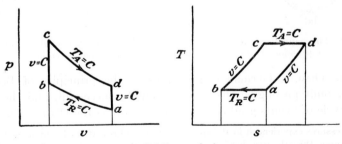

Fig. 17-5. Stirling cycle for a gas.

would be supplied by the cooling process $da$.   With all processes reversible, heat would need to be added only at $T_A$ (constant) and rejected at $T_R$ (constant); therefore, the thermal efficiency would equal that of the Carnot cycle.

Recently, interest in the Stirling cycle has been revived by the development of a high-speed engine by the Philips Research Laboratory (Eindhoven).   The cycle uses hydrogen at high pressures (140 kg/cm² max) and develops 40 bhp at 2,500 rpm. Minimum fuel (oil) consumption is 0.36 $lb_m$/bhp-hr at 1,500 rpm (about 38 per cent thermal efficiency); weight, 11 $lb_m$/bhp; 120 hp/liter displacement.

**17-2. The Otto Engine (1862-1876).** In 1862, Beau de Rochas proposed for the internal-combustion engine a sequence of operations that is, even today, typical of most spark-ignition engines:

1. An *intake stroke* to draw a combustible mixture into the cylinder of the engine (Fig. 17-6a) (a variable-mass process)

2. A *compression stroke* to raise the temperature of the mixture (Fig. 17-6b) (a constant-mass process)

3. Ignition and combustion of the mixture at the end of the compression stroke with the consequent liberation of energy raising the temperature and pressure of the gases, with the piston then descending downward on the *expansion* or *power stroke* (Fig. 17-6c)

4. An *exhaust stroke* to sweep the cylinder free of the burned gases (Fig. 17-6d) (a variable-mass process)

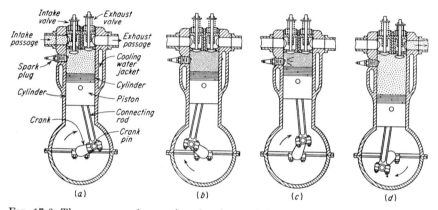

Fig. 17-6. The sequence of operations for the spark-ignition (SI) engine. (a) Intake stroke; (b)compression stroke; (c) expansion stroke; (d) exhaust stroke.

In 1876, Otto, a German engineer, using the principles of Beau de Rochas, built an engine that became highly successful, and the name of the cycle of events gradually became known as the Otto cycle.

The pressures experienced in the spark-ignition (abbreviated SI) engine at full load (at wide-open throttle) are shown by the *indicator diagram* of Fig. 17-7a. During the intake stroke the pressure in the cylinder is closely atmospheric, being less only because of fluid friction. On the following stroke, the mixture of gasoline vapor and air is compressed to the highest temperature that can be attained without spontaneous ignition of the mixture. [Compression ratios (Art. 11-7) of 6 to 10 are usual.] A spark then ignites the mixture, which rapidly burns while the piston comes to rest at the top dead-center position of the crank mechanism. During this period of little movement of the piston, combustion of the mixture takes place at essentially constant volume (as proposed by de Rochas).† The increase in temperature from burning the fuel

† De Rochas also suggested that the spark plug would be unnecessary because the fuel-air mixture would spontaneously ignite at the end of the compression stroke.

causes the pressure to increase, and thus high pressures are available to drive the piston downward on the power stroke. Although, if the greatest amount of work is to be obtained, expansion should proceed to atmospheric pressure, such a complete expansion would require a higher expansion ratio than compression ratio. Too, the size of the engine would be large. For these reasons, the exhaust valve is opened before the end of the expansion stroke to allow the pressure to drop to atmospheric before the exhaust stroke is begun. The exhaust stroke then purges the remaining gas from the cylinder with the exception of the small amount contained within the clearance space (combustion chamber).

In the SI engine, a spark can ignite only a combustible mixture. A fairly definite relationship of fuel and air (approximately 12 to 16 parts of air to 1 part of fuel by mass) must therefore be present in all parts of the chamber at all loads if a flame is to

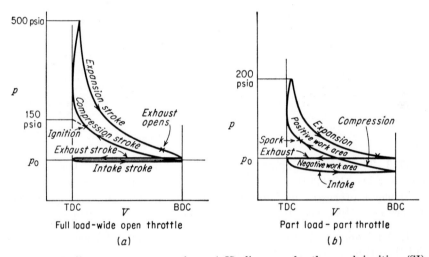

Fig. 17-7. Indicator or pressure-volume ($pV$) diagrams for the spark-ignition (SI) engine.

travel through the mixture. The turning effort applied to the crankshaft depends on the mass of mixture burned in the cylinder, and this effort is controlled by restricting the amount of mixture (but not primarily the air-fuel ratio) entering the cylinder on the intake stroke. This load control is accomplished by using a valve, called the *throttle*, to obstruct the passageway that leads to the cylinder. On the intake stroke, if the throttle is not open, the pressure in the cylinder is reduced below atmospheric, and a reduced mass of mixture will enter the cylinder with correspondingly lower compression pressures and combustion pressures (Fig. 17-7b). Note that the work obtained from the partially loaded engine is reduced, not only because less fuel (and air) is inducted than at full load, but also because the negative work of the engine cycle is increased. The negative work area of Fig. 17-7b is a result of the wasteful throttling process that is used as the method of governing.

Although combustion of the fuel should take place at the highest possible temperature, the Otto cycle is restricted to relatively low compression ratios. The reason for this can be understood by studying Fig. 17-8. After passage of the ignition spark, a flame sweeps out from the spark plug and transforms the mixture into products of combustion. The energy released will be proportional to the mass of charge burned, and this release can be accomplished in an orderly manner by properly controlling the speed of the flame. As the flame progresses across the chamber, the unburned mixture is compressed, and its temperature therefore increases. If the temperature should reach a certain critical stage, the unburned mixture will ignite, as at *a* in Fig. 17-8, without depending on the advancing flame. There will then result an energy release sufficient to produce a momentary high and localized pressure that causes an audible "knock" to be heard. The *knock*, or *detonation*, is caused by the almost instantaneous explosion of a part of the mixture. The resulting impact on the engine structure may cause failure; for this reason the compression ratio of the Otto engine must be kept at a low value (or else expensive fuels, which can resist compression ignition, are required). In other words, combustion of the fuel in the Otto engine must be fast but not explosive, because no mechanical structure can long resist impact.

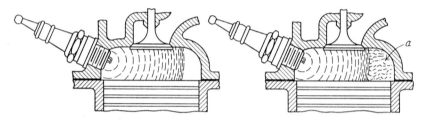

FIG. 17-8. Detonation in the SI engine.

**17-3. The Diesel Engine (1893).** The failure of Dr. Diesel's first engine (Art. 17-1) caused him to abandon, as ideals, the impractical isothermal processes in favor of adiabatic compression and expansion processes. His successful engine embodied the following events:

1. An *intake stroke* to induct air alone into the cylinder without wasteful throttling (a variable-mass process)

2. A *compression stroke* to raise the air to a high temperature, a temperature higher than the ignition point of the fuel

3. Injection of the fuel during the first part of the expansion stroke at a rate such that combustion maintains the pressure constant, followed by expansion to the initial volume of the cylinder

4. An *exhaust stroke* to remove the burned gases from the cylinder (a variable-mass process)

The success of the diesel engine can be attributed to the fact that only air is compressed in the engine; therefore, the compression ratio can be much higher than that of the Otto engine, which compresses a combustible mixture.

Typical pressures encountered in the compression-ignition (abbreviated CI) engine at full load are shown in Fig. 17-9. During the intake stroke, air alone is drawn into the cylinder, and the pressure is closely atmospheric. The air is then compressed to a pressure of 350 to 500 psia (compression ratios of 12 to 18 are in use) before the fuel is injected into the high-temperature air. An early method of injecting the fuel was to use a blast of compressed air to carry the fuel into the engine through a small orifice called the *nozzle*. This method gave good atomization and good control of the combustion process, as can be recognized in Fig. 17-9a, which shows, closely, a constant-pressure combustion. However, *air injection* is rarely used because an air compressor becomes a necessary and expensive auxiliary. The modern method of injection is to compress and spray the fuel alone into the cylinder through the nozzle and depend upon a high injection pressure (2,000 psia) for atomizing the fuel. If the injection period is long and the speed of the engine is slow, indicator cards similar to Fig. 17-9a can be obtained. However, although Dr. Diesel insisted upon operating his engines with essentially constant-pressure combustion, constant-volume combustion is thermodynamically more desirable because the chemical energy is liberated at the very beginning of the expansion stroke. For this reason, the indicator cards of modern engines resemble Fig. 17-9b; here the first part of the fuel to be injected burns essentially at constant volume while the remainder burns essentially at constant pressure.

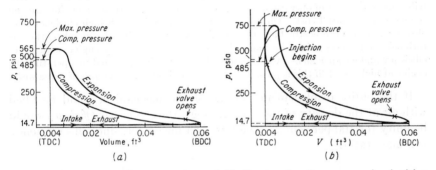

FIG. 17-9. Indicator or pressure-volume (*pV*) diagram for the compression-ignition (CI) engine. (*a*) Air injection (or late and slow mechanical injection); (*b*) solid (mechanical) injection.

The CI engine is not completely free from "knock" or detonation: When the first particles of liquid fuel are injected into the hot compressed air, an interval of time will elapse while atomization, vaporization, distribution, and initial chemical reaction take place. During this *delay* period, more and more fuel enters the chamber. Because of this accumulation of fuel, when combustion does begin, the reaction tends to be uncontrollably fast, and therefore a high localized pressure rise is experienced. This violent pressure rise is an impact on the parts of the engine that can be quite destructive. Since the diesel engine has a high pressure before combustion, any additional pressure rise from combustion only increases the design troubles; probably for this reason, Dr. Diesel insisted upon constant pressure as the desideratum for his engine.

The load control of the CI engine is a simple matter because only the quantity of fuel need be regulated. Thus, at full load it is desired to inject a quantity of fuel such that all the air in the cylinder can be burned. Practically, this limit cannot be reached because it is not possible for the localized fuel spray to find all the air; rich and lean regions abound, and the engine exhaust gas is colored in appearance and pungent in odor. At part load, only a fraction of the full-load fuel quantity is injected. In localized regions, combustion of the fuel occurs at ratios of air to fuel of about 15:1, although the over-all air-fuel ratio is much higher than this. Thus, at full output of the engine, most of the air is used for reaction; at part load only a fraction of the air need be combined with the fuel, and, because of the localized combustion, the intake-air process need not be throttled at any time. Because of this free induction of air, the efficiency of the CI engine is not penalized at part loads by a wasteful throttling process such as that used by the SI engine. (Whether or not the fuel in the CI engine is throttled is unimportant because the work spent in pumping the liquid is relatively small when compared with that for a gas.)

**17-4. The Piston-Cylinder Mechanism.** The positive and negative work quantities for a piston-cylinder mechanism cannot be readily separated, and therefore the *work ratio* (Art. 15-1) is not used. If the mechanism is an engine, the *mean effective pressure* (mep) (Art. 6-7) is the index that relates the size to the work output.

One objective of the engineer is to use engines that have high efficiencies [Eq. (17-1)] and also high mean effective pressures [Eq. (6-6b)]. A high efficiency is a factor contributing to low cost of operation, while a high mep is a factor in small engine size and therefore lower initial cost. On the other hand, an engine with high efficiency may also demand a more costly fuel (Otto engine compared with the diesel or the gas turbine), and power might be more cheaply obtained from an engine with lower efficiency and cheaper fuel. A low mep might also be circumvented by operating the engine at high speed (if this is possible), since power is but the time *rate* of doing work.

If the entire displacement volume of the mechanism could be filled with air at the density of the surroundings, the *volumetric efficiency* would be unity (or 100 per cent). The term is a misnomer, since this efficiency is a mass ratio:

$$\eta_v = \text{"volumetric" efficiency} \equiv \frac{m}{m_D} \qquad (17\text{-}3a)$$

where $m$ = mass of fluid inducted or delivered per stroke

$m_D$ = mass of fluid to fill piston displacement volume under optimum (usually atmospheric) conditions of pressure and temperature

One reason for the name volumetric efficiency is that it is usually found by measuring volumes:

$$\eta_v = \frac{\text{capacity (volume inducted in cfm)}}{\text{displacement (piston displacement in cfm)}} \bigg]_{p,T} = \frac{C}{D} \bigg]_{p,T} \qquad (17\text{-}3b)$$

For an air compressor, the *compressor displacement* is the volume swept by the pistons in 1 min. (A double-acting piston displaces a volume twice that of a single-acting piston.) The volume of air delivered by a compressor, when measured at atmospheric temperature, pressure, and humidity, is called *free air*.

The work done by an internal-combustion engine depends on the amount of energy released when a mixture of air and fuel burns. If the engine does not induct the largest possible amount of air, the work output of the engine will be restricted, no matter how much fuel is added. (Fuel is usually liquid, and the injection or addition of liquid fuel into air requires little work.)    Therefore, a basic requirement for a high-output engine is a high volumetric efficiency.

The volumetric efficiency of the engine can be increased by *supercharging*, a *supercharger* being merely an air compressor that supplies air to the engine.    The gain in work output may not be proportional to the increase in mass of air supplied, mainly because a considerable portion of the gain must be used to drive the supercharger (unless the supercharger utilizes the "blowdown" energy of the exhaust; Chap. 12, Prob. 5).

The *brake* or *shaft work* of the mechanism differs from the *indicated work* because of mechanical friction (Art. 6-7).    The *mechanical efficiency* is defined as

$$\eta_m \equiv \frac{\text{brake work}}{\text{indicated work}} \qquad \text{(engine)} \qquad (17\text{-}4a)$$

$$\eta_m \equiv \frac{\text{indicated work}}{\text{shaft work}} \qquad \text{(compressor)} \qquad (17\text{-}4b)$$

Even though an engine has a low mep, a low volumetric efficiency, or a low mechanical efficiency, the power obtained from the engine may be relatively high, if high cyclic speeds are possible, because power is the time rate of doing work.    Modern internal-combustion engines are usually operated, not at the speed of maximum mep (for example), but, rather, at higher speeds where the mep is lower because of mechanical or fluid friction.    More power can be obtained because, although the work per mechanical cycle is decreased, the number of mechanical cycles per second is increased in a greater ratio.

The speed of the mechanism is given in revolutions per minute of the crankshaft.    The piston must be stopped and reversed in direction twice in each revolution of the crank, and the attendant accelerations and decelerations of the reciprocating mass create inertia forces.    These forces are dependent on acceleration and mass; hence small mechanisms or multicylinder mechanisms with light pistons can be operated at higher rpm than can large mechanisms.    The *mean piston speed* is the parameter usually specified:

$$\text{Mean piston speed} \equiv 2 \, (\text{rpm})(\text{stroke}) \qquad (\text{ft/min}) \qquad (17\text{-}5)$$

When speed or size is limited by design stresses, additional power can be obtained by a *two-stroke* system, which produces a power stroke in every revolution of the crank. Here the exhaust and intake strokes are eliminated by using a *blower* or *scavenging pump*.    When the exhaust valve opens near the end of the power stroke, the high-pressure products escape to the atmosphere; air is then blown into the cylinder by the blower to complete the scavenging and also to charge the cylinder.    The piston then returns on the second stroke of the cycle, the compression stroke.

The power of the internal-combustion engine depends on the mass rate of air flow into the engine, and this factor in turn is controlled by the size and number of cylinders, the cyclic speed (rpm), and the degree of supercharge.    To develop high power output, two methods are available: large engines operated at low rpm (this is CI practice) and small engines operated at high rpm (this is SI practice) either with or without supercharging.    In any case the size or complexity of the engine does not lend itself to the production of power in units above 10,000 kw, and smaller units are preferable.

**17-5. Reciprocating-piston Compressors.** The indicator diagram for a reciprocating-piston gas compressor would appear as in Fig. 17-10. In process 1-2, a mass of gas is compressed to the receiver pressure; in the displacement process 2-3, a portion of the gas is pushed into the receiver; in process 3-4, the residual portion in the clearance (Art. 5-8d) is expanded to the suction pressure; and along path 4-1, a new charge of gas enters the cylinder. The compression and reexpansion of a portion of the gas

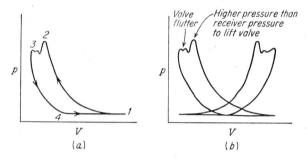

FIG. 17-10. Indicator cards for reciprocating-piston compressor.

causes a loss in the real machine because both processes are irreversible; hence it is desirable to use small clearances.

A usual value for the clearance with medium delivery pressures (100 psia) is 6 per cent of the piston displacement volume $V_D$. For higher delivery pressures, smaller clearances are desirable and necessary since the clearance volume controls the volumetric efficiency; if the clearance volume is large and the receiver pressure is high, it is entirely possible that no gas will be delivered by the compressor.

A *conventional* indicator card is one that idealizes each process. In Fig. 17-11 all processes are specified to be reversible, and therefore, the polytropic processes 1-2 and 3-4 must have identical values of $n$. (If the exponents were not the same, the temperature of the expanded gas at 4 would not be equal to the temperature of the incoming gas, and irreversible mixing would occur.) The reversible work can be found by Eq. (11-7b) or, for the isothermal compressor, by Eq. (11-8). Since Eq. (11-7b) was derived without picturing the mechanism for the compressor, clearance has no effect on the *reversible* work (and, too, the equations hold for *any* reversible compressor).

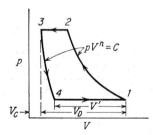

FIG. 17-11. Conventional (reversible) indicator diagram.

A *conventional volumetric efficiency* can be determined from the conventional card, and this efficiency represents the optimum value for the real compressor (Prob. 30):

$$\eta_{v \atop \text{conv}} = 1 + c - c \left(\frac{p_2}{p_1}\right)^{1/n} \qquad c = \frac{V_3}{V_D} \qquad (17\text{-}6)$$

**17-6. Multistage Compression.** The work required to compress a gas is controlled by three factors: (1) the compression process, (2) the pressure ratio, and (3) the initial temperature of the gas, although these factors cannot be readily varied. Any selection of the compression process is restricted by the mass flow rate, and therefore isothermal compression, while most desirable, is also most impractical. The pressure ratio cannot be changed because it is fixed by the process demands for the compressed fluid. However, the third factor, the initial temperature of the gas, can be lowered to effect a saving in work.

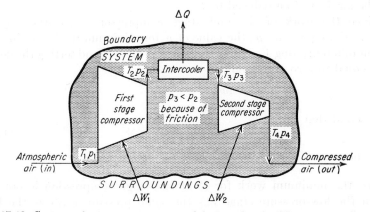

Fig. 17-12. System of two compressors and intercooler (multistage or series compression).

In the case of an air compressor, the inlet air is at atmospheric temperature, and therefore this temperature cannot be reduced by using lake or river water, which is the cheapest coolant, because the water would be at essentially the same temperature as the air. But, instead of using one compressor, suppose that two compressors are connected in series. This series combination will theoretically require the same amount of work as the single compressor, and, practically, it will require slightly more work owing to the increased friction caused by the more complex mechanism. However, the compressed air leaving the first compressor (the first *stage*) will be at a high temperature relative to the temperature of the available cooling water, and therefore effective cooling of the air is readily accomplished by an *intercooler* (Fig. 17-12). Thus, the air entering the second compressor (the second stage) is reduced in temperature, and for this reason less work will be required by the multistage compressor than if a single compressor had been used.

The work required for a real multistage compressor is readily evaluated by Eq. (5-7d) applied to the system of two compressors and intercooler (Fig. (17-12). For the *reversible* compressor, the ideal relationship, Eq. (11-7b), can be applied, in turn, to each compressor:

$$\Delta w_{\substack{rev \\ multistage}} = \frac{nRT_1}{1-n}\left[\left(\frac{p_i}{p_1}\right)^{\frac{n-1}{n}} - 1\right] + \frac{nRT_3}{1-n}\left[\left(\frac{p_4}{p_i}\right)^{\frac{n-1}{n}} - 1\right] \quad (17\text{-}7a)$$

Here it is assumed that the compression exponent is the same for both machines and that the outlet pressure ($p_i$) for the first compressor is equal to the inlet pressure for the second compressor ($p_i = p_2 = p_3$ in Fig. 17-12). The latter assumption is equivalent to stating that no pressure drop occurs in the intercooler. If the intercooling is *ideal* or *perfect*, no pressure drop will occur and, also, the outlet temperature of the air $T_3$ will be reduced to $T_1$.

Since the work of the multistage compressor is governed by the intermediate pressure $p_i$, the value of $p_i$ for minimum work is of interest. Upon differentiating Eq. (17-7a) with respect to $p_i$ and setting the derivative equal to zero (Prob. 33),

$$\frac{d\Delta w}{dp_i}\bigg]_{T_1 = T_3} = 0$$

it is found that

$$p_i = (p_1 p_4)^{\frac{1}{2}} \quad (17\text{-}7b)$$

and therefore

$$\frac{p_i}{p_1} = \frac{p_4}{p_i}$$

Thus, the minimum work for ideal two-stage compression is obtained when the low-pressure stage has the same pressure ratio as the high-pressure stage. With this division of the pressure ratio, each compressor will require the same amount of work because the inlet air temperature is the same for either stage.

The work saved by multistage compression can be illustrated on the *pv* diagram, Fig. 17-13. (Since clearance does not affect the reversible work, Fig. 17-13 can be construed to be for either a piston-type compressor without clearance or for a rotary-type compressor.) If compression were single-stage, the path for a polytropic compression process would be 1-5 and the work would be proportional to area 1579. If compression were isothermal and single-stage, the work would be proportional to area 1679. With multistage and polytropic compression,

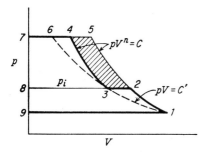

FIG. 17-13. Savings in work from two-stage compression with intercooler.

the work for the first stage is represented by area 1289, and the work for the second stage by area 3478. Thus, the work saved by two-stage polytropic compression, relative to single-stage polytropic compression, is proportional to the area 2345.

**17-7. The Gas Turbine.** The internal-combustion engine is a compact power plant, but it suffers from the limitation of a low mass-flow rate. This limitation can be removed by substituting continuous-flow machines for the reciprocating-piston engine while retaining the desirable feature of internal combustion of a fuel with the working fluid. Unfortunately, the flow compression process cannot be so efficiently conducted as non-flow compression. The usual† flow compressor first induces a high velocity in the fluid and then transforms the kinetic energy into pressure energy by means of a diffuser or diverging passageway; this is a difficult

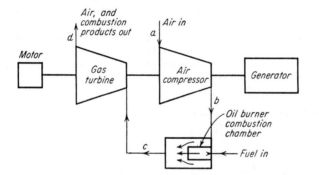

Fig. 17-14. Elements of a continuous-combustion gas-turbine system.

process to accomplish efficiently. On the other hand, a turbine can be used for the expansion process, and here the fluid can expand to the pressure of the atmosphere and thus attain an expansion ratio that cannot be readily approached by a reciprocating-piston engine. One other disadvantage is inherent in the flow process: the maximum temperature in the reciprocating-piston engine can be high because the process is intermittent; the maximum temperature in the continuous-flow machine must be relatively low because this temperature is constant in some part of the apparatus.

The basic elements of the continuous-combustion gas-turbine system are illustrated in Fig. 17-14. In this illustration is shown a compressor that compresses the air and forces it into and around the combustion chamber. Only a portion of the air is mixed with the fuel and burned at constant pressure. The high temperature of combustion is reduced by the main body of air passing around the combustion

† See Art. 5-8d.

chamber and mixing with the burned products. This mixture, at a temperature of about 1200°F, enters the turbine, which drives both the air compressor and a generator that absorbs the net power. (The small electric motor is used only for starting purposes.)

A speed governor regulates the fuel supply and thereby the turbine inlet temperature. The inlet temperature is restricted by metallurgical problems; for long life, a temperature of 1000 or 1200°F is usually recommended. Experimental work in testing full-scale turbine units and in developing high-temperature materials seems to predict that turbine inlet temperatures of 1500°F or better may be achieved in the near future. With these high temperatures, the gas turbine can compete for the high thermal efficiencies now developed by the CI engine. Even with lower inlet temperatures, the efficiency of the gas turbine may be acceptable for uses where simplicity and high mass-flow rates are desirable or in arid regions because a water supply is unnecessary. In general, fuels for the gas turbine can be cheaper than the fuels for internal-combustion engines because detonation is not a factor; experiments also indicate that powdered coal may prove successful.

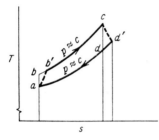

Fig. 17-15. Irreversible gas-turbine cycle.

The real gas-turbine system can be pictured as a cycle (Fig. 17-15). Here, by definition of $\eta_t$ [Eq. (4-6)] and of $\eta_c$ and $\eta_e$ [Eqs. (12-5)],

$$\eta_t = \frac{\Delta w_t + \Delta w_c}{\Delta h_{b'c}} = \frac{(h_c - h_d)\eta_e - (h_b - h_a)(1/\eta_c)}{h_c - h_{b'}}$$

This equation can be rearranged by assuming ideal gases so that $h = c_p T$ and by defining $x$ as

$$x = \frac{T_b}{T_a} = \frac{T_c}{T_d} = r_p{}^{(k-1)/k} \tag{17-8}$$

It then follows that

$$\eta_t = \frac{\left(\dfrac{T_c\eta_e}{x} - \dfrac{T_a}{\eta_c}\right)(x - 1)}{T_c - T_a\left(\dfrac{x - 1}{\eta_c} + 1\right)} \tag{17-9}$$

Equation (17-9), illustrated by Fig. 17-16, shows that an optimum pressure ratio exists for each turbine inlet temperature when compression and expansion efficiencies are fixed. Note that increasing the temperature from 800 to 1600°F more than doubles the optimum efficiency.

The pressure ratio is primarily fixed by the restriction offered by the nozzles of the turbine, and this restriction is controlled by the area of the nozzles and by the density

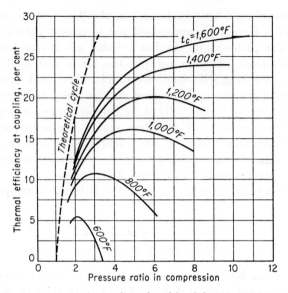

Fig. 17-16. Effects of compression ratio and turbine inlet temperature on the thermal efficiency of the nonregenerative Brayton cycle. ($t_c$ = turbine inlet temperature; $\eta_e = \eta_c = 0.85$; $t_a = 60°F$.)

of the fluid flowing. When the fuel quantity is reduced for part load, the combustion temperature decreases, and the density of the gas flowing through the fixed-area nozzles is increased (unless nozzle blocks can be shut off). This effect momentarily increases the amount of gas leaving the system, and the pressure before the nozzles falls (lower pressure ratio). But this drop in pressure decreases density and so partially compensates for the drop in temperature. The net effect is that the gas turbine operating at part load by temperature control also experiences a reduction in pressure ratio.

**17-8. The Regenerative Gas Turbine.** Inspection of Fig. 17-17 reveals that the temperature of the exhaust gases at state $d'$ is higher than the temperature of the compressed air at state $b'$, and, because of this fact, the gas turbine can be *regenerated*. The regeneration is accom-

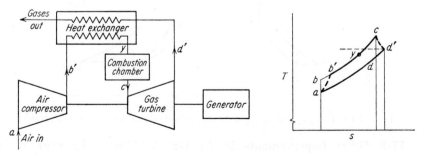

Fig. 17-17. Regenerative gas-turbine system.

plished by installing a heat exchanger in the flow system, as illustrated in Fig. 17-17. Here the compressed air is passed through a bundle of tubes or flat plates while the hot exhaust gas is circulated between the tubes or plates. In this manner the air is increased in temperature before reaching the combustion chamber, and thus less fuel need be burned to attain a specified turbine inlet temperature. The heat exchanger must be carefully designed; an excessive pressure drop of either the air or exhaust gas may nullify the anticipated gain in efficiency. Note that lowering the compression ratio increases the temperature difference available for regeneration and, at high compression ratios, regeneration becomes impossible.

The *efficiency* of the regenerator is defined (letters correspond to Fig. 17-17) as

$$\eta_r \equiv \frac{\text{actual temperature rise}}{\text{maximum temperature rise}} = \frac{t_y - t_{b'}}{t_{d'} - t_{b'}} \tag{17-10}$$

If the regenerator had infinite surface, $t_y$ and $t_{d'}$ would be equal. The size of the practical regenerator allows an $\eta_r$ of 0.50 to 0.75 without excessive pressure drop. The thermal efficiency of the regenerative cycle to simulate Fig. 17-17 is given by Eq. (17-9) when the term in the denominator involving $T_a$ is replaced by $T_y$:

$$\eta_t = \frac{\left(\dfrac{T_c \eta_e}{x} - \dfrac{T_a}{\eta_c}\right)(x - 1)}{T_c - T_y}$$

The temperature $T_y$ is found by Eq. (17-10):

$$T_y = \eta_r(T_{d'} - T_{b'}) + T_{b'}$$

and $T_{b'}$ can be expressed in terms of $T_b$, $T_a$, and $\eta_c$, while $T_{d'}$ is expressed in terms of $T_c$, $T_d$, and $\eta_e$.

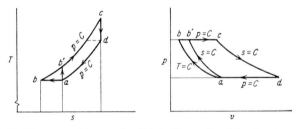

FIG. 17-18. The gas-turbine cycle with isothermal and isentropic compression.

**17-9. Other Improvements in the Gas Turbine.** The work ratio of the gas turbine can be raised if the isentropic compression process is replaced by isothermal compression. Figure 17-18 shows that this substitution increases the net work of the theoretical cycle (area $abcd$ is larger than area $ab'cd$) and decreases the thermal efficiency (heat added from $b$ to $b'$ is less effective than heat added from $b'$ to $c$ because a greater percentage is rejected). The decrease in thermal efficiency is more than removed when the cycle is regenerated because the energy available in the exhaust gases (area under $ad$) is sufficient to raise the temperature from $T_b$ to the limiting temperature $T_d$. Since isothermal compression is not a practical flow process, the real system employs intercooling and regeneration, as shown in Fig. 17-19.

The work ratio of the system can also be raised by injecting water into or around the combustion chamber. This procedure does not

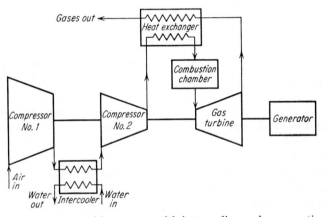

FIG. 17-19. Gas-turbine system with intercooling and regeneration.

affect the work of the compressor, but the steam that is formed from vaporizing the water increases the mass-flow rate to the turbine, lowers the $k$ value, and increases the specific gas constant of the mixture entering the turbine, and therefore increases† the work output. (The work spent in pumping the water, which is a liquid, into the high-pressure air is negligible.) This practice, of course, reduces the thermal efficiency because more fuel must be burned to raise the temperature and supply the latent heat of vaporization of the water. The water can also be injected into the air stream before compression or into the compressed air at a location preceding the regenerator; apparently, this last location is a patent claim by Lysholm. Note that water injected at any location that precedes the regenerator will probably result in a gain in thermal efficiency unless the energy available in the exhaust gases is insufficient to vaporize the water.

† See Art. 11-7 and Eq. (11-11).

The work ratio of the system can also be increased by reheating the air, after partial expansion in one turbine, before the air is admitted to a final turbine. In this instance, the thermal efficiency of the real system will probably be increased by the reheat because the work of the compressor, which affects the thermal efficiency, is not changed.

**17-10. Closed and Semiclosed Systems.** The open gas-turbine system is independent of a cooling source since it rejects hot waste gases to the atmosphere; this factor and the simplicity of the system have been major reasons for promoting development. A *closed system*, on the other hand, continuously circulates the same fluid and requires heat exchangers for the heat-addition and -rejection processes. Figure 17-20 depicts such a cycle. Combustion of the fuel takes place

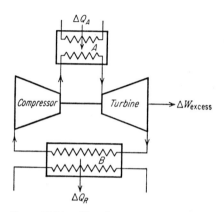

FIG. 17-20. Closed-system gas-turbine cycle.

in the air heater $A$ and not internally within the system, while water is used to cool the gases in the heat exchanger $B$. The advantages of the closed system are as follows:

1. Higher pressures can be used throughout the cycle, and therefore higher densities of the working fluid are obtained. This allows all parts of the system to be made smaller; smaller physical dimensions allow higher temperatures for a given stress limit.

2. The working fluid is clean, not contaminated with products of combustion; deposits on the turbine or compressor blades and wear or erosion of the turbine and compressor are reduced.

3. The working fluid can be a monatomic gas that has a more favorable heat-capacity ratio ($k$) than air.

4. Thermal efficiency at part load is high because part load can be secured by varying the density of the working fluid without varying the temperature. (Of course, a control system must be added to Fig. 17-20 if the density of the fluid is to be changed by adding or subtracting working fluid.)

5. A cheap fuel, such as coal, can be used.

Accompanying these advantages are the following disadvantages:

1. The efficiency of internal combustion has been eliminated by substituting a heat exchanger and an external furnace.

2. A coolant must be available.

3. Complexity and cost of the system have been increased.

**17-11. Jet Engines.**† In the analysis of jet engines, the methods of thermodynamics aid in visualizing why and how work can be obtained. A simple form of jet engine is the *ramjet* or *athodyd* (Fig. 17-21). Suppose that this device is traveling at a constant absolute speed $V_1$ through still air. To an observer on the ramjet, air enters the duct with relative speed $V_1$ and is decelerated in the diffuser with consequent rise in pressure. The change in kinetic energy for the reversible process is given by Eq. (6-3c), and since no work is done relative to the observer,

$$\Delta KE_{\substack{\text{compression}\\\text{in diffuser}}} = -m_{\text{air}} \int_{p_1}^{p_2} v\, dp \approx -m_{\text{air}} v_{\substack{\text{cold}\\(\text{av})}}(p_2 - p_1) \qquad (a)$$

Fuel is then burned with consequent increase in temperature and specific volume. Let it be assumed that this process is also at constant pressure. Now when the hot gases are discharged through the nozzle with relative speed $V_{r2}$, kinetic energy is regained, for a reversible process of amount

$$\Delta KE_{\substack{\text{expansion}\\\text{in nozzle}}} = -(m_{\text{air}} + m_{\text{fuel}}) \int_{p_2}^{p_1} v\, dp \approx (m_{\text{air}} + m_{\text{fuel}}) v_{\substack{\text{hot}\\(\text{av})}}(p_2 - p_1) \quad (b)$$

Since the specific volume (and mass flow) was increased by combustion of the fuel, a greater amount of kinetic energy is produced in the expansion process [Eq. (b)] than was supplied for the compression process [Eq. (a)], and the exit speed $V_{r2}$ will be greater than the entering speed $V_1$. Thus energy is available to drive the ramjet through the atmosphere (and work is done relative to an observer on the ground).

The same conclusion is reached by considering the reversible speed $V_{r2}$ developed by a nozzle (negligible velocity of approach):

$$V_{r2} = [2c_p T_{\text{combustion}}(1 - r_p^{(k-1)/k})]^{\frac{1}{2}} \qquad (13\text{-}11a)$$

and the inverse process in the diffuser:

$$V_1 = [2c_p T_{\text{compression}}(1 - r_p^{(k-1)/k})]^{\frac{1}{2}}$$

Hence, for this idealized case,

$$\frac{V_{r2}}{V_1} = \left(\frac{T_{\text{combustion}}}{T_{\text{compression}}}\right)^{\frac{1}{2}} = \left(\frac{T_{\text{combustion}}}{T_1 r_p^{(k-1)/k}}\right)^{\frac{1}{2}} \qquad (17\text{-}11)$$

A portion of the energy produced by expansion must necessarily be wasted, since residual kinetic energy remains in the leaving gas. Thus the exhaust gas has an absolute speed of $V_{r2} - V_1$ and the kinetic energy in the *leaving loss* is

$$\text{Leaving loss} = \frac{m_2}{2g_c}(V_{r2} - V_1)^2 \qquad (17\text{-}12)$$

† See, also, Art. 13-19.

The condition for zero leaving loss is that the speed of the ramjet $V_1$ should equal the relative exhaust speed $V_{r2}$, a condition which would correspond to zero fuel (heat) addition [Eq. (17-11)]. Hence if the ramjet is to operate, a leaving loss is a necessary accompaniment.

The work done by the ramjet will be calculated for an idealized case of steady flow (mass of fuel neglected). To an observer on the diffuser ($\Delta q$, $\Delta w$ zero),

$$\Delta q - \Delta w = \Delta h_{1x} + \Delta KE_{1x} \qquad (5\text{-}7d)$$

and, if the speed $V_x$ leaving the diffuser is negligible,

$$\Delta h_{\text{diffuser}} = h_x - h_1 = \frac{V_1{}^2}{2g_c} \qquad (c)$$

The energy added in the combustion chamber (considered as heat) is

$$\Delta q = \Delta h_{xy} = h_y - h_x \qquad (d)$$

For the nozzle (with $V_y$, like $V_x$, ignored),

$$\Delta h_{\text{nozzle}} = h_2 - h_y = -\frac{V_{r2}{}^2}{2g_c} \qquad (e)$$

An energy balance can be made on the surroundings:

Work = |heat added| − |leaving loss| − |sensible energy in exhaust|

$$= (h_y - h_x) - \frac{1}{2g_c} (V_{r2} - V_1)^2 - (h_2 - h_1)$$

And this reduces, with Eqs. (c) and (e), to

$$\text{Work} = \frac{V_1}{g_c} (V_{r2} - V_1) \qquad (17\text{-}13)$$

To check Eq. (17-13), note that the net thrust can be directly found from the momentum equation

$$F = \frac{\dot{m}_f}{g_c} (V_{r2} - V_1) \qquad (13\text{-}22a)$$

and the power from

$$P = FV = \frac{\dot{m}_f V_1}{g_c} (V_{r2} - V_1)$$

The locations of the forces (and their reactions) that produce the change in momentum may not always be obvious. Consider the flying ramjet in Fig. 17-21. Here the air entering the duct is slowed down in the diffuser, with consequent *increase* in pressure; fuel is sprayed into the air stream and burned, with consequent increase in velocity and *decrease* in pressure; the hot gases then increase in velocity in the nozzle as the pressure *decreases*. The forces exerted *by the fluid* on the walls of the ramjet are therefore (1) forces on the walls of the diffuser *in* the direction of motion, (2) friction forces on all the walls *opposite* to the direction of motion, and (3) forces on the walls of the nozzle *opposite* to the direction of motion. That the ramjet flies at all must arise from the forces on the walls of the diffuser being *greater* than the total of the other forces. Hence a "flying stovepipe" (a ramjet of constant internal diameter) cannot "fly" since there are no walls that the pressure of the fluid can act upon.

Fig. 17-21. Wall forces on a ramjet.

The ramjet can be converted into a slightly different form of jet engine by installing a gas turbine and compressor (such as that in Fig. 17-14). With this arrangement, air entering the engine is slowed down and compressed in the diffuser and then compressed to a higher pressure in the compressor. The fluid leaving the combustion chamber passes through the gas turbine on its way to the exit nozzle, and the work obtained from the turbine is used only to drive the air compressor. This combination of ramjet and gas turbine is called a *turbojet* engine.

The leaving loss from the ramjet or turbojet engine can be reduced (and even eliminated) by installing an exhaust-gas turbine. The output from the turbine can be used to drive a propeller. The usual arrangement is to employ the turbojet combination of units with a larger turbine. Then the power in excess of that required to drive the compressor drives the propeller. This variation of the turbojet is called a *turboprop* engine. Propulsion is obtained from the propeller and also from the reaction thrust of the leaving exhaust gas. In other words, a leaving loss is usually present, although the magnitude can be considerably reduced from that of a turbojet engine.

Examination of the turbojet and turboprop engines reveals that the same sequence of processes as in the ramjet should ideally be followed. These are (1) isentropic compression, (2) constant-pressure combustion, and (3) isentropic expansion. A hypothetical cycle employing these processes is the Brayton cycle (Fig. 17-3). This cycle represents the ideal or limiting performance for the jet engine, with maximum possible value for the thermal efficiency. To attain the Brayton-cycle efficiency, the kinetic energy in the exhaust gases must be fully utilized. The jet engine usually wastes a good portion of the available energy by discharging the exhaust gases at a high speed relative to the forward speed.

It should be carefully noted, however, that jet engines operate at much lower pressure ratios than reciprocating-piston engines and that the thermal efficiencies therefore tend to be inherently lower even if the leaving loss is negligible. Also, the limiting combustion temperature may demand a low pressure ratio for maximum efficiency (Fig. 17-16). This latter argument can be partly circumvented in jet design by employing high combustion temperatures that allow good performance (but short life). With an increase in pressure ratio to attain higher thermal efficiencies, the optimum velocity of the exhaust gas necessarily increases [Eq. (17-11)]. Thus jet engines must operate at high speeds if the efficiency is to even approach that of the reciprocating-piston engine. On the other hand, an advantage is gained by the high speed, since the reciprocating-piston engine employs a propeller and the propelling (propulsive) efficiency of the propeller may rapidly decrease as the flight speed exceeds about 300 mph. For this reason, when high speeds are demanded, the over-all efficiency of the jet engine can exceed that of the reciprocating-piston-engine–propeller combination because of the poor performance of the latter unit. The simple construction and light weight of the jet engine make it particularly desirable for high-speed aircraft operation.

## PROBLEMS

**1.** (*a*) Show all steps in the derivation of Eq. (17-2*a*). (*b*) Convert Eq. (17-2*a*) to Eq. (17-2*b*).

**2.** A Carnot cycle operates at temperatures of between 80 and 780°F with air as the gas. The pressure at the start of isothermal compression is 14.7 psia, and heat addition is 195 Btu/lb$_m$. Find the thermal efficiency, the imep, and the isothermal and isentropic compression ratios.

**3.** An Otto cycle with air has a compression ratio of 8, with the start of compression at 14.7 psia and 80°F. Heat addition is 195 Btu/lb$_m$. (*a*) Find the thermal efficiency and imep. (*b*) Compare the results with those of Prob. 2.

**4.** An engine operates on an air cycle. Air at 30 psia and 520°R is isentropically compressed to 450 psia. Heat is added at constant pressure until the temperature is 2000°R, and then an isentropic expansion proceeds to 30 psia. Heat is rejected at constant pressure, and the initial state is regained. Find for 1 lb$_m$ of air (*a*) the temperature at the end of compression, (*b*) the heat added at constant pressure, (*c*) the temperature at the end of expansion, (*d*) the heat rejected at constant pressure, (*e*) the volume at each point of the cycle, (*f*) the entropy change for each process; and plot (*g*) *Ts* and *pv* diagrams and (*h*) the cycle efficiency.

**5.** One pound mass of air in a Carnot cycle has a volume of 0.1 ft³ and a pressure of 3,000 psia at the start of the isothermal expansion stroke. If the thermal efficiency is 40 per cent and the isothermal expansion ratio is 10, find heat added and rejected, work done, mep, and the pressure at the end of each process.

**6.** A Carnot cycle with air as the fluid has a thermal efficiency of 50 per cent and an isothermal expansion of 2. If 64 Btu of heat is supplied to the cycle, which has a volume of 5 ft³ and a pressure of 100 psia at the start of isothermal expansion, find the work, mep, and pressure at the end of each process.

**7.** An Otto cycle using air with constant-volume addition of heat operates at a compression ratio of 6. Heat is supplied of amount 1280 Btu/lb$_m$ air. At the start of the cycle the temperature is 540°R and the pressure is 14.7 psia. Find the temperatures and pressures at key points of the cycle and the mep.

**8.** Find the thermal efficiency and mep for the cycle of Prob. 7.

**9.** For an Otto cycle with compression ratio of 4 but otherwise with the same data as Prob. 7, find the amount of heat rejected.

**10.** Plot thermal efficiency for the air-standard Otto cycle against compression ratio for values of the ratio from 1 to 10.

**11.** Compute the thermal efficiency of an Otto cycle that has a combustion-chamber volume that is 20 per cent of the piston displacement.

**12.** For the data of Prob. 7, determine the horsepower if the cycle contains $0.1$ lb$_m$ of air and the speed is 2,000 cycles per min.

**13.** Consider the data of Prob. 7 to be that of a diesel cycle with constant-pressure addition of heat. Find the pressures and temperatures at key points of the cycle, the mep, and the thermal efficiency. ($Q_A = 500$ Btu/lb$_m$.)

**14.** The compression ratio for a Diesel cycle is 14, the initial pressure is 14.7 psia, and the temperature is 60°F. The temperature at the end of the combustion process is 3000°F. Determine the thermal efficiency of the cycle if air is the fluid.

**15.** The initial temperature in a diesel cycle is 14 psia, and the temperature is 80°F. The temperature before combustion is 700°F, and after combustion it is 3000°F. Determine the thermal efficiency of the air-standard cycle.

**16.** Compare the Otto and diesel cycles on the $Ts$ and $pv$ diagrams for conditions of same maximum pressure and temperature. Determine which cycle is the more efficient.

**17.** Repeat Prob. 16 but for equal outputs from each cycle (compression ratios will differ).

**18.** A continuous-combustion turbine system operates with pressure ratio of 3. What will be the ideal thermal efficiency?

**19.** Repeat Prob. 18, assuming that the compression ratio is 3.

**20.** The turbine of Prob. 18 has compressor and turbine efficiencies of 85 per cent while the inlet temperature is 1200°F. Calculate the thermal efficiency and work ratio (atmospheric conditions, 14.7 psia and 60°F).

**21.** Repeat Prob. 20, assuming that a regenerator is added with efficiency of 75 per cent.

**22.** Repeat Prob. 21, assuming that the regenerator introduces a pressure drop of 2 psia between compressor and turbine (pressure ratio of turbine is now less than 3).

**23.** What will be the thermal efficiency and work ratio of a gas-turbine system if the inlet air is 100°F and 14.7 psia, the outlet from the compressor is 50 psia and 400°F, and the combustion temperature is 1200°F, falling to 840°F at exhaust?

**24.** Determine the compression and expansion efficiencies for the data of Prob. 23.

**25.** A reversible turbine with pressure ratio of 4 has an inlet temperature of 1200°F. Atmospheric conditions are 60°F and 14.7 psia. Determine the thermal efficiency and work ratio for (*a*) isentropic compression, (*b*) isothermal compression, (*c*) two-stage isentropic compression with perfect intercooling.

**26.** Repeat Prob. 25, assuming that a perfect regenerator is used in each instance. Compare results with those for Prob. 25.

**27.** A gas-turbine system has pressure ratio of 4 and compressor and engine efficiencies of 83 per cent, while the maximum temperature is 1300°F. Atmospheric conditions are 60°F and 14.7 psia. Determine the thermal efficiency and work ratio.

**28.** Calculate the effectiveness and loss of available energy for the regenerator of Prob. 22.

**29.** Calculate the effectiveness and loss of available energy for the cycle of Prob. 23.

**30.** Derive Eq. (17-6).

**31.** Calculate the mass of air in the cylinder during the compression and reexpansion processes of a reversible compressor if $n = 1.30$, $p_2 = 125$ psia, $p_1 = 14.7$ psia, $t_1 = 60°F$, $c = 0.04$ (based on 1 $lb_m$ delivered).

**32.** For the data of Prob. 31, calculate the heat transferred on each stroke.

**33.** Derive Eq. (17-7b).

**34.** Air is reversibly compressed from $p_1 = 14$ psia and $t_1 = 80°F$ to $p_2 = 60$ psia and $t_2 = 295°F$. If the conventional volumetric efficiency is 94 per cent, find the clearance.

**35.** A double-acting 7- by 7-in. air compressor, with clearance of 0.05, compresses air from $p = 14.7$ psia and $t = 70°F$ to $p = 100$ psia. Determine the (conventional) compressor capacity (300 rpm; $n = 1.35$).

**36.** For the data of Prob. 35, determine the heat transferred to the cooling water during a compression and, also, during an expansion stroke (for air on one side of the piston only).

**37.** What will be the area of the indicator card for the data of Prob. 35 if the spring scale is 100 psia/in. and the reducing motion is 2:1? Draw a conventional $pV$ card, and show true volumes.

**38.** A two-stage double-acting compressor takes in air at $p = 14.5$ psia, $t_1 = 70°F$ and delivers it to a receiver at 300 psia. If the intercooling is perfect, determine the conventional horsepower, the capacity of the compressor, the heat transferred in the intercooler, the displacement and dimensions of both cylinders (150 rpm, 12- by 12-in. low-pressure cylinder with 12-in. stroke of high-pressure cylinder; $c = 0.03$; $n = 1.3$ for both cylinders).

**39.** Find the isothermal compression efficiency for the data of Prob. 38.

**40.** Repeat the analysis leading up to Eq. (17-13) but for a turbojet engine.

## SELECTED REFERENCES

1. Obert, E. F.: "Internal Combustion Engines," International Textbook Company, Scranton, Pa., 1950.
2. Jennings, B. H., and W. L. Rogers: "Gas Turbines," McGraw-Hill Book Company, Inc., New York, 1953.
3. Vincent, E. T.: "The Theory and Design of Gas Turbines and Jet Engines," McGraw-Hill Book Company, Inc., New York, 1950.

# CHAPTER 18

# REFRIGERATION

He is the happiest, be he king or
peasant, who finds peace in his home.
*Goethe*

Refrigeration is the production and maintenance in a given space of a temperature lower than that of the surroundings; it is a necessary part of many industrial processes, as well as a very desirable kind of modern convenience.

**18-1. Definitions.** Not long ago, natural ice was the principal means of refrigeration, and therefore the refrigeration capacity was related to the latent heat of fusion of ice. A ton of refrigeration is, closely, the cooling effect or heat exchange equivalent to that obtained by melting 1 ton of ice at 32°F. Or precisely,† by arbitrary definition,

1 standard ton refrigeration $\equiv$ (2,000 lb$_m$)(144 Btu/lb$_m$) = 288,000 Btu

More often the ton of refrigeration is considered to be a rate:

$$1 \text{ standard commercial ton refrigeration} = 288,000 \text{ Btu/24 hr}$$
$$= 12,000 \text{ Btu/hr} = 200 \text{ Btu/min} \quad (18\text{-}1)$$

To obtain the cooling effect called *refrigeration*, work (or available energy) must be expended. The *coefficient of performance* CP is defined as the ratio of the refrigeration to the work supplied:‡

$$\text{CP} \equiv \frac{\text{refrigeration}}{\text{work added}} \bigg]_{\text{cycle}} \equiv \frac{\Delta Q_A}{-\Sigma \, \Delta W_A} \bigg]_{\text{cycle}} \quad (18\text{-}2)$$

The refrigeration $\Delta Q_A$ is the heat added to the working substance in the cycle, and $\Sigma \, \Delta W_A$ is the work or available energy used to drive the apparatus. The value of CP can be greater or less than unity.

Another gauge of the performance of the refrigerating machine is the *horsepower per ton of refrigeration*:

$$\text{hp/ton} = \frac{200}{\text{CP}(42.4)} = \frac{4.71}{\text{CP}} \quad (18\text{-}3)$$

† The latent heat of fusion of ice at 32°F is 143.35 Btu/lb$_m$.

‡ By the convention used in this text, work supplied to a system is a negative number; thus the CP defined by Eq. (18-2) is a positive number.

**18-2. The Carnot Refrigeration Cycle.** The Carnot heat pump (Art. 7-1) with its working substance, or *refrigerant*, confined to saturated and two-phase states is illustrated in Fig. 18-1. Here the cycle consists of the following processes:

*ab*  Work produced by isentropic expansion to the lower temperature $T_A$
*bc*  Heat added in the *evaporator* at the lower temperature $T_A$
*cd*  Work added in the isentropic compression to the higher temperature $T_R$
*da*  Heat rejected in the *condenser* at the higher temperature $T_R$

The *refrigeration* is proportional to the area *jbck*, the net work supplied is proportional to area *abcd*, and the heat rejected is proportional to area *jadk*. For this cycle, by Eq. (18-1),

$$\text{CP}_{\text{Carnot}} = \frac{\Delta Q_A}{-\Sigma \Delta W} = \frac{\Delta Q_{bc}}{-(\Delta Q_{bc} + \Delta Q_{da})} = \frac{T_A}{T_R - T_A} \qquad (18\text{-}4)$$

It has been demonstrated (Art. 7-1) that, *between fixed temperature limits*, all reversible engines have the same thermal efficiency, the thermal

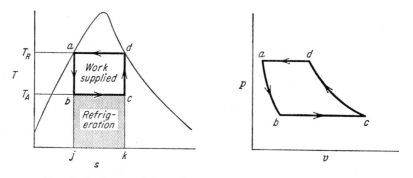

Fig. 18-1. Carnot refrigeration cycle (the refrigerant is a vapor).

efficiency of the Carnot cycle. Thus, the work of the Carnot cycle is the maximum work that can be obtained from transformation of heat energy in a heat-engine cycle. But if the Carnot cycle (or any reversible cycle) delivers the maximum work as a heat engine, it must therefore require the least work for the reversed operations as a heat pump. Thus, the coefficient of performance of a reversible engine is the optimum performance for the refrigeration machine. Moreover, all reversible heat pumps operating between the same temperature limits will have the same coefficient of performance no matter what fluid is used as the refrigerant or what pressures are experienced in the system. Then, the work or power required per ton of refrigeration will also be independent of the refrigerant and dependent only upon the temperatures of source and sink. These conclusions are valid only for the reversible cycle because the irreversibilities and the temperature limits of the real cycle will be affected in greater or less degree by the properties of the refrigerant (Table 18-1, page 352).

A *relative efficiency* can be defined by comparing the work of the real refrigeration cycle to that of the Carnot:

$$\eta_{rC} = \left. \frac{\Delta W_{\text{Carnot cycle}}}{\Delta W_{\text{refrigeration cycle}}} \right]_{Q_A = C} = \frac{CP_{\text{refrigeration cycle}}}{CP_{\text{Carnot pump}}} \qquad (18\text{-}5)$$

The relative efficiency is always less than unity. When this efficiency is computed, the work of the Carnot cycle should be based upon the temperatures of the refrigerated space and the cooling medium. If, instead, the relative efficiency is computed from the temperatures of heat addition and rejection within the cycle, then the system is not penalized for the irreversibility of the temperature differences that actually are present.

**Example 1.** A Carnot cycle is to operate between 80°F (atmospheric) and 40°F, with steam as the refrigerant. The quality is 0.80 at state $d$ (Fig. 18-1), and state $a$ is that of saturated liquid. Compute the CP, horsepower per ton, and the work of each process (all to be flow processes).

*Solution.* By Eq. (18-4),

$$CP = \frac{T_A}{T_R - T_A} = \frac{500}{540 - 500} = 12.5 \qquad Ans.$$

By Eq. (18-3),

$$\text{hp/ton} = \frac{4.71}{CP} = \frac{4.71}{12.5} = 0.378 \text{ hp/ton refrigeration} \qquad Ans.$$

The properties at each state of the cycle can be obtained from the Steam Tables and by calculation (letters correspond to Fig. 18-1),

| | | |
|---|---|---|
| $h_a = 48.02$ Btu/lb$_m$ | $x_a = 0.0$ | $v_a = 0.01608$ ft$^3$/lb$_m$ |
| $h_b = 46.6$ | $x_b = 0.0359$ | $v_b = 87.9$ |
| $h_c = 824.1$ | $x_c = 0.761$ | $v_c = 1,860$ |
| $h_d = 886.9$ | $x_d = 0.80$ | $v_d = 506.5$ |
| $s_a = 0.0932$ Btu/(lb$_m$)(°F) | | $p_a = 0.5069$ psia |
| $s_b = 0.0932$ | | $p_b = 0.12170$ |
| $s_c = 1.6474$ | | $p_c = 0.12170$ |
| $s_d = 1.6474$ | | $p_d = 0.5069$ |

The work of compression is

$$\Delta w_{cd} = -(h_d - h_c) = -(886.9 - 824.1) = -62.8 \text{ Btu/lb}_m \qquad Ans.$$

The work of expansion is

$$\Delta w_{ab} = -(h_b - h_a) = -(46.6 - 48.02) = +1.42 \text{ Btu/lb}_m \qquad Ans.$$

Note that the work obtained in the expansion process is but a small fraction of the work that must be supplied for the compression process.

This example indicates that steam would be a particularly undesirable refrigerant for these temperature limits. The pressures throughout the system are far below atmospheric so that the prevention of air leakage into the system would undoubtedly be a difficult if not impossible task. Compared with other refrigerants the specific volumes at $c$ and $d$ are large, and this would require a large compressor displacement. For these reasons it would be highly impractical if not impossible to use steam as the working fluid for this cycle (but see Art. 18-5).

**18-3. The Vapor-compression Refrigeration Cycle.** For most commercial purposes, the temperature range demanded of the refrigeration cycle is small, and therefore the work produced in the expansion process is negligible relative to the compression work (Example 1). Thus, the refrigeration machine can be simplified by substituting a throttling valve for the expansion turbine, with consequent savings in initial cost and

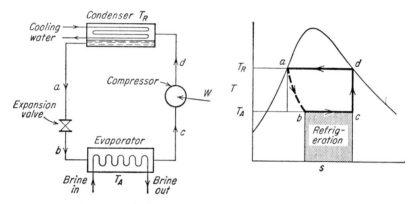

FIG. 18-2. Ideal vapor-compression refrigeration cycle.

maintenance. The elements of the ideal vapor-compression cycle are illustrated in Fig. 18-2.

  $ab$  Throttling at constant enthalpy to the lower temperature $T_A$
  $bc$  Heat added in the *evaporator* at the lower temperature $T_A$
  $cd$  Work added in the isentropic compression to the higher temperature $T_R$
  $da$  Heat rejected in the *condenser* at the higher temperature $T_R$

Note that in the throttling process the temperature of the refrigerant falls to the saturation temperature corresponding to the pressure maintained in the evaporator by the suction of the compressor.

In the real system, all processes are irreversible and temperature differences will be present. Figure 18-3 depicts some of these effects. The temperature $T_A$ at the end of the throttling process $ab$ must be lower than the temperature $T_A'$ of the room or space being refrigerated. The rate of vaporization of the refrigerant in the evaporator depends on this temperature difference (and on the surface offered by the evaporator). The refrigerant entering the compressor may be wet or dry vapor, the state being controlled by the location of the vapor take-off on the evaporator as well as by the mass-flow rate, which is governed by a control mechanism on the expansion valve. Dry compression is usually preferred because a greater refrigerating effect can be obtained per unit of mass flow and because the presence of large drops of liquid in the compressor may cause damage. Consequently, the final state $d$ lies, most often, in the superheat region at a temperature above $T_R$. The gas is cooled and condensed in the condenser because of a temperature difference between gas and coolant. The design of the condenser is to facilitate subcooling $ea$ to a temperature quite close to that of the cooling medium, $T_0$. Note that the pressure in the condenser for a selected refrigerant depends upon the temperature of the available coolant (tap water, usually).

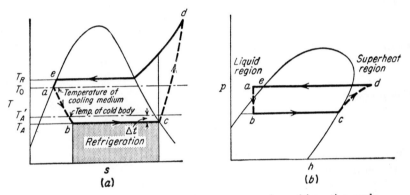

Fig. 18-3. Irreversibilities in the vapor-compression refrigeration cycle.

Suppose that the compressor is operated, but without flow of refrigerant. The pressure $p_{bc}$ then falls toward zero. Let refrigerant be admitted; since there is a "hot" body to be cooled, the refrigerant will absorb heat and vaporize, with consequent enormous specific volume because of the low pressure. Therefore, the capacity of the compressor is exceeded, and the pressure $bc$ rises. But, with the rise in pressure, the specific volume of the refrigerant decreases, and therefore a steady state is reached at that pressure (or temperature) dictated by the capacity of the compressor. [Thus, reciprocating-piston compressors can be used for small temperature ranges or for small refrigeration capacities, while centrifugal or rotary machines are usually demanded for low-pressure (low-temperature) and high-capacity demands.]

**Example 2.** For the same data as Example 1, let Freon, F-12, be the refrigerant, with saturated vapor leaving the evaporator and saturated liquid leaving the condenser. Compression efficiency is 80 per cent. Compute the coefficient of performance.

*Solution.* From Table B-14 for the states indicated in Fig. 18-3,

$$h_e = h_b = 26.28 \text{ Btu/lb}_m \qquad p_e = 98.76 \text{ psia}$$
$$h_c = 82.71 \text{ Btu/lb}_m \qquad p_c = 51.68 \text{ psia}$$

If compression were isentropic, state $d$ would be 6° superheated and with enthalpy of $86.80 + 0.95 = 87.75 \text{ Btu/lb}_m$ (linear interpolation of the data at 80°F for an entropy of 0.16833).

Hence $\qquad \Delta w_{\text{isen}} = -\Delta h = -(87.75 - 82.71) = -5.04 \text{ Btu/lb}_m$

$$\Delta w_{\text{actual}} = \frac{-5.04}{0.80} = -6.3 \text{ Btu/lb}_m$$

and therefore

$$h_d = 82.71 + 6.3 = 89.01 \text{ Btu/lb}_m$$

$$\text{CP} = \frac{\Delta q_A}{-\Delta w} = \frac{h_c - h_b}{h_d - h_c} = \frac{82.71 - 26.28}{89.01 - 82.71} = \frac{56.43}{6.3} = 8.95 \qquad Ans.$$

In analyzing the flow processes of the vapor-compression cycle, the following quantities may be calculated (all subscripts refer to Fig. 18-3):

*Expansion valve:*

$$\Delta q - \Delta w = \Delta h = 0$$
$$h_a = h_b = (h_f + x h_{fg})_b \tag{18-6}$$

*Evaporator:*

$$\Delta q - \Delta w = \Delta h \qquad \text{and} \qquad \Delta w = 0$$
$$\Delta q_A = h_c - h_b \tag{18-7}$$

*Compressor:*

$$\Delta q - \Delta w = \Delta h$$
$$\Delta w_{cd} = -(h_d - h_c) + \Delta q_{cd} \tag{18-8}$$

($\Delta q$ rejected is negative, by convention.)

If compression is isentropic,

$$\Delta w_{cd,\text{rev}} = -(h_d - h_c)_{s=C} \tag{18-9}$$

*Condenser:*

$$\Delta q - \Delta w = \Delta h \qquad \text{and} \qquad \Delta w = 0$$
$$\Delta q_R = h_a - h_d \tag{18-10}$$

*Coefficient of performance:*

$$\text{CP} = \frac{\Delta Q_A}{-\Delta W_A} = \frac{h_c - h_b}{(h_d - h_c) - \Delta q_{cd}} \tag{18-11}$$

*Mass-flow rate of refrigerant per ton of refrigeration:*

$$\dot{m}_f = \text{lb}_m/(\text{min})(\text{ton}) = \frac{200 \text{ Btu/(min)(ton)}}{\Delta q_A \text{ Btu/lb}_m} = \frac{200}{h_c - h_b} \tag{18-12}$$

*Compressor capacity per ton of refrigeration:*

$$C(\text{cfm/ton}) = \dot{m}_f[\text{lb}_m/(\text{min})(\text{ton})]v(\text{ft}^3/\text{lb}_m) = \dot{m}_f v$$
$$v = \text{specific volume of refrigerant at compressor inlet} \tag{18-13}$$

*Horsepower required per ton of refrigeration:*

$$\text{hp/ton} = \frac{12,000 \text{ Btu/(hr)(ton)}}{(2544 \text{ Btu/hp-hr})\text{CP}} = \frac{4.71}{\text{CP}} \tag{18-3}$$

**Example 3.** A refrigeration system uses ammonia as the refrigerant. The temperatures in the evaporator and condenser are, respectively, 5 and 86°F; the vapor entering the compressor is saturated, and the liquid entering the expansion valve is saturated; compression is isentropic. For these conditions find (a) the heat and work transfers for each process, (b) the CP, (c) horsepower per ton, (d) ideal CP, (e) relative efficiency, (f) mass-flow rate, (g) compressor capacity.

*Solution.* The properties at each state of the cycle will be found by means of Table B-16 and by calculation (subscripts correspond to Fig. 18-2, although in this problem state d lies in superheat region):

$$p_a = 169.2 \text{ psia} \quad t_a = 86°F \quad h_a = 138.9 \text{ Btu/lb}_m$$
$$p_c = 34.27 \quad t_c = 5 \quad h_c = 613.3$$
$$s_c = 1.3253 \text{ Btu/(lb}_m)(°R) \quad v_c = 8.150 \text{ ft}^3/\text{lb}_m$$

Also,
$$h_b = h_a = 138.9 \text{ Btu/lb}_m \quad (18\text{-}6)$$

Since the entropy at state d is equal to that at c, interpolation of the superheat values in Table B-16 shows that

$$p_d = p_a = 169.2 \text{ psia} \quad t_d = 211°F \quad h_d = 713.4 \text{ Btu/lb}_m$$

With these values, the work and heat transfers can be found:

(a)
$$\Delta q_A = \Delta q_{bc} = h_c - h_b = 613.3 - 138.9 = 474.4 \text{ Btu/lb}_m \quad Ans.$$
$$\Delta q_R = \Delta q_{da} = h_a - h_d = 138.9 - 713.4 = -574.5 \text{ Btu/lb}_m \quad Ans.$$
$$\Delta w_{cd} = -(h_d - h_c)_{s=C} = -(713.4 - 613.3) = -100.1 \text{ Btu/lb}_m \quad Ans.$$
or $$\Delta w_{cd} = \Delta q_A + \Delta q_R$$

(b) The coefficient of performance equals

$$CP = \frac{\Delta Q_A}{-\Delta W} = \frac{474.4}{100.1} = 4.74 \quad Ans.$$

(c) The horsepower per ton of refrigeration equals

$$hp/ton = \frac{4.71}{CP} = \frac{4.71}{4.74} = 0.995 \quad Ans.$$

(d) The Carnot or ideal CP equals

$$CP = \frac{T_A}{T_0 - T_A} = \frac{464.7}{545.7 - 464.7} = 5.74 \quad Ans.$$

(e) The relative efficiency equals

$$\eta_{rC} = \frac{CP_{actual}}{CP_{Carnot}} = \frac{4.74}{5.74} = 0.826 \quad Ans.$$

(f) The mass-flow rate per ton of refrigeration is

$$\dot{m}_f = \frac{200}{\Delta q_A} = \frac{200}{474.4} = 0.422 \text{ lb}_m/(\text{min})(\text{ton}) \quad Ans.$$

(g) The required compressor capacity is

$$C = \dot{m}_f v = 0.422(8.150) = 3.44 \text{ cfm/ton} \quad Ans.$$

These values differ somewhat from those shown in Table 18-1 because the two superheat states listed in Table B-16 do not allow precise evaluation of intermediate states.

## TABLE 18-1. COMPARISON OF REFRIGERANTS†

| Refrigerant | $M$ | Latent heat 5°F, Btu/lb$_m$ | $k$ | Mass flow rate, lb$_m$/(min)(ton) | Compressor capacity, cfm/ton | Evaporator pressure, psia | Condenser pressure, psia | Pressure ratio | CP | Power, hp/ton | $\eta_{rC}$ |
|---|---|---|---|---|---|---|---|---|---|---|---|
| Any fluid, Carnot cycle | ..... | ..... | ..... | ..... | ..... | ...... | ...... | ..... | 5.74 | 0.821 | 1.00 |
| Ammonia | 17.0 | 565 | 1.285 | 0.422 | 3.44 | 34.28 | 169.2 | 4.93 | 4.77 | 0.97 | 0.83 |
| Carbon dioxide | 44.0 | 115.3 | 1.304 | 3.68 | 0.99 | 339 | 1,054 | 3.12 | 2.56 | 1.84 | 0.45 |
| Freon, F-12 | 120.9 | 69.5 | 1.14 | 3.92 | 5.81 | 26.5 | 107.9 | 4.06 | 4.72 | 1.00 | 0.82 |
| Methyl chloride | 50.5 | 180.7 | 1.28 | 1.43 | 6.8 | 21.2 | 94.7 | 4.5 | 4.70 | 1.00 | 0.82 |
| Propane | 44.1 | 169.5 | 1.14 | 1.40 | 3.35 | 43.7 | 159.0 | 3.6 | 4.88 | 0.97 | 0.85 |
| Sulfur dioxide | 64.1 | 169.4 | 1.29 | 1.39 | 9.24 | 11.81 | 66.6 | 5.6 | 4.74 | 0.99 | 0.82 |

† For 1 ton of refrigeration: 5°F saturated vapor leaving evaporator; 86°F saturated liquid leaving condenser.

352

The real cycle contains an irreversible throttling process and, also, varying degrees of superheat. Since these variables are governed by the characteristics of the fluid and not directly by the evaporator or condenser pressure, the CP of different refrigerants will not be the same. In Table 18-1 comparison is made of different refrigerants.

**18-4. Properties of Refrigerants.** It is desirable from practical as well as theoretical considerations that the refrigerant should exhibit certain characteristics. The properties of the ideal refrigerant would show the following qualities (compare with those for an ideal heat-engine fluid; Art. 15-4):

1. The latent heat of vaporization should be large, and the heat capacity of the liquid should be small because then the mass-flow rate would be low. Note that, the smaller the heat capacity of the liquid, the less will be the vaporization during throttling and therefore the greater the amount of heat that can be abstracted from the cold source.

2. The critical point should be above the highest operating temperature, for then the fluid after compression is close to the two-phase region where condensation can take place at constant temperature; not only are heat-transfer rates better in the two-phase region, but also the irreversibility of a temperature difference is reduced by the constancy of temperature.

3. The vapor pressure in the condenser should not be high. High pressures increase design costs and maintenance.

4. The vapor pressure in the evaporator should be higher than atmospheric pressure. This would prevent air from leaking into the system and so increasing the amount of work that must be supplied to the compressor for a definite amount of refrigeration. Air in the system also adversely affects the rate of heat transfer. The humidity in the air is especially troublesome because the water tends to freeze in the smallest section of the system, the expansion valve.

5. The entropy of the saturated vapor should not change markedly with pressure, or else it should increase slightly as the pressure increases because then the refrigerant can enter the condenser as a wet or saturated vapor.

6. The properties of the fluid should be conducive to high rates of heat transfer in order that both surface areas and temperature differences can be small in the heat exchangers.

7. The refrigerant should be cheap in cost, stable, nonexplosive, noncorrosive under all conditions of operation, and nonpoisonous for safety of personnel.

No refrigerant is known that possesses all these properties, but certain fluids have qualities that are particularly suited for special applications. A few of these fluids are listed as follows and in Table 18-1.

*Anhydrous ammonia* is one of the oldest and most widely used refrigerants because of its high latent heat, moderate pressures, and small compressor capacity (Table 18-1). The evaporator pressure is above atmospheric in the usual installation where temperatures below −28°F are not demanded. On the debit side, ammonia, while noncorrosive to the ferrous metals, is corrosive to brass and bronze; ammonia is toxic and also irritating to the eyes, nose, and throat (Table B-16).

*Freons* make up the most important group of refrigerants: $CCl_3F$ (trichloromonofluoromethane)(Freon-11, Genetron-11); $CCl_2F_2$ (dichlorodifluoromethane)(Freon-12,† Genetron-12); $CClF_3$ (monochlorotrifluoromethane)(Freon-13); $CF_4$ (tetrafluoromethane)(Freon-14); $CHClF_2$ (monochlorodifluoromethane)(Freon-22, Genetron-141); $CCl_2F$-$CClF_2$ (trichlorotrifluoroethane)(Freon-113); $C_2Cl_2F_4$ (dichlorotetrafluoroethane)(Freon-114).

The Freons are nonflammable with low toxicity. Freon-11 is widely used with centrifugal compressors for medium- to low-temperature ranges. Freon-12 and -22 are for general refrigeration purposes (piston-type compressors) such as air conditioning. Freon-22 was developed for low temperatures (−20°F), while Freon-13 and -14 are for extremely low temperatures.

The properties of Freon-12 are listed in Table B-14. This is probably the most widely used of the Freons; although its latent heat is low, thus requiring a high massflow rate, the CP is essentially the same as for ammonia.

*Methyl chloride* ($CH_3Cl$) has properties quite similar to Freon-12 but is somewhat toxic. It has been used extensively in commercial and domestic installations (Fig. B-4).

*Carbon dioxide* ($CO_2$) has been used quite extensively as a refrigerant on shipboard because of its nontoxic properties. However, the high pressures in the system and the low CP have decreased its popularity with the advent of the Freons. Its use today is primarily in the manufacture of dry ice (Fig. B-3).

*Sulfur dioxide* ($SO_2$) was a popular refrigerant for the household refrigerator because of the low pressures in the cycle; it is somewhat more toxic than the other refrigerants.

## 18-5. Vacuum Refrigeration.

Water is, without doubt, the safest as well as the cheapest vapor refrigerant although the cycle temperatures must ordinarily be above 32°F. For certain processes, notably air conditioning, low temperatures are not needed and water can be used as the refrigerant although the pressures in the system are subatmospheric (Example 1) and the vapor volumes are large. To handle the large volume of refrigerant, an ejector is generally used although a centrifugal pump is a possible substitute.

In Fig. 18-4, relatively warm water is sprayed into a *flash chamber* that is maintained at a low pressure by an ejector or pump. A small portion of the water flashes into steam, and the latent heat of vaporization so demanded is supplied by the water (the flash chamber is insulated to reduce heat transfer from the surroundings). Thus, the water is cooled to the saturation temperature dictated by the pressure. The chilled water is then pumped to the point where it is required, and the warmed water returned to the flash chamber for cooling.

† This structural name indicates that two chlorine and two fluorine radicals have replaced hydrogen in the compound methane ($CH_4$).

The vapor withdrawn from the flash chamber by the ejector or pump is compressed and delivered to the steam condenser (Fig. 18-4). Here the pressure, as in the flash chamber, is far below atmospheric, the particular value being determined by the temperature of the available cooling water (Example 1).

The vacuum system, using steam-jet ejectors, has few moving parts because a mechanical compressor is eliminated. This simplification, along with the cheapness and nontoxicity of water, makes up for the inefficiency of the ejector. And if waste steam is available, at pressures above 5 psia (and, preferably, much higher), a water-vapor system becomes highly desirable.

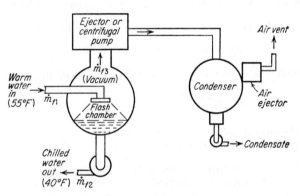

FIG. 18-4. Vacuum refrigeration system.

An energy balance can be made on the flash chamber to show the refrigeration:

$$\dot{m}_{f1}h_1 = \dot{m}_{f2}h_2 + \dot{m}_{f3}h_3 \qquad (\Delta Q,\ \Delta W = 0)$$

Since

$$\dot{m}_{f1} = \dot{m}_{f2} + \dot{m}_{f3}$$

then

$$\dot{m}_{f2}(h_1 - h_2) = \dot{m}_{f3}(h_3 - h_1)$$

and the refrigeration is

$$\Delta Q_A = \dot{m}_{f2}(h_1 - h_2) = \dot{m}_{f3}(h_3 - h_1) \qquad (18\text{-}14)$$

**Example 4.** A vacuum system produces 200 gpm of chilled water at 40°F with return water at 55°F. The vapor leaving the flash chamber has a quality of 0.98, and the temperature in the condenser is 90°F. Determine (a) the pressure in flash chamber and condenser, and pressure ratio, (b) the refrigeration capacity, (c) the amount of make-up water, and (d) the volume of vapor entering ejector.

*Solution.* (a) The vapor pressure of water at 40°F is the pressure in the flash chamber. From the Steam Tables,

$$p = 0.12170 \text{ psia} \qquad Ans.$$

The vapor pressure at 90°F is the pressure in the condenser:

$$p = 0.6982 \text{ psia} \qquad Ans.$$

The pressure ratio equals

$$r_p = \frac{p_{\text{condenser}}}{p_{\text{evaporator}}} = \frac{0.6982}{0.12170} = 5.74 \qquad Ans.$$

(b) The mass-flow rate equals

$$\dot{m}_{f2} = 200 \left(\frac{\text{gal}}{\text{min}}\right) \frac{1}{0.01602} \left(\frac{\text{lb}_m}{\text{ft}_3}\right) 0.1338 \left(\frac{\text{ft}^3}{\text{gal}}\right) = 1{,}670 \text{ lb}_m/\text{min}$$

And the refrigeration is

$$\Delta Q_A = \dot{m}_{f2}(h_1 - h_2) = 1670(15.02) = 25{,}100 \text{ Btu/min or } 125.5 \text{ tons} \qquad Ans.$$

(c) The mass of vapor entering the ejector is found by Eq. (18-14):

$$\Delta Q_A = \dot{m}_{f3}(h_3 - h_1) = 25{,}100 \text{ Btu/min}$$

where $h_3 = 8.05 + 0.98(1071.3) = 1058 \text{ Btu/lb}_m$, $h_1 = 23.07 \text{ Btu/lb}_m$. Hence,

$$\dot{m}_{f3} = \frac{25{,}100}{1{,}035} = 24.25 \text{ lb}_m/\text{min} \qquad Ans.$$

This is also the quantity of make-up water required.

(d) The specific volume of the vapor at 40°F is

$$v = 0.016 + 0.98(2{,}444) = 2{,}395 \text{ ft}^3/\text{lb}_m$$

and
$$C = \dot{m}_f v = 24.25(2{,}395) = 58{,}100 \text{ cfm} \qquad Ans.$$

## 18-6. Absorption Refrigeration.

It has already been remarked that the work necessary to compress a liquid is but a small fraction of that required to compress a gas (Art. 5-8d). Thus, the work supplied to the refrigeration system could be reduced if the refrigerant were pumped to the condenser pressure as a liquid rather than as a gas. A means of achieving this objective is offered by the *absorption* system (Fig. 18-5). Here as in the compression system the refrigerant passes from condenser to expansion valve to evaporator. But, unlike the compression system, the vapor issuing from the evaporator is dissolved in a cold solvent in the *absorber;* this liquid solution is then pumped into the high-pressure *generator* where the solution is heated.

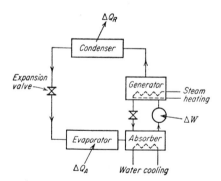

FIG. 18-5. Elements of the absorption refrigeration system (cycle).

The refrigerant is thus liberated from the solution and passes to the condenser while the solvent returns to the absorber. Of course, the solvent must be able to hold in the cold solution a greater amount of refrigerant than in the hot solution.

An *aqua-ammonia* solution is generally used (ammonia as the refrigerant, water as the solvent) in the absorption system. The ammonia vapor entering the absorber is dissolved in relatively cold (80 to 90°F) water. Heat is liberated in the process, and cooling coils are necessary to maintain the low temperature. The cold solution of water and

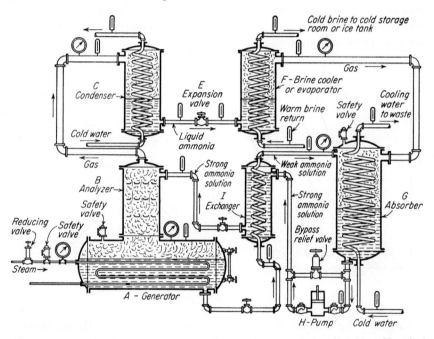

Fig. 18-6. An absorption refrigeration system. [From J. H. Perry (ed.), "Chemical Engineers' Handbook," 3d ed., McGraw-Hill Book Company, Inc., New York, 1950.]

ammonia (called the *strong aqua*) is pumped to the generator where it is heated (200 to 300°F). The hot solution cannot hold as much ammonia as the cold solution; hence, ammonia vapor is liberated and passes to the condenser. The hot and therefore weak solution of ammonia and water (called the *weak aqua*) is then throttled back to the absorber to be cooled and strengthened.

The simple system of Fig. 18-5 would deliver not only ammonia but also a large amount of water to the condenser. To improve the performance, a more complicated system must be used. Figure 18-6 shows the ammonia vapor being withdrawn from the evaporator *F*

because of the continuous removal of the ammonia vapor in the absorber
*G.* The strong aqua from the absorber is pumped (*H*) through a heat
exchanger *I* before entering the *analyzer B.* In the analyzer, the strong
solution is heated while the gas from the generator is cooled, and in this
cooling, water vapor more than ammonia vapor is condensed. The gas
leaving the analyzer thus has a high percentage of ammonia vapor.
Sometimes a second cooler, called the *rectifier* (not shown in Fig. 18-6),
is placed before the condenser. The rectifier is simply a precondenser
that by cooling the gases removes a greater percentage of water than of
ammonia. The condensate, or drip, from the rectifier is returned to the
generator via the analyzer. The gases finally entering the condenser *C*
are thus primarily ammonia.

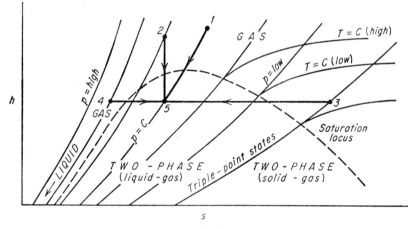

Fig. 18-7. Cooling processes.

Although the amount of mechanical energy supplied to the absorption
system is small, the amount of thermal energy greatly exceeds the
energy requirements of the compression system, and a larger amount of
cooling water is required. For these reasons the absorption system is
rarely used unless waste heat is available, say low-pressure exhaust steam
from the power plant.

**18-7. Liquefying and Solidifying Processes.** Suppose that the problem is to produce a liquid or a solid phase of a substance. One of the
following processes, illustrated in Fig. 18-7, could be selected:

1. Change of phase by cooling at essentially constant pressure (1-5)

2. Change of phase by essentially adiabatic expansion in either a
piston-type or a turbo expander (2-5)

3. Change of phase by compression and cooling (3-5)

4. Change of phase by throttling (4-5)

The throttling process has the merit of simplicity (no moving parts) for attaining low temperatures where lubrication may be a problem.

In the Linde process for liquefying gases such as oxygen or air, the apparatus appears as in Fig. 18-8. Gas enters at 1 and is throttled at $x$ to a lower pressure, with consequent fall in temperature, and the lower-pressure and lower-temperature gas counterflows to the exit, 2, thus cooling the incoming flow. Eventually, a steady state is reached, and a fraction $y$ of the incoming flow is liquefied (or solidified). For an adiabatic steady-flow process without work,

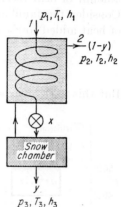

$$\Sigma \dot{m}_f \, \Delta h = 0$$
$$y h_3 + (1 - y) h_2 = h_1$$

And the solid or liquid fraction $y$ equals

$$y = \frac{h_2 - h_1}{h_2 - h_3} \qquad (18\text{-}15)$$

FIG. 18-8. Linde process.

The enthalpy $h_3$ is set by the exit pressure $p_3$ ($h_3 = h_f$ or $h_i$), the enthalpy $h_2$ is determined by the exit pressure and the pressure drop in the heat exchanger, while the initial enthalpy $h_1$ is set by the values of $p_1$ and $t_1$. Of all these variables, $p_1$ is most readily varied; hence, for maximum yield (maximum $y$), the foregoing equation shows that $h_1$ should be small.

**18-8. The Refrigeration Cycle as a Heat Pump.** The heating of buildings is an ever-recurring engineering problem. A building can be heated by burning fuels or by dissipating work, as, for example, when an electric current passes through a resistance heater. Electric-resistance heating, while convenient, is the ultimate degradation of energy, for here available energy is used to produce only heat. Consider that in the power plant a fuel is burned and work is obtained in amount seldom as much as 25 per cent of the heat of combustion of the fuel. Then, the work must have value at least four times that of the heat used to produce the work. Because of this fact, the average building can be heated more cheaply by direct firing of an expensive fuel in an inefficient furnace than by irreversibly using electrical energy that was produced by burning an inexpensive fuel in an efficient furnace.

Even when the electrical energy is produced by water power, the use of such energy for an irreversible heating purpose may be more expensive than direct firing of fuel. The cost of electrical energy includes not only the cost of any fuel used but also the fixed costs of the installation and the distribution costs.

The remedy is to replace the highly irreversible electric-resistance heating process with a process that can at least approach reversibility. Since work in the form of electrical energy can be derived from a heat-engine cycle, then the cycle can be reversed and heat obtained by supplying work. By this means the ratio of performance is reversed and the amount of heat received can be many times the amount of work added. Consider the familiar Carnot cycle of Fig. 18-9. Here, for every unit of heat added to the cycle, work is obtained of amount equal to

$$\Delta W = \eta_t \Delta Q_A = \frac{560 - 460}{560} (1) = 0.178 \text{ unit}$$

But this cycle can be reversed to act as a heat pump; work of amount 0.178 unit can be supplied, and 1 unit of heat will be received at the higher temperature. Thus, for the heat pump, the coefficient of performance is defined as

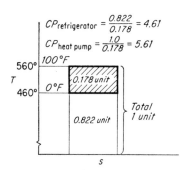

$$CP = \frac{\text{heat delivered}}{\text{work supplied}} = \frac{\Delta Q_R}{\Sigma \Delta W} \quad (18\text{-}16a)$$

And for Fig. 18-9,

$$CP = \frac{1}{0.178} = 5.61$$

FIG. 18-9. The coefficients of performance of the reversed Carnot cycle.

or 5.61 units of heat can be obtained by supplying 1 unit of work to the cycle. Compare this answer with the 1 unit of heat that would be received by direct conversion of electrical energy in a resistance heater.

Although the heat pump is a refrigeration cycle, the coefficients of performance differ. Comparison shows that

$$CP_{\text{heat pump}} = CP_{\text{refrigeration}} + 1 \quad (18\text{-}16b)$$

It is unfortunate that a heat pump can have two different coefficients of performance.

The reversed heat-engine cycle is called a *refrigerator* (and, also, a *heat pump*) when the evaporator is used for cooling purposes, as shown in Fig. 18-10a; the same cycle is called a *heat pump* (but not a refrigerator) when the condenser is used for heating purposes, as shown in Fig. 18-10b. A combined system that serves as a heat pump in the winter and a cooling system in the summer is illustrated in Fig. 18-11. Here the system consists of a heat exchanger A, a condenser B, an expansion valve C, an evaporator D, and a compressor (not shown). The refrigerant (Freon, F-12) is circulated through compressor, condenser, expansion valve, and evaporator. In the heating cycle the cooling water from the condenser passes through the heat exchanger alone and warms the supply air, which is also humidified by the spray humidifier.   The evaporator is supplied with water from a deep well, and therefore the temperature of this water is higher than that of the outside winter air.   This high-temperature water increases the coefficient of performance.   In the cooling cycle the well water is pumped through the evaporator and cooled to a low temperature; it then enters the heat exchanger, which is now a cooling section.   The water leaving the heat exchanger passes through the condenser before returning to the ground.   This is an example of a *water-to-water* design; water is used to heat the evaporator and also to cool the condenser.

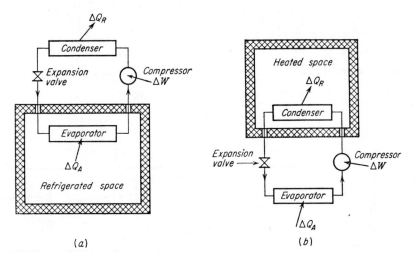

Fig. 18-10. The vapor cycle as a refrigeration cycle and as a heat pump cycle. (a) Refrigerating; (b) heating.

In the installation described in the previous paragraph, water passed over the evaporator of the heat pump and so transferred heat to the cycle. The temperature of the water is frequently above the temperature of the atmosphere because the temperature of the earth does not markedly change with changes in climatic conditions.   (In the Chicago area, a well 60 ft in depth will supply water at a temperature of about 50°F.)   Where well water is not available, or where withdrawal of water with consequent lowering of the water table is to be avoided, a heat exchanger can be buried in the earth.   The heat exchanger can be a vertical U tube running several hundred feet below the surface of the earth.   A small quantity of liquid can be circulated through the heat exchanger and the evaporator, the water being heated in the heat exchanger by the relatively warm earth and cooled in the evaporator by the cold refrigerant.

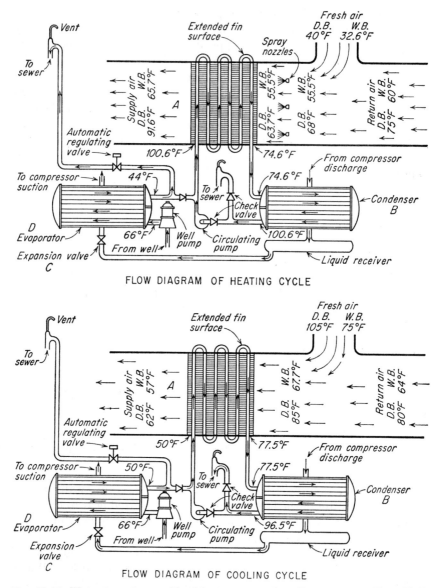

FLOW DIAGRAM OF HEATING CYCLE

FLOW DIAGRAM OF COOLING CYCLE

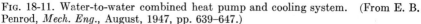

FIG. 18-11. Water-to-water combined heat pump and cooling system. (From E. B. Penrod, *Mech. Eng.*, August, 1947, pp. 639–647.)

In the winter the surface temperature of the earth decreases in pace with the air temperature as the winter progresses, but the temperature below the surface will lag the air temperature because of the heat capacity and thermal resistance of the earth. Thus, the lowest temperature reached at a depth of 20 ft may occur more than a month after the lowest air temperature of the winter has been experienced. Too, under strong sunlight the ground surface temperature may be much higher than the air temperature because of the absorption of radiant energy. Thus, heat exchangers buried at shallow depths may prove useful.

In regions with mild winters where the heating and cooling loads are approximately equal, an *air-to-air system*† can be used and the cost of wells or buried heat exchangers can be eliminated. In this heat pump the outside air is used as the source of heat while air is also directly

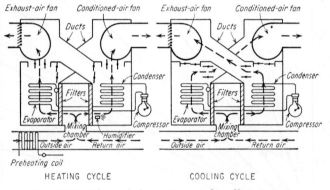

HEATING CYCLE       COOLING CYCLE

Fig. 18-12. Air-to-air heat pump and cooling system.

used to cool the condenser. The refrigeration cycle is simplified because air dampers serve as the means of control, as illustrated in Fig. 18-12. In regions with severe winters, this system is penalized because the capacity provided for winter heating must be greater than the capacity required for summer cooling; too, frost will form on the evaporator, thus reducing the rate of heat transfer and requiring a defrosting system.

Returning again to the cost factor of the heat pump, recall that the power cycle produces work from heat and that work in the form of electrical energy can be readily transported. Now if the power cycle were to be reversed, heat would be obtained at a very high temperature but the cost of the heat would be far greater than the original cost to the power plant because all real processes are irreversible and fixed

---

† Other systems are designated as follows: *earth-to-air* (the earth directly heats the evaporator and air cools the condenser); *water-to-air* (water heats the evaporator and air cools the condenser).

costs and distribution costs must be included. On the other hand, for home heating, heat need be supplied only at moderate temperatures, and therefore the CP of the heat pump will be larger than the CP of the reversed power cycle. In fact, when the thermal efficiency of the power plant is 30 per cent and the CP of the heat pump is 4, the heat obtained from the heat pump is equal to 120 per cent of the heat originally supplied to the power cycle. The heat pump appears to be especially advantageous when electricity is obtained from cheap water power or when costs of fuels are high. For these reasons, it is difficult to estimate whether the heat pump can show an economic gain over direct

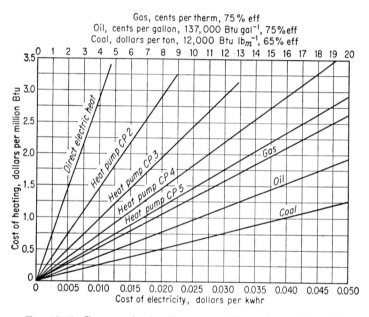

FIG. 18-13. Comparative heating costs. (From data of Penrod.)

firing; in most cases, direct firing is the cheaper method. An indication of the relative costs is given by Fig. 18-13. The heat pump, being a refrigeration cycle, however, offers the advantages of both winter heating and summer cooling, and this fact may well prove to be the deciding factor in future installations.

## PROBLEMS

**1.** A Carnot reversed cycle is used as a refrigeration system between the standard temperatures of 5 and 86°F. Determine the coefficient of performance and the horsepower per ton of refrigeration.

**2.** Repeat Example 3, assuming that carbon dioxide is the refrigerant.

**3.** Repeat Example 3, assuming that the liquid ammonia is subcooled 20°F before entering the expansion valve.

**4.** Determine the condenser and evaporator pressures for carbon dioxide, methyl chloride, and Freon-12 for the data of Example 1. Discuss.

**5.** Ammonia is used as the refrigerant between temperatures of −40 and 80°F. The liquid leaving the condenser is saturated, and the vapor entering the compressor is also saturated. Compression is isentropic. Determine the coefficient of performance, the horsepower per ton, and the capacity of the compressor.

**6.** Repeat Prob. 5, assuming that carbon dioxide is the refrigerant.

**7.** (a) For the data of Example 3, determine the compressor capacity for a refrigeration capacity of 25 tons.

(b) If a double-acting, reciprocating-piston compressor is used with clearance of 0.05 and speed of 100 rpm, what must be the displacement of the compressor (see Art. 17-5)?

**8.** Repeat Example 3, assuming that the isentropic compression efficiency is 0.80 and the process is adiabatic.

**9.** Repeat Example 3, assuming that the vapor entering the compressor contains 10 per cent moisture and compression is isentropic.

**10.** Determine the coefficient of performance and horsepower per ton for a vapor-compression system that uses methyl chloride as the refrigerant between temperatures of 0 and 80°F. The liquid leaving the condenser is subcooled to 70°F, and the vapor entering the compressor is 10°F superheated; isentropic compression efficiency is 85 per cent for the adiabatic compression.

**11.** A vapor-compression cycle uses methyl chloride as the refrigerant between evaporator and condenser temperatures of 20 and 100°F, respectively. The horsepower per ton is 1.08, and the liquid leaving the condenser is subcooled to 90°F. Determine the coefficient of performance and the quality of the vapor entering the evaporator.

**12.** Investigate the feasibility of installing an adiabatic expansion engine, with isentropic efficiency of 80 per cent, in the system of Prob. 10.

**13.** A vacuum system with ejector is to produce 50 tons of refrigeration by circulating at 45°F water that is warmed to 55°F. The available cooling water for the condenser can maintain a condensing temperature of 100°F. Determine (a) the mass flow rate of chilled water, (b) the amount of make-up, and (c) the volume of vapor removed from the flash chamber if the vapor is saturated.

**14.** Repeat Prob. 13 assuming that a centrifugal compressor (adiabatic) is to be used with isentropic compression efficiency of 65 per cent. Determine the coefficient of performance and horsepower per ton of refrigeration.

**15.** A Carnot heat pump supplies 80,000 Btu/hr of heat at 75°F when the outside temperature is 32°F. Determine the coefficient of performance, the power required to drive the pump, and the power required for electric-resistance heating (kilowatts).

**16.** A vapor-compression system with Freon-12 as the cycle fluid is to be used as a heat pump to supply 80,000 Btu/hr at 75°F. The outside temperature is 32°F, and a 10°F difference is present between evaporator and outside air and also between condenser and circulating air for heating. The vapor entering the compressor is saturated; the compression is isentropic; the liquid at the expansion valve is saturated. Determine the coefficient of performance and the cost of heating per hour if electricity is supplied at 5 cents per kilowatthour.

**17.** Suppose that the flash chamber of Fig. 18-4 has a total volume of 10 ft³ and contains 5 ft³ of saturated vapor and 5 ft³ of saturated liquid at 70°F. A vacuum pump lowers the pressure in the chamber (assumed adiabatic). Make reasonable assumptions and find the mass of ice that can be formed.

**18.** Carbon dioxide gas is compressed to 900 psia and cooled to 80°F. It then enters the Linde chamber, where it is throttled to 15 psia. The gas leaves the heat exchanger at 60°F and 15 psia.

(*a*) Calculate the mass of dry ice produced per pound of gas supplied.

(*b*) Calculate the temperature of the gas directly before throttling if the pressure is 890 psia.

**19.** Repeat Prob. 18, but precool the gas to 40°F while maintaining the 20° terminal difference.

## SELECTED REFERENCES

1. "Refrigerating Data Book," American Society of Refrigeration Engineers, New York.
2. Sparks, N. R., and C. C. Dillio: "Mechanical Refrigeration," McGraw Hill Book Company, Inc., New York, 1959.
3. Penrod, E. B.: A Review of Some Heat Pump Installations, *Mech. Eng.*, **69**(8): 639–647 (August, 1947).

# Introduction to Heat Transfer

Although the application of thermodynamic principles to a system gives both the amount and the direction of heat flow, no information is obtained on the factors which control the *rate* of heat flow. Methods for estimating the rate at which heat is transferred is the subject of the engineering science known as *heat transfer*.

In heat-exchanger design, as in the selection of a regenerator for a gas-turbine unit or in the arrangement of the coolant channels in the core of a nuclear reactor, the engineer is faced with the problem of attaining a maximum rate of heat transfer. To accomplish this, he must estimate the influence of exchanger geometry, fluid properties, exchanger materials, fluid velocity, and other variables upon the rate at which heat will be transmitted.

But variations of these parameters to increase heat transfer may lead to undesirable effects on other important factors, such as pumping power requirements and material temperatures and stresses. Thus the economic design must involve consideration of many disciplines: thermodynamics, heat transfer, fluid mechanics, properties of materials, and stress analysis.

In other situations, the engineer may find it necessary to minimize the transfer of heat as in the specification of materials for structural walls. Here the insulating properties of solids and fluids are of interest.

Frequently, the engineer must know the temperature distribution in a system. Such information may be required to avoid temperatures that lead to undesirable changes in the properties of materials or to destructive thermal stresses.

In all these situations, the basis of analysis is provided by the fundamental laws of thermodynamics. The First Law establishes energy balances, and the Second Law, availability balances (that dictate the direction of heat transfer). Thus the laws of thermodynamics bound the problem, but no information is provided on the factors which influence the rate of heat transmission. Means for obtaining such information are found through a study of the subject called heat transfer.

# CHAPTER 19

# FUNDAMENTALS OF HEAT TRANSFER

*Where it is a duty to worship the sun, it is certain*
*to be a crime to examine the laws of heat.*

*Voltaire*

The subject of heat transfer is not a part of thermodynamics; however, a knowledge of the laws that govern the transfer of heat is a necessary requirement in the training of an engineer.

Unfortunately, the word *heat* is not used in a strict thermodynamic sense in this new subject. Recall that from the thermodynamic viewpoint, an energy flow which takes place because of a temperature difference is identified as heat only if (1) the energy transfer occurred across a system boundary and (2) it was not transported by the mass flow. However, it is conventional in the subject of heat transfer to denote *all energy flows that arise because of a temperature difference as heat transfers.* For this reason, the subject of heat transfer might be more properly called thermal-energy transfer.

**19-1. The Modes of Heat Transfer.†** Heat may be transferred by three different mechanisms—*conduction, convection,* and *radiation.* These three modes are similar in that a temperature difference must exist and that the net energy transfer is in the direction of decreasing temperature. However, they are dissimilar in that the physical picture of each of the three phenomena and the laws controlling them differ.

*a. Conduction.* The usual means of heat transmission through opaque solids is by conduction. To illustrate, consider that a high-temperature energy source is attached to one end of a rod which has perfectly insulated sides. Let the other end of the rod be exposed to a region of lower temperature. After a period of time, measurements will show that the temperature of the unheated end has become greater than that of the surroundings. From this observation, it may be concluded that some of the energy added at the heated end has traveled the length of the rod, or in the terminology of this subject, *heat* has been *conducted* from one end of the rod to the other.

A simplified microscopic picture of conduction may be obtained by

† The reader will find it helpful to review the first two paragraphs of Art. 2-1 and all of Art. 2-6.

recalling that an increase in temperature is accompanied by an increase in the intensity of the motions of the many particles (electrons, atoms, molecules) making up the body. Because of greater temperature, the particles near the heated end are in a more violent state of agitation than those near the unheated end. However, these more active particles tend to excite their less active neighbors, who, in turn, tend to excite their neighbors, and so on, so that energy is conducted along the length of the rod.

Biot (1804) and Fourier (1822) are credited with formulating the following expression, commonly referred to as Fourier's heat-conduction equation:

$$\dot{Q} \equiv \frac{\partial Q}{\partial \tau} \equiv -kA\,\frac{\partial t}{\partial x} \tag{19-1}$$

In Eq. (19-1), $\dot{Q}$ is the instantaneous rate of heat transfer through area $A$ by conduction in the $x$ direction, $k$ is a transport property of the medium called the *thermal conductivity*, $A$ is an area perpendicular to $x$, and $\partial t/\partial x$ is the variation of temperature with distance in the $x$ direction, *the temperature gradient*. The partial derivative appears because temperature, in addition to being a function of $x$, may also be dependent upon time and other coordinate directions. For the insulated rod used to describe the conduction phenomena, $x$ would be the coordinate along the axis of the rod and $A$ would be the cross-sectional area of the rod. Usually, $x$ is measured positive in the direction of heat flow. Therefore, $\partial t/\partial x$ is a negative quantity because temperature decreases as $x$ increases. So, in order that $\dot{Q}$ may be a positive quantity, a minus sign is required in Eq. (19-1).

There are two primary types of conduction heat transfer—the unsteady and the steady state. For *unsteady-state conduction*, the temperature of the body may vary with both time and location, leading to a heat-transfer rate which changes with time and position. In *steady-state conduction*, the temperature of the body varies with location but not with time, resulting in a constant rate of heat transfer.

*b. Convection.* When heating or cooling fluids by passing them over solid surfaces, or when mixing hot and cold fluids, energy is transferred by convection. In convection, conduction occurs as adjacent particles interchange energy, while, superimposed upon the conduction process, large numbers of particles at some temperature level tend to circulate and thus exchange energy with regions in the fluid at other temperature levels. The added transport of energy by the migration of large numbers of particles distinguishes convection from conduction.

As an example of convection, suppose that the insulation were removed from the rod described in the preceding illustration. Let energy be added to the rod to maintain it at a temperature greater than that of the surrounding air. Energy would be transmitted from the surface of the rod into the air, where it would be diffused by conduction.†

But the heat transferred by conduction causes the temperature of the air in the vicinity of the rod to become higher than that of the fluid more remotely located. The pressure is essentially constant; therefore the density of the air near the rod is less than that of the air far from the rod. The resulting buoyant effects cause the warmed air to rise from the heated surface, thus permitting cooler air to flow into the vicinity of the rod. As the warmed air rises, it mixes with, and transfers energy to, the cooler air. The over-all process of conduction of heat from solid surface to fluid, with subsequent circulation of warmed fluid as it transports energy to other regions of the system, is characteristic of convection heat transfer.

The equation (Newton, 1701) for the calculation of convection between a solid surface and an adjacent fluid is

$$\dot{Q} \equiv h_c A \, \Delta t \tag{19-2}$$

In Eq. (19-2), $A$ is the surface area of the solid, $\Delta t$ is the difference between the surface temperature and the fluid temperature, and $h_c$ is the surface coefficient for heat transfer by convection, or simply the *convection coefficient*. Unlike thermal conductivity, $h_c$ is not a property of the solid or the fluid but is dependent on many parameters of the system: the geometry and the surface finish of the solid, the velocity of the fluid, and the properties of the fluid (density, viscosity, thermal conductivity, and temperature). As it gives little indication of the dependence of $h_c$ on many factors, Eq. (19-2) is best regarded as simply a definition of $h_c$ rather than as a law of nature.

REPRESENTATIVE VALUES OF THE CONVECTION COEFFICIENT

$h_c$, $Btu/(hr)(ft^2)(°F)$

| | |
|---|---|
| Natural or free convection (air) | 0.5–6 |
| Forced convection (air) | 2–100 |
| Forced convection (liquids, nonmetallic) | 40–1000 |
| Forced convection (liquid metals) | 500–20,000 |
| Boiling (water) | 200–10,000 |
| Condensation (steam) | 1000–20,000 |

Convection in which circulation is induced by buoyancy effects is referred to as *natural or free convection*.‡ In cases where the fluid is

---

† A significant fraction of the energy added to the rod might be lost from the surface of the rod by radiation (see part *c* of this article).

‡ Although conventionally used, the terms *natural* and *free* are not very descriptive of this type of convection. The distinguishing characteristic is that the fluid motion stems from gravitational attraction. Hence *gravitational attraction convection* or simply *gravity convection* seems to be a more appropriate name.

forced past the surface by mechanical means such as fans, pumps, or stirrers, or where the solid surface moves in the fluid, the heat transfer is known as *forced convection.* If the surface is moved through the fluid at low velocity, or if the mechanical means for forcing the flow are not strong, the convection is influenced by both buoyancy effects and the forced flow, and analysis is complex. Fortunately, in most practical situations, either the heat transfer is clearly natural convection or the mechanical means tending to give forced convection are of such intensity that the ever-present buoyant effects may be neglected.

The boiling of a liquid on a warm surface and the condensing of a vapor on a cold surface are normally classified as convection phenomena. Here again the fluid motion may be solely the result of density differences in the medium, or it may be caused by mechanical means.

*c. Radiation.* The means by which energy can be transmitted through space without the presence of an intervening medium is known as radiation. As an example of radiation heat transfer, consider a rod placed in an evacuated enclosure. To maintain the rod at a temperature greater than that of the enclosure walls, it would be found that energy must be continually supplied. The observed heat loss from the rod cannot be accounted for by conduction or convection for there is no medium between rod and walls. It is not implied that heat transfer by radiation occurs only when no intervening matter is present. Consider the heat transfers if the enclosure is filled with fluid. Now the rod surface loses heat to the fluid by convection, and, in addition, heat is interchanged by the rod surface and the enclosure walls by radiation.

The propagation of heat radiation obeys laws identical with those controlling the transmission of light; in fact, the wavelengths of light occupy a small portion of the heat-radiation spectrum. Thus heat radiation is transmitted through space or matter at the velocity of light and both the wave and corpuscular theories of physics are used to visualize and understand the process.

The heat transfer by radiation, $\dot{Q}_{12}$, between the rod (1) and the evacuated enclosure walls (2) is given by the following equation:

$$\dot{Q}_{12} \equiv \mathfrak{F}_{12}A_1\sigma(T_1{}^4 - T_2{}^4) \tag{19-3}$$

The factor $\mathfrak{F}_{12}$ (script F one to two) accounts for the geometry of the system and the radiating characteristics of each of the surfaces; $A_1$ is the surface area of the rod. Sigma ($\sigma$) is the Stefan-Boltzmann constant equal to $0.1714(10^{-8})$ Btu/(hr)(ft²)(°R⁴). The fourth-power relationship between radiation heat transfer and absolute temperature ($T_1, T_2$) was first described by Stefan (1879) and Boltzmann (1884). The factor $\mathfrak{F}$ was proposed by Hottel.[1]†

† See Selected References at the end of the next chapter.

It is sometimes convenient to describe radiation heat transfer by an equation similar to Eq. (19-2) for convection:

$$\dot{Q} \equiv h_r A (T_1 - T_2) \tag{19-4}$$

where $h_r$ is an equivalent surface coefficient for radiation, similar in form to the convection coefficient. From Eqs. (19-3) and (19-4), $h_r$ may be expressed as

$$h_r \equiv \frac{\mathfrak{F}_{12}\sigma(T_1^4 - T_2^4)}{T_1 - T_2} \tag{19-5}$$

If the evacuated enclosure were filled with air, the heat transfer from the rod would be the sum of convection [Eq. (19-2)] and radiation [Eq. (19-3)]. If air temperature

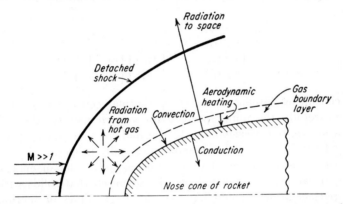

FIG. 19-1. Heat transfer and fluid flow about a body traveling at hypersonic speed.

and enclosure surface temperature were equal, the energy loss from the rod by convection and radiation might conveniently be expressed as

$$\dot{Q} = (h_c + h_r)A \, \Delta t = hA \, \Delta t \tag{19-6}$$

The term $h$ (equal to $h_c + h_r$) is the *combined surface coefficient* for heat transfer by convection and radiation.

Usually, the transfer of heat occurs by more than one mode. Consider the energy transfers for the nose cone of a rocket moving through the atmosphere, as shown in Fig. 19-1. (For simplification, Fig. 19-1 shows air velocities relative to the nose cone.) In the detached shock wave which precedes the hypersonic rocket, the air is suddenly compressed to high temperature and pressure and is partially dissociated and ionized. In the region behind the shock wave, heat is transmitted to the nose cone by convection and radiation from the hot gas. Friction between the gas layers in the boundary layer (*aerodynamic heating*) provides another source of heat, while recombination of the dissociated and ionized gases results in an energy release and thus provides still another heat source. Finally, a portion of the energy from the various sources is conducted into the nose cone while the remainder is radiated from the surface to the surroundings. Detailed knowledge of the several heat transfers is of great importance, for the energy accumulated in the nose cone may be sufficient to destroy it by melting or subliming. A certain amount of surface melting or subliming may be desirable (*ablation cooling*) because the change of phase process absorbs large amounts of energy.†

**19-2. Thermal Conductivity.** The thermal conductivity $k$, introduced in Eq. (19-1), is a property dependent upon the physical and chemical nature of the material, its temperature, and its pressure. Typical engineering units for $k$ may be developed by rearranging Eq. (19-1) as follows:

$$k = \frac{-\dot{Q}}{A\,(\partial t/\partial x)}$$

$$[k] \equiv \left[\frac{Btu}{hr\ ft\ °F}\right] = \left[\frac{Btu}{hr}\right]\left[\frac{1}{ft^2}\right]\left[\frac{ft}{°F}\right]$$

When consulting tabulated values of thermal conductivity, it is wise to note carefully the units specified for $k$. Frequently, the units are as given above, but occasionally the thermal conductivity is specified as Btu in./(hr)(ft²)(°F), which, of course, is equally proper.

Because conduction may be visualized as a transfer of energy from particle to particle in a body, it can be surmised that solids with their closely packed atoms should have the greatest values of thermal conductivity; gases with their widely separated molecules, the smallest values; and liquids, intermediate values. Figure 19-2 confirms these general trends. Values of $k$ for certain solids and fluids are tabulated in Table B-12. For many materials, the dependence of thermal conductivity on temperature may be expressed as

$$k = k_0(1 + at) \tag{19-7}$$

In Eq. (19-7), $k$ is the thermal conductivity at temperature $t$, $k_0$ is the thermal conductivity at $t = 0$, and $a$ is a constant adjusted to best fit data as in Table B-12.

**19-3. Further Remarks on Thermal Conductivity.** Although values of $k$ are of great utility for engineering calculations, thermal conductivity is a gross or macroscopic concept which gives little insight into the exact mechanism of heat conduction. To understand the conduction mechanism and to predict the values of $k$ for substances, knowledge of the microscopic structure of the substance must be sought. Once a reason-

---

† For a general discussion, see John W. Bond, Jr., Problems of Aerophysics in the Hypersonic Region, *Aero Digest*, **72**: 21–26 (June, 1956). For recommendations for calculating heat transfer, see Lester Lees, Laminar Heat Transfer over Blunt-nosed Bodies at Hypersonic Flight Speeds, *Jet Propulsion*, **26**: 259–269 (April, 1956).

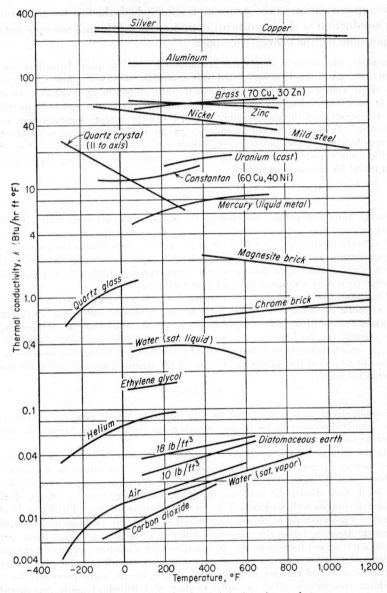

FIG. 19-2. The thermal conductivity of various substances.

able approximation to the structure is established, the key to understanding conduction is the identification and study of the microscopic carrier responsible for the energy transport.

In gases, the carrier is simply the gas molecule. By the most elementary kinetic theory, the gas molecule is postulated to be a minute, hard-shell sphere which undergoes elastic collisions with other, similar molecules. When energy is added to the gas, the translational velocity, and hence the kinetic energy of the molecule, increases. The temperature of the gas is simply a measure of the average kinetic energy of its molecules. Heat is conducted through the gas by the energy transfers which occur when molecules of higher kinetic energy collide with those of lower energy. Considering this simple model of the molecule to be the carrier, the following expression has been found for thermal conductivity:

$$k = \mu c_v$$

Although numerical values predicted by this equation do not agree well with those found by experiment, it is of interest because it shows that $k$, a transport property, is related to the product of another transport property, $\mu$, the coefficient of viscosity, and an equilibrium property, $c_v$. Furthermore, the equation correctly predicts that $k$ increases with temperature ($\mu$ increases with temperature) and that $k$ is independent of pressure (true over a range of low and moderate pressures).

It is known that a weak attraction force exists between widely separated molecules and a strong repulsion force exists between closely spaced molecules. Such forces influence the collision and energy-transfer processes, and thus the simple-kinetic-theory result can be improved by including the attraction-repulsion effect. However, no exact mathematical description of these forces is known; so recourse must be had to empirical formulations. By one of the means of approximating the forces between molecules

$$k = 2.5\mu c_v$$

The conductivity values predicted by this equation are in good agreement with those found experimentally for the monatomic gases, such as argon, helium, and neon, at low pressure.

The use of kinetic-theory concepts to predict conductivity values for polyatomic gases such as oxygen, nitrogen, the hydrocarbons, etc., has not been so successful. In these gases, the molecule can accept and give up energy by changes in translational, rotational, and vibrational motions; therefore, the energy-transfer process during collision is not so well understood as for the monatomic gas. Prediction is also difficult for gas mixtures or dissociating gases because of the problem of determining the energy transfer during the collision of different kinds of molecules. Large errors may be encountered if one attempts to calculate $k$ for a mixture of two or more gases by simply taking the mole fraction of one component times its $k$ value plus the mole fraction of a second component times its $k$ value plus the same procedure for the remaining components. For example, Lindsay and Bromley[†] quote experimental results for an ammonia-air mixture at 176°F as follows: pure ammonia, $k = 0.01742$; pure air, $k = 0.01659$; mixture of 0.41 mole fraction ammonia, $k = 0.01824$.[‡] The simple linear mixing law yields an incorrect value of 0.01693 for the $k$ of the mixture.

[†] A. L. Lindsay and L. A. Bromley, Thermal Conductivity of Gas Mixtures, *Ind. Eng. Chem.*, **42**: 1508-1509 (August, 1950).

[‡] If units are not specified for $k$, they may be assumed to be Btu/(hr)(ft)(°F).

For pressures up to about one-third of the critical pressure of a gas, $k$ increases with pressure, but the increase in $k$ from atmospheric pressure to one-third of the critical pressure is less than 10 per cent for most gases. At pressures greater than one-half of the critical pressure, $k$ changes rapidly with pressure, usually increasing with an increase in pressure. Reid and Sherwood[2] cite generalized correlations showing the effects of temperature and pressure on $k$ which are similar in form to the generalized compressibility chart.

For solids having an ordered, crystalline sort of structure,[†] as, for example, copper, iron, quartz crystal, diamond, and ice, two carriers for the transport of energy have been identified.[‡] The two carriers are free electrons, which diffuse and thus transport energy through the substance, and vibrations of the crystal lattice, which cause energy transfer from atom to atom.

The large values of $k$ for the good electrical conductors (pure metals; Fig. 19-2) stem from the availability of many free-electron carriers. The abundance of free electrons also accounts for the large values of electrical conductivity, and a direct proportion exists between the two conductivities of pure metals. Analysis shows that the energy transport by lattice vibrations contributes less than 1 per cent of the total heat conduction through these substances. Even a small amount of impurity in the metal impedes the mobility of the free electrons and markedly decreases $k$. For example, for nickel of 99.9 per cent purity at 212°F, $k = 48$, and for nickel of 99.2 per cent purity at 212°F, $k = 37$. Alloying of metals produces a similar effect, as shown by brass and constantan (Fig. 19-2).

Crystalline solids which are poor conductors of electricity have few free-electron carriers and exhibit much smaller values of $k$ than the good electrical conductors. Among such substances are quartz crystal, diamond, and ice. It is postulated that vibrations of the crystal lattice constitute the carrier for heat conduction in these solids. For purposes of visualization, it can be considered that the crystal lattice is similar to a three-dimensional mattress with the atoms connected by springs. The addition of energy causes the lattice to shiver, thus displacing the atoms from their normal position and transferring energy from one lattice to another. For these solids, as well as for the good electrical conductors, $k$ tends to decrease with increase in temperature. When lattice vibration is the predominant means of conduction, the value of $k$ depends upon the direction of conduction relative to the crystal orientation. For quartz crystal, $k$ parallel to the crystal axis is about two times that perpendicular to the axis. Once again, impurities or lattice defects hinder the energy-transport mechanism.

Noncrystalline (amorphous or glassy) solids are of random structure and have neither free-electron nor lattice-vibration carriers. Hence thermal-conductivity values for these substances are less than those for crystalline substances, as shown by the comparison of quartz crystal and quartz glass in Fig. 19-2. Also, $k$ tends to increase with temperature increases for these substances. Some refractories seem to exhibit the characteristics of both crystalline and amorphous substances.

The thermal conductivity of organic liquids is of the order of 0.1, while water is several times higher (Fig. 19-2). The effect of pressure on these values is practically insignificant. The structure of liquids is not so well understood as that of gases or solids. Hence the carrier responsible for energy transport has not been clearly iden-

---

† A crystal may be crudely pictured as a symmetrical group of atoms which are bound together by electrostatic forces.

‡ See C. Kittel, "Introduction to Solid State Physics," 2d ed., John Wiley & Sons, Inc., New York, 1956, for a more complete discussion.

tified and prediction of $k$ for liquids is inexact. Modern theories indicate that near the boiling point, liquids have a structure similar to gases, while near the freezing point, liquids have a structure similar to either (depending upon the liquid) crystalline or amorphous solids. The large $k$ values of the liquid metals stem from the free-electron carriers in these liquids.

In the field of heat insulation, the basic requirement is a material with small thermal conductivity. This end point is accomplished with porous, fibrous, or laminated materials of low density, because then the heat must be transferred across and around many air spaces. If these pores are small, convection will be reduced, and the heat will be transferred almost entirely by conduction through the gas and radiation across the pores. The given $k$ value for such materials includes all of these modes. Because the conductivity of gases is low, porous materials are good insulators and are often judged on the basis of their densities (note diatomaceous earth; Fig. 19-2).

For low-temperature insulation such as that for refrigeration systems, hair, felt, and cork are extensively used; for medium temperatures, say 200°F, corrugated asbestos paper is frequently used because of its low cost; and to insulate buildings, cork, rock wool, or glass wool may be applied.† At somewhat higher temperatures, say 500°F, a mixture of 85 per cent magnesia and 15 per cent asbestos is popular. For temperatures greater than 600°F, diatomaceous earth or asbestos may be supplied. The conductivities of all these materials are in the range of 0.025 to 0.05 Btu/(hr)(ft)(°F) and invariably increase with temperature.

**19-4. Steady-state Heat Conduction.** As previously defined, conduction is referred to as steady-state if the heat-transfer rate and the temperature do not vary with time. If, in addition, the conduction occurs through a homogeneous medium in one coordinate direction only (as, for example, the $x$ direction), Eq. (19-1) is written as

$$\dot{Q} = -kA\frac{dt}{dx} \tag{19-8}$$

By rearranging and integrating Eq. (19-8) from a coordinate position $x_1$ with corresponding temperature $t_1$ to a coordinate position $x_2$ at temperature $t_2$,

$$\dot{Q}\int_{x_1}^{x_2}\frac{dx}{A} = -\int_{t_1}^{t_2}k\,dt = \int_{t_2}^{t_1}k\,dt \tag{19-9}$$

Note that for the steady state, $\dot{Q}$, a constant, need not appear under the integral sign. Equation (19-9) can be integrated if:

1. The manner in which $k$ varies with $t$ is known [as in Eq. (19-7) for some materials]

2. A known relationship exists between the coordinate in the direction of heat flow, $x$, and the area perpendicular to heat flow, $A$

Consider first the evaluation of the integral on the right-hand side

† These materials are often used with aluminum foil (or other reflective materials) to minimize radiation and to prevent vapor transmission.

of Eq. (19-9). Let $k$ vary with $t$, as shown in Fig. 19-3. From this figure, it is seen that, for the given temperature range, a *mean thermal conductivity* $k_m$ may be defined:

$$k_m \equiv \frac{1}{t_1 - t_2} \int_{t_2}^{t_1} k \, dt \qquad (19\text{-}10)$$

Thus, by Eqs. (19-9) and (19-10),

$$\dot{Q} \int_{x_1}^{x_2} \frac{dx}{A} = k_m(t_1 - t_2) \qquad (19\text{-}11)$$

If the variation of thermal conductivity with temperature is linear, as in Eq. (19-7), $k_m$ may be found simply by determining $k$ at the arithmetic mean temperature. In other cases, the integration specified in Eq. (19-10) can be performed analytically or graphically.

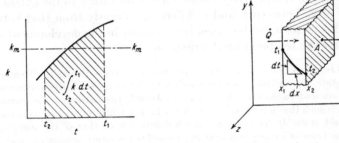

FIG. 19-3. Definition of $k_m$.        FIG. 19-4. Heat conduction through a plane wall.

The left-hand side of Eq. (19-11) can be simply evaluated for three cases of *unidirectional steady-state conduction in homogeneous bodies.*

a. *The Plane Wall.* Consider a plane wall oriented in $x$, $y$, and $z$ coordinates, as shown in Fig. 19-4. For heat to be conducted only in the $x$ direction, the wall dimensions in the $y$ and $z$ directions must be large compared with the wall thickness or the faces of the wall in these directions must be perfectly insulated. Of course, if there is no conduction in the $y$ and $z$ directions, there will be no temperature gradient in these directions. Thus the surfaces at $x_1$ and $x_2$ will be isothermal at temperatures $t_1$ and $t_2$.

Referring to Eq. (19-11), the area perpendicular to heat flow is an area parallel to the $yz$ plane. As this area does not vary with $x$, integration of Eq. (19-11) leads to

$$\frac{\dot{Q}}{A}(x_2 - x_1) = k_m(t_1 - t_2) \quad \text{and} \quad \dot{Q} = \frac{k_m A(t_1 - t_2)}{x_2 - x_1} \qquad (19\text{-}12)$$

If $k$ is assumed constant, the variation of temperature in the wall with $x$ may be determined by writing Eq. (19-12) between $x_1$ and any plane in the body at $x$ where the temperature is $t$.

$$\dot{Q} = \frac{k_m A (t_1 - t)}{x - x_1} \tag{a}$$

But for the steady state, the values of $\dot{Q}$ in Eqs. (19-12) and $(a)$ are identical. Thus these equations may be set equal to each other:

$$\frac{k_m A (t_1 - t_2)}{x_2 - x_1} = \frac{k_m A (t_1 - t)}{x - x_1} \tag{19-13}$$

Solving for the temperature distribution $t$,

$$t = t_1 - \frac{(t_1 - t_2)(x - x_1)}{x_2 - x_1} \tag{19-14}$$

If the thermal conductivity varies with temperature to the extent that the value of $k_m$ between $t_1$ and $t_2$ differs significantly from that between $t_1$ and $t$, this factor must be taken into account in the development of an equation for the temperature distribution.

It should be noted that $t_1$ and $t_2$ are surface temperatures of the wall. To maintain the steady state, the wall must be continually interchanging heat with its surroundings by convection and radiation. As will be explained, the nature of the convection phenomena is such that a temperature measured a few tenths of an inch from the wall surface would probably have a value much different from that of the surface. To obtain a true value of surface temperature, a small temperature-sensing element (such as a thermocouple) must be placed in the surface.

*b. The Hollow Cylinder (Pipe).* Consider a hollow cylinder of length $L$ oriented in cylindrical coordinates, as shown in Fig. 19-5. Assume that the inner and outer surfaces are isothermal so that conduction in both the $\theta$ and $z$ directions is negligible. The problem then is simply the transfer of heat radially through the pipe walls between $r_1$ at temperature $t_1$ and $r_2$ at temperature $t_2$. Note that, if this problem were considered in rectangular coordinates, conduction would not be unidirectional since energy would be transmitted in both the $x$ and $y$ directions.

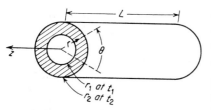

FIG. 19-5. Heat conduction through the walls of a cylinder.

Modifying Eq. (19-11) for this situation gives

$$\dot{Q} \int_{r_1}^{r_2} \frac{dr}{A} = k_m (t_1 - t_2)$$

Here the area perpendicular to conduction, $A$, is $2\pi rL$, a function of the direction $r$ of the heat transfer. Upon performing the integration and rearranging,

$$\dot{Q} = \frac{2\pi k_m L(t_1 - t_2)}{\ln{(r_2/r_1)}} \qquad (19\text{-}15)$$

By the method outlined in Eqs. (19-13) and (19-14), the temperature distribution in the cylinder wall (assuming constant thermal conductivity) is found to be

$$t = t_1 - \frac{(t_1 - t_2)\ln{(r/r_1)}}{\ln{(r_2/r_1)}} \qquad (19\text{-}16)$$

*c. The Hollow Sphere.* The case of a hollow sphere with isothermal inner and outer surfaces is treated in the same manner as in (*b*). Considering a spherical coordinate system, let $r_1$ be the inner radius of the sphere at surface temperature $t_1$ and let $r_2$ be the outer radius at $t_2$. For the steady state,

$$\dot{Q} = \frac{4\pi k_m(t_1 - t_2)}{1/r_1 - 1/r_2} \qquad (19\text{-}17)$$

Experimental determination of the thermal conductivity of a material normally involves the use of Eqs. (19-12), (19-15), or (19-17). A symmetrical model is made from the material to be tested, and the heat conducted, geometrical dimensions, and surface temperatures are carefully measured. The appropriate equation is then solved for $k$. Note that this method yields a mean value of $k$ ($k_m$) for the particular range of surface temperatures of the experiment, whereas a value of $k$ at a particular temperature is desired. To more nearly approach this, small temperature differences are employed, and the resulting heat flows are not large. Thus great precision in temperature measurement and calorimetry is required if accurate results are to be obtained. For model shapes similar to the plane wall or the hollow cylinder, care must be taken to account for the heat loss from the sides or ends. In measuring $k$ for liquids or gases, temperature differences must be small and fluid layers thin if convection effects are to be minimized.

**Example 1.** The thermal conductivity of a material is to be determined by fabricating the material into the shape of a hollow sphere, placing an electric heater at the center, and measuring the surface temperatures with thermocouples when steady state has been reached.

Experimental data: $r_1 = 1.12$ in., $r_2 = 3.06$ in.; for an electrical energy input at the rate of 11.1 watts to the heater, $t_1 = 203°F$ and $t_2 = 184°F$.

Determine (*a*) the experimental value of thermal conductivity and (*b*) the temperature at a point halfway through the sphere wall.

*Solution.* (*a*) Solving Eq. (19-17) for $k_m$,

$$k_m = \frac{(11.1)(3.413)(12/1.12 - 12/3.06)}{4\pi(203 - 184)} = 1.08 \text{ Btu/(hr)(ft)(°F)} \qquad Ans.$$

(*b*) Let $r_3 = 2.09$ in. and note that for steady state, $\dot{Q}$ remains equal to 11.1 watts.

Solving Eq. (19-17) for $t_1 - t_3$,

$$t_1 - t_3 = \frac{(11.1)(3\ 413)(12/1.12 - 12/2.09)}{(4\pi)(1.08)} = 13.8°F$$

Thus                    $t_3 = 203 - 13.8 = 189.2°F$     *Ans.*

The result for $t_3$ is only approximate, for if $k$ is a function of temperature, then the experimentally determined $k_m$ is strictly applicable only in the temperature range $t_1$ to $t_2$. Note that the rate of heat flow through the sphere walls was assumed equal to the electrical energy input to the heater. For the spherical arrangement, this may be essentially true because heat losses through heater and thermocouple leads can be made small. For this reason, the spherical test apparatus is desirable, but difficulties in fabrication and the time required for steady-state conditions frequently preclude its use.

**19-5. The Electrical Analogy and Conduction through Composite Bodies.** Consider that temperature is the *potential* that causes a heat transfer. This concept gives rise to an *analogy* between the transfer of heat and the transmission of electricity as expressed by Ohm's law. Thus current (heat transfer) is governed by the potential difference of voltage (temperature) and the electrical (thermal) resistance of the circuit.

$$\text{Current} = \frac{\text{potential difference}}{\text{electrical resistance}}$$

$$\text{Heat flow} = \frac{\text{potential difference}}{\text{thermal resistance}}$$

Denoting $R_t$ as the *thermal resistance to heat flow*,

$$\dot{Q} \equiv \frac{\Delta t}{R_t} \tag{19-18}$$

Thus, for conduction through the shapes considered in the previous section,

$$R_t \text{ for a plane wall} = \frac{x_2 - x_1}{k_m A} \tag{a}$$

$$R_t \text{ for a hollow cylinder} = \frac{\ln(r_2/r_1)}{2\pi k_m L} \tag{b}$$

$$R_t \text{ for a hollow sphere} = \left(\frac{1}{r_1} - \frac{1}{r_2}\right)\frac{1}{4\pi k_m} \tag{c}$$

In addition, for heat transfer from a surface by convection and radiation,

$$R_t \text{ for convection and radiation} = \frac{1}{hA} \tag{d}$$

The electrical analogy may be conveniently applied in the determination of steady-state unidirectional conduction through a *composite body* with surface temperature $t_1$ and $t_3$, as shown in Fig. 19-6. The *equivalent thermal circuit* for the composite wall is shown at the top of the figure. Note that heat is conducted in a *series path*, and therefore the total resistance to flow equals the sum of the individual resistances:

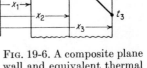

$$\Sigma R_t = R_{t12} + R_{t23}$$
$$= \frac{x_2 - x_1}{k_{m12}A} + \frac{x_3 - x_2}{k_{m23}A} \quad (19\text{-}19)$$

Thus the heat transferred through the wall will be equal to

$$\dot{Q} = \frac{t_1 - t_3}{\Sigma R_t} \quad (19\text{-}20)$$

FIG. 19-6. A composite plane wall and equivalent thermal circuit.

The temperature at the interface, $t_2$, may be found by noting that the energy conducted from $x_1$ to $x_2$ is equal to that conducted from $x_2$ to $x_3$:

$$\dot{Q} = \frac{t_1 - t_2}{R_{t12}} = \frac{t_2 - t_3}{R_{t23}}$$

Solving,
$$t_2 = \frac{R_{t12}t_3 + R_{t23}t_1}{\Sigma R_t} \quad (19\text{-}21)$$

**Example 2a.** Calculate the heat loss from an insulated (8 in. nominal diameter) pipe 100 ft. in length and covered with a composite insulation of two different materials. Other data:

Inside surface temperature of pipe, $t_1 = 625°F$

$$k_{12} = 25 \text{ Btu/(hr)(ft)(°F)} \qquad k_{23} = 0.05 \text{ Btu/(hr)(ft)(°F)}$$
$$k_{34} = 0.04 \text{ Btu/(hr)(ft)(°F)}$$

Outside surface temperature of insulation, $t_4 = 125°F$

$$r_1 = 4.035 \text{ in.} \qquad r_2 = 4.312 \text{ in.} \qquad r_3 = 5.312 \text{ in.} \qquad r_4 = 6.812 \text{ in.}$$

*Solution.* The resistance to heat transfer is (refer to Fig. 19-6)

$$\Sigma R_t = \frac{1}{(2\pi)100} \left[ \frac{\ln (4.312/4.035)}{25} + \frac{\ln (5.312/4.312)}{0.05} + \frac{\ln (6.812/5.312)}{0.04} \right]$$

$$\Sigma R_t = \frac{1}{200\pi} (0.0027 + 4.17 + 6.18) \approx 0.0165 \text{ hr °F/Btu}$$

The heat loss equals

$$\dot{Q} = \frac{t_1 - t_4}{\Sigma R_t} = \frac{625 - 125}{0.0165} = 30{,}300 \text{ Btu/hr} \qquad Ans.$$

Observe that the pipe wall, since it is metal, does not contribute much to the total thermal resistance. It is well to note that this method ignores the *contact resistance* (a minute air space, for example) that may exist at the two interfaces.

**Example 2b.** If the combined surface coefficient for heat transfer by convection and radiation from the insulated pipe surface to surrounding fluid and surfaces is equal to 1.7 Btu/(hr)(ft²)(°F), (a) sketch the equivalent thermal circuit for the heat flow from the inside surface of the pipe to the surrounding fluid and surfaces and (b) compute the temperature of the surrounding fluid and surfaces.

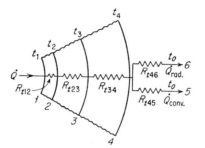

FIG. 19-7. Example 2b, equivalent thermal circuit.

*Solution.*   (a) Denote the surrounding fluid as 5 and the surrounding surfaces as 6, both at temperature $t_0$.   The equivalent circuit is shown in Fig. 19-7.   Note the parallel path of the heat transfer by convection and radiation at the surface.

(b)

$$\dot{Q} = 30,300 \text{ Btu/hr} = \dot{Q}_{\text{radiation}} + \dot{Q}_{\text{convection}} = hA_4(t_4 - t_0)$$

$$t_4 - t_0 = \frac{(30,300)(12)}{(1.7)(2\pi)(6.812)(100)} = 50°F$$

Thus

$$t_0 = 125 - 50 = 75°F \quad Ans.$$

## 19-6. Convection.

To better understand the combined fluid-dynamic and heat-transfer phenomenon known as convection, consider Fig. 19-8, which shows a fluid moving past a stationary wall. A common but complex engineering problem involves determining the heat transfer by convection between wall and fluid.

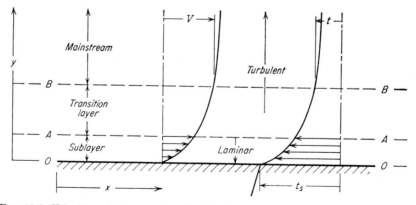

FIG. 19-8. Velocity and temperature distribution for the flow of a fluid parallel to a solid surface.

The bulk of the fluid at $y$ greater than $B$ (Fig. 19-8) is little influenced by the wall and is in *turbulent motion* (see Art. A-3). Although the velocity of a particular element of fluid in turbulent motion may, at a given time, be of any magnitude and direction, the net flow is parallel to the wall at an *average or mainstream velocity V*.

The particles of fluid in intimate contact with the wall tend to adhere to the wall and to have the wall velocity (zero in Fig. 19-8). Because of viscosity, those fluid particles near the stationary particles have a velocity only slightly greater than zero. These particles in turn cause a viscous drag on their neighbors, and so on, such that a thin layer of fluid near the wall (denoted as region $O$-$A$ in Fig. 19-8) is moving at a velocity substantially less than the mainstream velocity. Region $O$-$A$ is called the *laminar sublayer*, a thin layer of fluid in a state of well-ordered (laminar) motion.

In the laminar sublayer, the viscous force (drag) per unit area between adjacent fluid layers is

$$\frac{F}{A} = -\mu_f \frac{\partial V}{\partial y} \qquad (19\text{-}22a)$$

Equation (19-22a), known as Newton's viscosity law, declares that $F/A$ is proportional to the velocity gradient in the $y$ direction. All gases and the usual liquids obey this relationship and are classed as Newtonian fluids (Fig. 1-2).

The viscous drag at the solid surface normally represents a major portion of the pressure loss encountered in pumping fluids. Postulating that a sublayer in laminar motion exists next to the solid, the surface drag (also referred to as skin friction) can be expressed as

$$F = -\mu_f A \left.\frac{\partial V}{\partial y}\right)_{y=0} \qquad (19\text{-}22b)$$

where $(\partial V/\partial y)_{y=0}$ is the velocity gradient at the solid surface.

So that a useful analogy may be established between energy transport (by conduction only in region $O$-$A$) and momentum transport (by viscous drag only in region $O$-$A$), Eq. (19-22a) will be interpreted in a different manner.[†] Considering Fig. 19-8 again, note that the layer of fluid near $A$ has a certain velocity and hence momentum (mass times velocity) in the $x$ direction. Now it can be assumed that this layer of fluid imparts or transports some of its momentum to its adjacent layer in region $O$-$A$, thus keeping this layer in motion but at a smaller velocity. By such means it is imagined that $x$-direction momentum is transported through the laminar sublayer from fluid of high velocity at $A$ to fluid of low velocity near $O$. Thus velocity may be considered the driving force for momentum transport just as temperature is the driving force for energy (heat) transport. Furthermore, by Newton's second law of motion, the $F/A$ term in Eq. (19-22a) is proportional to $1/A$ multiplied by the time derivative of momentum, that is, the rate of momentum transport per unit area. Thus, by Eq. (19-22a), the momentum-transport rate is proportional to a velocity gradient in the same fashion that Eq. (19-1), Fourier's equation, declares that the energy-transport rate (conduction) is proportional to a temperature gradient. So it is

[†] The development of the momentum-transport concept follows that of Bird, Stewart, and Lightfoot.[3]

permissible and convenient to adopt the viewpoint that in the laminar sublayer viscous drag between adjacent fluid particles results in a momentum transport.

At $y$ equal to $A$ in Fig. 19-8, the velocity of the fluid has increased to a value such that there is a tendency for the flow to become turbulent. In region $A$-$B$, this tendency becomes more pronounced until, at $B$, the flow is highly turbulent and the mainstream velocity has been almost attained. In region $A$-$B$, momentum transport occurs partially by viscous drag and increasingly by transverse movements ($y$ direction) of fluid elements from regions of small (or large) velocity to regions of large (or small) velocity. In the mainstream region ($y$ greater than $B$), the momentum transport is predominantly caused by transverse movements of fluid elements.

The entire region $O$-$B$ is called the *hydrodynamic (velocity) boundary layer*, that is, a region where the fluid velocity differs from the mainstream velocity.† The thickness of the boundary layer is arbitrarily defined; one common definition is that the boundary layer extends from the wall to a place where the local velocity is 99 per cent of the mainstream velocity.‡ The concept of the boundary layer and its subdivisions is expedient from the standpoints of visualization and idealization to permit mathematical analysis. However, it should be recognized that the locations of these layers are not sharply defined, for in reality the transition from laminar to turbulent flow is continuous. Furthermore, flow-visualization studies show that swirls in the fluid cause the location of these layers to vary with time.

Though usually thin, the boundary-layer region $O$-$B$ in Fig. 19-8 can be of sufficient thickness to permit accurate measurement of velocity profiles. By such means, the existence of such a layer has been established beyond doubt. Region $O$-$A$, the laminar sublayer, must be extremely thin, and conclusive experimental evidence of its existence has not been presented. Some authorities even doubt that it exists.§ Knudsen and Katz[4] cite the following order of magnitude values for the thickness of the laminar sublayer for isothermal flow through a pipe: Reynolds number of 10,000, sublayer thickness one-hundredth of pipe diameter; Reynolds number of 1,000,000, sublayer thickness of one ten-thousandth of pipe diameter. These values are not

---

† In some cases, laminar flow may prevail throughout the entire boundary layer, as for flow parallel to a flat plate where the Reynolds number based on free-stream velocity and plate length is less than 80,000, or for flow through a circular pipe with Reynolds number less than 2,100. In the latter case, the entire pipe flow may be classed as a laminar boundary-layer type.

‡ Another concept of boundary-layer size is the displacement thickness, the distance the mainstream flow is shifted in the $y$ direction because of the presence of the boundary layer.

§ See Benjamin Miller, The Laminar Film Hypothesis, *Trans. ASME*, **71**: 357–367 (May, 1949), and for another viewpoint, consult R. G. Deissler, Investigation of Adiabatic Turbulent Flow in Smooth Tubes, *NACA TN* 2138, July, 1950.

experimentally determined but are thicknesses calculated from a postulated velocity distribution which has given results for skin friction and heat transfer in good agreement with experimental values.

As in the case of the velocity field, it is expedient to consider the *thermal boundary layer*, a region of fluid near the wall where the fluid temperature differs from the mainstream temperature. From the temperature distribution shown in Fig. 19-8 one could somewhat arbitrarily claim that the thermal boundary layer extended from $O$ to $B$.† In region $O$-$A$, little mixing of the fluid occurs and the heat transfer is primarily by conduction between neighboring layers of fluid. Many fluids, particularly gases, are poor conductors of heat; thus there is a large temperature gradient and large resistance to heat transfer in this region. In transition region $A$-$B$, heat is transferred both by conduction and by mixing induced by the turbulent motion. Mixing predominates in the mainstream region, and the temperature gradient is small.

Note the similarity between the momentum transport in the hydrodynamic boundary layer and the energy transport in the thermal boundary layer. In the laminar sublayer, momentum (energy) is transported by viscous drag (heat conduction) between adjacent layers of fluid. In the transition layer, some transport of momentum (energy) occurs by this mechanism, but increasingly, in proceeding from $A$ to $B$, momentum (energy) is transported by transverse movements and mixing of elements of fluid of differing velocities (temperatures). However, in the mainstream region, almost all of the momentum (energy) transport occurs by the mixing process. Jakob[5] and Schlichting[6] should be consulted for a much more comprehensive treatment of boundary layers.

The heat transfer at the surface may be expressed by Fourier's conduction equation:

$$\dot{Q} = -k_f A \left. \frac{\partial t}{\partial y} \right)_{y=0} \tag{19-23a}$$

In Eq. (19-23a), $k_f$ is the thermal conductivity of the fluid and $(\partial t/\partial y)_{y=0}$ is the temperature gradient in the fluid at the wall.‡ But, from Newton's

---

† Figure 19-8 shows the hydrodynamic and thermal boundary layers to be of the same thickness, as is approximately the case for gases. For nonmetallic liquids, it is found that the thermal boundary layer is much thinner than the hydrodynamic boundary layer, while for liquid metals, the thermal layer is much thicker than the hydrodynamic layer.

‡ Fourier's equation may be applied to the wall at its surface:

$$\dot{Q} = -k_s A \left. \frac{\partial t}{\partial y} \right)_{y=0} \tag{19-23b}$$

Here $k_s$ is the thermal conductivity of the wall and $(\partial t/\partial y)_{y=0}$ is the temperature gradient in the wall at the interface between wall and fluid. The heat flow given by Eq. (19-23b) represents the total heat transfer from the surface (both convection and radiation), whereas Eq. (19-23a) expresses the heat transfer to the fluid only. The two temperature gradients are normally not equal (see Fig. 19-8).

equation defining the convection coefficient,

$$\dot{Q} \equiv h_c A \, \Delta t \tag{19-2}$$

By Eqs. (19-23a) and (19-2),

$$h_c = \frac{-k_f}{\Delta t} \left.\frac{\partial t}{\partial y}\right)_{y=0} \tag{19-24}$$

Thus, if an expression for the fluid temperature distribution in the neighborhood of the wall can be found, the convection coefficient can be evaluated by Eq. (19-24).

Suppose that, while maintaining constant surface and mainstream temperatures, an increase in the convection coefficient were desired. One means of attaining it would be to increase the mainstream velocity and thus decrease the thickness of the laminar sublayer. Note that in this region the heat is transferred primarily by conduction and the resistance to heat flow is great; therefore, a decrease in the thickness of the laminar sublayer increases the convection coefficient. From Eq. (19-24), it can be seen that this increase in $h_c$ would be accompanied by an increase in the temperature gradient at the surface.

But how would the increase in mainstream velocity influence the viscous drag at the wall surface? Evidently the velocity gradient at the surface would increase as the laminar sublayer thickness decreases. Thus, by Eq. (19-22b), the viscous drag, and consequently the pressure loss, would increase. So it is concluded that an increase in the mainstream velocity is beneficial from the standpoint of heat transfer but is detrimental from the standpoint of pressure loss.[†] Therefore, the mainstream velocity is normally selected so as to strike an economic compromise between pressure loss (pumping costs) and convection coefficient (heat-exchange surface area required). Another means for increasing $h_c$ would be to roughen the wall surface, for this induces additional mixing in the boundary layer and thus aids conduction heat transfer. However, once again, drag will also increase.

In Fig. 19-8, it is assumed that the fluid in intimate contact with the wall has the velocity and temperature of the wall surface. Such an assumption is valid only as long as the fluid behaves as a continuum; that is, collisions of the molecules are so frequent that the fluid may be treated as a whole without considering the motion of each individual molecule. Values of the ratio of the mean free path ($\lambda$, average distance traveled by a molecule between collisions) to a characteristic surface length ($L$, length of flat plate, diameter of sphere, etc.) serve to indicate the

[†] O. A. Reynolds (1874) first declared that a direct proportionality exists between the viscous drag and convection heat transfer taking place between a surface and flowing fluid. In his honor, this generality is known as the *Reynolds analogy*. In many instances, experimentation to determine viscous drag is much simpler than experimental measurement of convection heat transfer. Thus, the concept of Reynolds analogy has been extensively applied to predict convection coefficients. Problem 28 further illustrates the use of this concept. Giedt[7] considers this analogy in greater detail.

validity of assuming continuum flow. This dimensionless ratio is called the *Knudsen number*. As a fluid becomes more rarefied (less dense), $\lambda$ increases. In air at 68°F and atmospheric pressure, $\lambda$ equals $2.5 \times 10^{-6}$ in. and there are $4.3 \times 10^{20}$ molecules per cubic inch. At an altitude of 90 miles in the earth's atmosphere $\lambda$ equals 10 ft although there are $10^{13}$ molecules per cubic inch.

Based upon the Knudsen number ($\mathbf{Kn} = \lambda/L$), the following approximate regimes of flow have been specified:[†] $\mathbf{Kn} < 0.01$, continuum flow; $\mathbf{Kn} = 0.01$ to 10, slip flow; $\mathbf{Kn} > 10$, free-molecule flow. For free-molecule flow, collisions are so rare that no boundary layer exists, and the fluid must be treated on the basis of the behavior of its individual molecules.[‡] Slip flow involves a combination of fluid behavior associated with the other two regimes and is characterized by the fact that the velocity and temperature of the fluid at a surface are not those of the surface.

**19-7. Calculation of Convection Coefficients.** The calculation of convection coefficients by theoretical means involves applying the following laws to an element of fluid in the system:

1. Conservation of mass (continuity equation)
2. Conservation of momentum (Newton's equations of dynamics)
3. Conservation of energy (First Law analysis)

In addition, three other relationships are required:[§]

4. An equation of state for the fluid
5. An equation relating viscous forces to velocity gradients (Newton's viscosity equation for Newtonian fluids)
6. An equation relating heat conduction and temperature gradient (Fourier's heat-conduction equation)

In some cases, application of these six relationships has led to explicit formulas for the velocity and temperature distribution in the fluid. From these results, the viscous drag and convection coefficient can be evaluated by Eqs. (19-22b) and (19-24), respectively. Because of a lack of knowledge concerning the exact mechanism of turbulent flow, theoretical approaches have been most successful when the entire flow is of the well-ordered laminar type.[¶]

---

[†] From H. Tsien, Superaerodynamics, Mechanics of Rarefied Gases, *J. Aeronaut. Sci.*, **13**: 653–664 (December, 1946). Considerable uncertainty still exists concerning the boundaries of these flow regimes. See Mac C. Adams and Ronald F. Probstein, On the Validity of Continuum Theory for Satellite and Hypersonic Flight Problems at High Altitudes, *Jet Propulsion*, **28**: 86–89 (February, 1958).

[‡] For methods of calculating heat transfer in this flow regime, see A. K. Oppenheim, Generalized Theory of Convection Heat Transfer in a Free-molecule Flow, *J. Aeronaut. Sci.*, **20**: 49–58 (January, 1953).

[§] Relationships 5 and 6 imply a Second Law analysis.

[¶] Jakob[5] describes several of these analyses.

However, even for turbulent flow, these six relationships are of value because they indicate those parameters (fluid velocity, properties, surface geometry, for example) which influence the fluid flow and convection heat transfer. With the parameters pertinent to the situation known, the method of *dimensional analysis* in combination with experimentation (see Art. A-2) can be utilized to develop empirical formulas for the convection coefficient.

When the method of *dimensional analysis* is applied to the variables present in problems of forced and free convection, certain *dimensionless groups* characteristically appear. Because of the frequency of appearance of these groups, names have been assigned that also serve to honor pioneer investigators in the field of heat transfer. A few of these groups are listed here.

| Name | Symbol | Dimensionless group |
|---|---|---|
| Reynolds number.......... | **Re** | $\dfrac{DV\rho}{\mu_m}$ |
| Nusselt number............ | **Nu** | $\dfrac{h_c D}{k}$ |
| Prandtl number........... | **Pr** | $\dfrac{c_p \mu_m}{k}$ |
| Grashof number........... | **Gr** | $\dfrac{D^3 \rho^2 g \beta\ \Delta t}{\mu_m{}^2}$ |
| Stanton number........... | **St** | $\dfrac{h_c}{c_p V \rho} = \dfrac{\textbf{Nu}}{\textbf{Re Pr}}$ |

Suppose that the convection coefficient of heat transfer $h_c$ is to be determined for fluid flowing through a pipe and being either heated or cooled without change in phase. The first step in the solution is to select those variables which are believed to be pertinent (and these may best be found from the laws controlling the process, as pointed out previously).

$V$ = velocity of fluid, ft/hr
$D$ = diameter of pipe, ft
$\rho$ = density of fluid, $lb_m/ft^3$
$\mu$ = viscosity of fluid, $lb_m/(ft)(hr)$ or $lb_f\ hr/ft^2$
$k$ = thermal conductivity of fluid, Btu/(hr)(ft)(°F)
$c_p$ = Btu/$(lb_m)$(°F)

and the unknown variable is

$h_c$ = convection coefficient, Btu/(hr)(ft²)(°F)

Inspection of these quantities shows that the unit conversion factor

$$g_c = 4.17 \times 10^8 \; \text{lb}_m \; \text{ft}/(\text{lb}_f)(\text{hr}^2)$$

should be included because of the presence of four basic mechanical units: $\text{lb}_m$, $\text{lb}_f$, hr, and ft. However, since the Btu unit has also been introduced, still another unit conversion factor, Joule's equivalent, may be necessary:

$$J = 778.16 \; \text{ft lb}_f/\text{Btu}$$

Then it can be premised that

$$h_c = f(V,D,\rho,\mu,k,c_p,g_c,J)$$

and, quite possibly,

$$h_c = (\text{constant}) \; V^a D^b \rho^c \mu^d k^e c_p{}^f g_c{}^g J^h$$

This equation is solved in the manner illustrated in Art. A-2, although several correct answers may be obtained, depending upon which exponents were chosen to remain unknown. (Of course, an answer can be converted from one form to another by substitution of dimensionally equivalent groups.) It may also be found that the unit conversion factors $g_c$ and $J$ are superfluous.

With a solution found from the dimensional analysis, the next step is to evaluate the constants and exponents that most probably will be necessary to portray the experimental data. But here a new difficulty arises: the temperature of the fluid may have a different value at each section of the heat exchanger and many of the variables will assume different values. Then an average temperature, say the arithmetic average between inlet and outlet of the exchanger, must be postulated for use in evaluating the many variables. Obviously, if different methods are used for determining this average temperature, different constants and exponents will result if correlation can be found at all, and several solutions may be present in the literature.

In many problems a *characteristic length* may be involved, and this length is designated in the various groups by the letter $D$. The characteristic length may be the *equivalent (hydraulic) diameter*, which is defined as

$$D_e \equiv 4 \times \frac{\text{cross-sectional area}}{\text{wetted perimeter}} \tag{19-25a}$$

For round pipes, completely filled,

$$D_e = \frac{4(\pi/4)D^2}{\pi D} = D \tag{19-25b}$$

A *double pipe exchanger* contains a tube within a tube; one fluid flows in the annular space while the second fluid is in the center tube. The equivalent diameter for the *annulus* is

$$D_e = \frac{4(\pi/4)(D_2{}^2 - D_1{}^2)}{\pi(D_2 + D_1)} = D_2 - D_1 \qquad (19\text{-}25c)$$

For ducts of square or rectangular section,

$$D_e = \frac{4D_1D_2}{2D_1 + 2D_2} = \frac{2D_1D_2}{D_1 + D_2} \qquad (19\text{-}25d)$$

The Prandtl number (**Pr**) can be calculated from the defining equation or, for perfect gases, from the semiempirical formula

$$\mathbf{Pr} = \frac{4}{9 - 5/(c_p/c_v)} \qquad (19\text{-}26)$$

The Grashof number contains the *coefficient of thermal expansion* $\beta$, which is defined by

$$\beta = \frac{1}{v}\left(\frac{\partial v}{\partial T}\right)_p \qquad (19\text{-}27a)$$

And when $RT/p$ is a valid substitution for $v$,

$$\beta = \frac{1}{T} \qquad (19\text{-}27b)$$

It may also be found that the geometric arrangement influences the solution, an effect noticeable in flow through short tubes where conditions at the entrance and at the center of the tube are quite different and, also, when flow is laminar. Most geometric arrangements may be characterized by the ratio $L/D$ of two characteristic lengths.

**19-8. Summary of Formulas for Convection Coefficients.**† A few of the many formulas that have been tested experimentally are listed in this section.‡ In most instances, the properties of the fluid are evaluated at the *mean of the mixing-cup or bulk temperatures:*

$$t \equiv \frac{t_{\text{inlet}} + t_{\text{outlet}}}{2} \qquad (19\text{-}28)$$

---

† Mainly from McAdams,[1] except as otherwise specified.

‡ No attempt has been made to compile a complete list of handy recipes for the calculation of convection coefficients. An effort has been made to select those expressions which are representative of the many available and thus to give the reader some insight into the problems of applying them. In the development of formulas to represent experimental data, differences of the order of plus or minus 20 per cent between experimental values and values predicted by the formulas have been considered satisfactory.

The *bulk temperature* ($t_{\text{inlet}}$ or $t_{\text{outlet}}$) is the equilibrium temperature attained by a complete sample of the fluid at a cross section when it is perfectly mixed in an adiabatic container. For turbulent flow of most fluids in ducts, this temperature is very nearly equal to a fluid temperature measured near the duct axis. The surface temperature, if required, must be estimated in most instances. The *mean film temperature* is estimated by

$$t_f \equiv \frac{t + t_s}{2} \tag{19-29a}$$

where the *mean surface temperature* is

$$t_s \equiv \frac{t_{s,\text{inlet}} + t_{s,\text{outlet}}}{2} \tag{19-29b}$$

The effect of radiation is not included in these formulas.

a. *Heating and Cooling of Fluids in Turbulent Flow through Long Pipes*

$$\mathbf{Nu} = 0.023\ \mathbf{Re}^{0.8}\ \mathbf{Pr}^{0.4} \tag{19-30}$$

$$10{,}000 < \mathbf{Re} < 120{,}000 \qquad 0.7 < \mathbf{Pr} < 120 \qquad \frac{L}{D} > 60$$

REMARKS

1. Physical properties are evaluated at mean bulk temperature $t$ [Eq. (19-28)].
2. For gases such as air, the $\mathbf{Pr}$ is essentially constant and can be dropped by changing the constant to 0.02 or, for cooling superheated steam without condensation, to 0.021.
3. Not valid for extreme shapes such as thin rectangles.
4. The equivalent diameter should be substituted for the characteristic length $D$ [Eqs. (19-25a), (19-25c), (19-25d)].
5. For short tubes ($L/D < 60$) having sharp-edged entrances, it is recommended that $h_c$ be calculated from Eq. (19-30) and then increased by multiplying by $1 + (D/L)^{0.7}$.

b. *Heating and Cooling of Viscous Fluids in Laminar Flow in Round Pipes*

$$\mathbf{Nu} = 1.86 \left( \mathbf{Re}\ \mathbf{Pr}\ \frac{D}{L} \right)^{\frac{1}{3}} \left( \frac{\mu}{\mu_s} \right)^{0.14} \tag{19-31}$$

$$\mathbf{Re} < 2{,}100 \qquad \frac{L}{D} > 2 \qquad \mu > 1 \text{ centipoise} \qquad D = D_i$$

REMARKS

1. Physical properties evaluated at mean bulk temperature $t$, except $\mu_s$, which is evaluated at the mean surface temperature $t_s$.
2. The term $\mu/\mu_s$ permits correlating heating and cooling data with one equation.
3. At small flow rates, natural convection effects may predominate and a more complex relationship involving $\mathbf{Re}$, $\mathbf{Pr}$, and $\mathbf{Gr}$ may be required.

*c. Heat Transfer to Fluids of Small Prandtl Numbers in Turbulent Flow through Pipes (Liquid Metals)*†

$$\mathbf{Nu} = 0.625(\mathbf{Re}\ \mathbf{Pr})^{0.4} \tag{19-32}$$

$$200 < \mathbf{Re}\ \mathbf{Pr} < 20{,}000 \quad \frac{L}{D} > 100$$

REMARKS

1. Properties evaluated at mean bulk temperature $t$.
2. Use equivalent diameter for characteristic length.
3. Equation verified experimentally for liquid metals with $\mathbf{Pr}$ less than 0.03.

*d. Heating and Cooling of Fluids Flowing Normal to Tubes or Pipes*

Single tube, liquids

$$\mathbf{Nu} = 0.35 + 0.56(\mathbf{Re})^{0.52} \tag{19-33}$$
$$0.1 < \mathbf{Re} < 200$$

Single tube, gases

$$\mathbf{Nu} = [0.35 + 0.47(\mathbf{Re})^{0.52}]\ \mathbf{Pr}^{0.3} \tag{19-34}$$
$$0.1 < \mathbf{Re} < 1{,}000$$
$$\mathbf{Nu} = 0.26\ \mathbf{Re}^{0.6}\ \mathbf{Pr}^{0.3} \tag{19-35}$$
$$1{,}000 < \mathbf{Re} < 50{,}000$$

Single tube for air, approximately

$$h_c = 0.026\ \frac{G^{0.6}}{D^{0.4}} \tag{19-36}$$

$$100°\text{F} < t_f < 300°\text{F} \quad 1{,}000 < \mathbf{Re} < 50{,}000$$

Tube banks for gases and liquids‡

$$\mathbf{Nu} = 0.244C\ \mathbf{Re}^{0.6}\ \mathbf{Pr}^{0.33} \tag{19-37}$$

$2{,}000 < \mathbf{Re} < 50{,}000$ ratio of tube pitch (minimum distance between adjacent tube centers) to tube diameter of 1.25 to 3.0

| Number of rows (banks) | 1 | 2 | 4 | 6 | 8 | 10 |
|---|---|---|---|---|---|---|
| Staggered tubes, $C$ | 1.00 | 1.11 | 1.31 | 1.45 | 1.51 | 1.54 |
| In-line tubes, $C$ | 1.00 | 1.10 | 1.24 | 1.34 | 1.40 | 1.43 |

REMARKS

1. Physical properties are evaluated at the mean film temperature [Eq. (19-29a)] except in Eq. (19-37), in which they are evaluated at the bulk temperature [Eq. (19-28)].

† Bernard Lubarsky and Samuel J. Kaufman, Review of Experimental Investigations of Liquid-metal Heat Transfer, *NACA Tech. Note* 3336, March, 1955.

‡ Equation (19-37) proposed by Giedt.[7] For a survey of more complex (and probably more accurate) relationships for tube banks, see McAdams.[1]

2. The significant length $D = D_o$, outside tube diameter.

3. Data are for wires and pipes.

4. Convection coefficients predicted by these equations are mean values for entire tube periphery. Larger local $h_c$ values occur in the neighborhood of the stagnation point and the separation point.

5. In Eq. (19-37), $\mathbf{Re} = GD_o/\mu_m$, where $G = \rho V$ is the mass velocity $[\mathrm{lb}_m/(\mathrm{hr})(\mathrm{ft}^2)]$ passing through the smallest opening between tubes.

### e. Heating and Cooling of Fluids Flowing Parallel to an Isothermal Flat Plate†

Laminar ($\mathbf{Re} < 80{,}000$ to $2{,}000{,}000$)

$$\mathbf{St}_x = 0.332\ \mathbf{Re}_x{}^{-0.5}\ \mathbf{Pr}^{-0.667} \tag{19-38‡}$$

$$\mathbf{St} = 0.664\ \mathbf{Re}^{-0.5}\ \mathbf{Pr}^{-0.667} \tag{19-39}$$

Turbulent ($\mathbf{Re} > 2{,}000{,}000$)

$$\mathbf{St}_x = 0.0296\ \mathbf{Re}_x{}^{-0.2}\ \mathbf{Pr}^{-0.667} \tag{19-40}$$

$$\mathbf{St} = 0.037\ \mathbf{Re}^{-0.2}\ \mathbf{Pr}^{-0.667} \tag{19-41}$$

REMARKS

1. Value of Reynolds number at which transition from laminar to turbulent flow occurs depends upon turbulence in free-stream fluid and plate roughness. Cooling stabilizes stream and tends to retard transition, while heating has the opposite effect. $\mathbf{Re} = 500{,}000$ is frequently quoted as a representative transition value.

2. Free-stream velocity is used in $\mathbf{St}$ and $\mathbf{Re}$ calculation, and all properties are evaluated at the film temperature.

3. Characteristic length used in Eqs. (19-39) and (19-41) is the plate length, and these equations give an average value of $h_c$. Equations (19-38) and (19-40) give point or local values of the convection coefficient $h_{cx}$ at characteristic length $x$, measured from the leading edge of the plate.

4. These equations have been found suitable for heat-transfer calculations at high subsonic and supersonic velocities if:§

† E. R. G. Eckert, Engineering Relations for Heat Transfer and Friction in High-velocity Laminar and Turbulent Boundary-layer Flow over Surfaces with Constant Pressure and Temperature, *Trans. ASME*, **78**: 1273–1283 (August, 1956). Also see Eckert and Drake.[8]

‡ Equation (19-38) was derived by Pohlhausen (1921) by theoretical means utilizing the six relationships discussed in Art. 19-7. See Jakob[5] for details of the solution. Equation (19-39) follows from Eq. (19-38), as shown in Prob. 26. Pohlhausen also showed that the boundary-layer thickness for this case is approximately $5.83x/(\mathbf{Re}_x)^{0.5}$.

§ At first thought, it seems obvious that the Mach number (the criterion for subsonic or supersonic flow) should be included in the dimensionless groupings to represent heat transfer at high subsonic and supersonic velocities. However, the Mach number is mainly a function of fluid velocity and temperature, and it is included implicitly in the usual $\mathbf{St}$, $\mathbf{Re}$, $\mathbf{Pr}$ groupings. So it has been found possible to include the Mach-number effect by evaluating fluid properties at a more complex reference temperature, denoted here as $t^*$. Because convection heat transfer is a boundary-layer phenomenon, convection is much more dependent on whether the boundary layer is laminar or turbulent (criterion is $\mathbf{Re}$) than on whether the main-stream flow is subsonic or supersonic (criterion is $\mathbf{M}$).

*a.* Aerodynamic heating effects are included by defining the convection coefficient as

$$h_c \equiv \frac{\dot{Q}}{A(t_s - t_{aw})} \tag{19-42}$$

$$t_{aw} \equiv t + r\frac{V^2}{2g_cJc_p} \tag{19-43}$$

The term $t_{aw}$ is the adiabatic wall temperature, that is, the temperature that would be attained by a perfectly insulated wall subjected to the stream. Free-stream static temperature and velocity are $t$ and $V$, respectively, and $r$ is a recovery factor. For laminar flow,

$$r = (\mathbf{Pr})^{\frac{1}{2}} \tag{19-44}$$

For turbulent flow,

$$r = (\mathbf{Pr})^{\frac{1}{3}} \tag{19-45}$$

*b.* All properties are evaluated at a reference temperature $t^*$:

$$t^* = t + 0.50(t_s - t) + 0.22(t_{aw} - t) \tag{19-46}$$

*f. Natural Convection from Vertical Planes, Vertical Pipes, and Horizontal Cylinders*

Vertical planes and vertical cylinders

$$\mathbf{Nu} = 0.59(\mathbf{Gr\ Pr})^{0.25} \tag{19-47}$$
$$10^4 < \mathbf{Gr\ Pr} < 10^9$$
$$\mathbf{Nu} = 0.13(\mathbf{Gr\ Pr})^{0.333} \tag{19-48}$$
$$10^9 < \mathbf{Gr\ Pr} < 10^{12}$$

Horizontal cylinders
In air

$$\mathbf{Nu} = 0.53(\mathbf{Gr\ Pr})^{0.25} \tag{19-49}$$

$$h_c = 0.27\left(\frac{\Delta t}{D_o}\right)^{0.25} \tag{19-50}$$

$$10^3 < \mathbf{Gr\ Pr} < 10^9$$

REMARKS

1. Physical properties evaluated at the mean film temperature $t_f$ and $\Delta t = t_s - t$.
2. In Eqs. (19-47) and (19-48), $D$ = height from the lower edge of the plate or cylinder.
3. $D = D_o$ in Eq. (19-49).
4. In Eq. (19-50), $D_o$ in feet, $\Delta t$ in °F gives $h_c$ in Btu/(hr)(ft²)(°F).

**Example 3.** Five hundred thousand pounds per hour of cooling water enters a steam condenser at a bulk temperature of 70°F, flows through the condenser inside of one hundred 1-in.-outside-diameter BWG No. 18 tubes arranged in parallel, with each tube 16 ft long, and leaves at a bulk temperature of 80°F. Steam condenses on the outer surface of each tube. Predict the coefficient for heat transfer by convection between the cooling water and the tube.

*Solution.* The water-side convection coefficient can be computed by the empirical relationships on page 393 if the variables in the problem are in a range for which an equation is available. This case might fit either item *a* or item *b* on page 393. To determine which, if either, is applicable, the Reynolds number $(DV\rho/\mu_m)$ is calculated.

$$D \text{ (inside diameter)} = 1 - 2(0.049) = 0.902 \text{ in.}$$

(Table B-3 gives 0.049 in. as the wall thickness of No. 18 tube.)   Evaluating water properties at a mean bulk temperature [Eq. (19-28)],

$$t = \frac{70 + 80}{2} = 75°F$$

$$\rho = 62.2 \text{ lb}_m/\text{ft}^3 \qquad\qquad \mu_m = 2.22 \text{ lb}_m/(\text{ft})(\text{hr})$$
$$k = 0.352 \text{ Btu}/(\text{hr})(\text{ft})(°F) \qquad c_p = 1 \text{ Btu}/(\text{lb}_m)(°F)$$

The velocity $V$ is calculated from the continuity equation

$$\frac{500,000 \text{ lb}_m/\text{hr}}{100 \text{ tubes}} = 5,000 \frac{\text{lb}_m/\text{hr}}{\text{tube}} = \frac{\pi(0.902)^2(V)(62.2)}{4(144)}$$

Solving,
$$V = 18,100 \text{ ft/hr}$$
$$\mathbf{Re} = \frac{(0.902)(18,100)(62.2)}{(2.22)(12)} = 38,100$$

(Here care must be taken to assure that consistent units have been used so that the Reynolds number is dimensionless.)   Although Eq. (19-30) is valid for this Reynolds number value, the other restrictions should be checked:

$$\mathbf{Pr} = \frac{c_p\mu_m}{k} = \frac{(1)(2.22)}{0.352} = 6.3$$
$$\frac{L}{D} = \frac{(16)(12)}{0.902} > 60$$

Thus Eq. (19-30) is applicable.

$$\mathbf{Nu} = 0.023(38,100)^{0.8}(6.3)^{0.4} = 0.023(4,620)(2.09) = 222$$

and    $$h_c = 222 \frac{k}{D} = \frac{(222)(0.352)(12)}{0.902} = 1040 \text{ Btu}/(\text{hr})(\text{ft}^2)(°F) \qquad Ans.$$

**Example 4.** A long horizontal pipe, 6 in. outside diameter, with an oxidized surface, passes through a large room. The surface temperature of the pipe is 200°F, and the surrounding air and solid surfaces are at 80°F. What percentage change in convection heat transfer would occur if the pipe surface temperature were increased to 400°F with surrounding air and surface temperatures remaining at 80°F?

*Solution.* Equation (19-49), or Eq. (19-50), is applicable for computing the convection coefficient if the **Gr Pr** product is in the proper range. Evaluating air properties at the mean film temperature defined by Eq. (19-29a) gives, for $t_s = 200°F$,

$$t_f = \frac{200 + 80}{2} = 140°F$$

Interpolating to determine air properties from Table B-15,

$$\rho = 0.0662 \text{ lb}_m/\text{ft}^3 \qquad c_p = 0.2408 \text{ Btu}/(\text{lb}_m)(°F)$$
$$\mu_m = 0.0483 \text{ lb}_m/(\text{hr})(\text{ft}) \qquad k = 0.0167 \text{ Btu}/(\text{hr})(\text{ft})(°F)$$

From Eq. (19-27b)

$$\beta = \frac{1}{T_f} = \frac{1}{460 + 140} = 0.00167°R^{-1}$$

$$Pr = \frac{(0.2408)(0.0483)}{0.0167} = 0.696$$

$$Gr = \frac{D^3\rho^2 g\beta \, \Delta t}{\mu_m{}^2} = \frac{(0.5)^3(0.0662)^2(32.17)(3,600)^2(0.00167)(200 - 80)}{(0.0483)^2}$$

$$Gr = 1.93 \times 10^7$$

$$Gr \, Pr = (1.93 \times 10^7)(0.696) = 1.34 \times 10^7$$

Hence Eq. (19-49) may be employed to calculate $h_c$ for the 200°F surface temperature. For $t_s = 400°F$,

$$t_f = \frac{400 + 80}{2} = 240°F$$

$$\rho = 0.0567 \text{ lb}_m/\text{ft}^3 \qquad c_p = 0.2419 \text{ Btu}/(\text{lb}_m)(°F)$$
$$\mu_m = 0.0542 \text{ lb}_m/(\text{hr})(\text{ft}) \qquad k = 0.0190 \text{ Btu}/(\text{hr})(\text{ft})(°F)$$

$$\beta = \frac{1}{460 + 240} = 0.00143°R^{-1} \qquad Pr = 0.69$$

$$Gr = \frac{(0.5)^3(0.0567)^2(32.17)(3,600)^2(0.00143)(400 - 80)}{(0.0542)^2} = 2.61 \times 10^7$$

$Gr \, Pr = 1.8 \times 10^7$; hence Eq. (19-49) may be used for the 400°F temperature. Computing the heat transfer by convection for the case of the 200°F surface temperature,

$$h_c = \frac{k}{D_o} (0.53)(Gr \, Pr)^{0.25} = \frac{0.0167}{0.5} (0.53)(1.34 \times 10^7)^{0.25}$$
$$= 1.07 \text{ Btu}/(\text{hr})(\text{ft}^2)(°F)$$

$$\left.\frac{\dot{Q}}{A}\right)_{t_s=200°F} = h_c \, \Delta t = 1.07(200 - 80) = 129 \text{ Btu}/(\text{hr})(\text{ft}^2)$$

For the 400°F surface temperature,

$$h_c = \frac{0.0190}{0.5} (0.53)(1.8 \times 10^7)^{0.25} = 1.31 \text{ Btu}/(\text{hr})(\text{ft}^2)(°F)$$

$$\left.\frac{\dot{Q}}{A}\right)_{t_s=400°F} = (1.31)(400 - 80) = 419 \text{ Btu}/(\text{hr})(\text{ft}^2)$$

The percentage change in convection heat transfer caused by increasing the surface temperature from 200 to 400°F is

$$\frac{(100)(419 - 129)}{129} = 225 \text{ per cent} \qquad Ans.$$

**Example 5.** The wing of a supersonic transport plane designed to fly at 1,200 mph in air at a temperature of 80°F has an average chord dimension of 40 ft. (a) Estimate the temperature of the wing surface (skin) assuming no heat loss from the skin. (b) If because of radiation, or possibly internal refrigeration, the skin loses energy and thus attains a steady temperature of 200°F, determine the direction and magnitude of the convection heat transfer.

*Solution.* (a) At this speed, aerodynamic heating effects cause the air in the boundary layer to increase in temperature and result in a significant convection heat transfer to the skin. If the skin is considered to lose no energy, then its temperature approaches the adiabatic wall temperature. For purposes of estimation, it will be assumed that the wing is equivalent to a flat plate of 40 ft length. From page 396, the adiabatic wall temperature can be computed if the recovery factor is known. The value of the recovery factor depends upon whether the boundary-layer flow is laminar or turbulent. To ascertain the nature of the flow, the Reynolds number is calculated. (As only an approximate value of **Re** is needed, fluid properties are evaluated at the free-stream temperature of 80°F from Table B-15.)

$$\mathbf{Re} = \frac{(1,200)(5,280)(40)(0.0735)}{0.0447} = 4.16 \times 10^8$$

Hence the boundary-layer flow is turbulent and

$$r = (\mathbf{Pr})^{\frac{1}{3}} = \left[ \frac{(0.2404)(0.0447)}{0.0152} \right]^{\frac{1}{3}} = 0.892$$

$$t_{aw} = t + r\frac{V^2}{2g_cJc_p} = 80 + \frac{(0.892)(1,200)^2(5,280)^2}{(2)(32.17)(778)(0.2404)(3,600)^2} = 310°F \qquad Ans.$$

Radiation heat transfer alone will cause the skin temperature to be other than 310°F.

(b) The convection coefficient may be found by Eq. (19-41), with property values evaluated at

$$t^* = 80 + (0.50)(200 - 80) + (0.22)(310 - 80) = 191°F$$

$$\mathbf{Re} = \frac{(1,200)(5,280)(40)(0.0610)}{0.0514} = 3.00 \times 10^8$$

$$\mathbf{Pr} = \frac{(0.2412)(0.0514)}{0.0179} = 0.693$$

$$\mathbf{St} = \frac{h_c}{c_pV\rho} = (0.037)(\mathbf{Re})^{-0.2}(\mathbf{Pr})^{-0.667}$$

$$h_c = (0.2412)(1,200)(5,280)(0.0610)(0.037)(3.00 \times 10^8)^{-0.2}(0.693)^{-0.667}$$
$$= 88.9 \text{ Btu/(hr)(ft}^2)(°F)$$

The convection heat transfer is

$$\frac{\dot{Q}}{A} = h_c(t_s - t_{aw}) = 88.9(200 - 310) = -9,790 \text{ Btu/(hr)(ft}^2) \qquad Ans.$$

The convection heat transfer is from air to plate. Note that sea-level pressure (14.7 psia) has been assumed. How would the results differ if the plane were flying at an altitude of 50,000 ft (approximately one-tenth sea-level pressure)?

**19-9. Thermal Radiation.** The modern theory of the mechanism of radiation premises that corpuscles of energy called *photons* are propagated through space as rays. Although the motion and position of each photon cannot be exactly specified, the movement of a swarm of photons can be described as an *electromagnetic wave*. This wave is propagated with unchanging frequency at a velocity equal to the speed of light. The relationship between velocity $V$, frequency $\nu$, and wavelength $\lambda$ is

$$V = \nu\lambda \tag{19-51}$$

Each photon, or *light quantum*, is conceived to have a definite energy of amount

$$E = h\nu \tag{19-52}$$

where $h$ is Planck's constant equal to $6.62377(10^{-27})$ erg sec. Thus the corpuscle-wave nature assigned to radiation dictates that the total energy is spread over a greater and greater area as the distance from the source increases (so that the inverse-square law of optics is upheld) and yet that each photon arrives at its final destination without loss of energy. If the photon is completely absorbed by a substance, it is annihilated, and its energy is taken up by the substance.

Consider a case in which the particles and subparticles of a substance occupy different energy levels. Then when energy is diminished (or increased) by one packet of energy (a photon), a particle experiences a transition from one energy level to another. Because the energy levels are not evenly spaced, radiation from a substance is composed of photons of various amounts of energy and therefore of various wavelengths [by Eqs. (19-51) and (19-52)]. Although in a microscopic sense radiation is not continuous, macroscopic measurements are too coarse to show the discontinuities.

The particle excitement that gives rise to radiation can be induced by many means, and certain methods produce radiation in very narrow wavelength (frequency) ranges. Arbitrarily, names have been assigned to certain frequency ranges of the electromagnetic spectrum (Table 19-1).

Radiation emitted by a substance solely because of its temperature is called *thermal radiation*. This radiation is distributed (macroscopically) over all wavelengths, but for the temperatures usually encountered in engineering, thermal radiation is mostly concentrated in the infrared region of the spectrum (Table 19-1). Examples of radiation not induced by thermal means include that emitted by a substance experiencing radioactive decay and that emitted by a phosphor under electronic bombardment (fluorescent lamp). The remaining discussion will consider only thermal radiation, and the term *radiation* will imply thermal radiation.

Matter emits radiation at all attainable temperatures, and the direction of radiant flux is not determined by the temperature difference. The preceding statement stems from the Prevost law of exchanges (1792). To illustrate, consider an isolated system consisting of a sphere in an evacuated enclosure. If at first the enclosure is hotter than the sphere, equality of temperature will result from transfer of heat by radiation from enclosure to sphere. But if the enclosure is cooled, the transfer of heat proceeds from sphere to enclosure. From these observations, it might be concluded that the sphere was radiating energy to the enclosure in the first case as well as in the second case. Following the same line of reasoning when sphere and enclosure are at the same temperature, it is seen that an equal exchange of radiation must take place to maintain the equilibrium. Thus radiation interchanges are continually occurring in the direction of both increasing and decreasing temperature, although,

TABLE 19-1. THE ELECTROMAGNETIC SPECTRUM
(Speed of propagation is 186,000 miles/sec in a vacuum)

| Name | Origin | Representative wavelength, cm | Approximate energy per photon, electron volts[†] |
|------|--------|------|------|
| Cosmic ray........ | Unknown | $10^{-13}$ | $10^9$ |
| Gamma ray....... | Emitted by radioactive substances | $10^{-10}$ | $10^6$ |
| X ray............ | Secondary emission from materials bombarded with electrons | $10^{-7}$ | $10^3$ |
| Ultraviolet ray.... | High activity of outer electrons of molecule, usually by increasing the temperature | $10^{-6}$ | $10^2$ |
| Light: | | | |
| Violet.......... | ............................ | $4 \times 10^{-5}$ | |
| Green.......... | ............................ | $5 \times 10^{-5}$ | 10 |
| Red............ | ............................ | $7 \times 10^{-5}$ | |
| Infrared.......... | Vibrations of atoms in molecules | $10^{-3}$ | $10^{-1}$ |
| FM radio and television | From flow of electricity through systems with capacitance and inductance | $10^3$ | $10^{-7}$ |
| AM radio | | $10^5$ | $10^{-9}$ |

[†] 1 electron volt $\approx 1.5 \times 10^{-22}$ Btu (Art. A-1).

of course, the *net* transfer of heat must always be from hotter to colder regions.

**19-10. The Concept of the Black Body.** When radiation falls on a surface, some of the incident energy may be reflected and the remainder either absorbed or transmitted through the material. This division of incident radiation can be shown as three fractional parts:

$$\alpha + \rho + \tau = 1 \tag{19-53}$$

wherein $\alpha$ = absorptivity (fraction of incident radiation absorbed)
$\rho$ = reflectivity (fraction of incident radiation reflected)
$\tau$ = transmissivity (fraction of incident radiation transmitted)
The absorption and emission of radiation can be considered to be a surface effect for most solids (involving, at most, 0.1 in. depth). Engineering materials are usually sufficiently thick so that they can be considered opaque, that is, $\tau = 0$ (glass appears to be an exception, but see Art. 19-15).

Black substances are effective absorbers of radiation in the wavelengths that are encountered in heat transfer. From this observation, the name *black body* is assigned to a perfect absorber of radiation.

**A black body has an absorptivity of unity for all incident radiation.**

The characteristics of a black body can be closely approached by a construction first proposed by Kirchhoff. A small opening is made in a hollow sphere (or other hollow body) with its inner surface coated with (for example) lampblack ($\alpha \approx 0.95$, $\rho \approx 0.05$) (Fig. 19-9). Here the

FIG. 19-9. A practical black body.

small opening approaches black-body behavior since radiation entering it will be largely absorbed. Note, in Fig. 19-9, that an incident ray will have a multiplicity of ever-weaker reflections on the lampblack surface, and only an extremely small portion of the incident radiation will ever again encounter the opening.†

In addition to being a perfect absorber, the black body is also a *perfect emitter;* that is, at a given temperature it emits more radiation than any other body. Suppose that a black body is placed inside a hot oven. The body would absorb all incident radiation and attain the oven temperature (but its temperature could not exceed this value or the Second Law would be violated). Now substitute a non-black body for the black body. A longer interval of time will elapse before the non-black body reaches the oven temperature since not all of the incident radiation is absorbed (a slower absorption rate). The conclusion for each body at thermal equilibrium is that the body must be emitting radiation at exactly the same rate that it is absorbing radiation. Then it follows that since the black body is the perfect absorber, it must therefore be the perfect emitter.

With this reasoning, consider again Fig. 19-9. Since the small hole is (essentially) a perfect absorber, it is also a perfect emitter. The inner walls of the cavity are continually exchanging radiation and therefore the enclosure is filled with radiant flux with density dictated by the equilibrium temperature. Thus black-body radiation is continually escaping through the small hole to the surroundings. Ideally, all of the escaping radiation originates from emission (not reflection) by the inner walls. Devices such as that of Fig. 19-9 are used as black-body

FIG. 19-10. Emission from a black surface.

sources for research in radiation and for calibration of radiation instruments.

**19-11. The Wavelength Distribution of Black-body Radiation.** In Fig. 19-10, the black surface‡ $A$ is radiating energy in all directions of its enclosing hemispherical angle. An expression for the rate at which radiation is emitted by unit area of $A$ throughout the hemisphere for one

---

† The smaller the opening, the better the approximation to black-body behavior. To approach black-body behavior closely enough for most experimentation, the opening need not be as small as one might expect; for example, a hole of 1 in. diameter in the end of a hollow cylinder 10 in. long and 3 in. in diameter suffices.

‡ Black surface will be used as a synonym for black-body surface.

particular wavelength ($\lambda$) was derived from theoretical considerations by Planck (1901):[†]

$$W_{b\lambda} = \frac{C\lambda^{-5}}{e^{c/\lambda T} - 1} \qquad \left[ \frac{\text{Btu}}{\text{hr ft}^2 \text{ cm}} \right] \qquad (19\text{-}54)$$

where $\lambda$ = wavelength, cm
$\quad C = 1.1870(10^{-8})$ Btu cm$^4$/(hr)(ft$^2$)
$\quad c = 2.5896$ cm °R

The quantity $W_{b\lambda}$ is called the *monochromatic hemispherical emissive power* of a black surface.[‡]

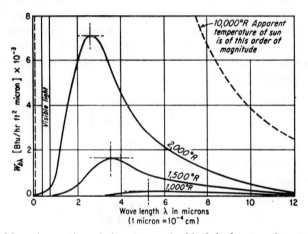

FIG. 19-11. Monochromatic emissive power of a black body at various temperatures.

Equation (19-54) is plotted in Fig. 19-11 with the wavelength as the abscissa. The total rate of radiation per unit area of black surface for a specified temperature is equal by definition to

$$W_b = \int_0^\infty W_{b\lambda} \, d\lambda \qquad (19\text{-}55)$$

Hence $W_b$, the *total hemispherical emissive power* per unit area of black surface for a particular temperature, is represented by the area under that particular temperature curve in Fig. 19-11. The maxima of the

[†] Prior to Planck, the use of Newtonian mechanics failed to produce an expression for the distribution of black-body radiation that agreed with experimental findings. In obtaining a valid relationship, Planck was led to the viewpoint that matter cannot emit or absorb energy of arbitrary amount, but must interchange energy in discrete amounts—one photon, two photons, etc. [Eq. (19-52)]. Planck's concept was revolutionary for it led to *quantum mechanics*, a cornerstone of twentieth-century science. See Jakob[5] for a description of the earlier formulations and further information on Planck's contribution.

[‡] Radiation constants are those recommended by N. W. Snyder, A Review of Thermal-radiation Constants, *Trans. ASME*, **76**: 537–539 (May, 1954).

curves shift to shorter wavelengths (higher frequencies) as the temperature increases; these maxima can be obtained by differentiating Eq. (19-54) or from *Wien's displacement law* (derived in another manner by Wien prior to the introduction of Planck's equation):

$$\lambda_{max} T = 0.52156 \text{ cm } °R \tag{19-56}$$

The increase in radiation with temperature is illustrated by the larger areas lying under the higher temperature curves. Note that because of the high temperature of the sun, a significant amount of its emission is in the visible light band (0.4 to 0.8 micron). For the relatively low temperatures encountered in engineering, however, most of the energy will be concentrated in the longer wavelengths—the infrared region (Table 19-1)—and relatively little energy is emitted as visible light.

Substituting Eq. (19-54) into Eq. (19-55) and performing the integration yields the *Stefan-Boltzmann law:*†

$$W_b = \sigma T^4 \quad \left[ \frac{\text{Btu}}{\text{hr ft}^2} \right] \tag{19-57}$$

where $\sigma = 0.1714(10^{-8}) \text{ Btu}/(\text{hr})(\text{ft}^2)(°R^4)$
$T$ = absolute temperature of emitter, °R
It is convenient to use Eq. (19-57) in the form

$$W_b = 0.1714 \left( \frac{T}{100} \right)^4$$

**19-12. The Spatial Distribution of Radiation.** The distribution of black-surface radiation throughout space will be investigated. Figure 19-12 shows a small black surface $dA_1$ (emitter) and a black-body radiation collector‡ $dA_2$ which can be placed at various angular locations about $dA_1$. Maintaining the radius $r$ constant, the locus of locations is a hemisphere. It will be found that the collector measures a maximum amount of radiation (heat) when it is at the position normal to the emitter and that the amount progressively decreases with increase in $\phi$, becoming zero at $\phi$ of 90°. Also, the radiation will be symmetrical about the normal axis.

† For a detailed account of the experimentation and thermodynamic reasoning that led to the laws of black-body radiation, refer to J. K. Roberts, "Heat and Thermodynamics," 4th ed., Blackie & Son, Ltd., Glasgow, 1955. As consideration of black-body radiation led to formulation of the quantum theory, most textbooks on modern physics consider black-body radiation in some detail; see, for example, F. K. Richtmyer, E. H. Kennard, and T. Lauritsen, "Introduction to Modern Physics," 5th ed., McGraw-Hill Book Company, Inc., New York, 1955.

‡ The collector might be crudely thought of as a small mass, with its temperature rise during a measured time period serving as an indicator of the incident radiation.

The radiation impinging on (and absorbed by) the collector is emitted by $dA_1$ through a constant solid angle† $d\omega$ equal to $dA_2/r^2$. For a fixed position on the hemispherical surface, the quantity of radiation absorbed by such a collector is proportional to the intensity of emission ($I$) of $dA_1$, defined for black bodies as

$$W_b = \int_0^{2\pi} I \, d\omega = \sigma T^4 \quad (19\text{-}58)$$

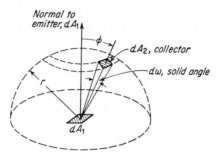

FIG. 19-12. Determination of the angular distribution of radiation emitted by a black surface.

The dimensions of $I$ are radiant flux per unit area of emitter and unit solid angle [Btu/(hr)(ft²)(solid angle)].

The collector would measure a variation of black-body intensity with angular position $\phi$, in accord with *Lambert's cosine law:*

**The intensity of radiation in a direction $\phi$ from the normal to a plane black emitter is proportional to cosine $\phi$.**

Let $I_n$ designate the normal intensity and $I_\phi$ denote the intensity at an angle $\phi$ from the normal. Then, according to Lambert's cosine law,

$$I_\phi = I_n \cos \phi \quad\quad\quad\quad\quad (19\text{-}59)$$

By Lambert's cosine law,‡ the radiation emitted by $dA_1$ which impinges on $dA_2$ may be expressed as

$$dA_1 \, W_{1\to2} = \frac{I_{n1} \cos \phi_1 \, dA_1 \, dA_2}{r^2} \quad\quad (19\text{-}60)$$

Further, if the normal to $dA_2$ is oriented at an angle $\phi_2$ to a radius connecting $dA_1$ and $dA_2$ such that the projected area $dA_2 \cos \phi_2$ is seen when looking toward $dA_2$ from $dA_1$ (note that $\phi_2 = 0$ in Fig. 19-12), then

$$dA_1 \, W_{1\to2} = \frac{I_{n1} \cos \phi_1 \, dA_1 \cos \phi_2 \, dA_2}{r^2} \quad\quad (19\text{-}61)$$

† A solid angle is defined as the spherical area divided by the radius squared. The area of a hemisphere is $2\pi r^2$; hence a hemisphere is a solid angle of $2\pi$ or contains $2\pi$ solid angles.

‡ The concept that the intensity (brightness) of a plane surface varies with the direction of view is contrary to our observations. If a "red-hot" surface is viewed from its normal direction and then from a direction other than normal at angle $\phi$, our senses report that the surface appears equally intense (bright) in both cases. However, in defending this viewpoint, it must be remembered that if an area $A$ was seen from the normal direction, then a smaller projected area $A \cos \phi$ was seen from the other direction. But whether one adopts the concept that intensity varies with $\phi$ or the seemingly more sensible viewpoint that intensity is constant and the area seen varies with $\phi$, the same result [Eq. (19-60) or (19-61)] is obtained for the exchange between emitter and collector.

A relationship between the normal intensity $I_n$ and the total emission of a plane black surface of unit area may be obtained by reference to Fig. 19-13. In Fig. 19-13, the collector $dA_2$ is now a symmetrical band about the normal axis.

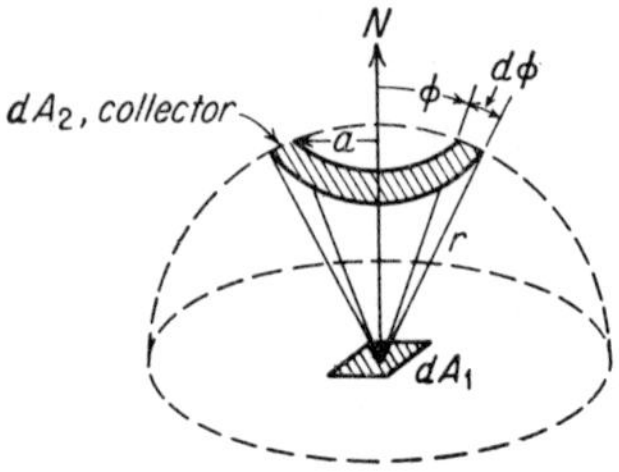

FIG. 19-13. Determination of the relationship between the normal intensity and the total energy emitted by a plane black surface.

$$W_b = \sigma T_1{}^4 = \int_0^{2\pi} I \, d\omega$$

$$= \int_{A_2} I_n \cos \phi \, \frac{dA_2}{r^2} \quad (19\text{-}62a)$$

Note that
$$dA_2 = 2\pi a (r \, d\phi)$$
$$= 2\pi r \sin \phi (r \, d\phi) = 2\pi r^2 \sin \phi \, d\phi$$

Substituting for $dA_2$ in Eq. (19-62a),

$$W_b = \sigma T_1{}^4$$

$$= \int_0^{\pi/2} I_n \cos \phi \, 2\pi \sin \phi \, d\phi = \pi I_n \quad (19\text{-}62b)$$

Thus the normal intensity of radiation from a plane black surface is

$$I_n = \frac{W_b}{\pi} = \frac{\sigma}{\pi} T_1{}^4 \quad \left[ \frac{\text{Btu}}{(\text{hr})(\text{ft}^2)(\text{solid angle})} \right] \quad (19\text{-}63)$$

Employing Eqs. (19-61) and (19-63), a useful relationship for evaluating the amount of radiation leaving $dA_1$ and striking $dA_2$ is obtained:

$$dA_1 \, W_{1 \rightarrow 2} = \frac{\sigma T_1{}^4 \cos \phi_1 \cos \phi_2 \, dA_1 \, dA_2}{\pi r^2} \quad (19\text{-}64)$$

Integration of Eq. (19-64) over finite areas $A_1$ and $A_2$ permits evaluation of the radiation interchange between plane surfaces bearing various geometrical relationships to each other. Specific means of applying this technique to certain configurations will be considered in Art. 19-17.

**19-13. The Concept of the Gray Body.** The concept of the black body gives rise to an ideal absorber or emitter, but the materials of engineering exhibit only a fraction of this behavior. This fraction is called the *emissivity*, which is defined as the *ratio of the total emissive power of a material (surface) to that of a black body at the same temperature.*

$$\epsilon \equiv \frac{W}{W_b} \bigg]_T \quad (19\text{-}65)$$

A gray body can now be defined.

**A gray body has a constant emissivity less than unity at all temperatures and wavelengths.**

The radiation from a gray body can be shown in Fig. 19-11 by constructing a curve with ordinates a constant fraction of those for the black body. Although the gray body is an ideal concept since the emissivities of many materials vary markedly with temperature and wavelength, it is still a convenient assumption that may be closely true for many engineering problems.

The departure of *real materials* from the gray ideal is shown by variations in the emissivity with temperature and wavelength. Thus the monochromatic emissivity is defined as the ratio of the monochromatic emissive power at wavelength $\lambda$ and temperature $T$ to that of a black body at the same wavelength and temperature:

$$\epsilon_{\lambda T} \equiv \frac{W_\lambda}{W_{b\lambda}}\bigg]_T \qquad (19\text{-}66)$$

Emissivity is frequently based upon the total energy radiated in all directions into or through an imaginary hemisphere above the radiating surface, and such emissivity is denoted as *hemispherical emissivity*. In some instances, the ratio is measured only in a direction normal to the radiating surface, and this results in a directional emissivity called *normal emissivity*. For a black body or for any other surface having emissions specified by Lambert's cosine law, the hemispherical and normal emissivity values are identical, as is nearly so for real nonmetallic materials which deviate only slightly from the cosine law. The intensity of radiation from polished metallic surfaces does not obey the cosine law, and hemispherical emissivities of these substances may be as much as 20 per cent greater than the normal emissivities. See Eckert and Drake[8] and Jakob[5] for additional information.

**19-14. Kirchhoff's Law.** Consider a black body placed in an evacuated enclosure which is at a definite temperature. When thermal equilibrium is reached, the body is radiating exactly the same amount of energy as it is absorbing. Since the emissivity and absorptivity of the black body are unity, obviously

$$\epsilon_b = \alpha_b = 1.0$$

Now let a gray body of emissivity 0.8 be placed in the enclosure. At equilibrium the emitted radiation must again exactly equal the absorbed radiation. But since the gray body is emitting exactly 0.8 of the amount emitted by the black body, it must also be absorbing exactly 0.8 of the amount absorbed by the black body; that is,

$$\epsilon_g = \alpha_g \qquad (19\text{-}67a)$$

Similar procedure and logic can be used for monochromatic emissivity and absorptivity when the enclosure is premised to be emitting only monochromatic radiation:

$$\epsilon_{\lambda_1 T_1} = \alpha_{\lambda_1 T_1} \qquad (19\text{-}67b)$$

Here the subscripts $T_1$ and $\lambda_1$ emphasize that body and enclosure are at the same temperature and that the same wavelength of radiation is being emitted and absorbed.   Such a restriction is necessary in considering real materials, because when the enclosure and real body are not at the same temperature, the equality need not exist.   In other words, $\epsilon_{\lambda_1 T_1}$ need not be equal to $\alpha_{\lambda_1 T_2}$ (or to $\alpha_{\lambda_2 T_1}$).   Note further that the emissivity of a body depends solely upon the body's characteristics, such as kind of material, surface condition, and temperature.   But absorptivity, in addition to being a function of material, surface condition, and temperature, is also dependent upon the wavelength distribution of the radiation incident upon the surface.   *Kirchhoff's law* can now be formally stated:

**The emissivity and absorptivity of a real surface are equal for radiations with identical temperatures and wavelengths.†**

**19-15. The Emissivities and Absorptivities of Real Materials.**   The emissivity is quite strongly a surface characteristic, since the surface controls the reflectivity, the complement to absorptivity for opaque materials, and $\epsilon_{\lambda_1 T_1} = \alpha_{\lambda_1 T_1}$.   Values of the emissivities for various materials are shown in Table 19-2.   Note that rough surfaces have larger emissivities (and therefore larger absorptivities) than smooth surfaces, while oxidizing a surface greatly increases the emissivity.   In general, emissivity increases with surface temperature.

Caution must be used in interpreting the values listed in Table 19-2.   Thus white paper at room temperature has high emissivity (0.93) and therefore high absorptivity (0.93) for radiation of the same (room) temperature.   This is proved by Kirchhoff's law.   Inspection of Fig. 19-11 shows that such low-temperature radiation is primarily concentrated in the long wavelengths.   Now when the white paper, at room temperature with emissivity of 0.93, is exposed to the rays of the sun, its absorptivity will be found to be very small because the high-temperature radiation is concentrated in short wavelengths and the white color is a good reflector.   An extreme example of this was demonstrated in the atomic-bomb explosions.   Here people with black clothing were severely burned while those with white garments were shielded; the high-temperature radiation was absorbed in the one case and reflected in the other.   Thus white clothing is preferred in tropical climates.   But if the ordinary house radiator is to be painted, the choice of color is not critical because the emissivity for low-temperature radiation has little to do with the visible wavelengths (Fig. 19-11), being approximately the same for all colors (Table 19-2).

Table 19-3 illustrates the variation of absorptivity of certain surfaces as the wavelength distribution of the incident radiation is changed.   Here

† For a black or gray body, Kirchhoff's law may be paraphrased as:   The emissivity and absorptivity of a black or gray surface are equal for all radiations at all temperatures and wavelengths.

TABLE 19-2. NORMAL EMISSIVITIES OF VARIOUS SURFACES†

| Surface | °F | $\epsilon$ |
|---|---|---|
| Aluminum: | | |
| Highly polished plate............ | 440–1070 | 0.039–0.057 |
| Rough plate.................... | 78 | 0.055 |
| Oxidized plate................. | 390 | 0.11 |
| Roofing....................... | 100 | 0.216 |
| Iron and steel: | | |
| Polished iron.................. | 800–1800 | 0.144–0.377 |
| Oxidized iron.................. | 212 | 0.736 |
| Oxidized steel................. | 390–1100 | 0.79 |
| Rough, oxidized steel plate........ | 100–700 | 0.94–0.97 |
| Molten....................... | 2370–3270 | 0.28 |
| Asbestos paper................. | 100–700 | 0.93–0.945 |
| Red rough brick................ | 70 | 0.93 |
| Paints, lacquers, varnishes: | | |
| White enamel on rough iron....... | 73 | 0.906 |
| Black lacquer on iron........... | 76 | 0.875 |
| Oil paints.................... | 212 | 0.92–0.96 |
| Aluminum paint, 26% Al......... | 212 | 0.3 |
| Paper, thin................... | 66 | 0.93 |
| Plaster, rough lime............. | 50–190 | 0.91 |
| Roofing paper................. | 69 | 0.91 |
| Water....................... | 32–212 | 0.95–0.963 |

† From J. H. Perry (ed.), "Chemical Engineers' Handbook," 3d ed., McGraw-Hill Book Company, Inc., New York, 1950. The table was contributed by H. C. Hottel with the following comments:

1. Normal and hemispherical emissivities can be considered equal except for well-polished metal surfaces where the latter is 15 to 20 per cent greater than the normal value.

2. When two temperatures and two emissivities are given, linear interpolation is permissible for intermediate values.

black-body radiation emitted by a source strikes the substance. The source temperature is varied, as shown in the table, thus changing the wavelength distribution of the radiation incident upon the substance (Fig. 19-11). The substance is maintained at a constant temperature of 70°F.

The difficulties of evaluating the absorbing and emitting power of surfaces by visual examination may be illustrated by considering the radiating characteristics of snow and glass. For the visible wavelengths, snow is a very efficient reflector; however, its total emissivity at 32°F is 0.98. Glass is obviously a very good transmitter of radiation in the visible range of the spectrum; yet its emissivity at 72°F is of the order of 0.9.

TABLE 19-3. VARIATION OF ABSORPTIVITY WITH TYPE OF INCIDENT RADIATION†

| Substance at 70°F | Source temperature | | |
|---|---|---|---|
| | 70°F (530°R) | 5000°R | 10,000°R (approx. solar radiation) |
| Slate composition roofing: | | | |
| $\epsilon$............ | 0.92 | 0.92 | 0.92 |
| $\alpha$............ | 0.92 | 0.90 | 0.90 |
| Concrete: | | | |
| $\epsilon$............ | 0.88 | 0.88 | 0.88 |
| $\alpha$............ | 0.88 | 0.67 | 0.61 |
| Red brick: | | | |
| $\epsilon$............ | 0.93 | 0.93 | 0.93 |
| $\alpha$............ | 0.93 | 0.41 | 0.54 |
| Flat black paint: | | | |
| $\epsilon$............ | 0.90 | 0.90 | 0.90 |
| $\alpha$............ | 0.90 | .... | 0.97–0.99 |
| Flat white paint: | | | |
| $\epsilon$............ | 0.88 | 0.88 | 0.88 |
| $\alpha$............ | 0.88 | .... | 0.12–0.26 |
| Aluminum paint: | | | |
| $\epsilon$............ | 0.43 | 0.43 | 0.43 |
| $\alpha$............ | 0.43 | 0.22 | 0.20 |
| Polished aluminum: | | | |
| $\epsilon$............ | 0.04 | 0.04 | 0.04 |
| $\alpha$............ | 0.04 | 0.23 | 0.28 |

† Mostly from L. S. Marks (ed.), "Mechanical Engineers' Handbook," 6th ed., McGraw-Hill Book Company, Inc., New York, 1958. Table contributed by H. C. Hottel.

**19-16. Heat Transfer by Radiation between Parallel Planes.** Consider, first, two parallel black planes at different temperatures and of equally large area $A$, so that end effects can be neglected. The radiant energy emitted by the hot $(T_1)$ plane will equal

$$W_{1 \to 2} = \sigma T_1{}^4$$

while the radiant energy received from the cold $(T_2)$ plane will equal

$$W_{2 \to 1} = \sigma T_2{}^4$$

The net heat transferred from the hot plane is†

$$\frac{\dot{Q}}{A_1} = W_{1 \to 2} - W_{2 \to 1} = \sigma(T_1{}^4 - T_2{}^4) \qquad (19\text{-}68)$$

† The concept that the net heat transfer is the difference of radiation streams moving with and opposite to the direction of decreasing temperature is consistent with the Prevost law (Art. 19-9).

Now let the hotter plane be gray and the colder plane black, and evaluate the net heat transfer. This can be done by summing up the net heat exchange *for one plane* (the picture is clearer if radiation is considered to be corpuscular in nature). Thus the first step is emission from the gray plane, and by definition of the emissivity,

$$W_{1\to2} = \epsilon_1 \sigma T_1{}^4$$

This packet of energy strikes the black plane, and none of the energy is reflected. Meanwhile, the black plane is emitting packets that are partially absorbed by the gray plane:

$$W_{2\to1} = \alpha_1 \sigma T_2{}^4$$

(The part reflected will be absorbed by the black plane.)   Adding these two equations together (and since $\epsilon = \alpha$),

$$\dot{Q}_{12} = A_1(W_{1\to2} - W_{2\to1}) = \epsilon_1 A_1 \sigma (T_1{}^4 - T_2{}^4) \qquad (19\text{-}69)$$

Although Eq. (19-69) was derived for parallel planes, it may be used for other cases in which similar interpretations can be placed on the heat exchange. Thus a small body in a large enclosure will radiate energy to the walls of the enclosure, and *only a small fraction of the reflected energy will ever again encounter the small body*. But this condition, that no reflected energy will reach the gray body, is exactly the condition imposed in the derivation. The same remarks would apply to the case of a hot pipe in a large room or a thermocouple in a large duct. Thus for a small body $A_1$ of emissivity $\epsilon_1$ in a large enclosure $A_2$, Eq. (19-69) applies.

Consider next the case of two large parallel gray planes. Here the problem is complicated by the infinite number of reflections that will occur at each surface, but when the net energy is summed, it will be found to equal

$$\dot{Q}_{12} = \frac{\sigma}{1/\epsilon_1 + 1/\epsilon_2 - 1} A_1(T_1{}^4 - T_2{}^4) \qquad (19\text{-}70)$$

**19-17. The General Problems of Heat Transfer by Radiation.**   In the calculation of heat transfer by radiation, it is usually necessary to approximate real-body behavior by the gray-body idealization. The assumption that emissivity is always equal to absorptivity (strictly valid only for black or gray bodies) is frequently required if the problem is to be solved. Even as simple a situation as the one resulting in Eq. (19-70) becomes quite complex if real-body behavior is considered and the resulting convenience of replacing absorptivity with emissivity is lost. A second difficulty in considering real-body behavior is the lack of sufficient data. To properly account for real-body behavior, extensive tabulations similar to the data in Table 19-3 would be required for all types of incident radiation on engineering substances.

The usual problem of heat transfer by radiation is further complicated not only because the surfaces are nonblack but also because the configuration of the areas involved is not simple and reflecting surfaces may be present to augment the direct exchange. However, a number of solutions have been made and expressed in terms of certain parameters by Hottel.[1] These parameters are $F$, *the geometric configuration factor;* $\bar{F}$, *the geometric configuration-reflection factor; and* $\mathfrak{F}$, *the geometric configuration-reflection-emissivity factor.*

The geometric factor $F$ depends only upon the physical configuration of the radiator and receiver areas and is defined by

$$F_{12} = \frac{\text{direct radiation from surface 1 incident upon surface 2}}{\text{total radiation from surface 1}} \quad (19\text{-}71)$$

As the name implies, factor $F$ is the fraction of the total radiation leaving the radiating surface and directly falling upon a second surface. In all real radiation problems, the radiating surface will be of finite area; therefore it will be enclosed by many surfaces, real or imaginary (*an enclosure*). The total unit radiation leaving the radiating surface is made up of the fractional parts sent to each confining surface of the enclosure:

$$F_{11} + F_{12} + F_{13} + \cdots = 1 \quad (19\text{-}72)$$

In cases where body 1 cannot "see itself," $F_{11}$ is zero.

The configuration factor $F_{12}$ can be used to determine the net heat exchange between two black bodies since, in an enclosure of black surfaces, no reflection is possible. Consider the radiation leaving one surface and falling on a second surface:

$$A_1 W_{1\to2} = A_1 F_{12} \sigma T_1^4$$

Since the second surface is also radiating energy,

$$A_2 W_{2\to1} = A_2 F_{21} \sigma T_2^4$$

But when $T_1 = T_2$, the two exchanges must be equal, and since $F$ is not a function of temperature but only of geometric configuration,

$$A_1 F_{12} = A_2 F_{21} \quad (19\text{-}73)$$

Thus the net heat transfer between 1 and 2 is

$$\dot{Q}_{12} = A_1 F_{12} \sigma (T_1^4 - T_2^4) = A_2 F_{21} \sigma (T_1^4 - T_2^4) \quad (19\text{-}74)$$

where only $F_{12}$ or $F_{21}$ need be evaluated, depending upon which is more easily determined. Since Eq. (19-74) may be made general for the enclosure of black surfaces, the net heat transferred from the radiator is

$$\dot{Q}_{1\,net} = A_1 F_{12} \sigma (T_1{}^4 - T_2{}^4) + A_1 F_{13} \sigma (T_1{}^4 - T_3{}^4) + \cdots \quad (19\text{-}75)$$

**Example 6.** A small, 2-in.-outside-diameter sphere with a surface temperature of 540°F is located at the geometric center of a large, hollow, 10-in.-inside-diameter sphere with an inner surface temperature of 40°F. Assuming both bodies approach black-body behavior, determine the fraction of the energy emitted by the inner surface of the large sphere which is absorbed by the outer surface of the smaller sphere.

*Solution.* Denote the smaller sphere as 1 and the larger sphere as 2. As all the radiation incident upon the small sphere is absorbed, the fraction desired is $F_{21}$. All radiation emitted by the small sphere is incident upon (and absorbed by) the inner surface of the larger sphere. Thus from Eq. (19-71), $F_{12} = 1$. But from Eq. (19-73),

$$F_{21} = \frac{A_1 F_{12}}{A_2} = \frac{(4\pi)(1)^2(1)}{(4\pi)(5)^2} = 0.04 \qquad Ans.$$

Thus only 4 per cent of the radiation emitted by surface 2 is incident upon the small sphere. From Eq. (19-72), it can be seen that

$$F_{22} = 1 - F_{21} = 1 - 0.04 = 0.96$$

The remaining 96 per cent of the energy emitted by the larger sphere is absorbed by the inner surface of the larger sphere.

Equation (19-64) may be applied to determine the geometric configuration factor $F$ for two surfaces of finite area (say $A_1$ and $A_2$) of known geometrical relationship. Indicating the integration of Eq. (19-64) and using the definition of $F$ [Eq. (19-71)],

$$A_1 W_{1\to 2} = A_1 F_{12} \sigma T_1{}^4 = \int_{A_1} \int_{A_2} \frac{\sigma T_1{}^4 \cos \phi_1 \cos \phi_2 \, dA_1 \, dA_2}{\pi r^2}$$

Thus $$F_{12} = \frac{1}{\pi A_1} \int_{A_1} \int_{A_2} \frac{\cos \phi_1 \cos \phi_2 \, dA_1 \, dA_2}{r^2} \quad (19\text{-}76)$$

As premised earlier, it can be seen that $F$ is a function of geometric configuration only. Values for $F$ obtained from expressions similar to Eq. (19-76) are shown in Figs. 19-14 and 19-15.[†]

**Example 7.** Calculate the net exchange of radiation between two parallel rectangles, 2 by 4 ft, and 3 ft apart, if the temperatures are 1800 and 70°F and both surfaces behave as black bodies.

*Solution.* The fraction of the total radiation leaving the hotter and falling upon the colder rectangle is found from Fig. 19-14, curve 3:

$$\text{Ratio} = \frac{\text{shorter side}}{\text{distance apart}} = \frac{2}{3} \qquad F = 0.165$$

[†] Factors for other geometric arrangements (and by H. C. Hottel) can be found in L. S. Marks (ed.), "Mechanical Engineers' Handbook," 6th ed., McGraw-Hill Book Company, Inc., New York, 1958. Also see D. C. Hamilton and W. R. Morgan, Radiant-interchange Configuration Factors, *NACA Tech. Note* 2836 (December, 1952).

Thus only 16.5 per cent of the radiation leaving the radiator falls upon the sink. By Eq. (19-74),

$$\dot{Q}_{12} = A_1 F_{12}\sigma(T_1{}^4 - T_2{}^4)$$
$$= 8(0.165)0.1714 \left[ \left( \frac{2,260}{100} \right)^4 - \left( \frac{530}{100} \right)^4 \right]$$
$$= 59,400 \text{ Btu/hr} \quad Ans.$$

But in most problems there will be confining walls between the radiating (source) and the receiving (sink) surfaces that will reflect and thus

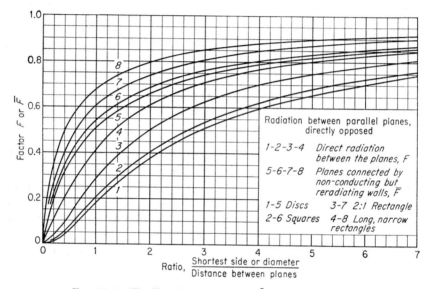

FIG. 19-14. The Hottel factors $F$ and $\bar{F}$ for parallel planes.

increase the incident radiation to the sink. A factor that includes both *direct radiation* ($F$) and *reflected radiation* is defined by $\bar{F}$.

$$\bar{F}_{12} = \frac{\text{direct and reflected radiation from surface 1 incident upon surface 2}}{\text{total radiation from surface 1}}$$

$$(19\text{-}77)$$

It is assumed that the reflecting (*refractory*) walls reflect all the incident radiation. For the same reasons discussed for $F$,

$$\bar{F}_{11} + \bar{F}_{12} + \bar{F}_{13} + \cdots = 1 \qquad (19\text{-}78)$$
$$A_1\bar{F}_{12} = A_2\bar{F}_{21} \qquad (19\text{-}79)$$

The net heat transferred from the black radiating surface $A_1$ to the black surfaces $A_2, A_3, \ldots$ is

$$\dot{Q}_{1\,net} = A_1\bar{F}_{12}\sigma(T_1{}^4 - T_2{}^4) + A_1\bar{F}_{13}\sigma(T_1{}^4 - T_3{}^4) + \cdots \quad (19\text{-}80)$$

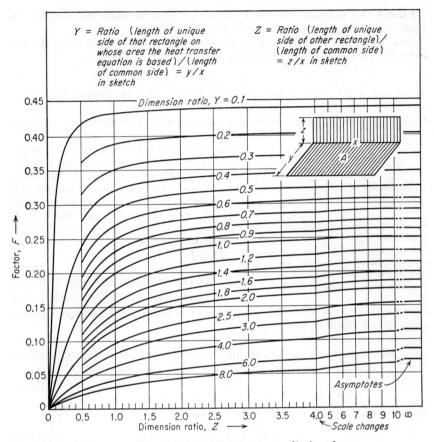

FIG. 19-15. The Hottel factor $F$ for perpendicular planes.

Values for $\bar{F}$ for some cases are shown in Fig. 19-14, where the reradiating (refractory) walls connecting the radiating planes are straight and reflect in a *diffuse manner* (Fig. 19-16). The reflection from most engineering materials approximates the diffuse type, although reflection from polished metal surfaces is more nearly *specular* (*mirrorlike*).†

† In engineering problems involving sources, sinks, and refractory surfaces, the assumption of perfect reflection ($\rho = 1$) by the refractory surfaces appears invalid. Frequently the refractory is exactly that (such as a refractory wall in a furnace), and from Table 19-2, the emissivity for such substances is of the order of 0.9. Applying Kirchhoff's law and Eq. (19-53), $\rho$ must be about 0.1, quite different from the value of unity assumed in using the $\bar{F}$ concept. A viewpoint that permits the use of the perfect-reflector idealization is that the refractory walls are well insulated compared with the source and sink, and so the heat conducted through them is negligible compared with the total heat transfer between source and sink. Thus, although the refractory walls may absorb much of a beam of incident radiation, they subsequently

**Example 8.** Calculate the net exchange of radiation when the rectangles in Example 7 are connected by refractory walls.

*Solution.* The fraction of direct and reflected radiation incident upon the cold rectangle is found from Fig. 19-14, curve 7:

$$\bar{F} = 0.51$$

$$\dot{Q}_{12} = A_1\bar{F}_{12}\sigma(T_1{}^4 - T_2{}^4) = 8(0.51)0.1714(22.6^4 - 5.3^4) = 184{,}000 \text{ Btu/hr} \qquad Ans.$$

Of the total emission from the hotter rectangle, 51 per cent went directly and indirectly to the colder rectangle: 16.5 per cent $(F_{12})$ went directly and 34.5 per cent was initially incident upon the refractory and was ultimately diffusely reflected to the colder rectangle. The remaining 49 per cent [Eq. (19-78)] must have been diffusely reflected back to the hotter rectangle. What would have been the energy accounting had the refractory reflected specularly rather than diffusely?

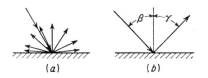

FIG. 19-16. Diffuse and specular reflection. (a) Diffuse reflection (distribution of reflected radiation ideally obeys Lambert's cosine law); (b) specular reflection (angle of incidence $\beta$ equals angle of reflection $\gamma$).

To better approximate the behavior of radiation enclosures consisting of sources, sinks, and refractories, the third Hottel factor, $\mathfrak{F}$, assumes that the sources and sinks are gray surfaces.

$$\mathfrak{F}_{12} = \frac{\text{direct and reflected radiation from surface 1 absorbed by surface 2}}{\text{total radiation from surface 1 if it emitted as a black body}}$$

$$(19\text{-}81)$$

By reasoning similar to that for the $F$ factor,

$$\mathfrak{F}_{11} + \mathfrak{F}_{12} + \mathfrak{F}_{13} + \cdots = \epsilon_1 \qquad (19\text{-}82a)$$

$$A_1\mathfrak{F}_{12} = A_2\mathfrak{F}_{21} \qquad (19\text{-}82b)$$

$$\dot{Q}_{1\text{ net}} = A_1\mathfrak{F}_{12}\sigma(T_1{}^4 - T_2{}^4) + A_1\mathfrak{F}_{13}\sigma(T_1{}^4 - T_3{}^4) + \cdots \qquad (19\text{-}82c)$$

For some enclosures consisting of one gray source, one gray sink, and any number of refractory surfaces, Hottel found that

$$\mathfrak{F}_{12} = \frac{1}{\dfrac{1}{\bar{F}_{12} \text{ or } F_{12}} + \left(\dfrac{1}{\epsilon_1} - 1\right) + \dfrac{A_1}{A_2}\left(\dfrac{1}{\epsilon_2} - 1\right)} \qquad (19\text{-}83)$$

In applying Eq. (19-83), use $\bar{F}_{12}$ if refractories are present and $F_{12}$ if refractories are not involved. In evaluating $\bar{F}_{12}$ (or $F_{12}$), the source and sink may be assumed black.

---

emit most of the beam, which, from the over-all picture (and for easy visualization), is as if all of the incident beam had been perfectly reflected. This viewpoint is what is meant by the term "nonconducting but reradiating walls" in Fig. 19-14.

**Example 9.**  Determine the net heat transfer for the data of Examples 7 and 8 if the absorptivity of the cold rectangle is 0.60 while the emissivity of the hot rectangle is 0.9.  (Assume gray-body conditions.)

*Solution.*  With these data,

$$\mathcal{F}_{12} = \frac{1}{\dfrac{1}{0.51} + \left(\dfrac{1}{0.9} - 1\right) + \dfrac{1}{1}\left(\dfrac{1}{0.6} - 1\right)} = 0.378$$

$$\dot{Q}_{12} = A_1\mathcal{F}_{12}\sigma(T_1{}^4 - T_2{}^4) = 8(0.378)0.1714(22.6^4 - 5.3^4)$$
$$= 136{,}000 \text{ Btu/hr} \qquad Ans.$$

Although Eq. (19-83) does not apply to all situations involving one source and one sink (both gray) and any number of refractory surfaces (see Art. 20-6), this equation may be used as a convenient starting point for obtaining specific solutions for simple cases.  To illustrate, consider the case of an object radiating energy to a surrounding enclosure. Here $F_{12} = 1.0$ because all the energy leaving the radiator is intercepted by the enclosure.  Substituting this value in Eq. (19-83),

$$\mathcal{F}_{12} = \frac{1}{\dfrac{1}{\epsilon_1} + \dfrac{A_1}{A_2}\left(\dfrac{1}{\epsilon_2} - 1\right)} \qquad (19\text{-}83a)$$

The condition that $F_{12} = 1.0$ is also valid for two large parallel **gray** planes, and here $A_1 = A_2$; therefore,

$$\mathcal{F}_{12} = \frac{1}{\dfrac{1}{\epsilon_1} + \dfrac{1}{\epsilon_2} - 1} \qquad (19\text{-}83b)$$

which is directly related to Eq. (19-70).  When the enclosure $A_2$ is large compared with $A_1$, Eq. (19-83a) reduces to

$$\mathcal{F}_{12} = \epsilon_1 \qquad (19\text{-}83c)$$

which can be compared with Eq. (19-69) and the comments of Art. 19-16.

**19-18. Radiation from Gases and Flames.**  Most gases are perfect transmitters of radiation (*diathermous*) for the wavelengths of visible light.  Because of this observation, it was long thought that gases did not absorb thermal radiation of any wavelengths.  Studies in furnaces subsequently showed that certain gases do absorb (and thus by Kirchhoff's law, emit) significant amounts of radiation in the infrared region. Among these absorbing gases are certain combustion products produced in burning fuels, such as carbon dioxide, water vapor, carbon monoxide, the hydrocarbon gases, sulfur dioxide, and other heteropolar gases such as the alcohols and ammonia.  No significant absorption of radiation in the wavelengths of thermal radiation has been observed in gases with symmetrical molecules, such as hydrogen, oxygen, and nitrogen. Thus, in considering the radiant interchange between surfaces separated by dry air, the assumption that air is a perfect transmitter of thermal radiation is justified. However, in air at high temperatures, such as behind the shock wave of the nose cone described in Art. 19-1, dissociation and ionization of the molecules produce species which emit radiation. Present indications are that the radiation from the "air" is significant but represents only a small fraction of the heat transfer by convection for return through the earth's atmosphere from an initial entry speed of satellite velocity or less (about 26,000 ft/sec). However, gas radiation will probably be the predominant mode of heat transfer for a direct return through the earth's atmosphere from an interplanetary flight with an atmospheric entry speed of escape velocity (36,000 ft/sec) or greater.

Gases are selective emitters; that is, emission (and absorption) of energy occurs only in certain discrete wavelengths of the radiation spectrum. The absorption in a layer of gas exposed to radiation streaming normal to it may be pictured as follows: As radiation of certain wavelengths passes through the layer, it is partially absorbed and partially transmitted. Radiation of other wavelengths is transmitted without absorption. The amount of absorption in the gas layer depends upon the type of gas, the number of molecules of gas present (a function of the gas pressure and temperature), and the thickness of the layer. Thus the absorption and emission is a volume phenomenon rather than a surface occurrence as in solids.

Luminous flames, such as those obtained in the burning of powdered coal and atomized oil sprays, emit significant amounts of radiation. It is premised that the emission is primarily from glowing particles of soot and dust.

In a modern, coal-fired steam generator, the steam-generating tubes and superheater receive energy by radiation from the luminous flame and by both radiation and convection from the hot combustion gases. Normally, radiation accounts for a large fraction of the total heat transfer.

The calculation of heat transfer by radiation for surfaces separated by absorbing gases or luminous flames has proved difficult. Hottel[1] discusses some means of solving the problem, and Oppenheim has shown that the network method (Art. 20-6) can be applied to simplify the analysis.

**Example 10.** For the situation described in Example 4, determine the percentage change in total heat transfer from the pipe surface.

*Solution.* In addition to convection, a transfer of heat by radiation takes place between the pipe and surrounding surfaces. Considering the pipe (1) to be a small body in a large enclosure (2), Eq. (19-69) can be used to calculate the radiation heat transfer, with the emissivity of the pipe surface, 0.79, selected from Table 19-2.

$$\frac{\dot{Q}_{12}}{A_1} = (0.79)(0.1714)\left[\left(\frac{660}{100}\right)^4 - \left(\frac{540}{100}\right)^4\right] = 142 \text{ Btu/(hr)(ft}^2) \qquad \text{for 200°F surface}$$

$$\frac{\dot{Q}_{12}}{A_1} = (0.79)(0.1714)(8.60^4 - 5.40^4) = 630 \text{ Btu/(hr)(ft}^2) \qquad \text{for 400°F surface}$$

Thus the total heat transfer from the surface is $129 + 142 = 271$ Btu/(hr)(ft²) at 200°F and $419 + 630 = 1049$ Btu/(hr)(ft²) at 400°F.

The percentage change in total heat transfer in increasing the pipe surface temperature is

$$\frac{(100)(1049 - 271)}{271} = 287 \text{ per cent} \qquad Ans.$$

Some important generalizations may be made on the basis of this solution.

1. The radiation heat transfer accounts for more than half the total heat transfer in both cases. This is often the case in situations where radiation and natural convection comprise the total heat transfer from a surface. In forced convection with coefficients several times those for natural convection, radiation may not contribute appreciably to the total heat transfer.

2. The radiation contribution represents a greater portion of the total heat transfer as the temperature of the surface is increased. Because of the fourth-power temperature difference in the radiation equation, this is usually the case unless very large decreases occur in the emissivity of the hot emitter as its temperature increases (and this is unlikely; see Table 19-2).

**Example 11.** It is planned to measure the temperature of a hot gas (diathermous) flowing through a large duct with a small thermocouple located centrally in the duct. Data for typical conditions are:

Duct diameter = 24 in.
Wall temperature of duct $(t_w) = 613°F$
Temperature indicated by thermocouple probe $(t_p) = 797°F$
Emissivity of thermocouple probe $(\epsilon_p) = 0.8$
Convection coefficient between gas and thermocouple = 7 Btu/(hr)(ft²)(°F)

Is the temperature indicated by the thermocouple probe equal to the temperature of the hot gas $t_g$?

*Solution.* At steady state, the thermocouple loses as much energy as it receives, as expressed by

$$\dot{Q}_{\text{convection from gas}} = \dot{Q}_{\text{radiation to walls}} + \dot{Q}_{\text{conduction}}$$

The conduction loss may be minimized by using small thermocouple lead wires and will be neglected in this approximate analysis. By Eqs. (19-2) and (19-69),

$$h_c A_p (t_g - t_p) = \epsilon_p A_p \sigma (T_p{}^4 - T_w{}^4)$$

From the data given,

$$t_g - t_p = \frac{(0.8)(0.1714)(12.57^4 - 10.73^4)}{7} = 226°F \qquad Ans.$$

The error in thermocouple probe reading is 226°F, and the true gas temperature is approximately 1023°F.† The value of the calculated error should be regarded as only approximate, for the convection coefficient and emissivity are not known precisely. Thus when a large error is predicted, modifications should be made in the temperature-measuring arrangement. The energy-balance equation serves to point out modifica-

---

† For a convenient means of solving this type of problem, see W. M. Rohsenow, A Graphical Determination of Unshielded Thermocouple Thermal Correction, *Trans. ASME,* **68:** 195–198 (April, 1946). For shielded thermocouples, consult W. M. Rohsenow and J. P. Hunsaker, Determination of the Thermal Correction for a Single-shielded Thermocouple, *Trans. ASME,* **69:** 699–704 (August, 1947).

tions that lead to better accuracy in gas-temperature measurement. Some or all of the following improvements might be made:

1. Increase $h_c$, perhaps by increasing the gas velocity in the vicinity of the thermo-couple probe.

2. Decrease the emissivity of the probe.

3. Insulate the duct wall such that the wall temperature more nearly approaches the gas temperature.

4. Provide a shield or shields around the probe so that it does not radiate directly to the cold wall. Problem 45 shows the beneficial effect of one shield.

**19-19. Heat Transfer between Fluids.** Consider the transfer of heat between two fluids separated by a wall, as shown in Fig. 19-17; the

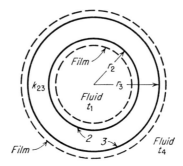

Fig. 19-17. Heat transfer through a boundary layer, a wall, and a second boundary layer.

resistance to flow of heat is offered by a film (boundary layer) on the inner side of the pipe, by the metal of the pipe, and by a film on the outer side of the pipe. The resistance of this series path is equal to the sum of the individual resistances. As considered in Art. 19-5,

$$\Sigma R_t = \frac{1}{h_{12}A_2} + \frac{\ln{(r_3/r_2)}}{2\pi k_{23}L} + \frac{1}{h_{34}A_3} \tag{19-84a}$$

and the heat transfer equals

$$\dot{Q} = \frac{t_1 - t_4}{\Sigma R_t} \tag{19-84b}$$

**Example 12a.** Determine the heat transfer between two fluids separated by a copper condenser tube of $\frac{3}{4}$-in. outside diameter, 6-ft length, and 0.1-in. wall thickness if the outer (steam) temperature is 212°F and the inner (water) temperature is 60°F. Assume that the film coefficient is 280 Btu/(hr)(ft²)(°F) on the water side and 2000 Btu/(hr)(ft²)(°F) on the steam side, while the $k$ value for copper is 220 Btu/(hr)(ft)(°F).

*Solution.* The resistance of the path is found by Eq. (19-84a):

$$A_2 = \pi \frac{(0.55)(6)}{12} = 0.866 \text{ ft}^2$$

$$A_3 = \pi \frac{(0.75)(6)}{12} = 1.18 \text{ ft}^2$$

$$\Sigma R_t = \frac{1}{280(0.866)} + \frac{\ln (0.75/0.55)}{(2\pi)(220)(6)} + \frac{1}{2000(1.18)}$$

$$= \underset{\text{(water side)}}{0.004124} + \underset{\text{(tube)}}{0.000372} + \underset{\text{(steam side)}}{0.000424} = 0.00459$$

and

$$\dot{Q} = \frac{\Delta t}{\Sigma R_t} = \frac{152}{0.00459} = 33,125 \text{ Btu/hr} \qquad Ans.$$

The solution of Example 12a should be carefully examined because several important generalizations become apparent.

1. The resistance to heat flow arises mainly from the films and not from the metal in the wall.

2. The resistance of the condenser tube could have been neglected with a resulting error of less than 2 per cent.

3. The total resistance is most strongly controlled by the film with lowest coefficient. Thus little gain would be obtained were the steam-side coefficient to be increased, but a large gain would result from an increase in the water-side coefficient.

The fluids in industrial heat exchangers are often contaminated and form deposits on the exchanger surfaces, thus imposing another resistance to heat transfer. For certain situations, the effect of the deposit on heat transfer may be approximated by using the *fouling factors* listed in Table 19-4. A continuation of Example 12 illustrates the method of using these factors.

TABLE 19-4. FOULING FACTORS
(Standards of the Heat Exchange Institute)

| Situation | Fouling factor, $Btu/(hr)(ft^2)(°F)$ |
|---|---|
| Organic liquids; refrigerating fluids; brine; clean recirculating oil; machinery and transformer oils; city, well, or treated water under 125°F; sea water over 125°F | 1000 |
| Distilled water or sea water below 125°F | 2000 |
| River water under 125°F, velocity over 3 ft/sec | 500 |
| River water over 125°F, velocity under 3 ft/sec | 250 |
| Crude oil | 200 |

**Example 12b.** Use the data in Example 12a, but include the effect of the deposit formed on the inside of the tube if river water flows through the tube at a velocity of 5 ft/sec.

*Solution.* From Table 19-4, the fouling factor is 500 Btu/(hr)(ft²)(°F); thus the additional resistance imposed by the deposit is

$$R_{t_{\text{deposit}}} = \frac{1}{(500)(0.866)} = 0.002309$$

and the total resistance is now

$$\Sigma R_t = 0.00459 + 0.002309 = 0.006899$$

and
$$\dot{Q} = \frac{152}{0.006899} = 22{,}000 \text{ Btu/hr} \qquad Ans.$$

**19-20. The Over-all Coefficient.** Without investigating the multiple films and conductivities that might be present in a heat exchanger, it can be premised that

$$\dot{Q} \equiv U_2 A_2 (t_4 - t_1) = U_3 A_3 (t_4 - t_1) \qquad (19\text{-}85)$$

$$\left[ \frac{\text{Btu}}{\text{hr}} \right] = \left[ \frac{\text{Btu}}{\text{hr ft}^2 \text{ °F}} \right] [\text{ft}^2][\text{°F}]$$

Equation (19-85) defines the *over-all coefficient of heat transfer U*, also named the *over-all conductance*. Inspection of this equation and of Fig. 19-17 shows that such an over-all coefficient can be based upon either

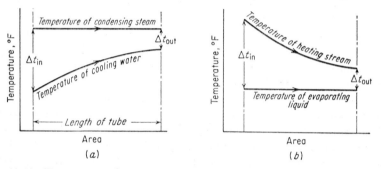

Fig. 19-18. Temperature differences in condensation and evaporation. (a) Condensation; (b) evaporation.

the inner or outer area of the surface, and therefore the thermal resistance equals

$$\Sigma R_t = \frac{1}{A_2 U_2} = \frac{1}{A_3 U_3} \qquad (19\text{-}86)$$

**Example 13.** Determine the over-all coefficient of heat transfer, based on the outer diameter of the tube, for the data of Example 12.

*Solution.* Equation (19-86) and the data in Example 12 yield

$$\frac{1}{1.18 U_3} = 0.00459 \qquad \text{thus } U_3 = 185 \text{ Btu/(hr)(ft}^2)(\text{°F}) \qquad Ans.$$

**19-21. The Log Mean Temperature Difference.** It has been tacitly assumed that the temperature of each fluid was constant throughout the heat exchanger; actually, this condition is not encountered in practice. Consider steam condensing on the outside of a condenser tube with water flowing inside the tube as the coolant. The fluid temperature on the outside of the tube would be constant, as shown in Fig. 19-18a, but the coolant temperature would progressively, but not necessarily linearly, rise from the inlet to the outlet of the tube. The inverse condition exists when a hot fluid is used for evaporating another fluid. Here the temperature of the hot stream continuously decreases in passing from inlet to outlet, as illustrated in Fig. 19-18$b$.

In many instances, one fluid travels along one wall of the exchanger, while the second fluid travels in the reverse direction along the other side of the wall; this is an example of a *counterflow*, or *countercurrent*, heat exchanger (Fig. 19-19$a$). If both fluids travel in the same direction, then *parallel flow* results (Fig. 19-19$b$). Usually, counterflow heat exchangers can transfer more heat than similar, but parallel-flow, exchangers.

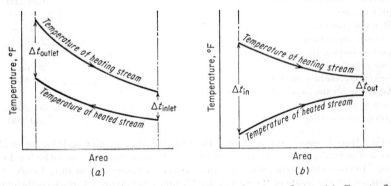

Fig. 19-19. Temperature differences in parallel and counterflow. (a) Counterflow; (b) parallel flow.

For all these simple cases, an expression can be derived[†] for the average or mean temperature difference between the two fluids in terms of the terminal temperature differences if it can be assumed that:

1. The coefficient $U$ is constant throughout the exchanger.
2. The heat capacities of the fluids are constant.
3. Condensation or evaporation does not occur in particular parts of the exchanger (but if these events occur throughout the exchanger, Fig. 19-18, the development is valid).
4. The exchanger is perfectly insulated externally, and steady-flow conditions exist.

When the heat capacities are constant, the change in temperature of either fluid is a linear function of the rate of heat addition or rejection per unit length:

$$\frac{d\dot{Q}}{dL} = \dot{m}_f c_p \frac{dt}{dL}$$

$$\frac{d\dot{Q}}{dt} = \dot{m}_f c_p \qquad \text{(a constant if } c_p \text{ is constant)} \qquad (a)$$

[†] Derivation suggested by that of McAdams.[1]

TABLE 19-5. REPRESENTATIVE VALUES OF THE OVER-ALL COEFFICIENT $U$†

| Situation | $U$, $Btu/(hr)(ft^2)(°F)$ |
|---|---|
| Brick wall, 8 in. thick (air to air) | 0.48 |
| Frame wall construction, face-brick veneer exterior, ½-in. gypsum-board sheathing, 3⅝-in. air space (between studding), ½-in. insulating-board interior (air to air) | 0.24 |
| Same with 3-in. thickness of insulation [$k$ equal to 0.022 Btu/(hr)(ft)(°F)] in air space (air to air) | 0.065 |
| Liquid to liquid: | |
| Free convection, water | 25–60 |
| Forced convection, water | 150–300 |
| Liquid to air: | |
| Free convection | 1–3 |
| Forced convection | 2–10 |
| Liquid to boiling liquid: | |
| Free convection, water | 20–60 |
| Forced convection, water | 50–150 |
| Condensing steam to liquid: | |
| Free convection | 50–200 |
| Forced convection | 150–800 |

† First three items of table from the "Heating, Ventilating, Air Conditioning Guide" published annually by the American Society of Heating and Air Conditioning Engineers. In these three items, still air with a surface coefficient of 1.46 Btu/(hr)(ft²)(°F) is assumed for the inside of the wall and a 15-mph wind giving a surface coefficient of 6.00 Btu/(hr)(ft²)(°F) is assumed for the outside of the wall.

Remainder of table from J. H. Perry (ed.), "Chemical Engineers' Handbook," 3d ed., McGraw-Hill Book Company, Inc., New York, 1950.

In Fig. 19-20$a$ the temperature of each fluid is plotted versus the quantity of heat transferred, that is, the heat-transfer rate integrated from the inlet of the exchanger ($L = 0$) to an intermediate position ($L$). For example, the integrated heat-transfer rate $\dot{Q}_{total}$ for the entire exchanger is

$$\dot{Q}_{total} = \int_{inlet}^{outlet} \frac{d\dot{Q}}{dL} dL$$

In accord with Eq. ($a$), a linear relationship is obtained in Fig. 19-20$a$. It follows that the temperature difference between fluids $\Delta t$ is also linear. The slope of $\Delta t$ in Fig. 19-20$b$ is seen to be

$$-\frac{d(\Delta t)}{(d\dot{Q}/dL) dL} = \frac{\Delta t_1 - \Delta t_2}{\dot{Q}_{total}}$$

But by definition of the conductance,

$$\left(\frac{d\dot{Q}}{dL}\right) dL = U \Delta t \frac{dA}{dL} dL$$

Eliminating $(d\dot{Q}/dL)\,dL$ from the above equations,

$$\int_{\Delta t_2}^{\Delta t_1} \frac{d(\Delta t)}{\Delta t} = \frac{\Delta t_1 - \Delta t_2}{\dot{Q}_{total}} \int_{inlet}^{outlet} U \frac{dA}{dL}\,dL$$

or, for constant $U$,

$$\dot{Q}_{total} = U A \frac{\Delta t_1 - \Delta t_2}{\ln\,(\Delta t_1/\Delta t_2)}$$

And, by definition of the over-all conductance,

$$\dot{Q}_{total} = U A\,\Delta t_m$$

Then the average temperature difference, the *logarithmic mean temperature difference*, is equal to

$$\Delta t_m = \frac{\Delta t_1 - \Delta t_2}{\ln\,(\Delta t_1/\Delta t_2)} \qquad (19\text{-}87a)$$

which is easier to remember in the form

$$\Delta t_m = \frac{\Delta t_{in} - \Delta t_{out}}{\ln\,(\Delta t_{in}/\Delta t_{out})} \qquad (19\text{-}87b)$$

Equation (19-87b) applies to countercurrent flow as well as to parallel flow and, also, to any exchanger wherein one of the fluid temperatures

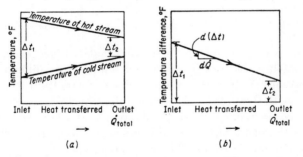

FIG. 19-20. (a) Temperatures and (b) temperature differences in a parallel-flow exchanger (when $c$ is constant).

remains constant (Fig. 19-18). When $\Delta t_{in}$ and $\Delta t_{out}$ in Eq. (19-87b) approach the same value as may occur in a counterflow exchanger, the arithmetic mean temperature difference may be more readily calculated (with less possibility of error):

$$\Delta t_a = \frac{\Delta t_{in} + \Delta t_{out}}{2} \qquad (19\text{-}88)$$

**19-22. Mean Temperature Differences for Complex Heat Exchangers.**[†] The flow in commercial heat exchangers does not usually follow a simple pattern. In Fig. 19-21 is shown a typical heat exchanger which consists of a *shell, baffles* to direct the flow of the shell-side fluid, and *tubes* to carry the second fluid. Note that the shell fluid makes *two shell passes* because of the presence of the longitudinal baffle. The tube fluid makes *four tube passes* because the flow is passed four times through the shell by the tube arrangement. This can be called a *two-shell-pass, four-tube-pass, reversed-current* exchanger. Here it should be noted that the flow is a combination of parallel, countercurrent, and *crossflow.* The latter term applies when one fluid flows perpendicularly to the tube arrangement, a condition achieved in Fig. 19-21 by the cross baffles.

The analysis of the mean temperature difference in real heat exchangers becomes quite complex, but theoretical analyses have been published,

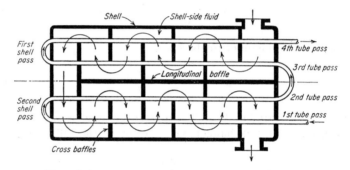

Fig. 19-21. Two-shell-pass, four-tube-pass, reversed-current heat exchanger.

in the form of correction factors for various types of heat exchangers, to be used to modify the logarithmic mean temperature difference. The same assumptions as those that were outlined in Art. 19-21 were made in the analyses, with the additional condition that the fluids were well mixed; that is, at any one section of a pass, the temperature of the fluid is uniform. Then the real temperature difference has been arbitrarily related to a $\Delta t_m$ computed as if counterflow were present:

$$\Delta t_m = Y \, \Delta t_{mcf} \qquad (19\text{-}89)$$

where $Y$ is obtained from Fig. 19-22. The parameters $X$ and $Z$ are based upon

$t_1''$ entering cold fluid $\qquad$ $t_1'$ entering hot fluid
$t_2''$ leaving cold fluid $\qquad$ $t_2'$ leaving hot fluid

[†] Kays and London[9] describe a method, the effectiveness–number of exchanger heat transfer units (NTU) approach, which, in some cases, simplifies the calculations for complex exchangers.

The steep gradient of the curves in Fig. 19-22 at low values of $Y$ may cause extreme errors in design. Since the theoretical development rests upon certain restricting assumptions, values of $Y$ less than 0.8 are also more susceptible to error and therefore are not recommended by the authors of Fig. 19-22.

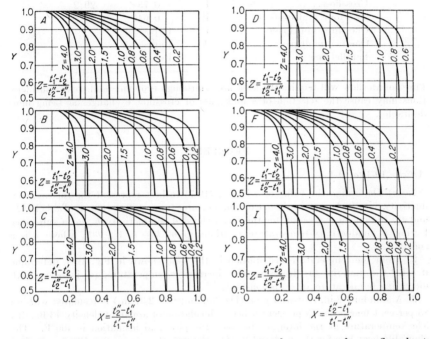

FIG. 19-22. Mean temperature difference in reversed-current and crossflow heat exchangers. ($A$) 1 shell pass and 2, 4, 6, etc., tube passes. ($B$) 2 shell passes and 4, 8, 12, etc., tube passes. ($C$) 3 shell passes and 6, 12, 18, etc., tube passes. ($D$) 4 shell passes and 8, 16, 24, etc., tube passes. ($F$) 1 shell pass and 3, 6, 9, etc., tube passes. ($I$) Crossflow, 2 tube passes, shell fluid flows over second and first passes in series. [R. A. Bowman, A. C. Mueller, and W. M. Nagle, *Trans. ASME*, **62**: 283–294 (May, 1940).]

**Example 14.** What style of heat exchanger will require the least area for conditions of heating a cold fluid from 160 to 260°F while cooling a hot fluid from 300 to 200°F? (Assume same $U$ for all exchangers.)

*Solution.* The log mean temperature difference for counterflow is

$$\Delta t_{m.f} = \frac{(300 - 260) - (200 - 160)}{\ln (40/40)}$$

or, by inspection, 40°F. The parameters $X$ and $Z$ are

$$X = \frac{260 - 160}{300 - 160} = 0.715 \qquad Z = \frac{300 - 200}{260 - 160} = 1.0$$

With these data and Fig. 19-22,

| Exchanger | Chart | $Y$ | $\Delta t_m$, °F |
|---|---|---|---|
| Countercurrent flow........ | ... | 1.0 | 40.0 |
| Four-eight multipass....... | D | 0.90 | 36.0 |
| Crossflow................ | I | 0.73 | 29.2 |
| Two-four multipass........ | B | 0.645 | 25.8 |
| One-three multipass........ | F | Impossible | |
| One-two multipass......... | A | Impossible | |
| Parallel flow............. | ... | Impossible | |

Pure counterflow will require the least area (although economic and space considerations may cause the selection of a multipass or crossflow exchanger as being most desirable).

## PROBLEMS

*Articles* 19-1 *to* 19-4

**1.** (a) Determine the value of $a$ in Eq. (19-7) to best represent the thermal-conductivity data for copper in a range of temperatures from −100 to 1000°F (Table B-12), and (b) obtain an expression for the mean thermal conductivity $k_m$ for a substance having a conductivity-temperature relationship of $k = k_0(1 + bt^2)$, where $b$ is a constant.

**2.** From the temperature distribution in the solid shown in Fig. 19-4, determine if $k$ for the material (a) decreases with increasing temperature, (b) increases with increasing temperature, or (c) is constant.

**3.** A standard 2-in. cast-iron pipe (ID 2.07 in., OD 2.37 in.) is insulated with an 85 per cent magnesia, 15 per cent asbestos insulation of apparent density 13 $lb_m/ft^3$. The temperature at the interface between the pipe and insulation is 400°F. The allowable heat loss for the material flowing through the pipe is 200 Btu/hr per foot of pipe length, and for safety the temperature of the outside surface of the insulation must not exceed 150°F. Determine (a) the minimum thickness of insulation required and (b) the temperature of the inside surface of the pipe.

**4.** Heat is conducted unidirectionally in the $x$ direction through a large plane wall of thickness $x_1$ with surface temperatures $t_0$ at $x = 0$ and $t_1$ at $x = x_1$. The thermal conductivity of the wall material varies with $x$ as follows:

$$k = \frac{k_0}{1 + Bx/x_1}$$

where $k_0$ is the thermal conductivity at $x = 0$ and $B$ is a positive constant. (a) Derive an expression for the steady-state conduction heat transfer per unit area of wall surface in terms of $B$, $k_0$, $t_0$, $t_1$, and $x_1$. (b) At what location $x$ does the temperature gradient attain a maximum absolute value?

**5.** It is sometimes convenient to consider steady-state, unidirectional heat conduction through the walls of cylinders and spheres by means of an equation similar to the plane-wall conduction equation

$$\dot{Q} = \frac{k_m A_m \, \Delta t}{\text{wall thickness}}$$

where $A_m$ is a mean area of the cylinder or sphere that gives the same value of $\dot{Q}$ as that obtained from Eq. (19-15) or (19-17). Determine $A_m$ for a cylinder of surface area $A_1$ and surface temperature $t_1$ at inner radius $r_1$ and of surface area $A_2$ and temperature $t_2$ at outer radius $r_2$.

Ans. $A_m = (A_2 - A_1)/\ln(A_2/A_1)$ denoted as the logarithmic mean area.

**6.** Repeat Prob. 5 for a hollow sphere of inner radius $r_1$, area $A_1$, temperature $t_1$ and of outer radius $r_2$, area $A_2$, temperature $t_2$.

**7.** During the morning, the sun shines on the outer surface of a 6-in.-thick concrete wall. Thermocouples embedded in the wall at various locations along the thickness of the wall ($y$ direction) indicate that, at one particular time of day, the temperature distribution in the $y$ direction may be represented as $t = 100 - 60y + 24y^2 - 16y^3 + 64y^4$, where $t$ is in degrees Fahrenheit and $y$ is in feet. Determine (a) the outer and inner surface temperatures of the wall at this time, (b) the rate of heat transfer per unit wall area at the inner and outer surfaces, (c) the rate at which energy is being stored in, or released by, the wall at this time.

*Article* 19-5

**8.** Derive an equation for the heat transfer through a composite plane wall of two different materials without using the electrical analogy. Solution procedure: (a) Set up Eq. (19-12) for each section of the wall, and note that $\Delta t$ has different subscripts. (b) Solve each equation for $\Delta t$. (c) Add the equations together, and note that $\Delta t_{12} + \Delta t_{23} = \Delta t_{13}$. (d) Simplify.

**9.** Repeat Prob. 8, but assume a composite cylinder of two different materials.

**10.** Determine the heat loss per square foot of furnace wall when the wall is made up of 6 in. of firebrick [$k_m = 0.8$ Btu/(hr)(ft)(°F)], 6 in. of insulating material [$k_m = 0.075$ Btu/(hr)(ft)(°F)], and 2 in. of concrete [$k_m = 0.44$ Btu/(hr)(ft)(°F)] and the inside and outside surface temperatures are, respectively, 1900 and 120°F.

**11.** An 8-in.-diameter (nominal size) steam line carrying saturated steam at 250 psia is to be insulated with a 1-in. thickness of material with $k_m = 0.05$ Btu/(hr)(ft)(°F). If the outside surface temperature is 90°F, determine the heat loss for a length of 200 ft.

**12.** A furnace wall of magnesite brick is 12 in. thick and is to be covered with diatomaceous earth insulation (apparent density 18 lb/ft³) of sufficient thickness to limit the heat loss to 500 Btu/(hr)(ft²). Assuming that the outer surface temperature of the insulation will be about 80°F, while the inner surface temperature of the furnace is known to be 2200°F, calculate the required thickness of the insulation.

**13.** A large plane wall is composed of an 8-in. layer of refractory brick [assume $k_m$ constant and equal to 0.75 Btu/(hr)(ft)(°F)] and a 2-in. layer of insulating material with thermal conductivity which varies with temperature as $k = 0.02 + 0.0001t$, where $t$ is in degrees Fahrenheit and $k$ is in Btu/(hr)(ft)(°F). If the surface temperatures of brick and insulating material are 2000 and 100°F, respectively, determine the temperature at the interface between brick and insulating material.

*Articles* 19-6 *to* 19-8

**14.** Assume for a particular case of convection heat transfer between a surface and a fluid, as shown in Fig. 19-8, that the variation in temperature with $y$ ($t_y$ in degrees Fahrenheit) through the boundary-layer region $O$-$B$ of thickness $B$ may be expressed as

$$t_y = t_s - \frac{400y}{B} + \frac{400y^3}{B^3} - \frac{200y^4}{B^4}$$

Utilizing this expression for the temperature distribution in air when $t_s = 300°F$ and

$B = 0.05$ in., determine (a) the air temperature at $y = B$, (b) a value for the convection heat transfer per unit area between surface and air, and (c) the value of the convection coefficient $h_c$.

**15.** Complete the derivation discussed in Art. 19-7, and obtain an expression of the same form as Eq. (19-30).

**16.** By means of dimensional analysis, determine an expression for the height a liquid will rise in a vertical capillary tube. The height may be assumed to be a function of the fluid density, inner tube diameter, fluid temperature, and fluid surface tension. Compare your expression with the equation for this phenomenon, which may be found in most elementary physics textbooks.

**17.** Convert Eq. (19-30) to the form in which Stanton number equals a function of Reynolds and Prandtl numbers, and evaluate the necessary constants and exponents. Is the equation in terms of Stanton number of any greater utility than Eq. (19-30)?

**18.** Substitute the average values of the properties of air at room temperature into Eq. (19-49), and reduce to Eq. (19-50).

**19.** Determine the average convection coefficient for the flow of water with velocity of 5 ft/sec through 1-in.-outside-diameter No. 16 BWG tubes. The water enters the tubes at 60°F and leaves at 70°F. Data for water at 65°F: $c_p = 0.999$ Btu/$(lb_m)$(°F); $\rho = 62.3$ $lb_m/ft^3$; $\mu_m = 2.54$ $lb_m/(ft)(hr)$; $k = 0.346$ Btu/$(hr)(ft)$(°F).

**20.** If it were desired to increase the convection coefficient calculated in Prob. 19, would it be better to maintain the same velocity and halve the tube diameter or to use the same size tube and double the water velocity?

**21.** Superheated steam at 20 psia and 400°F flows through the annular space of a double-pipe heat exchanger at the rate of 240 $lb_m/hr$. The inner diameter of the outer tube is 2 in., and the outer diameter of the inner tube is 1 in. If the steam leaves at a temperature of 250°F, determine the convection coefficient.

**22.** Water is flowing over an eight-row bank of staggered tubes of 1 in. outside diameter with a mean velocity through the smallest opening of 6,000 ft/hr. If the average film temperature is 65°F, calculate the average convection coefficient. (See Prob. 19 for property data.)

**23.** Air at atmospheric pressure and an average temperature of 300°F flows through a duct 10 in. in diameter at a rate of 100,000 $ft^3/hr$. Two tubes are placed at right angles to the direction of flow and far enough apart so that one does not interfere with the other. The outside diameters of the two tubes are 2 and ½ in., and a fluid at 100°F evaporates as it flows through the tubes. Find (a) the ratio of $h_c$ for the small tube to $h_c$ for the large tube and (b) the ratio of the convection heat transfer from the small tube to that from the large tube.

**24.** Show that at small velocities the reference temperature $t^*$ given by Eq. (19-46) reduces to the film temperature defined by Eq. (19-29a).

**25.** A flat plate 4 in. long is moving in air at atmospheric pressure and a temperature of 200°F. For what plate velocity would no external heating or cooling be required to maintain the plate at a temperature of 300°F?

**26.** Equations (19-38) and (19-40) give the values of the convection coefficient at particular distances $(x)$ from the leading edge of a thin, flat plate, and Eqs. (19-39) and (19-41) yield the average values of the convection coefficient over the length of the plate. (a) Show that the average convection coefficient for laminar flow over a plate of length $L$ is equal to a constant times the local convection coefficient at $x = L$, and determine the numerical value of the constant. (b) What is the value of the local convection coefficient at the plate's leading edge $(x = 0)$? By consideration of the expected change in boundary-layer thickness along the length of the plate, show that the numerical value given by the equation at $x = 0$ is logical, though not attainable in practice.

**27.** Prepare a sketch of the cross section of a short tube having a sharp-edged entrance. Visualize the expected change in boundary-layer thickness along the length of the tube (similar to that on a flat plate), and show that the multiplication factor given in item 5, part $a$, of Art. 19-8 tends to influence the value of $h_c$ in a proper manner.

**28.** Both experiment and theory show that the drag force $F_D$ caused by the skin friction between a flat plate of unit width and length $L$ and a gas flowing in laminar motion along its length is

$$F_D = C_f \frac{\rho V_\infty^2 L}{2g_c}$$

where $V_\infty$ is the free-stream velocity and $C_f$ is the average value of the skin friction coefficient, which may be expressed as $C_f = 1.328(LV_\infty\rho/\mu_m)^{-0.5}$. Compare the expression for the skin friction coefficient with Eq. (19-39), assuming a Prandtl number of unity (nearly valid for most gases). Obtain a simple relationship between St and $C_f$ in support of Reynolds' analogy (footnote, page 388). Now using the relationship between St and $C_f$ and Eq. (19-41), obtain an expression for the average value of $C_f$ for the case of a turbulent boundary layer on a flat plate. The experimental value of $C_f$ for this case is known to be equal to $0.074(LV_\infty\rho/\mu_m)^{-0.2}$.

*Articles* 19-9 *to* 19-18

**29.** Verify Wien's displacement law by finding the maximum value of Eq. (19-54). The final step is by trial.

**30.** An icehouse is to be built. What color would you paint it? Explain.

**31.** Two pieces of wood are placed in sunlight; one piece is painted white and the other black. Which piece will absorb more radiation? Explain. The same two pieces of wood at room temperature are laid on the ground at night in midwinter. Which piece will cool faster? Explain. The same two pieces of wood are placed before an infrared heat lamp. Which piece will absorb more radiation? Explain.

**32.** Repeat the derivation of Eq. (19-69), but sum the net radiation leaving the black plane and find $\dot{Q}_{21}$.

**33.** Repeat the derivation of Eq. (19-69), but assume that the one plane is not gray because the absorptivity for incident radiation at $T_2$ is not equal to the emissivity of the plane at $T_1$.

**34.** Compute the heat lost by radiation from a cast-iron oxidized furnace door at 400°F; the furnace is located in a large room with a temperature of 80°F. The dimensions of the door are 2 by 3 ft. If the absorptivity of the door for low-temperature radiation is only three-quarters of that for high-temperature radiation, will the answer be changed?

**35.** In a room 12 ft wide, 16 ft long, and 9 ft high, one of the 9- by 12-ft walls is a radiant heating panel. What per cent of the radiant energy emitted by the panel is absorbed by (a) the floor, (b) the ceiling, (c) the two adjacent walls, (d) the opposite wall? Assume that all surfaces have an emissivity and absorptivity of unity.

**36.** Calculate the heat-transfer rate between a radiant panel in the ceiling of a small furnace and the floor if the dimensions are 2 by 2 ft with a height of 2 ft. The temperature and emissivity of the radiant panel are 1900°F and 0.80, while for the floor the corresponding values are 400°F and 0.78. At the end of several hours, the floor temperature is 1000°F and emissivity is 0.79. What is the heat-transfer rate for these conditions?

**37.** A liquid at $-100°F$ is placed in the inner container of a vacuum bottle which can be considered to be a sphere of 5 in. diameter. The outer shell of the sphere has a diameter of 6 in. and temperature of 60°F. If the emissivity of the container is 0.04, determine the radiant heat exchange.

**38.** A 2-in. oxidized steel pipe is located centrally in a concrete tunnel of square cross section 2 ft on each side. If the temperature of the walls of the tunnel is 80°F and that of the pipe is 300°F, determine the radiant heat transfer per foot of length by an approximate method and by a more exact method.

**39.** Repeat Prob. 38, but assume that the emissivity of the walls of the tunnel is only 0.10.

**40.** If the pipe in Prob. 38 is painted with aluminum paint, what will be the radiation heat transfer per foot?

**41.** Calculate the radiation heat transfer in the studding of the walls of a house if the brick surface is at $-10°F$ and the plaster surface is at 80°F. Repeat, but assume that both surfaces are covered with aluminum foil with an emissivity of 0.10 and that the temperatures remain essentially as before.

**42.** It has been suggested that the walls of refrigerators might be made thinner by lining the inside of the outer metal wall with aluminum foil (sheet $a$), lining the outside of the inner wall with foil (sheet $c$), and supporting a third sheet of foil (bright on both sides) between them (sheet $b$). The entire space between outer and inner walls would then be evacuated to minimize convection heat transfer. Assume the emissivity of foil to be 0.10 and determine ($a$) the temperature of sheet $b$ if sheet $a$ is at 80°F and sheet $c$ is at 0°F, ($b$) the radiation heat transfer per unit surface area of wall, and ($c$) the thickness of cork needed to provide the same insulation.

**43.** A mercury-in-glass thermometer, 0.35 in. diameter and 1 ft long, is suspended in a large, electrically heated drying oven containing still, dry air. The convection coefficient between air and thermometer is 1.0 Btu/(hr)(ft²)(°F). The thermometer indicates a temperature of 600°F, and the walls are at a uniform temperature of 620°F. Determine the true value of the air temperature.

**44.** A thermocouple is housed in a ½-in.-diameter oxidized steel tube which is placed at right angles to the direction of gas flow in a 3-ft-diameter duct. When the gas velocity is 15 ft/sec and the duct wall temperature is 600°F, the thermocouple indicates a temperature of 900°F. Neglecting conduction heat transfer, calculate the true gas temperature, and recommend two or more modifications of the system which would reduce the thermocouple error. Gas properties are $c_p = 0.25$ Btu/$(lb_m)(°F)$; $\mu_f = 0.20 \times 10^{-9}$ $lb_f$ hr/ft²; $k = 0.022$ Btu/(hr)(ft)(°F); $\rho = 0.04$ $lb_m/ft^3$.

**45.** In Example 11, the true gas temperature was computed as 1023°F from an unshielded thermocouple reading of 797°F. Assuming the true gas temperature to be

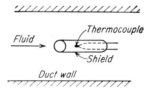

FIG. 19-23. Problem 45, a shielded thermocouple.

1023°F, what temperature would the thermocouple indicate if it were shielded from radiation interchange with the duct walls at 613°F by a single shield, as shown in Fig. 19-23? The shield is made of thin metal ($\epsilon = 0.2$) and is 1 in. in diameter and

5 in. long. The convection coefficient between the air and the shield is 5 Btu/ (hr)(ft²)(°F). SUGGESTED SOLUTION: (a) Prepare an energy balance for the shield, discard a small term (or terms), and solve for the shield temperature. (b) Make an energy balance for the thermocouple assuming the thermocouple "sees" only the shield, and determine the thermocouple temperature.          $Ans.\ t_p = 964°F.$

NOTE: More than one shield may be necessary to attain accurate readings. King[†] reports on an experiment in which gas at 1400°F passed through a duct at about 1000°F. The temperature measured by an unshielded thermocouple was about 1310°F; a 1-in.-diameter, 3-in.-long single shield reduced the error to approximately 50°F; and four concentric shields were necessary to decrease the error to about 10°F.

**46.** An electrically heated wire of polished aluminum, 0.125 in. diameter, is supported horizontally in a galvanized iron duct 24 in. square. The outer surface of the duct is well insulated, and air at a pressure of 0.90 atm absolute and a temperature of 70°F flows through the duct normal to the wire at a velocity of 30 ft/sec. Calculate the energy input to the wire to maintain it at a temperature of 500°F if (a) convection heat transfer only is considered and (b) both radiation and convection heat transfer are considered.

**47.** A mercury-in-glass thermometer, 0.35 in. in diameter and 12 in. long, is suspended in a large refrigerated room containing still, dry air. The refrigerant pipes are located in the walls of the room, and it may be assumed as a first approximation that the room walls are at constant temperature. The emissivity of the walls is 0.9, and the convection coefficient between the air and the thermometer is 0.5 Btu/ (hr)(ft²)(°F). (a) Explain whether the true value of the air temperature is less or greater than that indicated by the thermometer. (b) Compute the true value of the air temperature when the wall surface temperature is −20°F and the thermometer reads a temperature of 0°F.

*Articles* 19-19 *to* 19-22

**48.** Derive Eq. (19-84a) without using the electrical analogy. (Follow the procedure described in Prob. 8.)

**49.** Water or steam flows through the inside of tubes in a steam generator and receives energy by convection and radiation heat transfer to the outer surfaces of the tubes. Such tubes tend to burn out when deposits form on the inside of the tubes. Why?

**50.** Assuming Fig. 19-17 represents the cross section of a hollow sphere, obtain an expression of the form of Eq. (19-84a) for a thermal resistance. Also find an expression for $U_3$.

**51.** Repeat Example 12, but assume that the steam-side coefficient was increased by 50 per cent. Repeat the same problem, but assume that the water-side coefficient is increased by 50 per cent.

**52.** Repeat Example 12, but assume that the water side has a scale deposit (and, of course, the usual fluid film), with $h$ of 250 Btu/(hr)(ft²)(°F).

**53.** Repeat Example 13 but find the over-all coefficient based on the inner area of the tube.

**54.** Determine the over-all heat-transfer coefficient for a 1-in.-outside-diameter condenser tube [$k = 80$ Btu/(hr)(ft)(°F)] with a wall thickness of 0.049 in. if the steam- and water-side coefficients are, respectively, 1000 and 600 Btu/(hr)(ft²)(°F).

**55.** Determine $U$ for a brick wall 8 in. thick with still air on the inside and a 15-mph

† W. J. King, Measurement of High Temperatures in High-velocity Gas Streams, *Trans. ASME*, **65**: 421–431 (July, 1943).

wind on the outside.   Compare with the value for $U$ given in the table in Art. 19-21.
See footnote to Table 19-5 for surface coefficient values.

**56.** Determine $U$ for a window consisting of a single pane of glass $\frac{1}{8}$ in. thick with
still air on the inside and a 15-mph wind on the outside.   Now assume that a well-
fitted storm window of $\frac{1}{8}$-in. glass is installed over the window such that a 2-in. air
space is formed between windows.   Compute $U$ for this arrangement.   The "Heating, Ventilating, Air Conditioning Guide" referred to in Art. 19-21 gives $U$ for the
single pane of glass as 1.13 Btu/(hr)(ft²)(°F) and $U$ for the two-window combination
as 0.53 Btu/(hr)(ft²)(°F).   Explain the discrepancy between the computed values
and the values recommended by the guide.

**57.** Derive Eqs. (19-87) for a counterflow heat exchanger.

**58.** Draw diagrammatic sketches of each of the heat exchangers mentioned in Fig.
19-22.

**59.** Explain why, in Example 14, certain types of heat exchangers could not be used
for the given data.

**60.** A heat exchanger is to be built to heat water from 90°F at the rate of 30,000 lb$_m$/
hr while cooling 20,000 lb$_m$/hr of water from 190 to 150°F.   The over-all coefficient $U$
is 300 Btu/(hr)(ft²)(°F).   Determine the area necessary for parallel and counterflow
operation and by using a one-two reversed-current exchanger.   What does the computed area represent?

**61.** Hot gases at 800°F are to be used to heat 100,000 lb$_m$/hr of water from 100 to
300°F while the gas temperature decreases to 400°F.   For an over-all coefficient of
10 Btu/(hr)(ft²)(°F), compute the area of exchanger necessary for (*a*) parallel flow,
(*b*) counterflow, (*c*) four-eight reversed-current exchanger.   Precisely what area
has been calculated?

**62.** A double-pipe heat exchanger contains 10 ft² of heat-transfer surface.   Steam
condensing in the annular space at 250°F heats water as it passes through the inner
pipe.   When first placed in service, water in the amount of 1,000 lb$_m$/hr was heated
from 100 to 200°F by the condensing steam.   After several weeks of operation, it was
found that the 1,000 lb$_m$/hr of water could be heated only from 100 to 175°F by the
condensing steam.   Assuming that dirt deposit on the tube has caused the change,
compute the fouling factor that has developed.

**63.** A double-pass heat exchanger is to condense without subcooling 2,000 lb$_m$/hr
of steam initially saturated at 1.5 in. Hg absolute pressure.   The cooling (river)
water enters at 60°F and leaves at 70°F through 1-in.-outside-diameter No. 16 BWG
tubes (70 per cent copper, 30 per cent zinc).   For an effective length of condenser of
10 ft, calculate the area and number of tubes required if the steam-side coefficient is
1000 Btu/(hr)(ft²)(°F), and determine the over-all coefficient based upon the outside
tube area.   (See Prob. 19 for water-property data.)

**64.** Repeat Prob. 63 using most of the data, except that the water velocity is
specified to be 5 ft/sec.

**65.** In Prob. 21, water is flowing through the inner tube, which has a wall thickness
of 0.049 in. and $k$ of 60 Btu/(hr)(ft)(°F).   What length of pipe is necessary to heat
compressed water from 150 to 225°F by (*a*) counterflow and (*b*) parallel flow?   How
much water can be thus heated?

# CHAPTER 20

# ADVANCED TOPICS IN HEAT TRANSFER

*The nation that has the schools has the future.*
*Bismarck*

The principles of heat transfer have been used to analyze many systems of engineering importance. A few such cases are described in the following articles.†

**20-1. Steady-state Conduction with Internal Heat Generation.** In many situations of practical importance, heat is generated within a solid, conducted to the surface, and dissipated to the surroundings by convection and radiation. Examples are the passage of electricity through a conductor with a consequent Joulean heat generation and the fission heat generated within a nuclear reactor. In these circumstances, the heat transfer from the solid surface and the temperature distribution in the solid are of interest. Frequently the temperature distribution is of major concern, because the allowable rate of heat generation is usually controlled by a temperature limitation, which, if exceeded, might result in failure of the solid.

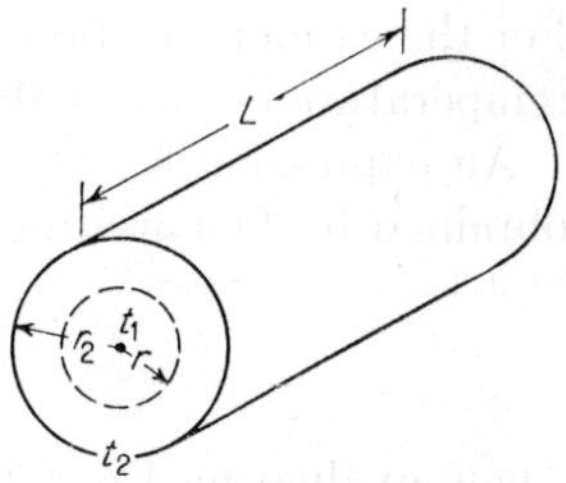

Fig. 20-1. Radial conduction in a rod with internal heat generation.

Consider a rod of length $L$ and outer radius $r_2$, as shown in Fig. 20-1. Let $\omega$ be the rate at which heat is generated per unit volume of rod [Btu/(hr)(ft³)], and assume that conduction occurs only in the radial direction. The total heat generated in the rod per unit time and, for steady-state conditions, the rate of heat transfer at the surface, $\dot{Q}_2$, is

$$\dot{Q}_2 = \int_0^{r_2} \omega 2\pi r L \, dr \qquad (20\text{-}1)$$

Considering an arbitrary radius $r$ in the solid, the heat generated in

† In this chapter, each article is a separate and independent unit. After completing the preceding chapter, these articles may be studied in any order, or if desired, each of these articles may be taken up at appropriate times during the study of the preceding chapter. For example, Art. 20-1 may be assigned upon completion of Art. 19-4.

the volume from 0 to $r$ is equal to the conduction heat transfer at $r$:

$$\dot{Q}_r = -k2\pi rL \frac{dt}{dr} = \int_0^r \omega 2\pi rL \, dr \tag{20-2}$$

The manner in which the rate of heat generation varies with radius must be known to evaluate Eqs. (20-1) and (20-2).

As a simple case, assume that $\omega$ is constant. Evaluating Eq. (20-2),

$$-k \, dt = \frac{\omega r \, dr}{2} \tag{20-3}$$

Integrating Eq. (20-3) between $t_1$ at $r = 0$ and $t$ at $r$,

$$k_m(t_1 - t) = \frac{\omega r^2}{4} \tag{20-4}$$

Solving for $t$, the temperature distribution in the rod, gives

$$t = t_1 - \frac{\omega r^2}{4k_m} \tag{20-5}$$

For the case of constant $\omega$, Eq. (20-5) shows that the maximum value of temperature occurs at the geometric center of the rod.

An expression for $\dot{Q}_2$, the rate of heat transfer at the surface, can be obtained by first writing Eq. (20-4) for the surface ($r = r_2$):

$$k_m(t_1 - t_2) = \frac{\omega r_2^2}{4} \tag{20-6}$$

Then, evaluating Eq. (20-1) for constant $\omega$,

$$\omega = \frac{\dot{Q}_2}{\pi L r_2^2} \tag{20-7}$$

Finally, by Eqs. (20-6) and (20-7),

$$\dot{Q}_2 = 4\pi k_m L(t_1 - t_2) \tag{20-8}$$

By the same means, expressions similar to Eq. (20-8) can be obtained for other shapes of simple geometry assuming a constant rate of heat generation.

a. *The Plane Wall (Conduction in the x Direction Only)*

$$\dot{Q}_2 = \frac{2k_m A (t_1 - t_2)}{x_2 - x_1} \tag{20-9}$$

The wall is perfectly insulated at $x_1$ [if surface and surrounding temperatures are identical on both sides of the wall, then, by symmetry, the midplane of the wall can be considered a perfectly insulated (adiabatic) surface]. Conduction occurs through the constant area $A$ from $x_1$ to $x_2$. The surface at $x_2$ loses heat at the rate $\dot{Q}_2$ to the surroundings.

b. *The Solid Sphere (Conduction in the r Direction Only)*

$$\dot{Q}_2 = 8\pi k_m r_2 (t_1 - t_2) \tag{20-10}$$

Outer radius $r_2$ of the sphere is at temperature $t_2$, and $t_1$ is the temperature at the geometric center.

**Example 1.** A 66,000-volt copper electrical transmission line is carrying a current of 840 amp. The diameter of the line is 0.811 in., and the electrical resistance of the copper conductor is 0.1196 ohm/mile. Assuming that the surroundings are at 100°F and that the combined convection and radiation coefficient for heat transfer from the wire surface to the surroundings is 2.5 Btu/(hr)(ft²)(°F), determine (a) the surface temperature of the transmission line, (b) the rate of heat generation per unit volume of wire, and (c) the maximum temperature in the line.

*Solution.* (a) Under steady-state conditions, the heat generated in the wire must be dissipated from the wire surface by convection and radiation to the surroundings. Thus by Eq. (19-6)

$$(840)^2(0.1196) = 84,300 \text{ watts/mile} = 54.5 \text{ Btu/(hr)(ft)} = 2.5\frac{\pi(0.811)}{12}(t_2 - 100)$$

Solving, $t_2$ (wire surface temperature) $= 103 + 100 = 203°F$    *Ans.*

(b) As a first approximation, assume that the line is of solid copper and that heat is generated at a uniform rate over the wire cross section. By Eq. (20-7),

$$54.5 \text{ Btu/(hr)(ft)} = \frac{\pi(\omega)(0.811)^2(1)}{(4)(144)}$$

Solving, $\omega = 15,200 \text{ Btu/(hr)(ft}^3)$    *Ans.*

(c) Assuming a constant rate of heat generation, the maximum temperature in the line will occur at the geometric center of the wire and may be computed by Eq. (20-6):

$$k_m(t_1 - t_2) = \frac{\omega r_2^2}{4}$$

Evaluating $k_m$ for copper (Table B-12) at an estimated mean temperature of 210°F, $k_m = 220 \text{ Btu/(hr)(ft)(°F)}$.

$$t_1 = \frac{(15,200)(0.405)^2}{(4)(144)(220)} + 203 = 0.02 + 203 = 203.02°F \quad \textit{Ans.}$$

The small difference between surface and center temperature results from the relatively small heat-generation rate and the high thermal conductivity of copper. To be strictly correct, $k_m$ should be again evaluated at a temperature of 203.01°F; however, the computed temperature changes only slightly with the corrected thermal-conductivity value.

**20-2. Extended Surfaces.**† Figure 20-2a shows one type of *extended surface*, in this instance a rod attached to a plate at temperature $t_1$ and extending into a fluid at a lesser temperature $t_o$. Heat from the hot surface is conducted through the rod and transferred by convection and

† If desired, this article may be assigned upon completion of Art. 19-17.

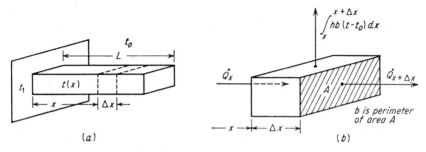

FIG. 20-2. An extended surface.

radiation from the sides of the rod to the surrounding fluid and solid surfaces.

Extended surfaces (commonly called *fins*) are frequently placed on heat-exchanger tubes to increase the total area for heat transfer; this results in an exchanger of smaller dimensions (a *compact heat exchanger*). However, the gain in heat transfer with fins is not directly proportional

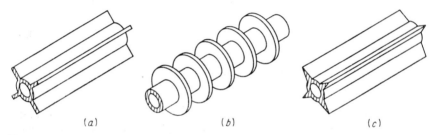

FIG. 20-3. Finned surfaces. (*a*) Straight fins of rectangular profile; (*b*) circular fins of rectangular profile; (*c*) straight fins of triangular profile.

to the increase in surface area, because the sides of the fins are at a temperature less than $t_1$. Hence the temperature difference for convection and radiation from the exposed fin surface is less than that for the bare tube. A few of the many types of fins are shown in Fig. 20-3.

Consider again the fin of total length $L$ (Fig. 20-2). From an energy balance on the elemental fin volume shown in Fig. 20-2$b$, the variation of fin temperature $t$ with length $x$ can be found. The temperature of the fin, $t$, is assumed constant over any cross-sectional area $A$, and thus conduction is considered to occur in the $x$ direction only.†

† At this point, the careful reader inquires: If $t$ varies only with $x$, how does the heat get out of the sides of the fin? The only possible viewpoint is that $t$ really varies both with $x$ and with directions perpendicular to $x$, so that in reality the problem is a complex one of multidimensional conduction. However, one reasons (or perhaps merely hopes) that the predominant variation of $t$ is along the $x$ direction, establishes a mathematical model on this basis, and achieves a reasonably simple solution. For-

Assuming steady-state conditions, an energy balance shows that the rate of heat conduction into the incremental volume at $x$ is equal to the rate of heat conduction out of the volume at $x + \Delta x$ plus the rate of surface heat transfer by convection and radiation:

$$\dot{Q}_x = \dot{Q}_{x+\Delta x} + \int_x^{x+\Delta x} hb(t - t_o)\,dx \qquad (20\text{-}11)$$

But, by the mean value theorem of the differential calculus,

$$\dot{Q}_{x+\Delta x} = \dot{Q}_x + \frac{d\dot{Q}_x}{dx}\bigg|_{\bar{x}} \Delta x \qquad (20\text{-}12)$$

where $d\dot{Q}_x/dx\big|_{\bar{x}}$ is evaluated at $\bar{x}$, a point between $x$ and $x + \Delta x$. In addition, by the mean-value theorem of the integral calculus,

$$\int_x^{x+\Delta x} hb(t - t_o)\,dx = hb\,\Delta x(t - t_o)\big|_{\bar{\bar{x}}} \qquad (20\text{-}13)$$

where $h$, $b$, and $(t - t_o)$ are evaluated at $\bar{\bar{x}}$, a point (not necessarily at the same location as $\bar{x}$) between $x$ and $x + \Delta x$.

From Eqs. (20-11), (20-12), and (20-13),

$$\frac{d\dot{Q}_x}{dx}\bigg|_{\bar{x}} + hb(t - t_o)\bigg|_{\bar{\bar{x}}} = 0 \qquad (20\text{-}14)$$

But $\Delta x$ is an arbitrary length, and hence, as $\Delta x$ is allowed to approach zero, $d\dot{Q}_x/dx\big|_{\bar{x}}$ approaches $d\dot{Q}_x/dx$ at $x$ while $hb(t - t_o)\big|_{\bar{\bar{x}}}$ approaches $hb(t - t_o)$ at $x$. Thus Eq. (20-14) becomes

$$\frac{d\dot{Q}_x}{dx} + hb(t - t_o) = 0$$

However, $\dot{Q}_x$ (conduction) equals $-kA\,(dt/dx)$:

$$\frac{d}{dx}\left(kA\,\frac{dt}{dx}\right) - hb(t - t_o) = 0 \qquad (20\text{-}15)$$

Considering a fin of constant thermal conductivity with constant cross-sectional area $A$ and constant perimeter $b$ (as in Figs. 20-2 and 20-3a), it is seen that

$$\frac{d^2t}{dx^2} - \frac{hb}{kA}(t - t_o) = \frac{d^2t}{dx^2} - a^2(t - t_o) = 0 \qquad (20\text{-}16)$$

---

tunately, results based upon the simplified mathematical model agree quite well with those obtained from experiment for the usual long thin fin. But, upon understanding the limitations of the mathematical model, one expects large deviations between theory and experiment for stubby thick fins.

Here $hb/kA$ is denoted as $a^2$. Equation (20-16) is the differential equation for the temperature distribution in a fin of constant $A$, $b$, and $k$. If $A$ varies with $x$ (as in the circular fin of Fig. 20-3$b$), and/or if $k$ cannot be considered constant, the temperature distribution is given by Eq. (20-15).

If $h$ is assumed constant,† the form of the solution to the differential equation (20-16) is

$$t = t_o + M \sinh ax + N \cosh ax \qquad (20\text{-}17)$$

sinh and cosh are the hyperbolic sine and hyperbolic cosine, respectively. The terms $M$ and $N$ are constants to be evaluated from the boundary conditions, that is, two given relationships between the dependent and independent variables. One obvious boundary condition is

$$t = t_1 \qquad \text{at } x = 0$$

A second convenient boundary condition is

$$\frac{dt}{dx} = 0 \qquad \text{at } x = L$$

The secondary boundary condition implies that the tip of the fin is perfectly insulated—a condition that may not be true in practice. For many fins the tip-surface area is quite small compared with the fin-side area; thus the tip heat transfer is small, and the approximation introduced by the boundary condition results in little error.

Substituting the two boundary conditions in Eq. (20-17) and simplifying results in

$$M = (t_o - t_1) \tanh aL$$
$$N = t_1 - t_o$$

With these values of $M$ and $N$, the temperature distribution is

$$t = t_o + (t_1 - t_o)(\cosh ax - \tanh aL \sinh ax)$$

† The assumption of constant $h$ means that the surface heat transfer is considered proportional to $t - t_o$, and so introduces some approximation into the solution. In truth, both convection and radiation are more complex functions of $t - t_o$. But if $h$ is considered a function of $t - t_o$, the differential equation becomes nonlinear and solution is much more difficult. The few results published for the case of variable $h$ show that the assumption of constant $h$ may not introduce much error. For example, I. C. Hutcheon and D. B. Spalding [Prismatic Fin with Non-linear Heat Loss Analysed by Resistance Network and Iterative Analogue Computer, *British Journal of Applied Physics*, vol. 9, pp. 185–191 (May, 1958)] considered the surface heat transfer proportional to $(t - t_o)^{5/4}$ as in natural convection. For "short" fins ($\eta > 80$ per cent), they found the fin efficiencies for constant and for variable $h$ essentially equal, but for "long" fins ($\eta < 50$ per cent) the fin efficiency for variable $h$ was as much as 10 per cent less than that for constant $h$. Fin efficiency $\eta$ is defined by Eq. (20-20).

which may be simplified by the hyperbolic identities to

$$t = t_o + (t_1 - t_o) \frac{\cosh a(L - x)}{\cosh aL} \qquad (20\text{-}18)$$

The total heat transferred by the fin per unit time, $\dot{Q}_{\text{total}}$, can be determined by noting that the rate of heat conduction at the root of the fin, $x = 0$, is equal to the rate of heat transfer by convection and radiation at the exposed surface of the fin:

$$\dot{Q}_{\text{total}} = -kA \frac{dt}{dx}\Big|_{x=0}$$

Differentiating Eq. (20-18) and evaluating $dt/dx$ at $x = 0$ gives

$$\dot{Q}_{\text{total}} = kA(t_1 - t_o)a \tanh aL \qquad (20\text{-}19)$$

The *fin efficiency* $\eta$ is a criterion for judging the relative merits of fins of differing geometries or materials:

$$\eta = \frac{\text{heat transferred by fin}}{\text{heat that would be transferred if fin were all at } t_1}$$

Thus

$$\eta = \frac{h(\text{surface area})(t_m - t_o)}{h(\text{surface area})(t_1 - t_o)} = \frac{\Delta t_m}{t_1 - t_o} \qquad (20\text{-}20)$$

The term $\Delta t_m$ represents the mean temperature difference for convection and radiation heat transfer between the fin surface and its surroundings. From the definition of fin efficiency and Eq. (20-19), the efficiency for fins of constant $A$, $b$, and $k$ may be expressed as

$$\eta = \frac{kA(t_1 - t_o)a \tanh aL}{hbL(t_1 - t_o)} = \frac{\tanh aL}{aL} \qquad (20\text{-}21)$$

From the results of a more exact analysis for the fin with a bare tip, the approximation involved in the second boundary condition can be decreased by replacing $L$ in the preceding equations with a corrected length equal to $L + A/b$.

Circular fins (Fig. 20-3b) and triangular fins (Fig. 20-3c) have been investigated by the same procedure, starting with an expression similar to Eq. (20-15). However, the solution of the differential equation for the temperature distribution is more complicated for these fins than for the fin of constant $A$, $b$, and $k$. Values of fin efficiency for various circular fins of rectangular profile are presented in Fig. 20-4, as given by Gardner.[†] Consideration of extended surfaces of other geometries and discussions of such problems as the optimum fin configuration for minimum weight

[†] K. A. Gardner, Efficiency of Extended Surfaces, *Trans. ASME*, **67**: 621–631 (November, 1945).

can also be found in the literature. Schneider[10] describes several such analyses.

According to Eckert and Drake,[8] the use of fins on surfaces is justified if the ratio $kb/hA$ is greater than 5. Thus fins almost always result in a worthwhile increase in heat transfer between tubes and gases, because $k$ of the metallic fin is normally large compared with $h$. But fins are rarely used on tubes immersed in liquids, because $h$ is normally great

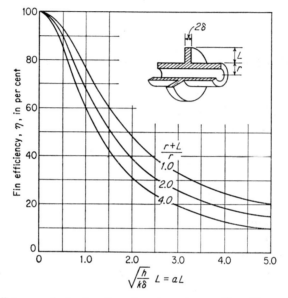

FIG. 20-4. Efficiency of circular fins of rectangular profile. NOTE: Upper curve [$(r + L)/r$ of 1.0] may be used to determine the efficiency of fins of constant $A$, $b$, and $k$ [Eq. (20-21)].

enough relative to $k$ that the change in heat transfer does not justify the added cost of finned surfaces.

**Example 2.** An air heater is composed of 1-in.-outside-diameter steel tubes on which circular steel fins, 0.50 in. long and 0.025 in. thick, are mounted. There are five fins per inch of tube, and $h$ is estimated to be 10 Btu/(hr)(ft²)(°F). The air temperature is 170°F, and the tube surface temperature is 430°F. Determine (a) the heat loss per hour from 1 ft of bare tube, (b) the heat loss per hour from 1 ft of finned tube, (c) an approximate temperature for the tip of the fin, (d) the heat loss per hour from 1 ft of finned tube if the fins are made of aluminum.

*Solution.* (a) The heat rate per foot of bare tube is

$$\dot{Q} = \frac{(10)(\pi)(1)(1)(430 - 170)}{12} = 681 \text{ Btu/hr} \qquad Ans.$$

(b) The heat rate per foot of finned tube may be calculated with the aid of Fig. 20-4:

$$\frac{r + L}{r} = \frac{0.5 + 0.5}{0.5} = 2.0$$

Select $k$ as 26 Btu/(hr)(ft)(°F) from Table B-12:

$$L \left(\frac{h}{k\delta}\right)^{\frac{1}{2}} = \frac{0.50}{12} \left[\frac{(10)(12)}{(26)(0.0125)}\right]^{\frac{1}{2}} = 0.80$$

From Fig. 20-4, $\eta = 0.77$, and by Eq. (20-20),

$$\Delta t_m = \eta(t_1 - t_o) = (0.77)(430 - 170) = 200°F$$

Since the surface area of each fin is $(2)(\pi)(2^2 - 1^2)/4 = 4.71$ in.², the hourly heat loss from the 60 fins on 1 ft of tube is

$$\dot{Q} = \frac{(10)(60)(4.71)(200)}{144} = 3920 \text{ Btu/hr}$$

The hourly heat loss from the surface of the 1 ft of tube not covered by fins is

$$\dot{Q} = \frac{(10)(\pi)(1)[12 - (60)(0.025)](430 - 170)}{144} = 596 \text{ Btu/hr}$$

and the total hourly heat loss per foot of finned tube is

$$3920 + 596 = 4516 \text{ Btu/hr} \qquad Ans.$$

(c) An approximate value of the temperature at the tip of the fin may be found by using Eq. (20-18), although it is strictly valid only for fins of constant $A$ and $b$.

$$aL = L \left(\frac{hb}{kA}\right)^{\frac{1}{2}} \approx L \left(\frac{h}{k\delta}\right)^{\frac{1}{2}} = 0.80$$

$$t_L = 170 + (430 - 170) \frac{1}{\cosh 0.80} = 364°F \qquad Ans.$$

(From the exact analysis for circular fins, $t_L = 351.5°F$.)

(d) For aluminum fins, select $k$ as 132 Btu/(hr)(ft)(°F) (Table B-12). Determine $\eta$ from Fig. 20-4:

$$aL = \frac{0.50}{12} \left[\frac{(10)(12)}{(132)(0.0125)}\right]^{\frac{1}{2}} = 0.36$$

At this value, $\eta = 0.94$ and

$$\Delta t_m = 0.94(430 - 170) = 244°F$$

The heat rate from the 60 fins is 4780 Btu/hr, and the total hourly heat loss from the 1 ft of finned tube is 5376 Btu/hr. *Ans.*

The utility of the extended-surface concept is not limited to the design of finned-surface heat exchangers. The following example problem illustrates one of the many other cases in which this concept is applicable.

**Example 3.**† A 3,600-rpm electric motor drives a centrifugal pump which circulates a liquid metal at a temperature of 1000°F. The motor is connected to the pump impeller by a horizontal, oxidized steel shaft 1 in. in diameter. If the temperature of the electric motor is limited to a maximum value of 125°F, what length of shaft should be specified between the motor and pump? Air and surrounding surfaces about the shaft are at 80°F.

*Solution.* The shaft can be considered as an extended surface conducting heat from the pump toward the motor while losing energy from its surface by radiation and convection. As the shaft is of constant $A$ and $b$, Eq. (20-18) represents the temperature distribution where $x = 0$ at the pump and $x = L$ at the motor. Thus the shaft temperature at the motor, $t_L$, may be expressed as

$$t_L = t_o + (t_1 - t_o) \frac{\cosh a(L - L)}{\cosh aL} = t_o + \frac{t_1 - t_o}{\cosh aL}$$

As a conservative estimate, assume that the shaft temperature at the pump, $t_1$, is equal to the liquid-metal temperature in the pump, 1000°F; assume that the shaft temperature at the motor is 125°F, and $t_o$ is 80°F. Thus

$$\frac{1}{\cosh aL} = \frac{t_L - t_o}{t_1 - t_o} = \frac{125 - 80}{1000 - 80} = 0.049$$

Solving, $aL = 3.71$, where $L$ is the necessary shaft length and $a = (hb/kA)^{1/2}$. From the given data,

$$b = \frac{(\pi)(1)}{12} = 0.262 \text{ ft} \qquad A = \frac{(\pi)(1)^2}{(4)(144)} = 0.00546 \text{ ft}^2$$

$k$ (Table B-12) = 25 Btu/(hr)(ft)(°F).

The term $h$ is the sum of the surface coefficient for radiation $h_r$ and the surface coefficient for convection $h_c$. By assuming the shaft to be a small body in a large enclosure, $\mathfrak{F} = \epsilon$ of the shaft [Eq. (19-83c)]. Selecting $\epsilon$ for the oxidized shaft as 0.79 from Table 19-2, and using an arithmetic-mean surface temperature of the shaft, $(1000 + 125)/2 = 562°F = 1022°R$, $h_r$ is found from Eq. (19-5) to be

$$h_r = \frac{(0.79)(0.1714)[(10.22)^4 - (5.4)^4]}{1022 - 540} = 2.85 \text{ Btu/(hr)(ft}^2)(°F)$$

(Note that, because the $T^4$ difference is involved, the $h_r$ predicted by using the arithmetic-mean surface temperature is probably too small. However, if $h_r$ is smaller than the true value, the computed length of shaft will be larger than that actually required. Thus the use of the arithmetic-mean surface temperature appears to be an approximation leading to a safe design estimate.)

The evaluation of $h_c$ poses a new problem. A value for the convection coefficient for a rotating horizontal shaft is needed, and a survey of Art. 19-8 reveals that no information is presented for this situation. It would seem that Eq. (19-49) for natural convection from horizontal cylinders is most like the rotating-shaft case, but in natural convection the flow is induced by buoyancy effects, while for the rotating shaft, flow should be influenced by both buoyancy and centrifugal effects. A search of the Selected References shows that only Refs. 8 and 11 consider the case of rotating

† In addition to illustrating another use of the extended-surface concept, this lengthy example points out that frequently the current literature must be searched to find needed data and that the data so found must be critically inspected to ensure that they are applicable to the problem under consideration.

surfaces. Eckert and Drake[8] list two papers on the subject, but it is clear that these concern rotating plates. Kreith[11] presents the main results found by Anderson and Saunders.† Although Ref. 11 summarizes their results, it is well to study the original paper. Anderson and Saunders present experimental data for cylinders of 1.0, 1.8, and 3.9 in. diameter, with maximum rotational speeds of 550 rpm. Although their results might be extrapolated to the conditions of the present problem, an attempt should be made to find a more recent paper because (1) it may contain data more applicable to the present problem, and (2) it should contain a critical discussion of, and a comparison with, past work.

Upon consulting "The Engineering Index" for 1958 (published by Engineering Index, Inc., New York) under the heading Heat Transmission, reference to a paper by Kays and Bjorklund‡ is found. This paper contains a summary of the results of previous investigators and new results for a cylinder 2.26 in. in diameter rotated at speeds of 50 to 4,500 rpm. The following equation is proposed:

$$\mathbf{Nu} = 0.135\{[0.5(\mathbf{Re}_p)^2 + (\mathbf{Re}_s)^2 + \mathbf{Gr}]\mathbf{Pr}\}^{1/3}$$

This equation represented their experimental results for heat transfer from a rotating cylinder within plus or minus 15 per cent when the value of the terms in the braces was between $3(10^7)$ and $10^9$. All properties are evaluated at the film temperature, Eq. (19-29a); $\mathbf{Re}_p$ is a Reynolds number based upon cylinder diameter and peripheral velocity, and $\mathbf{Re}_s$ is a Reynolds number based upon cylinder diameter and the velocity of the gas being blown across the cylinder. Evaluating air properties at a film temperature of $(562 + 80)/2 = 321°F$ from Table B-15, the dimensionless numbers are calculated as

$$\mathbf{Re}_p = 2,040 \qquad \mathbf{Re}_s = 0 \qquad \mathbf{Gr} = 1.1(10^5) \qquad \mathbf{Pr} = 0.685$$

The braced term is found to be

$$[0.5(2,040)^2 + 0 + 1.1(10^5)]0.685 = 1.5(10^6)$$

Unfortunately, the present problem does not fall within the range of validity of the proposed equation. However, the paper also contains much of the experimental data, and a plot of $\mathbf{Nu}$ versus the braced term shows that $\mathbf{Nu}$ equals approximately 18 when the braced term is equal to $1.5(10^6)$.

$$h_c = \frac{\mathbf{Nu}\ k}{D} = \frac{(18)(0.0208)(12)}{1} = 4.5 \text{ Btu/(hr)(ft}^2)(°F)$$

[Neglecting the effect of rotation, Eq. (19-49) for natural convection gives 2.2 as the value of $h_c$.]

With a reasonable value for $h_c$ finally determined, $a$ may be found as

$$a = \left(\frac{hb}{kA}\right)^{1/2} = \left[\frac{(4.5 + 2.85)(0.262)}{(25)(0.00546)}\right]^{1/2} = 3.75 \text{ ft}^{-1}$$

But $aL = 3.71$; hence

$$L = \frac{3.71}{3.75} = 0.99 \text{ ft} \qquad Ans.$$

† J. T. Anderson and O. A. Saunders, Convection from an Isolated Heated Horizontal Cylinder Rotating about Its Axis, *Proc. Roy. Soc. (London)*, **A217**: 552–562 (1953).

‡ W. M. Kays and I. S. Bjorklund, Heat Transfer from a Rotating Cylinder with and without Crossflow, *Trans. ASME*, **80**: 70–77 (January, 1958).

**20-3. Unsteady-state Heat Conduction.**†   Consider that volume $A \, \Delta x$ shown in Fig. 20-5 is part of a larger mass through which heat is conducted in an *unsteady-state manner*. From Art. 19-1, recall that for unsteady-state conduction, temperature and rate of heat conduction are dependent upon both coordinate position and time. For simplicity, consider conduction in the $x$ direction only; hence temperature varies with coordinate position $x$ and time. Assume that heat is generated within the volume (by electrical resistance heating, for example) at a volumetric rate of $\omega$ [typical units, Btu/(hr)(ft$^3$)].

Application of the conservation of energy principle to volume $A \, \Delta x$ for a certain time interval $\Delta \tau$ shows that the heat conducted into the volume through the plane at $x$ plus the heat generated in the volume equals the heat conducted out of the volume through the plane at $x + \Delta x$ plus the change in internal energy of the volume.

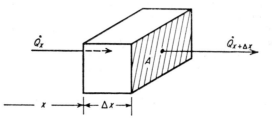

FIG. 20-5. Unsteady-state conduction.

During the time $\Delta \tau$, the difference in the heat conducted into and out of the volume is

$$\int_{\tau}^{\tau + \Delta \tau} (\dot{Q}_x - \dot{Q}_{x+\Delta x}) \, d\tau$$

By the mean-value theorem of the integral calculus, the preceding expression may be set equal to

$$(\dot{Q}_x - \dot{Q}_{x+\Delta x}) \Big]_{\bar{\tau}} \Delta \tau$$

The difference in conduction rates is evaluated at a mean time $\bar{\tau}$ such that the two preceding expressions are equal. $\bar{\tau}$ must lie within the time interval $\tau$ to $\tau + \Delta \tau$.

The heat generated within the volume during the time $\Delta \tau$ is

$$\omega A \, \Delta x \, \Delta \tau$$

The change in internal energy of the volume during the time $\Delta \tau$ equals

$$\int_{x}^{x+\Delta x} [(\rho u)_{\tau + \Delta \tau} - (\rho u)_\tau] A \, dx = [(\rho u)_{\tau + \Delta \tau} - (\rho u)_\tau]_{\bar{x}} A \, \Delta x$$

† If desired, this article may be assigned upon completion of Art. 19-8.

The terms $u$ and $\rho$ are specific internal energy and density, respectively. The bracketed term on the right-hand side of the expression is the difference in internal energy per unit volume at a mean location $\bar{x}$ between $x$ and $x + \Delta x$.

Summing the preceding expressions in accord with the conservation of energy principle,

$$(\dot{Q}_x - \dot{Q}_{x+\Delta x})_{\bar{\tau}}\, \Delta\tau + \omega A\, \Delta x\, \Delta\tau = [(\rho u)_{\tau+\Delta\tau} - (\rho u)_\tau]_{\bar{x}} A\, \Delta x \qquad (a)$$

But, by the mean-value theorem of the differential calculus,

$$\dot{Q}_{x+\Delta x,\bar{\tau}} = \dot{Q}_{x,\bar{\tau}} + \left(\frac{\partial \dot{Q}}{\partial x}\right)_{\bar{x},\bar{\tau}} \Delta x$$

$$(\rho u)_{\tau+\Delta\tau,\bar{x}} = (\rho u)_{\tau,\bar{x}} + \left[\frac{\partial(\rho u)}{\partial\tau}\right]_{\bar{\tau},\bar{x}} \Delta\tau$$

Thus Eq. ($a$) becomes

$$-\left(\frac{\partial \dot{Q}}{\partial x}\right)_{\bar{x},\bar{\tau}} \Delta x\, \Delta\tau + \omega A\, \Delta x\, \Delta\tau = \left[\frac{\partial(\rho u)}{\partial\tau}\right]_{\bar{\tau},\bar{x}} A\, \Delta x\, \Delta\tau$$

But $\Delta x$ and $\Delta\tau$ are of arbitrary value, and as they approach zero, $\bar{x}$ and $\bar{x}$ approach $x$, and $\bar{\tau}$ and $\bar{\tau}$ approach $\tau$:

$$-\frac{\partial \dot{Q}}{\partial x} + \omega A = \left[\frac{\partial(\rho u)}{\partial\tau}\right] A$$

The term $\dot{Q}$ (conduction) is equal to $-kA\, \partial t/\partial x$. If the internal energy is a function of temperature only, $u = ct$, where $c$ is the heat capacity. With these substitutions, the preceding equation may be written as

$$\frac{\partial}{\partial x}\left(k\frac{\partial t}{\partial x}\right) + \omega = \frac{\partial(\rho ct)}{\partial\tau}$$

Including conduction in the $y$ and $z$ directions,

$$\frac{\partial}{\partial x}\left(k\frac{\partial t}{\partial x}\right) + \frac{\partial}{\partial y}\left(k\frac{\partial t}{\partial y}\right) + \frac{\partial}{\partial z}\left(k\frac{\partial t}{\partial z}\right) + \omega = \frac{\partial(\rho ct)}{\partial\tau} \qquad (20\text{-}22)$$

If properties $k$, $\rho$, and $c$ are assumed constant,

$$\frac{\partial^2 t}{\partial x^2} + \frac{\partial^2 t}{\partial y^2} + \frac{\partial^2 t}{\partial z^2} + \frac{\omega}{k} = \frac{\rho c}{k}\frac{\partial t}{\partial\tau} \qquad (20\text{-}23)$$

Equation (20-23), or Eq. (20-22) for variable properties, is known as the *general differential equation for the temperature distribution for unsteady-state conduction.*

For a cylindrical coordinate system with coordinates $r$, $\theta$, and $z$ (see Fig. 20-6$a$), Eq. (20-23) may be transformed to

$$\frac{\partial^2 t}{\partial r^2} + \frac{1}{r}\frac{\partial t}{\partial r} + \frac{1}{r^2}\frac{\partial^2 t}{\partial \theta^2} + \frac{\partial^2 t}{\partial z^2} + \frac{\omega}{k} = \frac{\rho c}{k}\frac{\partial t}{\partial \tau} \qquad (20\text{-}24a)$$

For a spherical coordinate system with coordinates $r$, $\phi$, and $\psi$ (see Fig. 20-6$b$), Eq. (20-23) takes the form

$$\frac{\partial^2 t}{\partial r^2} + \frac{2}{r}\frac{\partial t}{\partial r} + \frac{1}{r^2}\frac{\partial^2 t}{\partial \psi^2} + \frac{1}{r^2 \tan \psi}\frac{\partial t}{\partial \psi} + \frac{1}{r^2 \sin^2 \psi}\frac{\partial^2 t}{\partial \phi^2} + \frac{\omega}{k} = \frac{\rho c}{k}\frac{\partial t}{\partial \tau} \qquad (20\text{-}24b)$$

The solution of these equations for the temperature distribution involves finding an expression for temperature as a function of coordinate directions, time, rate of heat generation, and properties $k$, $\rho$, and $c$.

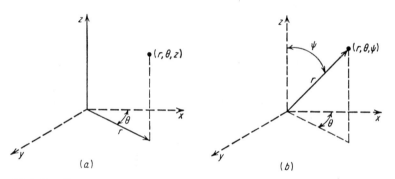

FIG. 20-6. Coordinate systems. ($a$) Cylindrical coordinates; ($b$) spherical coordinates.

Having found a valid expression for $t$ for a particular problem, the temperature gradient may be found by differentiation, and hence the rate of heat conduction through any plane in the body can be found from Fourier's equation.

The expression for $t$ must be a valid solution of the differential equation for the temperature distribution, and in addition, it must also satisfy the boundary and initial conditions of the problem under consideration. The usual boundary conditions are $t$ or gradient of $t$ equal to a certain value at a particular coordinate position, and initial conditions are specified by giving the value of $t$ or its gradient at a certain time, as, for example, $t = 0$ at $\tau = 0$.

When used as the starting point in a particular conduction problem, Eq. (20-22) can frequently be simplified. For example, if properties can be assumed constant, Eq. (20-23) may be used instead of Eq. (20-22). If steady state, the term on the right-hand side of the equation is zero. For situations without heat generation, the term containing $\omega$

may be discarded, and if conduction occurs in only two directions or one direction, one or two of the second-order partial terms may be discarded. The following example problem illustrates the means of applying Eq. (20-22) to simple conduction problems.

**Example 4.** The steady-state conduction solutions obtained previously by integration of Fourier's equation (Arts. 19-4 and 20-1) may be obtained by simplification and solution of Eq. (20-22).    To illustrate, consider the case of steady-state, unidirectional conduction through a plane wall, as shown in Fig. 19-4.    By integration of Fourier's equation, Eqs. (19-12) for the rate of heat transfer and Eq. (19-14) for the temperature distribution were obtained.    Assuming $k$ constant, and noting that $t$ depends upon $x$ only, Eq. (20-22) simplifies to

$$\frac{d^2t}{dx^2} = 0$$

By integration, the solution of this equation is

$$\frac{dt}{dx} = C_1 \quad\text{and}\quad t = C_1 x + C_2$$

Integration constants $C_1$ and $C_2$ are evaluated from the boundary conditions.    For the particular problem under consideration, the boundary conditions are

$$t = t_1 \quad \text{at } x = x_1$$
$$t = t_2 \quad \text{at } x = x_2$$

Thus $\qquad\qquad t_1 = C_1 x_1 + C_2 \quad\text{and}\quad t_2 = C_1 x_2 + C_2$

Solving for $C_1$ and $C_2$,

$$C_1 = \frac{t_2 - t_1}{x_2 - x_1} \quad\text{and}\quad C_2 = \frac{t_1 x_2 - t_2 x_1}{x_2 - x_1}$$

With these expressions for $C_1$ and $C_2$,

$$t = \frac{(t_2 - t_1)(x)}{x_2 - x_1} + \frac{t_1 x_2 - t_2 x_1}{x_2 - x_1}$$

This expression for the temperature distribution can be rearranged to be identical to Eq. (19-14).    By Fourier's equation, $\dot{Q} = -kA\, dt/dx$, and from the equation for the temperature distribution,

$$\frac{dt}{dx} = \frac{t_2 - t_1}{x_2 - x_1}$$

Hence $\qquad\qquad \dot{Q} = \frac{kA(t_1 - t_2)}{x_2 - x_1}$

which is identical to Eqs. (19-12).

To further illustrate the means of applying the general differential equation for the temperature distribution for unsteady-state conduction, consider that heat is generated at a uniform rate in a sphere made of material having a constant $k$ value.    At steady state, let $t_1$ be the temperature at the geometric center of the sphere and let $t_2$ be the temperature at the outer surface of the sphere of radius $r_2$.    The heat transfer from the surface of such a sphere was given in Art. 20-1 as

$$\dot{Q}_2 = 8\pi k r_2 (t_1 - t_2)$$

Noting that $t$ is a function of $r$ only, Eq. (20-24$b$) simplifies to

$$\frac{d^2t}{dr^2} + \frac{2}{r}\frac{dt}{dr} + \frac{\omega}{k} = 0 = \frac{1}{r^2}\frac{d}{dr}\left(r^2\frac{dt}{dr}\right) + \frac{\omega}{k}$$

Integrating twice, the solution for $t$ is

$$t = -\frac{C_1}{r} - \frac{\omega r^2}{6k} + C_2$$

As $r$ approaches zero, $t$ must remain finite, and hence $C_1 = 0$. At $r$ equals $r_2$, $t$ equals $t_2$; thus

$$C_2 = t_2 + \frac{\omega r_2{}^2}{6k}$$

So the expression for the temperature distribution is

$$t = t_2 + \frac{\omega}{6k}(r_2{}^2 - r^2)$$

As expected, the maximum temperature in the sphere occurs at $r = 0$ and may be expressed as

$$t_1 = t_2 + \frac{\omega}{6k}r_2{}^2$$

To obtain an expression for $\dot{Q}_2$, note that

$$\omega = \frac{\dot{Q}_2}{\frac{4}{3}\pi r_2{}^3}$$

Thus $\qquad\qquad t_1 = t_2 + \dfrac{3\dot{Q}_2}{24\pi r_2{}^3 k}r_2{}^2$

and, as before, $\qquad \dot{Q}_2 = 8\pi k r_2(t_1 - t_2)$

The concept of thermal resistance provides a convenient means for classifying unsteady-state conduction problems. To illustrate the utility of the resistance concept, consider that a body is immersed in a fluid. Initially, let the fluid and the solid be at the same temperature. Then assume that the temperature of the fluid is suddenly increased. Now, energy will be transferred to the surface of the solid by convection and radiation, where it will be conducted into and stored in the body. The manner in which temperatures in the solid vary with coordinate position and time depends upon the relative values of two resistances—the resistance to conduction heat transfer within the body and the resistance to convection and radiation heat transfer at the surface. If the internal resistance (conduction) is small compared with the surface resistance (convection and radiation), only small temperature gradients occur in the solid. Consequently, temperatures vary little with coordinate position, and to good approximation, the temperatures in the body may be considered functions of time only. At the other extreme, if the internal resistance is large compared with the surface resistance, the temperature of the body varies with both position and time. But the surface temperature closely approaches the fluid temperature, and with little error, surface temperature may be assumed equal to fluid temperature. In either case, the mathematical treatment of the problem is simplified compared with the general case in which the internal and surface resistances are of about equal importance.

Considering the case in which internal resistance is negligible compared with surface resistance, Eq. (20-23) may be reduced to

$$\omega = \rho c \frac{dt}{d\tau} \tag{20-25}$$

The terms describing the variation of temperature with coordinate position have been discarded by reasoning that $t$ is a function of time only.

Assume that a body of surface area $A$, volume $V$, density $\rho$, thermal conductivity $k$, and specific heat $c$ is immersed in a fluid. Consider that initially the solid and the fluid are at the same temperature $t_i$ but that at $\tau = 0$ the fluid temperature is suddenly changed to, and maintained at, a different temperature $t_o$. Denoting $h$ as the surface coefficient for heat transfer between the fluid and the body, $\omega$ may be regarded as an equivalent surface source (or sink) of heat per unit volume of solid:

$$\omega = \frac{hA(t_o - t)}{V}$$

Upon substituting this value of $\omega$ into Eq. (20-25),

$$\frac{hA(t_o - t)}{V} = -\rho c \frac{d(t_o - t)}{d\tau} \tag{20-26}$$

Rearranging and integrating Eq. (20-26), with $t$ and $\tau$ as the variables, results in

$$\ln (t_o - t) = -\frac{hA\tau}{\rho c V} + C_1 \tag{20-27}$$

The integration constant $C_1$ is evaluated from the initial condition that $t = t_i$ at $\tau = 0$:

$$C_1 = \ln (t_o - t_i)$$

With this value of $C_1$, Eq. (20-27) may be written as

$$\frac{t_o - t}{t_o - t_i} = \exp \left( -\frac{hA\tau}{\rho c V} \right) \tag{20-28}$$

The term exp represents $e$ (base of the natural logarithm system) raised to the exponent indicated in parentheses. Solving Eq. (20-28) for $t$,

$$t = t_o - (t_o - t_i) \exp \left( -\frac{hA\tau}{\rho c V} \right) \tag{20-29}$$

Equation (20-29) describes the variation of temperature with time when the internal resistance within the body is small compared with the surface resistance.

From analyses of cases where both internal and surface resistance were considered, a criterion has been established to show under what conditions internal resistance may be neglected without introducing much error. For bodies with shapes similar to plates, cylinders, or spheres, temperatures in the body will differ by less than 5 per cent at any one time, and hence neglecting internal resistance will introduce little error if

$$\frac{hL}{k} \leq 0.1 \tag{20-30}$$

Here $h$ is the surface coefficient for heat transfer, $k$ is the thermal conductivity of the solid, and $L$ is a characteristic length equal to the ratio of the volume of the solid to its surface area.

The dimensionless term $hL/k$ is known as the Biot number (**Bi**). The numerical value of the Biot number is an indication of the ratio of internal resistance to surface resistance. Although the Biot number is of the same form as the Nusselt number, there is an important difference between the two. In the Nusselt number, $k$ is the thermal conductivity of the fluid, while in the Biot number, $k$ is the thermal conductivity of the solid.

**Example 5.** A copper-constantan thermocouple probe is made such that the thermocouple tip is a sphere 0.1 in. in diameter. The probe is to measure the temperature of air at atmospheric pressure flowing at a velocity of 10 ft/sec. Initially, the probe and the air are at a temperature of 80°F. The air temperature is suddenly changed to, and maintained at, 440°F. Approximately how many seconds will be required for the thermocouple to indicate a temperature of 400°F?

*Solution.* By assuming that the internal resistance is much less than the surface resistance, Eq. (20-29) can be solved to obtain an approximate value for the required time. In Eq. (20-29), $t = 400°F$, $t_o = 440°F$, $t_i = 80°F$, $V/A = 0.1/(6)(12)$, and $\tau = $ time required. The properties of the thermocouple probe are needed. For copper, $\rho = 559$ lb$_m$/ft³, $c = 0.0915$ Btu/(lb$_m$)(°F), and $k \approx 220$ Btu/(hr)(ft)(°F); for constantan, $\rho = 557$ lb$_m$/ft³, $c = 0.098$ Btu/(lb$_m$)(°F), and $k \approx 14$ Btu/(hr)(ft)(°F). Calculating average values for the tip, $\rho = 558$ lb$_m$/ft³ and $c = 0.0947$ Btu/(lb$_m$)(°F).

A value for $h$ is required to complete the solution. McAdams[1] recommends that the following equation be used to calculate an average value of the convection coefficient for the case of air flowing past a sphere:

$$\mathbf{Nu} = 0.37(\mathbf{Re})^{0.6}$$

The characteristic length is the sphere diameter, all properties are evaluated at the film temperature, and the equation has been found valid for a range of Reynolds numbers from 20 to 70,000.

From Table B-15, the air properties at a film temperature of 260°F are $\rho = 0.0551$ $lb_m/ft^3$, $\mu_m = 0.0553$ $lb_m/(hr)(ft)$, and $k = 0.0194$ Btu/(hr)(ft)(°F).

$$\mathbf{Nu} = \frac{(h_c)(0.1)}{(12)(0.0194)} = (0.37)\left[\frac{(10)(3,600)(0.1)(0.0551)}{(12)(0.0553)}\right]^{0.6}$$

$$\mathbf{Nu} = (0.37)(300)^{0.6} = 11.3$$

and $$h_o = 26.4 \text{ Btu}/(hr)(ft^2)(°F)$$

Note that the Reynolds number value of 300 falls within the recommended Reynolds number range. Although the film temperature varies as the probe temperature changes, the effect of the film-temperature variation on $h_c$ is small.

Checking the value of the Biot number to determine if the approach using Eq. (20-29) was valid:

$$\mathbf{Bi} = \frac{hL}{k} = \frac{hV}{kA} = \frac{(26.4)(0.1)}{(14)(6)(12)} < 0.1$$

Thus Eq. (20-29) may be used. [The thermal conductivity of the copper-constantan thermocouple probably has a value between that of constantan (14) and that of copper (220). In calculating the Biot number, $k = 14$ was used as the least value of thermal conductivity likely for the tip, thus yielding the largest value of the Biot number.]

Equation (20-29) may now be solved for $\tau$:

$$400 = 440 - (440 - 80) \exp \frac{-(26.4)(6)(12)(\tau)}{(0.1)(558)(0.0947)}$$

and $$\tau = 22 \text{ sec} \qquad Ans.$$

Radiation from the probe and conduction along the thermocouple lead wires may cause the actual time requirement to be somewhat greater than that calculated.

A characteristic time of response known as the *time constant* is defined for instruments which respond exponentially to a sudden change [as in Eq. (20-29)]. The time constant is defined as that time required to yield a value of unity for the exponent of $e$. Thus the time constant for the thermocouple considered in Example 5 is given by

$$\frac{hA\tau}{\rho c V} = 1$$

or $$\text{Time constant} = \frac{\rho c V}{hA} = \frac{(558)(0.0947)(0.1)(3,600)}{(26.4)(6)(12)} = 10 \text{ sec}$$

At 10 sec, the thermocouple indicates a temperature of

$$t = 440 - (440 - 80) \exp(-1) = 308°F$$

In a period of time equal to its time constant, the thermocouple has attained $1 - 1/e$ or 63.2 per cent of the sudden temperature change.

When internal and surface resistance are of equal importance, or when the internal resistance is large compared with the surface resistance, the temperature distribution in the body is a function of both coordinate position and time, and the mathematical solution is lengthy. However, many solutions for such cases have been obtained, and for convenience, several of these solutions are available in graphical form similar to Fig. 20-7. (See McAdams[1] and Schneider[10] for other examples.)

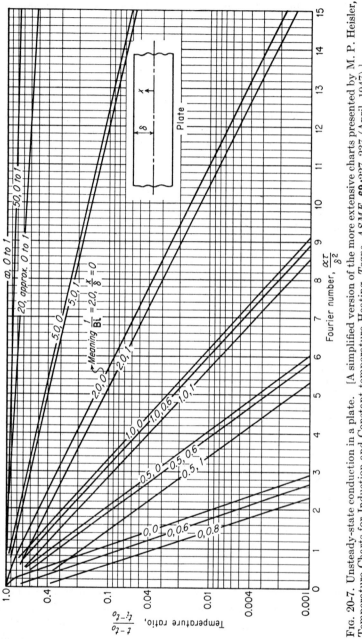

FIG. 20-7. Unsteady-state conduction in a plate. [A simplified version of the more extensive charts presented by M. P. Heisler, Temperature Charts for Induction and Constant-temperature Heating, *Trans. ASME*, **69**:227–237 (April, 1947).]

For the plate considered in Fig. 20-7, unsteady-state conduction occurs in only one coordinate direction without internal heat generation. If, in addition, constant properties are assumed, Eq. (20-22) reduces to

$$\frac{\partial^2 t}{\partial x^2} = \frac{\rho c}{k} \frac{\partial t}{\partial \tau} = \frac{1}{\alpha} \frac{\partial t}{\partial \tau} \tag{20-31}$$

The parameter $\alpha$ is equal to $k/\rho c$ and is a property known as the thermal diffusivity. Values of $\alpha$ for a few substances are given in the following table.

TABLE 20-1. THERMAL DIFFUSIVITIES OF VARIOUS SUBSTANCES AT 70°F

| Substance | $\alpha$, $ft^2/hr$ |
|---|---|
| Aluminum (pure) | 3.67 |
| Cast iron | 0.67 |
| Copper (pure) | 4.35 |
| Constantan | 0.24 |
| Steel (1% C) | 0.45 |
| Asbestos (36 $lb_m/ft^3$) | 0.013 |
| Wood | 0.0045 |
| Water (saturated liquid) | 0.0055 |
| Air | 0.82 |
| Hydrogen | 5.84 |

The relative magnitude of the thermal diffusivity is an approximate measure of the rapidity with which a substance responds to a sudden temperature change in its surroundings. Note that metals and gases have relatively large values of $\alpha$ and hence respond rapidly to temperature changes, while nonmetallic solids and liquids have relatively small values of $\alpha$ and thus respond slowly to temperature changes. (In the case of liquids and gases, convection effects may decrease the time of response to temperature change.)

In obtaining the results shown in Fig. 20-7, it is assumed that the plate at a uniform initial temperature $t_i$ is suddenly, at $\tau = 0$, exposed to a fluid stream at constant temperature $t_o$. Thus the initial condition for the solution of Eq. (20-31) is

$$t = t_i \qquad \text{at } \tau = 0, \ -\delta \le x \le \delta$$

The rate of heat transfer between the fluid and the plate at the plate surfaces is

$$h(t_o - t_{x=\pm\delta}) = -k \left( \frac{\partial t}{\partial x} \right)_{x=\pm\delta}$$

The preceding expression may be rearranged to give one of the boundary conditions:

$$\frac{\partial t}{\partial x} = \frac{h}{k} (t - t_o) \qquad \text{at } x = \pm\delta, \ \tau > 0$$

By symmetry, no conduction occurs across the mid-plane of the plate; thus the second boundary condition is

$$\frac{\partial t}{\partial x} = 0 \qquad \text{at } x = 0, \tau > 0$$

An expression for the temperature in the plate at any point $x$ and any time $\tau$ may be obtained by solving Eq. (20-31) and applying the initial condition and the two boundary conditions. The values obtained from this solution are presented in graphical form in Fig. 20-7. The following example problem illustrates the method of using Fig. 20-7.

**Example 6.** During the manufacture of plastic sheets 4 in. thick, the sheets are brought to a uniform temperature of 350°F and then allowed to cool to a surface temperature of 125°F in air at 100°F before further processing. Properties of the plastic material are

$$\rho = 80 \text{ lb}_m/\text{ft}^3 \qquad c = 0.4 \text{ Btu}/(\text{lb}_m)(°\text{F}) \qquad k = 0.167 \text{ Btu}/(\text{hr})(\text{ft})(°\text{F})$$

(a) How long a cooling period will be required if natural convection cooling is employed with an average surface coefficient of 2 Btu/(hr)(ft²)(°F)? (b) What is the temperature at the center of the plastic sheet when the surface temperature has reached 125°F?

*Solution.* (a) The Biot number should first be calculated to determine the relative values of surface and internal resistance:

$$\text{Bi} = \frac{h\delta}{k} = \frac{(2)(\frac{1}{6})}{0.167} = 2$$

Note that $\delta$ is one-half the plate thickness. For this value of Biot number, the internal resistance cannot be neglected, and Fig. 20-7 must be used. Calculating the temperature ratio,

$$\frac{t - t_o}{t_i - t_o} = \frac{125 - 100}{350 - 100} = \frac{25}{250} = 0.1$$

For the surface of the plate, $x/\delta = 1$. Reading the value of $\alpha\tau/\delta^2$ from Fig. 20-7 at a temperature ratio of 0.1, 1/Bi of 0.5, and $x/\delta = 1$ yields $\alpha\tau/\delta^2$ of approximately 1.5. Solving for $\tau$,

$$\tau = \frac{(1.5)(\frac{1}{6})^2(80)(0.4)}{0.167} = 7.98 \text{ hr} \qquad Ans.$$

The parameter $\alpha\tau/\delta^2$ is a dimensionless time known as the *Fourier number* (**Fo**).

(b) At the center of the sheet, $x/\delta = 0$. From Fig. 20-7 at **Fo** = 1.5, 1/**Bi** = 0.5, and $x/\delta = 0$, the temperature ratio is 0.21. Thus the center temperature is computed as

$$t_{\text{center}} = (0.21)(350 - 100) + 100 = 152°\text{F} \qquad Ans.$$

**20-4. The Critical-radius Concept.†**   In some instances, the addition of an insulating material causes not a decrease but an *increase in the total heat transfer.*   A method for predicting such an effect when the combined coefficient of convection and radiation may be considered constant will be developed.   Figure 20-8 shows a long cylinder of length $L$, with radius $r_1$ at temperature $t_1$. Surrounding this cylinder is an annular section of insulation with thermal conductivity $k_{12}$, thickness $r_2 - r_1$, and outer surface temperature $t_2$. Assuming steady-state conditions and $t_1$ greater than $t_2$, heat is conducted through the insulation and then transferred by convection and radiation from the surface to the surroundings at $t_o$.

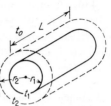

Fig. 20-8. Cylinder for establishing the critical radius criterion.

By the resistance concept (Art. 19-5),

$$\dot{Q} = \frac{t_1 - t_o}{R_{t12} + R_{t2o}}$$

and $$R_{t12} + R_{t2o} = \frac{\ln r_2 - \ln r_1}{2\pi k_{12}L} + \frac{1}{2\pi r_2 L h}$$

For fixed values of $t_1$ and $t_o$, the rate of heat transfer will be a maximum when the total resistance is a minimum.   To establish a criterion for maximum heat-transfer rate, let it be assumed that $t_1$, $t_o$, $r_1$, $L$, $k_{12}$, and $h$ are constant while the thickness of insulation is varied, thus changing $r_2$ and $t_2$.   As $r_2$ is increased, the surface temperature $t_2$ will decrease, but at the same time, the surface area $2\pi r_2 L$ will increase.

Using these assumptions, the criterion for minimum resistance is found as follows:

$$\frac{d(R_{t12} + R_{t2o})}{dr_2} = \frac{1}{2\pi r_2 k_{12}L} - \frac{1}{2\pi r_2^2 L h} = 0 \qquad \text{(for a minimum)}$$

Thus the *critical radius* for minimum resistance ($r_2 = r_{2c}$) is

$$r_{2c} = \frac{k_{12}}{h} \tag{20-32}$$

In a given situation, if $r_1$ is less than the critical radius given by Eq. (20-32), then the addition of insulation of the particular $k$ value will lead to increasing heat transfer until the outer radius of the insulation becomes equal to $r_{2c}$.   If $r_1$ is greater than the predicted $r_{2c}$, then the addition of insulation will decrease the heat transfer.   In practical cases, the phenomenon of increase in heat transfer is most likely to occur when

† If desired, this article may be assigned upon completion of Art. 19-19.

insulating materials of poor quality are applied to cylinders of small radius.

This effect is sometimes used to advantage in the insulation of electrical wires. The prime purpose of the insulation is to provide protection from electrical hazard, but by using the proper thickness, the ability of the insulated wire to dissipate heat may be made greater than that for the bare wire.

**Example 7.** Suppose that insulation of mean thermal conductivity equal to 0.1 Btu/(hr)(ft)(°F) were to be applied to the copper transmission line considered in Example 1. Would the ability of the bare wire to dissipate heat be decreased or increased by the addition of insulation? Data from Example 1 are $r_1 = 0.811/2 = 0.4055$ in. and $h = 2.5$ Btu/(hr)(ft²)(°F).

*Solution.* By Eq. (20-32), $r_{2c} = 0.1/2.5 = 0.04$ ft $= 0.48$ in. $r_{2c}$ is greater than $r_1$, and thus a thin layer of insulation would increase the heat dissipation from the wire and so permit some increase in the current-carrying capacity of the transmission line. *Ans.*

Another situation in which the addition of "insulation" increases heat transfer may be illustrated by considering the ducts used in warm-air heating and air-conditioning systems. The bare metal duct has a shiny surface, consequently a low emissivity (Table 19-2); thus heat transfer by radiation is small. Now if the duct is covered with a thin sheet of asbestos paper or similar material, the temperature of the surface more nearly approaches that of the surrounding air and (for practical duct sizes) convection decreases. But the emissivity of the outer surface is now several times greater than before (Table 19-2), and the heat transfer by radiation is increased severalfold. The net effect may well be that the "insulation" causes an increase in heat transfer (but a decrease in noise).

**20-5. Convection Heat Transfer with Change of Phase.†** In Art. 19-8, relationships were presented for computing the convection-heat transfer rate between a solid surface and a single-phase fluid. In this article, consideration is given to the problem of calculating convection *between a surface and a fluid which changes phase*—as in condensation, which involves a change from vapor to liquid phase, or in evaporation, where the inverse change in phase occurs.

*a. Condensation.* When a vapor encounters a cold surface, its temperature is reduced until the condensation temperature for the pressure is reached, and the surface is then covered with a thin liquid film or boundary layer of condensate. Vapor meeting the condensate film loses its latent heat by conduction through the film, while the film grows in thickness to a value dictated, for example, by the viscosity and the effect of gravity in displacing the liquid. This is *film-type condensation.* Since the thermal conductivities of liquids are small, film-type condensation causes an appreciable temperature difference to exist between the cold surface and the vapor. Note that an increased rate of condensation

† If desired, this article may be assigned upon completion of Art. 19-8.

increases the thickness of the liquid film and thus increases the temperature difference, although the surface coefficient of heat transfer will be lowered by the thicker film.

When the condensation surface is smooth and polished or when substances are present that prevent the condensate from wetting the surface, *dropwise condensation* may occur. Here the condensate forms in small drops that rapidly grow in size and run off the surface without forming a film. Thus the heat transfer is not restricted by the insulating effect of a liquid film, and local coefficients as great as 75,000 Btu/(hr)(ft²)(°F) have been reported. Certain oils, called *promoters,* can be added to achieve dropwise condensation, but this relatively unstable form of condensation may not materially change the over-all coefficient because some other resistance may be found to be the controlling factor.

The presence of a noncondensable gas in the vapor will greatly reduce the rate of heat transfer. When a mixture of gas and vapor encounters a cold surface, the vapor, but not the gas, will be condensed. Thus a layer will be formed immediately adjacent to the condensate film that contains more air (and less steam) than the main body of the mixture. The condensation process is retarded because the vapor must first diffuse through the gas layer before reaching the liquid film and condensing surface; consequently, the film coefficient of heat transfer is greatly reduced.

The Reynolds number assumes a special form for condensation processes. When the condensation surface is a vertical plate, of width $b$ and with film of thickness $x$ [refer to Eq. (19-25a)],

$$\mathbf{Re} = \frac{DV\rho}{\mu_m} = \frac{GD_e}{\mu_m} = \frac{G4(bx/b)}{\mu_m} = \frac{4Gx}{\mu_m}$$

But the runoff $\Gamma$ is defined as the mass rate of flow per foot of width $b$: $\Gamma = Gx$. Therefore,

$$\mathbf{Re}_{\text{vertical surfaces}} = \frac{4\Gamma}{\mu_m} \tag{20-33}$$

Equation (20-33) is also valid for vertical pipes (and here the width is $\pi D_o$).

CONDENSATION (FILM-TYPE) RELATIONSHIPS.[†] The physical properties of the fluid in the condensate film are evaluated at a mean film temperature:

$$t_f = t_{\text{sat vapor}} - 0.75 \, \Delta t \tag{20-34}$$

The average surface temperature ($t_s$) must be estimated, and

$$\Delta t = t_{\text{sat vapor}} - t_s$$

[†] Recommended by McAdams.[1]

where $t_{\text{sat vapor}}$ is the saturation temperature of the vapor. Physical properties are evaluated at $t_f$, as defined by Eq. (20-34), regardless of whether the vapor is initially saturated or superheated.

Vertical surfaces and tubes

$$h_m = 1.13 \left(\frac{k^3\rho^2g\lambda}{L\mu_m \,\Delta t}\right)^{0.25} = 1.88(\mathbf{Re})^{-0.33}\left(\frac{k^3\rho^2g}{\mu_m{}^2}\right)^{0.33} \qquad \mathbf{Re} < 1,800$$

$$(20\text{-}35)$$

$$h_m = 0.0077(\mathbf{Re})^{0.4}\left(\frac{k^3\rho^2g}{\mu_m{}^2}\right)^{0.33} \qquad \mathbf{Re} > 1,800 \qquad (20\text{-}36)$$

Horizontal tubes

$$h_m = 0.725\left(\frac{k^3\rho^2g\lambda}{ND_o\mu_m \,\Delta t}\right)^{0.25} \qquad \frac{2\Gamma'}{\mu_m} < 2,100 \qquad (20\text{-}37)$$

where $\lambda$ = latent heat of condensation $(h_{fg})$

$N$ = number of vertical tiers (banks) of horizontal tubes

$\Gamma'$ = mass flow of condensate per foot of length from lowest tube in tier, $\text{lb}_m/(\text{hr})(\text{ft})$; this is total flow from all tubes in any one vertical plane because condensate from upper tubes will drip or flow onto lower tubes

*b. Evaporation.* Suppose that saturated liquid, in equilibrium with saturated vapor, is contained over a surface that can be gradually raised in temperature. When the temperature of the surface is raised, the temperature of the adjacent liquid is increased over the saturation point, and the liquid becomes *superheated.* An examination of the liquid would reveal, at first, no visible change, although vaporization is occurring, as shown by a decrease in the liquid level. In this instance, heat is being transferred by conduction and convection through the liquid. When the temperature difference between the source and liquid reaches a value of about 4°F, bubbles of steam appear and rise through the liquid, thus greatly increasing the convection of heat. This is the beginning of the stage called *nucleate boiling.* As the temperature difference is increased, the number of bubbles also increases and a violent boiling takes place. But now, if a still greater temperature difference is allowed, an unstable vapor film will first gather over the heating surface and then break away, forming large bubbles, while another film attempts again to blanket the hot surface. Further increase in the temperature difference stabilizes the film in place over the heating surface, and therefore *film boiling* is said to be present. The duration of these stages of boiling is illustrated in Fig. 20-9, which shows the relationship between the coefficient $h_c$ and the temperature difference $\Delta t$. In the first stage, the heat-transfer rate is low because the conductivity of the liquid is small and convection is slight. But the rate rapidly rises as nucleate boiling increases until a maximum point is reached. (For water at 1 atm pressure, this critical temperature difference is around 50°F.) When film boiling occurs, the heat transfer decreases because of the insulating effect of the vapor film. Note that, with continued heating, the curve in Fig. 20-9 reaches a minimum value and then starts to climb again because of heat transferred by radiation from the extremely hot source; here failure of the heater or the heating surface by melting (*burnout*) may occur.

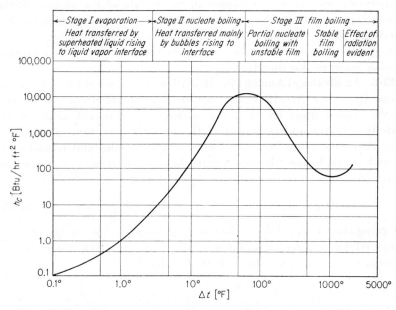

Fig. 20-9. Boiling of water at atmospheric pressure on a chromel C heating wire. [Data of E. A. Farber and R. L. Scorah, Heat Transfer to Water Boiling under Pressure, *Trans. ASME,* **70:** 369–384 (May, 1948).]

The following conclusions appear in the literature[†] for tests made with water and various metals for the heating surface:

1. The boiling curves for all the metals tested were similar but gave different numerical values. At atmospheric pressure and $\Delta t$ of 10°F, the $h_c$ values were: chromel C, 100; tungsten, 500; chromel A, 3300; nickel, 20,000.

2. The boiling curve is radically affected by pressure: chromel C and $\Delta t$ of 10°F; $h_c = 100$ for 0 psig, 1200 for 50 psig, 8000 for 75 psig, and 90,000 for 100 psig.

[†] E. A. Farber and R. L. Scorah, Heat Transfer to Water Boiling under Pressure, *Trans. ASME,* **70:** 369–384 (May, 1948).

3. The mechanism of boiling is different for different parts of the boiling curve, and the same heat-transfer rate can be obtained with three different values of $\Delta t$: chromel C at 0 psig, $\dot{Q}/A = 100,000$ Btu/(hr)(ft²) for $\Delta t$ values of 30, 430, and 1650°F, with corresponding values of $h_c$ of 3330, 232, and 60.5.

From these observations, it is evident that the rate of heat transfer in boiling depends in a complex manner upon many factors. Furthermore, accurate experimental results have been difficult to obtain because of the large values of $h_c$ for boiling and because of the consequent small temperature differences. Changes in the nature of the surface during boiling experimentation have also introduced errors. For these reasons, equations which correctly predict boiling convection coefficients over wide ranges of system variables have not been developed, but progress is being made in this direction.†

**20-6. An Electrical Analogy for Radiation‡.** As in conduction through composite bodies (Art. 19-5), the concept of a thermal circuit, or *radiation network*, based upon an electrical analogy for radiation heat transfer leads to a better understanding of the multiple reflections and absorptions that take place in an enclosure of gray bodies and refractories.§ In such an enclosure, the net radiant heat interchange between a gray source (1) and a gray sink (2) may be expressed as

$$\dot{Q}_{12} = A_1\mathfrak{F}_{12}\sigma(T_1^4 - T_2^4) = A_1\mathfrak{F}_{12}(W_{b1} - W_{b2}) \qquad (20\text{-}38)$$

Adopting the viewpoint from the electrical analogy that the heat transfer is caused by a *potential difference* $(W_{b1} - W_{b2})$ acting across a total thermal resistance $(\Sigma R_t)$, Eq. (20-38) becomes¶

$$\dot{Q}_{12} = \frac{W_{b1} - W_{b2}}{\Sigma R_t} \qquad \text{where } \Sigma R_t = \frac{1}{A_1\mathfrak{F}_{12}} \qquad (20\text{-}39)$$

Now the problem is to specify the items that contribute to the total resistance to radiation heat transfer between the two surfaces. Evidently resistance to radiation interchange is present at each of the two gray surfaces, for they are imperfect emitters and absorbers. Further-

† For example, see J. W. Westwater, Boiling Heat Transfer, *American Scientist*, **47**: 427–446 (September, 1959).

‡ If desired, this article may be assigned upon completion of Art. 19-18.

§ Formulated by A. K. Oppenheim, Radiation Analysis by the Network Method, *Trans. ASME*, **78**: 725–734 (May, 1956).

¶ Here the potential for heat transfer is temperature, or, more specifically for radiation, temperature to the fourth power. However, it is conventional to retain the Stefan-Boltzmann constant $(\sigma)$ in the potential-difference term rather than in the resistance; thus the potential difference becomes the difference in the hemispherical emissive powers of the surfaces considered as black bodies.

more, other resistances exist in the space between the source and the sink. These resistances are caused by the geometric configuration of source, sink, and refractories as it affects (1) the direct exchange of radiation (the $F_{12}$ factor) and (2) the indirect interchange by reflection from refractory surfaces (the $\bar{F}_{12} - F_{12}$ factor).† In Fig. 20-10a, items 1 and 2 constitute a parallel resistance ($R_{t12}$ and $R_{t1R} + R_{tR2}$) in series with the resistances caused by the gray-body behavior of the source and sink ($R_{t1}$ and $R_{t2}$). The parallel resistances have been replaced in Fig. 20-10b by a single equivalent resistance ($R_{t\overline{12}}$). From Fig. 20-10, the total resistance to radiation heat transfer is

$$\Sigma R_t = R_{t1} + R_{t\overline{12}} + R_{t2} = R_{t1} + \frac{1}{1/(R_{t1R} + R_{tR2}) + 1/R_{t12}} + R_{t2}$$

$$(20\text{-}40)$$

For evaluation of the resistances arising from gray-body behavior ($R_{t1}$ and $R_{t2}$), two new terms are defined:

$J$ (*radiosity*) = radiation flux density leaving a surface (emission and reflection), Btu/(hr)(ft²)

$G$ (*irradiation*) = radiation flux density striking a surface, Btu/(hr)(ft²)

By their definitions, $J$ and $G$ bear the following relationships to each other:

$$J = \epsilon W_b + \rho G = \epsilon W_b + (1 - \epsilon)G \qquad \text{for opaque, gray surface}$$

Thus
$$G = \frac{J - \epsilon W_b}{1 - \epsilon} \qquad (20\text{-}41)$$

From Fig. 20-10 and the definitions of $J$ and $G$,

$$\dot{Q}_{12} = \frac{W_{b1} - W_{b2}}{\Sigma R_t} = \frac{W_{b1} - J_1}{R_{t1}} = A_1(J_1 - G_1) \qquad (20\text{-}42)$$

A substitution of Eq. (20-41) into Eq. (20-42) leads to

$$R_{t1} = \frac{1 - \epsilon_1}{\epsilon_1 A_1} \qquad \text{and similarly} \qquad R_{t2} = \frac{1 - \epsilon_2}{\epsilon_2 A_2} \qquad (20\text{-}43)$$

Thus, in general, the resistance caused by gray-body behavior is

$$\frac{1 - \epsilon}{\epsilon A}$$

The resistance due to geometric configuration and refractory reflection may be obtained (without any loss of generality) by assuming for the

† If the gas between surfaces 1 and 2 is an absorber (and emitter), it causes an additional resistance. The present discussion considers only those cases in which the enclosure is filled with a gas that does not radiate significantly, such as air at the temperatures normally encountered in engineering. See Art. 19-18 for further discussion of this point.

present that source 1 and sink 2 are black, so that $R_{t1}$ and $R_{t2}$ are zero, $W_{b1} = J_1$, and $W_{b2} = J_2$. Referring to Eqs. (19-80) and (20-39),

$$\dot{Q}_{12} = \frac{W_{b1} - W_{b2}}{\Sigma R_t} = \frac{J_1 - J_2}{R_{t\overline{12}}} = A_1 \bar{F}_{12}(J_1 - J_2)$$

or
$$R_{t\overline{12}} = \frac{1}{A_1 \bar{F}_{12}} \qquad (20\text{-}44)$$

In general, the resistance caused by geometric configuration and refractory reflection is equal to $1/A\bar{F}$.†

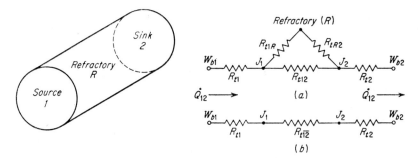

FIG. 20-10. Radiation networks for a source-sink-refractory enclosure.

Determining the total thermal resistance by Eqs. (20-40), (20-43), and (20-44),

$$\Sigma R_t = \frac{1 - \epsilon_1}{\epsilon_1 A_1} + \frac{1}{A_1 \bar{F}_{12}} + \frac{1 - \epsilon_2}{\epsilon_2 A_2} \qquad (20\text{-}45)$$

From Eqs. (20-39) and (20-45),

$$\mathfrak{F}_{12} = \frac{1}{1/\bar{F}_{12} + 1/\epsilon_1 - 1 + (A_1/A_2)(1/\epsilon_2 - 1)} \qquad (20\text{-}46)$$

Equation (20-46) is identical with that found by Hottel [Eq. (19-83)].

† If there are no refractories in the enclosure, $R_{t\overline{12}} = R_{t12}$ (see Fig. 20-10a) and $R_{t12} = 1/A_1 F_{12}$. Furthermore, from Eq. (20-40), note that $R_{t\overline{12}} = 1/A_1\bar{F}_{12}$ is equivalent to

$$\frac{1}{1/(R_{t1R} + R_{tR2}) + 1/R_{t12}}$$

Utilizing the fact that $R_t = 1/AF$ and simplifying leads to

$$A_1\bar{F}_{12} - A_1F_{12} = \frac{1}{1/A_1F_{1R} + 1/A_RF_{R2}}$$

Thus the $\bar{F}$ factor between sources and sinks may be found from the direct radiation factors $F$ between sources, sinks, and refractories. Problem 31 further illustrates this point.

As shown in Art. 19-17, Eq. (20-46) may be reduced to obtain specific solutions for simple cases. From this observation, it might be concluded that Eq. (20-46) is generally valid for enclosures of one gray source, one gray sink, and refractories. But, as previously pointed out by Hottel and, more recently (with the aid of the radiation network), by Oppenheim, Eq. (20-46) is strictly correct only when:

1. The source-sink system is symmetrical, as, for example, a small sphere located centrally inside of a larger sphere, or a small thermocouple located at the center of a duct. A general rule is that a system is symmetrical when all points on the source (or sink) have the same view of the sink (or source).

2. If the source-sink system is not symmetrical, Eq. (20-46) is still correct if source and sink are both black, but it becomes less valid as the emissivities of the source and sink become smaller. (As emissivity becomes less, reflection from the source and sink accounts for a greater portion of the radiant flux in the enclosure.)

The concepts of. radiosity and irradiation assist in making these limitations on Eq. (20-46) understandable. In Fig. 20-10, a single value of radiosity was ascribed to each gray surface. The radiosity of such surfaces is uniform only if the surface is at uniform temperature and the impinging irradiation is uniformly distributed (the case for symmetrical systems). For unsymmetrical systems, the radiosity approaches uniformity only as the emissivity of the constant-temperature sinks and sources approaches unity. Thus, in unsymmetrical enclosures, accurate results may not be obtained if sources and sinks of small emissivity are established only on the basis of selecting surfaces at uniform temperature. To approach uniform irradiation (and thus uniform radiosity), further subdivision (yielding a more complex network) may be necessary if accurate results are needed.

To illustrate the effects of deviations from symmetry and emissivity on the accuracy of Eq. (20-46), Jakob[5] reports on a small sphere, 1, at uniform temperature inside of a large sphere, 2, at uniform temperature, with $D_2/D_1$ equal to 4. When sphere 1 is centrally located, Eq. (20-46) leads to a correct value for heat transfer for all values of $\epsilon_1$ and $\epsilon_2$. When sphere 1 is located eccentrically such that it almost touches sphere 2, Eq. (20-46) results in a value about 7 per cent too great for $\epsilon_1 = 1$ and $\epsilon_2 = 0.5$, and about 3 per cent too great for $\epsilon_1 = \epsilon_2 = 0.75$. Although these inaccuracies are not great, much more serious errors may be encountered in other geometrical arrangements and for surfaces of smaller emissivities.

By means similar to those just described, thermal circuits may be drawn for more complex enclosures consisting of many sources, sinks, and refractories. In such cases, the networks are not so simple as the one considered, and specification of proper $F$ factors may be much more difficult. The following example problem illustrates the application of the network method to a complex enclosure.

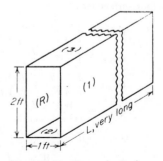

Fig. 20-11. Example 8, four-surface enclosure.

**Example 8.** Figure 20-11 shows an enclosure consisting of four surfaces having the following characteristics: surface 1: black-body surface maintained at a temperature of 1000°R; surface 2: gray-body surface, emissivity of 0.5, maintained at a temperature of 100°R; surface $R$: a well-insulated surface; surface 3: a completely transparent window which sees only outer space. Outer space is taken to have an effective radiating temperature of 0°R. It is desired (a) to make appropriate idealizations of the problem and sketch the radiation network and (b) to find numerical values for the net radiant-heat-transfer rate at surfaces 1, 2, and 3 ($\dot{Q}_{1\,net}$, $\dot{Q}_{2\,net}$, and $\dot{Q}_{3\,net}$, respectively) and for the temperature of surface $R$ ($T_R$).

*Solution.* (a) In idealizing the problem, it is convenient to assume that the well-insulated surface $R$ is an adiabatic (refractory) surface; hence $\dot{Q}_{R\,net} = 0$. Furthermore, the transparent surface which views outer space may be replaced by a black-body surface at temperature absolute zero. The substitution involves no approximation, for the transparent surface transmits all radiation impinging on it and emits no radiation, while the black surface at absolute zero absorbs all radiation falling on it and emits nothing. With these idealizations, the radiation in the enclosure shown in Fig. 20-11 can be represented by the network shown in Fig. 20-12.

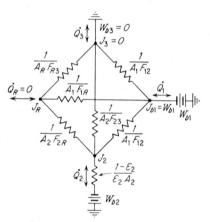

FIG. 20-12. Example 8, radiation network.

(b) For the network in Fig. 20-12, the following equations can be written for the four-surface enclosure:

$$\dot{Q}_{1\,net} = \frac{J_1 - J_2}{1/A_1F_{12}} + \frac{J_1 - J_R}{1/A_1F_{1R}} + \frac{J_1 - J_3}{1/A_1F_{13}}$$

$$\dot{Q}_{1\,net} = \frac{W_{b1} - J_1}{(1 - \epsilon_1)/\epsilon_1 A_1} \qquad \text{hence } W_{b1} = J_1 = \sigma T_1{}^4$$

$$\dot{Q}_{2\,net} = \frac{J_2 - J_1}{1/A_1F_{12}} + \frac{J_2 - J_R}{1/A_2F_{2R}} + \frac{J_2 - J_3}{1/A_2F_{23}}$$

$$\dot{Q}_{2\,net} = \frac{W_{b2} - J_2}{(1 - \epsilon_2)/\epsilon_2 A_2}$$

$$Q_{R\,net} = 0 = \frac{J_R - J_1}{1/A_1F_{1R}} + \frac{J_R - J_2}{1/A_2F_{2R}} + \frac{J_R - J_3}{1/A_RF_{R3}}$$

$$\dot{Q}_{R\,net} = 0 = \frac{W_{bR} - J_R}{(1 - \epsilon_R)/\epsilon_R A_R} \qquad \text{hence } W_{bR} = J_R = \sigma T_R{}^4$$

$$\dot{Q}_{3\,net} = \frac{J_3 - J_1}{1/A_1F_{13}} + \frac{J_3 - J_2}{1/A_2F_{23}} + \frac{J_3 - J_R}{1/A_RF_{R3}}$$

$$\dot{Q}_{3\,net} = \frac{W_{b3} - J_3}{(1 - \epsilon_3)/\epsilon_3 A_3} \qquad \text{where (by nature of surface 3) } W_{b3} = J_3 = 0$$

From Fig. 19-14, line 4, for long, narrow rectangles read $F_{1R}$ equals 0.62 at a ratio of 2 and $F_{23}$ equals 0.24 at a ratio of 0.5. The other values of $F$ can be obtained by Eqs. (19-72) and (19-73):

$$F_{12} = F_{13} = F_{R3} = 0.19 \qquad F_{2R} = 0.38$$

Other needed numerical values are

$$J_1 = W_{b1} = \sigma T_1{}^4 = 1{,}714 \qquad W_{b2} = \sigma T_2{}^4 = 0.1714$$
$$A_1 = A_R = 2L \qquad\qquad A_2 = A_3 = L$$

With these numerical values, the equations for the enclosure become

$$\left(\frac{\dot{Q}_1}{L}\right)_{net} = 0.38(1{,}714 - J_2) + 1.24(1{,}714 - \sigma T_R{}^4) + 0.38(1{,}714)$$

$$\left(\frac{\dot{Q}_2}{L}\right)_{net} = 0.38(J_2 - 1{,}714) + 0.38(J_2 - \sigma T_R{}^4) + 0.24(J_2)$$

$$\left(\frac{\dot{Q}_2}{L}\right)_{net} = \frac{0.5}{1 - 0.5}\,(0.1714 - J_2)$$

$$\left(\frac{\dot{Q}_R}{L}\right)_{net} = 0 = 1.24(\sigma T_R{}^4 - 1{,}714) + 0.38(\sigma T_R{}^4 - J_2) + 0.38(\sigma T_R{}^4)$$

$$\left(\frac{\dot{Q}_3}{L}\right)_{net} = 0.38(-1{,}714) + (0.24)(-J_2) + 0.38(-\sigma T_R{}^4)$$

Solving the system of five equations for the five unknown quantities,

$$\left(\frac{\dot{Q}_1}{L}\right)_{net} = 1774 \text{ Btu/(hr)(ft)} \qquad \left(\frac{\dot{Q}_2}{L}\right)_{net} = -547 \text{ Btu/(hr)(ft)}$$

$$\left(\frac{\dot{Q}_3}{L}\right)_{net} = -1227 \text{ Btu/(hr)(ft)} \qquad J_2 = 548 \text{ Btu/(hr)(ft}^2)$$

$$T_R = 909°\text{R} \qquad Ans.$$

**20-7. Radiation Instrumentation.†** An application of the principles of radiation has resulted in the development of two useful instruments, the *optical pyrometer* and the *total-radiation pyrometer*. Representative types of these devices are shown in Fig. 20-13. The prime advantage of these instruments is that they can be used to measure the temperature of bodies emitting radiant energy without the necessity of physical contact between the instrument and the body.

*a. Instrument Description.* The optical pyrometer measures the effective black-body temperature of hot sources, such as molten metals or luminous flames, which emit a significant amount of radiation in the visible range. As shown in Fig. 20-13a, the instrument consists of a viewing system that includes a light filter and a small filament which may be made to glow by electrical resistance heating. As the instrument is sighted upon the hot source, the filter permits the passage of only a small

† If desired, this article may be assigned upon completion of Art. 19-17. See Refs. 5 and 12 for additional information on this topic. Consult also T. R. Harrison, "Radiation Pyrometry and Its Underlying Principles of Radiant Heat Transfer," John Wiley & Sons, Inc., New York, 1960.

portion (normally red) of the visible spectrum. The filament in the viewing system is heated electrically until it appears to be of the same brightness as the filtered radiation from the source. The heating current required to bring the wire to a matching brightness is measured by the milliammeter, and from a previous calibration with a black-body source, the current reading is related to an effective black-body temperature. The meaning of the effective temperature is that, for a small portion of the visible spectrum, the source at its true temperature is emitting as much energy as a black body would emit at the effective temperature.

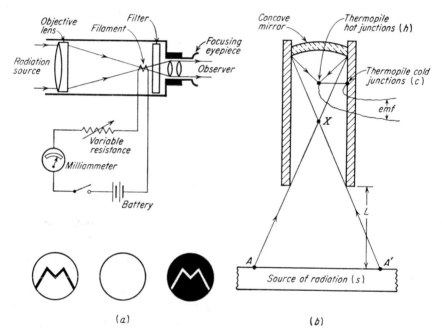

FIG. 20-13. Radiation pyrometers. (a) Optical pyrometer. Left circle—source brightness greater than filament brightness; middle circle—source brightness equal to filament brightness, proper adjustment of variable resistance for determining temperature; right circle—source brightness less than filament brightness. (b) Total-radiation pyrometer.

The optical pyrometer is widely utilized for the measurement of high temperatures where devices such as thermocouples are subject to physical damage. Good accuracy is obtained when measuring the temperature of a flame or a molten metal through a small hole in a furnace wall because (as discussed in Art. 19-10) the opening approaches black-body behavior.

In the total-radiation pyrometer of Fig. 20-13b, all wavelengths of radiant energy emitted in certain directions by the source are collected and focused upon the hot junction of a thermocouple such that a temperature difference, and hence a detectable voltage difference (emf), exists between the hot and cold junctions. Relating the measured emf to known calibration data permits the evaluation of the source temperature.

The part of the hot junction facing the mirror is usually blackened to assure effective absorption of the collected radiation. To minimize the influence of stray radiation on the junctions, the inside of the tube may be blackened, and the outside of the tube, the cold junction, and that portion of the hot junction not exposed to the reflected radiation from the mirror may be made bright.

Ideally, the temperature difference between the hot and cold junctions is solely the result of radiation from the source impinging upon, and being absorbed by, the hot junction. Practically, conduction of heat between the hot and cold junctions and different rates of heat transfer by convection and radiation from the two junctions prevent the attainment of the ideal. To approach the ideal situation, it is desirable to restrict the amount of radiation received by the hot junction from the source to an amount such that only a small difference in temperature exists between the two junctions. But if only two junctions are used, the emf produced by the little temperature difference is so small that it is difficult to measure accurately. To produce an easily measurable, large emf from a small temperature difference, many thermocouples (perhaps 100 or more) are arranged in series, and the emf produced by each hot and cold junction combination is additive. Such a device is referred to as a *thermopile*.

The minimum surface temperature that can be detected by an optical pyrometer is that temperature (approximately 1400°F) at which the body begins to emit a significant amount of energy in the visible range of the spectrum. The total-radiation pyrometer is not subject to this limitation; so it is used extensively to detect remotely the temperature of surfaces in this lower range. The total-radiation pyrometer is also suitable as the sensing device for control systems, and it has been employed to measure the emissivity of surfaces.

*b. Theory of Operation.* Consider that an optical pyrometer is focused on a black body at a temperature $T_b$. The radiant flux density per unit wavelength emitted by the black body is given by Planck's equation (Art. 19-11):

$$W_{b\lambda} = \frac{C\lambda^{-5}}{e^{c/\lambda T_b} - 1} \qquad (19\text{-}54)$$

The wavelength of the radiation passed by the usual pyrometer filter is approximately $6.5 \times 10^{-5}$ cm (0.65 micron), so that the first term of the denominator of Eq. (19-54) is much greater than unity. Thus with little error

$$W_{b\lambda} = C\lambda^{-5}e^{-c/\lambda T_b} \qquad (20\text{-}47)$$

[Eq. (20-47) is known as *Wien's law of radiation*.] The current required to bring the filament to a brightness equal to the emission from the black body is recorded, and the correct temperature, $T_b$, is read from calibration tables.

Now let the optical pyrometer be sighted upon a nonblack emitter at temperature $T$. The flux density per unit wavelength emitted by the non-black body is

$$W_\lambda = \epsilon_\lambda C\lambda^{-5}e^{-c/\lambda T}$$

where $\epsilon_\lambda$ [Eq. (19-66)] is the emissivity of the body for the wavelength of visible light passed by the pyrometer filter.

If the current required to attain matching brightness is now noted and used to obtain a temperature from the calibration tables, the obtained temperature, $T_b$, is the temperature of a black body which emits the same amount of energy in wavelength $\lambda$ as emitted in wavelength $\lambda$ by the non-black body at temperature $T$.

Thus

$$W_{b\lambda} = W_\lambda \qquad \text{or} \qquad e^{-c/\lambda T_b} = \epsilon_\lambda e^{-c/\lambda T}$$

Rearranging,
$$\frac{1}{T} - \frac{1}{T_b} = \frac{\lambda \ln \epsilon_\lambda}{c} \qquad (20\text{-}48)$$

Thus, if its emissivity is known, the *true temperature* of a non-black body may be determined with an optical pyrometer and calibration tables based on black-body behavior. Conversely, this procedure may be used to measure $\epsilon_\lambda$ if $T$ can be determined by other means (Prob. 34). Note that $\epsilon_\lambda$ is the *emissivity at a particular wavelength* and may differ in magnitude from average emissivity values as given in Table 19-2.

Consider that the total-radiation pyrometer shown in Fig. 20-13b is sighted upon a source of radiation at a uniform temperature $T_s$. Let the temperature of the hot junction of black surface area $A_h$ be $T_h$, and let the temperature of the cold junction be $T_c$. In a properly designed pyrometer, $T_c$ should be almost equal to the temperature of the optical pyrometer tube and surrounding air, and $T_h$ should be only slightly greater than $T_c$.

It is first assumed that the radiation source emits as a black body. Then the following energy balance may be written for the hot junction:

$$C(T_h - T_c) = F_{sh}A_s\sigma(T_s{}^4 - T_h{}^4) \qquad (20\text{-}49)$$

The left-hand side of this equation represents the energy leaving the hot junction by conduction to the cold junction, convection to the gas in the pyrometer, and radiation to all surfaces except the source. The term $C$ is a combined heat-transfer coefficient to represent all of these effects. The right-hand side represents the net radiation interchange between the hot junction and source (both assumed black). The emf produced by the thermopile is proportional to $T_h - T_c$, and hence

$$\text{emf} \propto \frac{F_{sh}A_s\sigma}{C}(T_s{}^4 - T_h{}^4)$$

or
$$\text{emf} = K(T_s{}^4 - T_h{}^4)\dagger \qquad (20\text{-}50)$$

From this equation, the value of $K$ for a particular total-radiation pyrometer may be obtained by calibration with a black-body source (Prob. 37, test 1).

Now assume that the source is a gray body with an emissivity $\epsilon_s$. From an energy balance on the hot junction (assumed black), including reflection from $s$ to $h$ of radiation initially emitted by $h$ (second term, right-hand side of equation) and reflection from $s$ to $h$ of radiation initially emitted by surroundings at $T_c$ (third term),

$$C(T_h - T_c) = \epsilon_s F_{sh}A_s\sigma T_s{}^4 + (1 - \epsilon_s)F_{sh}A_h\sigma T_h{}^4 + (1 - \epsilon_s)F_{sh}A_s\sigma T_c{}^4$$
$$- F_{hs}A_h\sigma T_h{}^4$$

Rearranging and noting that the second term on the right-hand side of the equation is negligibly small,

$$C(T_h - T_c) = F_{sh}A_s\sigma[\epsilon_s T_s{}^4 + (1 - \epsilon_s)T_c{}^4] - F_{hs}A_h\sigma T_h{}^4 \quad (20\text{-}51)$$

But $T_c$ is almost equal to $T_h$ and $F_{hs}A_h = F_{sh}A_s$ [Eq. (19-73)]:

$$C(T_h - T_c) = \epsilon_s F_{sh}A_s(T_s{}^4 - T_h{}^4)$$
and
$$\text{emf} = \epsilon_s K(T_s{}^4 - T_h{}^4) \qquad (20\text{-}52)$$

If emissivity is known, the temperature of the radiating source may be determined from this equation and an emf measurement with a total-radiation pyrometer of known calibration factor $K$. Conversely, if $T_s$ is measured by other means, the emissivity may be determined by use of the total-radiation pyrometer. Most of the emissivities given in Table 19-2 were determined by such means. Normally, $T_s$ to the fourth power is enough greater than $T_h$ to the fourth that negligible error is encountered by either (1) substituting the more readily known $T_c$ for $T_h$ or (2) neglecting the $T_h$ to the fourth-power term completely (Prob. 37, test 2).

† Results of tests of 22 different total-radiation pyrometers by the National Bureau of Standards (G. K. Burgess and P. D. Foote, Characteristics of Radiation Pyrometers, *Natl. Bur. Standards Scientific Paper* 250, 1915) showed that a better form of this equation is emf $= K(T_s{}^b - T_h{}^b)$, where $b$ was found to vary from 3.28 to 4.26, with an average value of 3.89.

An interesting and useful feature of the total-radiation pyrometer, shown in Fig. 20-13$b$, is that the emf generated by the thermopile, and hence the measured source temperature or emissivity, is not influenced by the distance $L$ from instrument to radiation source, as long as the area of the radiation source is at least as great as the base of a cone with vertex at $X$ (cone $AXA'$ in Fig. 20-13$b$). This can be understood by noting that, although the intensity of the radiation from any point on the source decreases with the square of the distance, the area seen by the instrument increases with the square of the distance. An excessive temperature rise in the instrument as the hot source is brought quite close to it can place a restriction on the minimum permissible value of $L$ (Prob. 37, test 3).

## PROBLEMS

*Article* 20-1

**1.** Derive Eq. (20-9).

**2.** Derive Eq. (20-10).

**3.** Heat is generated at a uniform rate of 20,000 Btu/(hr)(ft³) in a solid sphere of radioactive material, 2 ft. in diameter. The surrounding air temperature is 70°F, $k$ for the sphere material is 20 Btu/(hr)(ft)(°F), and the surface coefficient for heat transfer between air and sphere is 10 Btu/(hr)(ft²)(°F). Assuming steady-state conditions, determine (*a*) the location and numerical value of the maximum temperature in the sphere and (*b*) the location and numerical value of the maximum temperature gradient in the sphere.

**4.** Assuming that the rate of heat generation increases linearly with rod radius as $\omega = ar$, where $a$ is a positive constant, obtain expressions similar to Eqs. (20-5) and (20-8) for the temperature distribution and surface heat-transfer rate.

*Article* 20-2

**5.** Calculate and compare the fin efficiencies of the following straight fins of cylindrical cross section: (*a*) copper fins, 2 in. long; (*b*) steel fins, 2 in. long; (*c*) copper fins, 5 in. long; (*d*) steel fins, 5 in. long. All the fins are ¼ in. in diameter, and the surface heat-transfer coefficient is 2.0 Btu/(hr)(ft²)(°F).

**6.** Obtain an expression for the temperature distribution in a fin of constant $A$, $b$, and $k$ if the fin is so long that its tip ($x = L$) has the same temperature as the surrounding fluid.

**7.** Two large tanks containing liquid nitrogen at a temperature of 100°R are connected by a number of cylindrical steel supporting bars 2 in. in diameter and 2 ft long. Assume that $k$ for the bars is 24 Btu/(hr)(ft)(°F) and that the surface coefficient between the bars and surrounding air at 60°F is 4 Btu/(hr)(ft²)(°F). (*a*) Determine the rate of heat transfer for each bar. (*b*) In order to decrease the heat transfer, would it be preferable to halve the length or halve the diameter of each bar? (Neglect the change in surface coefficient caused by changing the bar diameter.)

**8.** Figure 20-14 shows four possible methods for measuring the surface temperature of a solid. In (*a*), the thermocouple is placed on the surface and covered with a piece of putty or similar material. In (*b*), the thermocouple tip is just barely embedded in the surface. In (*c*), the thermocouple tip and a length of the lead wires are barely embedded in the surface. The method shown in (*d*) is identical to that in (*c*), except that the lead wires are brought through the back side of the solid. Noting that the lead wires tend to be extended surfaces and that the thermal boundary layer in the fluid is thin, comment on the suitability of each of these methods for obtaining a true value of surface temperature.

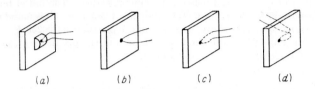

(a)          (b)          (c)          (d)

FIG. 20-14. Problem 8, methods for measuring surface temperature.

**9.** A copper thermometer well is inserted radially into a duct 12 in. in diameter. Hot air flows through the duct to a drying oven at atmospheric pressure and at a velocity of 400 ft/min. The well is 0.5 in. OD, 0.25 in. ID, and 6 in. long. The end of the well at the pipe wall is at 220°F, while the bottom of the well is at 300°F, as indicated by the thermometer. Heat conduction along the mercury-in-glass thermometer is to be neglected. (a) Assuming that radiation heat transfer is negligible, determine the air temperature. (SUGGESTED SOLUTION: Treat the thermometer well as an extended surface. In order to evaluate $h$, it is necessary to first assume an air temperature, calculate $h$, compute the air temperature and compare with the assumed air temperature, and proceed by trial.) (b) Would radiation tend to increase or decrease the error in thermometer reading? (c) List several means by which the error caused by conduction along the well could be reduced.

*Article 20-3*

**10.** Simplify and solve Eq. (20-24a) to obtain expressions for the temperature distribution and rate of heat transfer for steady-state conduction through the walls of a hollow cylinder of inner radius $r_1$ at temperature $t_1$, outer radius $r_2$ at temperature $t_2$, and length $L$. Compare the results with those found from integration of Fourier's equation in Art. 19-4.

**11.** Obtain Eq. (20-8), Art. 20-1, by simplifying and solving Eq. (20-24a).

**12.** In Example 5, how much would the time required for the thermocouple to attain a temperature of 400°F be changed if (a) the velocity of the air were doubled or (b) the diameter of the thermocouple tip were halved?

**13.** What values of Biot number and relative position $(x/\delta)$ apply to the ordinate of Fig. 20-7? What is the nature of the unsteady-state conduction case represented by these values?

**14.** What region of Fig. 20-7 corresponds to the situation in which Eq. (20-28) is valid?

**15.** What measures might be taken to reduce the cooling time calculated in Example 6?

**16.** The following experiment is performed to determine the value of the surface heat-transfer coefficient for a small metal object. The solid is heated to a uniform temperature of 200°F and then suddenly exposed to a low-speed air stream at 75°F. After cooling for 10 min, a single thermocouple embedded in the solid records a temperature of 90°F. The solid weighs 0.25 lb and has a surface area of 0.03 ft². Properties of the solid are: $\rho = 400$ lb$_m$/ft³, $k = 200$ Btu/(hr)(ft)(°F), $c = 0.1$ Btu/(lb$_m$)(°F). (a) From these data, determine a value for the surface coefficient $h$. (SUGGESTED SOLUTION: Assume that the surface resistance is much greater than the internal resistance, calculate $h$, and then check the validity of the assumption by referring to the Biot number criterion.) (b) Comment upon the worth of this method for experimentally determining $h$.

**17.** The thermal diffusivity of a plastic material is determined by clamping a 2-in.-thick sheet of the material between two metal plates. The initial temperature of the plastic material is 100°F, the metal plates are at 280°F, and it is found that, after 1 hr of heating, the temperature at the center of the plastic sheet is 250°F. Assuming perfect thermal contact between the metal plates and the plastic sheet, determine a value for the thermal diffusivity of the plastic.

*Article* 20-4

**18.** Assuming a constant value for the surface heat-transfer coefficient, is it possible that the addition of "insulation" to a plane wall will result in an increase in heat-transfer rate?

**19.** Show that the critical radius for a sphere is equal to $2k/h$.

**20.** Tubes carrying a refrigerant at 0°F pass through air at 45°F. A concentric layer of frost $[k = 0.25 \text{ Btu}/(\text{hr})(\text{ft})(°\text{F})]$ forms on the tubes. Assuming a constant $h$ of 2.0 $\text{Btu}/(\text{hr})(\text{ft}^2)(°\text{F})$ between air and frost (or bare tube) surface, will the frost layer tend to increase or decrease the heat loss for (a) tubes of 1 in. OD and (b) tubes of 4 in. OD?

**21.** Continuing Prob. 20, plot the surface temperature of the frost and the heat loss per foot of tube for various thicknesses of frost on the 1-in.-diameter tube. Plot in $\frac{1}{2}$-in. increments of frost thickness until the frost-surface temperature reaches 32°F.

**22.** Same as Prob. 21 for the 4-in.-diameter tube.

*Article* 20-5

**23.** Derive an equation, by dimensional analysis, in the form of Eq. (20-35). Justify your selection of variables.

**24.** Draw the cross section of a horizontal tube with steam condensing on the outside surface. Note that the thickness of the condensate will be small on the top of the tube and will progressively increase as the fluid flows over the sides. At the bottom of the tube, the flow from each side will meet, and one stream of fluid will leave the tube. Compute the **Re** for this flow, noting that $\Gamma = 2Gx$.

*Ans.* **Re** $= 2\Gamma/\mu$.

**25.** Determine the mean steam-side convection coefficient for a bank of condenser tubes, six tiers deep, that are to condense saturated steam at 1 in. Hg pressure with water flowing through the tubes and changing in temperature from 60 to 70°F. Tubes are 1 in. OD and 8 ft in length. The tube bundle is (a) horizontal and (b) vertical.

**26.** Repeat Prob. 25, but assume that the pressure is atmospheric.

**27.** Forty pounds per hour of saturated steam at 142°F is to be condensed on a 2-in.-OD BWG No. 14 condenser tube (or tubes) with the tube surface temperature constant at 138°F. Determine the necessary length of tube (or tubes) required if (a) a single vertical tube is used; (b) a single horizontal tube is used; (c) two vertical tubes are used (tubes do not interfere with each other); (d) two horizontal tubes are used, with one placed directly above the other.

**28.** Plot the heat-transfer rate $\dot{Q}/A$ versus $\Delta t$ for the data of Fig. 20-9.

**29.** High-pressure steam is rarely used to evaporate water at atmospheric pressure. Explain why not.

*Article* 20-6

**30.** Considering an enclosure consisting of $k$ surfaces, show that the following equations may be written for the net heat transfer at any surface $i$ of the $k$-surface enclosure:

$$\dot{Q}_{i\,\text{net}} = \sum_k A_i F_{ik}(J_i - J_k)$$

$$\dot{Q}_{i\,\text{net}} = \frac{A_i \epsilon_i}{1 - \epsilon_i}(W_{bi} - J_i)$$

Apply these equations to the enclosure considered in Example 8 and obtain the equations given in part *b* of Example 8.

**31.** A particular furnace may be approximated as a box 12 ft wide, 24 ft long, and 4 ft high. Considering the ceiling (12 by 24 ft) of the furnace to be a black-body source (denote as 1), the floor to be a black-body sink (denote as 2), and the side walls to be refractory surfaces (denote as $R$), evaluate $\bar{F}_{12}$ by (*a*) the $\bar{F}$ factor shown in Fig. 19-14 and (*b*) the $F$ factor of Fig. 19-14 and Eqs. (19-72) and (19-73). (See footnote, page 464.)

**32.** An enclosure of the same geometry as that shown in Fig. 20-11 is made up of four black surfaces. The two larger surfaces are at temperatures of 1000 and 500°R, respectively, and the two smaller surfaces are both at a temperature of 750°R. Sketch the radiation network for the enclosure, and determine the net heat-transfer rate at each surface.

**33.** Repeat Prob. 32, but assume that all of the surfaces are gray with an emissivity of 0.5.

*Article 20-7*

**34.** An optical pyrometer, calibrated for black-body radiation, indicates a temperature of 1540°F when sighted upon a body with known surface temperature of 1600°F. Determine the emissivity of the surface. What is the significance of the calculated emissivity value if (*a*) the surface behaves as a gray body, (*b*) the surface exhibits real body behavior?

**35.** When sighted upon an operating fluorescent lamp, an optical pyrometer indicates a temperature of several hundred degrees, yet the lamp is only slightly warm to the touch. Explain.

**36.** Equation (20-47), first derived by Wien on the basis of classical physics (see footnote, page 403), represents the monochromatic emissive power of a black body (Fig. 19-11). Although no flaw could be found in Wien's application of physical principles, his equation gave results which deviated greatly from some of the experimental data shown in Fig. 19-11. Under what conditions do the results from Wien's equation differ greatly from experimental findings (and common sense)? To further the confusion of scientists at that time (about 1900), Rayleigh and Jeans had made an apparently equally sound application of another set of physical principles to obtain the equation

$$W_{b\lambda} = K\lambda^{-4}T \qquad (K \text{ is a constant})$$

Their equation gave results at certain wavelength ranges in great disagreement with Wien's equation and with experimental data.

**37.** The data presented in this problem were obtained during the testing of a particular total-radiation pyrometer of the type shown in Fig. 20-13*b*.

1. In the first test, a blackened plate (emissivity of almost unity) was located so that the pyrometer could "see" only the front surface of the plate. A heater was placed behind the plate so that the plate temperature could be varied. The plate surface temperature was measured with small thermocouples mounted in the plate.

The following results were obtained:

| Plate temperature, °F | Emf produced by pyrometer thermopile, mv |
|:---:|:---:|
| 396.0 | 14.08 |
| 346.0 | 10.78 |
| 292.0 | 7.40 |
| 220.0 | 4.40 |
| 164.0 | 2.18 |
| 145.9 | 1.82 |

The thermopile construction was such that an emf of 1 mv was produced for each 1.8°F temperature difference between the hot and cold thermocouple junctions. In addition, it was observed during this and the following tests that room temperature was 78°F and that, when sighted upon the background, an emf of 0.1 mv was produced by the thermopile. Thus, as an approximation to account for background effects, the emf data might be corrected by subtracting 0.1 mv. From the given data, determine (a) if the test substantiated the Stefan-Boltzmann law, that is, if the emf is proportional to $T_s^4$, $- T_h^4$, and (b) a suitable value for the pyrometer calibration constant $K$ (Art. 20-7b). [SUGGESTED SOLUTION: (1) Convert plate temperature to absolute temperature $T_s$. (2) Determine the temperature of the hot junction of the thermopile as $T_h = 78 + 460 + 1.8$ (corrected emf). (3) Evaluate $T_s^4 - T_h^4$ and determine whether the variation of this quantity is directly proportional to the variation of the corrected emf.]

2. In the second test, a painted plate of emissivity considerably less than unity was substituted for the blackened plate. The following results were obtained:

| Plate temperature, °F | Emf produced by pyrometer thermopile, mv |
|:---:|:---:|
| 381.1 | 5.00 |
| 292.0 | 2.90 |
| 219.9 | 1.68 |
| 156.0 | 0.80 |

From these data, determine the emissivity of the painted plate. Compare the emissivity value with those given in Table 19-2. What type of paint might have been applied to the plate surface?

3. In the third test, the distance between the blackened plate and the total-radiation pyrometer was varied, while the plate was maintained at a constant temperature of about 330°F. As the distance between plate and pyrometer was changed, the following variation in the thermopile emf was measured:

| Distance, in. | Emf produced by pyrometer thermopile, mv |
|:---:|:---:|
| 2.5 | 9.9 |
| 3.0 | 9.9 |
| 4.0 | 9.9 |
| 6.0 | 9.5 |
| 10.0 | 7.5 |
| 15.0 | 5.4 |
| 20.0 | 3.5 |
| 25.0 | 2.3 |
| 30.0 | 1.6 |
| 35.0 | 1.15 |

Plot corrected thermopile emf versus distance on log-log paper and explain the slope of the resulting curve at (a) small distances (less than 6 in.) and (b) large distances (greater than 20 in.).

## SELECTED REFERENCES

1. McAdams, W. H.: "Heat Transmission," 3d ed., McGraw-Hill Book Company, Inc., New York, 1954.
2. Reid, R. C., and T. K. Sherwood: "The Properties of Gases and Liquids," McGraw-Hill Book Company, Inc., New York, 1958.
3. Bird, R. B., W. E. Stewart, and E. N. Lightfoot: "Transport Phenomena," John Wiley & Sons, Inc., New York, 1960.
4. Knudsen, J. G., and D. L. Katz: "Fluid Dynamics and Heat Transfer," McGraw-Hill Book Company, Inc., New York, 1958.
5. Jakob, M.: "Heat Transfer," vols. I and II, John Wiley & Sons, Inc., New York, 1949, 1957.
6. Schlichting, H.: "Boundary Layer Theory," 4th ed., McGraw-Hill Book Company, Inc., New York, 1960.
7. Giedt, W. H.: "Principles of Engineering Heat Transfer," D. Van Nostrand Company, Inc., Princeton, N.J., 1957.
8. Eckert, E. R. G., and R. M. Drake, Jr.: "Heat and Mass Transfer," 2d ed., McGraw-Hill Book Company, Inc., New York, 1959.
9. Kays, W. M., and A. L. London: "Compact Heat Exchangers," McGraw-Hill Book Company, Inc., New York, 1958.
10. Schneider, P. J.: "Conduction Heat Transfer," Addison-Wesley Publishing Company, Reading, Mass., 1955.
11. Kreith, F.: "Principles of Heat Transfer," International Textbook Company, Scranton, Pa., 1958.
12. American Institute of Physics: "Temperature, Its Measurement and Control in Science and Industry," vols. I and II, Reinhold Publishing Corporation, New York, 1941, 1955.

# APPENDIX A

# REFERENCE MATERIAL

**A-1. Other Dimensional Systems and Units.** The English system of units in the $ML\tau$ system has a metric prototype called more often the *absolute cgs* system. Here the unit of length is the *centimeter,* the unit of mass is the *gram,* and the unit of time is the *second.* The force unit, like the poundal, is expressed in derived dimensions:

$$F = ma$$

$$(1 \text{ dyne } [ML\tau^{-2}]) = (1 \text{ g } [M]) \frac{1 \text{ cm } [L]}{1 \text{ sec}^2 \, [\tau^2]}$$

*The dyne is defined as the force required to accelerate 1 gram mass at the rate of 1 centimeter per (second)².* In using this system, it must be remembered that

$$1 \text{ dyne} \equiv 1 \frac{\text{g cm}}{\text{sec}^2}$$

The unit of energy in this system is called an *erg* and is defined as the product of unit force acting through unit distance,

$$1 \text{ erg} \equiv 1 \text{ dyne} \times 1 \text{ cm} = 1 \text{ dyne cm}$$

A larger unit of energy is called the *joule,*

$$10^7 \text{ ergs} \equiv 1 \text{ joule}$$

The unit of power, which is the energy expended in unit time, is the *watt,*

$$1 \text{ watt} \equiv 1 \frac{\text{joule}}{\text{sec}}$$

The basic unit of charge is that carried by the *electron* or by the positive electron called a *positron* (which has only a transitory free existence),

$$1 \text{ unit of charge} \equiv \text{that carried by the electron}$$

A much large quantity of charge is called the *coulomb,*

$$1 \text{ abs coulomb} = 6.24273(10^{18}) \text{ unit charge}$$

A still larger quantity of charge is called the *faraday* (**F**),

$$1 \text{ faraday} = 6.02544(10^{23}) \text{ unit charge}$$
$$= 96,519 \text{ abs coulombs}$$

479

As a matter of interest, one faraday corresponds to:

1. The charge carried by one gram-atomic weight of monovalent ion
2. The charge carried by $N_0$ monovalent ions, where

$$N_0 \text{ (Avogadro's number)} = 6.02544(10^{23}) \frac{\text{atoms}}{\text{g atom}} \text{ or } \frac{\text{molecules}}{\text{g mole}}$$

3. The charge carried by one *gram-equivalent weight* of ion (the atomic weight divided by the valence)

The displacement of charge ($q$) gives rise to an electric current ($I$) which is defined as the time rate of flow of charge,

$$I \equiv \frac{\partial q}{\partial \tau}\bigg)_L$$

The unit of current called the *ampere* is defined as

$$1 \text{ amp} \equiv 1 \frac{\text{coulomb}}{\text{sec}}$$

It follows that

$$1 \text{ faraday} = 26.811 \text{ amp hr}$$

Because of the attraction or repulsion between charges, a force field is set up which, for stationary charges, is called an *electric field*. Energy is required, in general, to move charge from one position to another in an electric field. Each position in the electric field can be assigned a definite value of potential energy which would correspond to the energy required to bring unit charge from infinity to the point in question. The potential energy per unit charge at a point in the electric field is called the *potential, V*. When a potential difference exists, positive charge (by convention) can flow through a conductor from the region of high potential (energy) to the region of low potential (energy). The unit of potential is the *volt*, which, since it is a specific energy quantity, has units of joules (energy) per coulomb (charge),

$$1 \text{ volt} \equiv 1 \frac{\text{joule}}{\text{coulomb}}$$

Since a coulomb is a quantity of charge equivalent to $6.24273(10^{18})$ electrons,

$$1 \text{ volt} = 1.60186(10^{-19}) \frac{\text{joule}}{\text{electron}}$$

Hence, a new energy term is the *electron volt* (ev),

$$1 \text{ ev} = 1.60186(10^{-19}) \text{ joule} = 1.51829(10^{-22}) \text{ Btu}$$

When the flow of charge encounters *resistance R*, the potential of the charge is dissipated in heating the conductor. The unit of resistance is the *ohm*. A resistance of 1 ohm exists when a potential difference of 1 volt is required for the steady flow of 1 ampere of current. This relationship is defined by *Ohm's law*,

$$\Delta V_{ab} \equiv RI$$

To create a potential difference, energy in some form must be converted into electrical energy. Thus, electrical energy is obtained from chemical energy in a storage battery, from mechanical energy in a generator, from heat energy when thermoelectric effects are present. The capability of a device for *ideal* energy conversion is indicated by the *electromotive force* (emf, or $\mathcal{E}$). *The emf, by definition, is the ideal (reversible) work done in converting one form of energy into another form per unit of charge passing through the device.* (And the name is misleading since emf is not a force.) Since work and energy have the same dimensions, both emf and potential can be measured with the same unit—the volt. Note, however, that the concepts of emf and potential are not equivalent.

**A-2. Dimensional Analysis.** Since a dimension describes a measurable quantity, an equation made up of dimensional terms can be visualized as a description or summation of a more complicated event. A dimensional system is invariably used to ensure that equations are dimensionally homogeneous, and a check of the correctness of an equation is to test this homogeneity. If an event is known to be dependent on certain variables, it should be possible to predict the arrangement of these variables from the principle of dimensional homogeneity of equations.

*It should be carefully noted that dimensional analysis can do no more than systematically arrange the variables that presumably govern the event under study. If one or more of these variables are omitted, the answer found will be significant only to the degree that the missing variables are insignificant.*

**Example 1.** Find an expression for the force exerted by the air on the wing of an airplane.

*Solution.* The factors involved appear to be as in the following table:

| Variables | Symbol | Dimensional formula |
|---|---|---|
| Area of wing............. | $A$ | $[L^2]$ |
| Velocity of wing......... | $V$ | $[L\tau^{-1}]$ |
| Density of air............ | $\rho$ | $[ML^{-3}]$ |
| Viscosity of air........... | $\mu_f$ | $[F\tau L^{-2}]$ |
| Resistance of wing........ | $F$ | $[F]$ |

Since resistance of the wing is the factor to be investigated, the desired solution will have the general form

$$F = f(A, V, \rho, \mu_f, g_c)$$

where $g_c$ must be included since the $FLM\tau$ system is used. Dimensionally for any form of the unknown function $f$, the following relationship must be satisfied:

$$[F] = [L^2]^a[L\tau^{-1}]^b[ML^{-3}]^c[F\tau L^{-2}]^d[MLF^{-1}\tau^{-2}]^e$$

For dimensional homogeneity, the exponents of $FLM\tau$ on both sides of the equation must be the same:

$$\Sigma F: \quad 1 = \qquad\qquad\qquad d \; - e$$
$$\Sigma L: \quad 0 = 2a \; +b \; -3c \; -2d \; + e$$
$$\Sigma \tau: \quad 0 = \qquad -b \qquad\quad + d \; -2e$$
$$\Sigma M: \quad 0 = \qquad\qquad\quad c \qquad\quad + e$$

When these simultaneous equations are solved,

$$a = \tfrac{1}{2} - \tfrac{1}{2}e \qquad b = 1 - e \qquad d = 1 + e \qquad c = -e \qquad e = e$$

and the solution is

$$F = (A^{1/2} V \mu_f) f \left( \frac{\mu_f g_c}{A^{1/2} V \rho} \right)$$

This can be rearranged (to obtain a more familiar equation),

$$F = A^{1/2} V \mu_f \left( \frac{A^{1/2} V \rho}{\mu_f g_c} \right) f \left( \frac{A^{1/2} V \rho}{\mu_f g_c} \right) = \frac{A V^2 \rho}{g_c} f \left( \frac{A^{1/2} V \rho}{\mu_f g_c} \right) \qquad Ans.$$

Upon examination of this answer, it is noticed that two dimensionless groups have been obtained,

$$\frac{F g_c}{A V^2 \rho} \qquad \text{and} \qquad \frac{A^{1/2} V \rho}{\mu_f g_c}$$

The answer obtained by dimensional analysis will always contain an unknown function that must be experimentally determined. Over a limited range of variables, it is often possible to describe the experimental data with a simple exponential function. Thus, the solution to Example 1 can be premised to be

$$F = C \frac{A V^2 \rho}{g_c} \left( \frac{A^{1/2} V \rho}{\mu_f g_c} \right)^e$$

and this is a valid solution if values for the constant $C$ and the exponent $e$ can be found that will correctly evaluate the known results. (A series of terms, each with a different exponent and constant, could undoubtedly be found to describe a greater range.)

The dimensionless number

$$\frac{A^{1/2} V \rho}{\mu_f g_c} = C \left( \frac{D V \rho}{\mu_m} \right)$$

appears with great regularity whenever a dimensional analysis is made of fluid flow. In honor of Osborne Reynolds and his classic studies of fluid flow, this group has been named the *Reynolds number*. The Reynolds number is abbreviated **Re** and is defined as

$$\mathbf{Re} \equiv \frac{D V \rho}{\mu_m} \equiv \frac{D V \rho}{\mu_f g_c} \tag{A-1}$$

Experimentally it is found that, if two different fluids in flow under apparently dissimilar conditions have the same **Re**, the flows are dynamically similar.

A number of rules are available to facilitate dimensional analysis:

1. *The number of dimensionless groups obtained will equal the number of factors less the number of fundamental dimensions.*

2. *If two of the simultaneous equations give the same answer, an additional dimensionless group will be obtained over the number predicted by rule 1.*

3. (a) *If the list of variables contains both [F] and [M], the dimensional constant $g_c$ should be included, for then the answer is general for any system of dimensions (Table 1-1).* (b) *However, if the dimensions [F] and [M] both appear more than once, the constant may be superfluous.*

*Directional dimensional analysis*[5] can be used to obtain, in some cases, a more informative analysis. In this system, the direction of a force $F_x$, a length $L_y$, a velocity $dL_x/d\tau$, or a heat flow $dQ_x/dL_x$ is shown by a subscript (in this instance, for cartesian coordinates). Then a length $L_y$, for example, will not cancel a length $L_x$, since the dimensions are different. From this viewpoint, the Reynolds number is not dimensionless but has the dimensions of a tangent: $L_x/L_y$.

For example, viscosity is defined (Art. 1-8) as

$$\mu = \frac{F/A}{\dfrac{dV}{dz}}$$

and with the conventional analysis

$$\mu = \left[\frac{M}{L\tau}\right]$$

which is physically meaningless. In directional dimensions,

$$F = ma = m\frac{d^2L_x}{d\tau^2} \text{ or } \left[M\frac{L_x}{\tau^2}\right]$$

$$A = L_xL_y$$

$$\frac{dV}{dz} = \frac{d}{dL_z}\left(\frac{dL_x}{d\tau}\right) \text{ or } \left[\frac{L_x}{L_z\tau}\right]$$

Hence, 

$$[\mu] = \left[\frac{ML_xL_z1}{L_xL_y\tau^2L_x}\right] = \left[\frac{ML_z}{L_xL_y\tau}\right] = \left[\frac{MV_z}{A_{xy}}\right]$$

Here viscosity is shown to be the *momentum* $MV_z$ transferred normal to the flow ($x$ direction) per unit of cross-sectional area parallel to the direction of flow. (This is a nice physical picture for the viscosity of gases, but for liquids it is unrealistic since it ignores entirely the cohesion between molecules, which is probably a primary effect.)

## A-3. Reynolds Number.

The concept of one-dimensional flow (Art. 5-1) implies an ideal fluid (Fig. 1-2) since the viscosity of a real fluid is responsible for two distinct types of flow. When the velocity is low, definite *macroscopic* elements of the fluid move in a fixed layer, or stratum, guided in direction by the boundaries of the system. At the boundary, the fluid has zero velocity, but each succeeding stratum away from the boundary slips past its neighbor at a faster speed. For example, in a round pipe, a very thin annular layer of fluid next to the walls has zero velocity, but each succeeding annulus of fluid has a correspondingly greater velocity until the center of the pipe is reached. The distribution of velocity across the pipe is parabolic (Fig. A-1a), and the fluid motion is called *laminar, streamline,* or *viscous flow.*

As the mass velocity is increased, radial components of velocity may be imparted to elements of fluid from wall projections or other disturbances, and eddies appear—the flow is *turbulent* (Fig. A-1b). In turbulent flow, the velocity distribution tends to be uniform except near the walls, where the velocity must necessarily approach zero within a very thin *laminar sublayer*. For flow in pipes, the velocity distribution in turbulent flow is such that the average velocity is about 0.8 of the maximum velocity. [For flow through a nozzle, however, the walls are polished, and the ratio of average to maximum velocity is closely unity, and the flow is closely one-dimensional (Fig. A-1c).]

The transition from turbulent to laminar flow depends on the Reynolds number, **Re**. For circular pipes, **Re** $\approx$ 2,000 is considered to be the *lower critical* **Re**. Below this value, turbulent flow cannot be maintained—laminar flow is stable to disturbances.

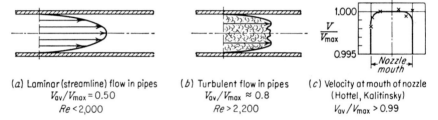

(a) Laminar (streamline) flow in pipes
$V_{av}/V_{max} = 0.50$
$Re < 2,000$

(b) Turbulent flow in pipes
$V_{av}/V_{max} \approx 0.8$
$Re > 2,200$

(c) Velocity at mouth of nozzle
(Hottel, Kalitinsky)
$V_{av}/V_{max} > 0.99$

Fig. A-1. Velocity-distribution diagrams in round conduits.

Laminar flow will persist to high Reynolds numbers if a long "stilling time," a smooth approach, and no impulses are present. Reynolds numbers of 40,000 and higher have been reported for laminar flow when extraordinary precautions were taken to avoid disturbances to the flow as the velocity was raised. Thus, the *upper critical* **Re** appears to be indeterminate; that is, the disturbance necessary to induce turbulent flow simply decreases in magnitude with increase in **Re**. (However, in commercial piping, disturbances are always present, and **Re** $\approx$ 2,200 is considered to be a relatively sharp boundary between laminar and turbulent flow.)

The **Re** is clearly a ratio of two forces: impulse and viscous (lb$_f$ sec/ft$^2$) forces. Thus, **Re** $\approx$ 2,200 (for circular pipes) divides† the flow regime into a region dominated by inertia forces and another region dominated by viscous forces.

**A-4. Velocity of Sound.** It is known from experience that pressure changes are propagated through substances at high speeds. To investi-

† Whenever **Re** is specified for other than circular passages, inquiry must be made as to the dimension represented by $D$ (a characteristic length).

gate the phenomenon, consider the plane pressure discontinuity in Fig. A-2a, which is moving at high velocity $V_1$ into a fluid at rest. Assume that directly behind the front of the pressure increase, the fluid moves at a lesser velocity $V_2$. Let the observer be located on the plane discontinuity (Fig. A-2b), so that to this observer the system is steady-flow with fluid entering at high velocity and leaving at a lower velocity (but

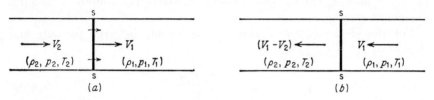

FIG. A-2. (a) Plane pressure discontinuity s-s moving at velocity $V_1$ relative to an external observer; (b) the plane pressure discontinuity s-s relative to an observer located on the plane.

with higher values of temperature and pressure). The equations of continuity and momentum can be applied, with wall forces being neglected. Thus, one of Eqs. (13-4b) is rearranged and reduced to ($A_1 = A_2$)

$$\rho_2 \frac{V_r^2}{g_c} + p_2 = \rho_1 \frac{V_1^2}{g_c} + p_1$$

Similarly, for the continuity equation [Eq. (5-6a)],

$$\rho_1 V_1 = \rho_2 V_r$$

Then, by substitution,

$$V_1 = \sqrt{g_c \frac{\rho_2}{\rho_1} \frac{p_2 - p_1}{\rho_2 - \rho_1}} = v_1 \sqrt{g_c \frac{p_2 - p_1}{v_1 - v_2}} \qquad (A-2a)$$

Equation (A-2a) relates the velocity of a "shock front" to the corresponding pressure rise and density change. The stronger the shock front ($p_2 - p_1$), the faster will the disturbance travel.

The phenomenon called *sound* travels as a longitudinal wave of alternate pressure increases and decreases (since it originates from a vibrating source), but each pressure pulse can be treated by Eq. (A-2a) if certain modifications are made. It can be surmised that various speeds could be assigned to various sounds, depending on the intensity. However, as $p_2$ approaches $p_1$, the ratio $\rho_2/\rho_1$ approaches unity and $\Delta p/\Delta \rho$ should approach a limit. Unfortunately, the value of the limit depends on the frequency of sound. At low frequencies, it is found from experiment that the limit corresponds to an adiabatic process, while, with increase in frequency, the limit for gases tends toward an isothermal process.[6]

It is customary, however, to define arbitrarily the concept of the *acoustic*, or *sonic*, *velocity* as

$$a \equiv \sqrt{g_c \frac{\partial p}{\partial \rho}\bigg)_s} \equiv v \sqrt{-g_c \frac{\partial p}{\partial v}\bigg)_s} \qquad (A\text{-}2b)$$

since this limiting value is quite closely approached by sounds in the aural frequency range.

For the ideal gas, the isentropic relationship between pressure and density is $pv^k = C$. Therefore,

$$a = \sqrt{\frac{g_c k R_0 T}{M}} \qquad (A\text{-}2c)$$

and, for air ($k = 1.4$),

$$a_{air} = 49.1 \sqrt{T} \qquad (A\text{-}2d)$$

Note that the sonic velocity increases with temperature increase of the gas. Also, gases with small molecular weights $M$ or large $k$ values have relatively larger sonic velocities.

For the concept of an incompressible fluid, Eq. (A-2b) shows that the speed of sound would be infinite. In liquids, changes of density with pressure are small relative to gases, and sound velocities are correspondingly larger (roughly, a factor of 5 for atmospheric conditions). On the other hand, the equation of state for a liquid has the form

$$v = f(T)p + \text{(terms not involving pressure)}$$

and since compression of a liquid changes its temperature little,

$$\frac{\partial p}{\partial v}\bigg)_s \approx \frac{\partial p}{\partial v}\bigg)_T \approx \text{constant} = \frac{\Delta p}{\Delta v}$$

Hence, quite closely, large or small disturbances in liquids travel at essentially the same speed.

**A-5. Mach Number.** An important parameter to distinguish certain flow regimes is the *Mach number*,

$$\mathbf{M} \equiv \frac{V}{a} \qquad (A\text{-}3)$$

which is the ratio of the actual velocity to the sonic velocity computed for the state where the velocity $V$ exists. When the Mach number is less than 1, the flow is *subsonic;* for Mach numbers greater than 1, the flow is *supersonic*, or *hypersonic* (for, say, $\mathbf{M} > 3.5$). The properties of the fluid at $M = 1$ or any measurement pertinent to this state are usually designated by an asterisk (\*),

$$A^* \qquad V^* = a^* \qquad p^* \qquad T^* \qquad \rho^* \qquad \text{(etc.)}$$

**A-6. Calculus Fundamentals.**† Mathematics, like thermodynamics, is a discipline based upon postulates and definitions; if the postulates and definitions are not thoroughly understood, little progress will be made in either subject. In this review, emphasis is laid upon certain aspects of the calculus which are often misunderstood and which are of importance in the study of thermodynamics. (But it should be remarked that *the pure mathematician would quarrel with several of the statements to follow on the grounds that they are too restrictive for his purposes.*)

*a. Set. A collection of particular things.*

In this text *set* will refer specifically to numbers: set of numbers between 3 and 5, for example.

*b. Variable. A quantity which can take on any of the numbers of some set.*

Thus, the name *variable* may designate a "point" quantity such as $x$, or an incremental quantity such as $\Delta x$, or a quantity given by a complicated formula.

*c. Function. A rule (or formula) which yields a single‡ value y for each value x of a set D.*

The variable $x$ is called the *independent variable* since its value is selected at will from the set $D$; the variable $y$ is called the *dependent variable* since its value depends upon the value assigned to $x$. The set $D$ is called the *domain of the function* (or the *range of the variable x*). The *range of the variable y* is the set of values yielded by the function.

The dependence of $y$ upon $x$ dictated by the function has led to the practice of loosely calling $y$ a function of $x$: *A variable y is said to be a function of another variable x (or of several other variables, $x_1$, $x_2$, . . .) if a single definite value (of y) is obtained for each value assigned to the variable x (or for each set of values assigned to the several variables, $x_1$, $x_2$, . . .).*

Note that the definition of a function agrees with what the pure mathematicians call a "single-valued function." But as Menger[2] points out:

The general concept of a many-valued function is physically useless; moreover, it is also unimportant in the elements of mathematics. In calculus we differentiate and integrate exclusively one-valued functions and even in the theory of complex functions we study only a very limited class of many-valued functions.

† This review is necessary since there is no single text that can be used as a reference (and many of the current texts in applied mathematics are misleading).

‡ And where two or more values might appear (as in, for example, $y^2 = x$), the restriction is preserved by specifying the range of the variable $y$ under study, for example, $y = +\sqrt{x}$. Thus, it is redundant (but not undesirable) to specify single-valued functions.

The results of most experimental work yield functions given by tabulated data (for example, the Steam Tables, or the temperature-time report in the newspaper) although in the classroom functions are usually presented in equational form (such as $y = x^2$). The functional relationship in all cases can be symbolically indicated by an equation,

$$y = f(x)$$

Here $y$ is the dependent variable since its value is fixed by the function $f$ operating upon the independent variable $x$; conversely, $x$ is the independent variable since its value is arbitrarily assigned. Another symbolism is

$$y = y(x)$$

wherein the symbol $f$ is replaced by $y$ for the function. The inverse function is shown by the notation

$$x = F(y) \qquad \text{or} \qquad x = x(y)$$

and the roles of dependent and independent variables are interchanged (and by so doing a restriction may need to be added; see the second footnote on page 487).

It should also be noted that the mathematician uses different symbols for different functions,

$$V = f(r) \qquad V = F(L)$$

whereas the engineer and the physicist use the same symbol for functions of identical meaning,

$$V = V(r) \qquad V = V(L)$$

In thermodynamics, the variables include quantities such as temperature, pressure, and specific volume. Then, when we say that a property is a *function* of other properties, we mean that when values are assigned, say to specific volume and temperature, one and only one value is found for pressure,

$$p = f(v,T) \qquad \text{or} \qquad p = p(v,T)$$

*d. Limit.* The symbols

$$\lim_{x \to a} f(x) = A \qquad \text{or} \qquad f(x) \to A \text{ as } x \to a$$

stand for the statement: The limit of $f(x)$ as $x$ approaches $a$ (by any arbitrary sequence) is $A$. This means that the values of $f(x)$ can be made to be as close to $A$ as desired, provided that $x$ is taken to be sufficiently close to $a$.

For example, if the function is $x^2$ and $x$ approaches 2, the limiting value of 4 is approached by $x^2$ as $x$ approaches 2 from either direction. In this instance, the function actually attains the value of 4 (when $x = 2$). On the other hand, a function such as $(\sin x)/x$ has the limit 1 as $x$ approaches zero, but strictly speaking, it can never reach this value since, at $x = 0$, $(\sin x)/x = \%$ and division by zero has no meaning.

*e. Continuity.* A function $f(x)$ is *continuous* at a point $x = a$ if:

1. The limit of $f(x)$ as $x$ approaches $a$ exists (assumes some unique finite value).

2. This limit is equal to the value of $f(a)$,

$$\lim_{x \to a} f(x) = f(a) = \text{finite number}$$

Throughout this text, only regions where functions are everywhere continuous (and single-valued) are studied. Continuity in a region implies that the curve or graph of the function in that region is an unbroken line or surface (and the derivative of the continuous function will also be continuous in that region if the unbroken line or surface is *smooth*—has no sharp point or cusp where the derivative fails to exist).

*f. Infinity.* A function such as $1/x^2$ is discontinuous at $x = 0$ because, as $x$ approaches this value, the value of the function $1/x^2$ increases without limit. The function is said to *become infinite* or *to approach infinity*.

Infinity is a word with several implications. One meaning is that given any number, say $10^9$, a function such as $1/x^2$ gets and stays bigger than $10^9$ when $|x|$ is sufficiently small. The other meaning (which really deserves a separate name) describes the number of objects in a set, say the set of terms in a series which is not finite: there is an *infinity* (or an *infinite collection*) of terms.

To the practical engineer, *infinity represents the realm beyond the precision of his measurements.* Thus, when theory demands an infinity of terms in a series or an infinity of heat reservoirs to enable heat to be reversibly transferred at, say, constantly increasing temperature, we can inquire as to the required precision. Then we need supply only a few terms of a series (two or three are not unusual) or a few heat reservoirs to obtain answers valid within the precision of our measurements.

*g. Infinitesimal.* An infinitesimal is a variable which at some stage of the discussion is allowed to approach zero as its limit.

Note that an infinitesimal is *not*, necessarily, a very small quantity. A variable $\Delta x$, for example, may be called an infinitesimal even though its value is relatively large; the crux of the matter is that *at some stage of the discussion* $\Delta x$ must be allowed to approach zero.

*h. Derivative.* A derivative is *not* the division of $\Delta y$ by $\Delta x$ when $\Delta x = 0$ since division by zero has no meaning; it is the limiting value of the ratio $\Delta y/\Delta x$ when $\Delta x$ approaches zero by any arbitrary sequence.

*i. Maxima, Minima, and Inflection Points.* Draw a graph of $y$ versus $x$ with maxima, minima, and inflection points. Note that points where $dy/dx = 0$ locate maxima, minima, and certain inflection points (locate points where the tangent becomes horizontal). Points where $d^2y/dx^2 = 0$ *may* be inflection points (points where the tangent reverses the direction in which it turns) (decide whether this requirement, a *necessary* requirement, is also *sufficient;* test, for example, $y = x^4$).

*j. Differential.* The increment of the dependent variable is a *function* of increments of the independent variables (usually a complicated function involving powers higher than the first). But *the differential of the dependent variable (called an exact, total, or complete differential) when it exists is defined as the linear part of the function (the principal part of the increment),*

$$dy \equiv \frac{dy}{dx} \Delta x \qquad (A\text{-}4a)$$

$$dz \equiv \frac{\partial z}{\partial x}\bigg)_y \Delta x + \frac{\partial z}{\partial y}\bigg)_x \Delta y \qquad (A\text{-}5a)$$

Nothing in the definitions requires $\Delta x$ or $\Delta y$ to be small! The differential is a linear function of $\Delta x$ (or $\Delta x$ and $\Delta y$) and is particular to a point—the point where $dy/dx$ [or $\partial z/\partial x)_y$ and $\partial z/\partial y)_x$] is evaluated—and Eqs. (A-4a) and (A-5a) describe the various functions at different points.

As further definitions, the *differentials of the independent variables* are declared equal to their increments; thus Eqs. (A-4a) and (A-5a) are usually written as

$$dy = \frac{dy}{dx} dx \qquad (A\text{-}4b)$$

$$dz = \frac{\partial z}{\partial x}\bigg)_y dx + \frac{\partial z}{\partial y}\bigg)_x dy \qquad (A\text{-}5b)$$

It is a *theorem* that Eqs. (A-4b) and (A-5b) hold whether or not $x$ in Eq. (A-4a) and $x$, $y$ in Eq. (A-5a) are the independent variables [that is, whether $dx$ (or $dx$ and $dy$) is the differential of a dependent or an independent variable]. But this fact does not dispute the fundamental definitions [Eqs. (A-4a) and (A-5a)]. [The student usually concludes, quite erroneously, that Eqs. (A-4b) and (A-5b) are something pertaining to infinitely small changes, quite forgetting that Eqs. (A-4a) and (A-4b) are the exact equivalents and with $\Delta x$ and $\Delta y$ assuming values as large as we please.]

*k. Relationships among Partial Derivatives.* In many instances, the independent variables are themselves functions of other variables.

Suppose that $z = z(x,y)$, while $x = x(u,w)$ and $y = y(u,w)$; then an increment $\Delta u$, at constant $w$, will have corresponding increments in $z$, $x$, and $y$,

$$\Delta z = \frac{\partial z}{\partial x}\bigg)_y \Delta x + \frac{\partial z}{\partial y}\bigg)_x \Delta y + \epsilon \Delta x + \eta \Delta y \qquad (a)$$

Equation (a) can be divided by $\Delta u$,

$$\frac{\Delta z}{\Delta u} = \frac{\partial z}{\partial x}\bigg)_y \frac{\Delta x}{\Delta u} + \frac{\partial z}{\partial y}\bigg)_x \frac{\Delta y}{\Delta u} + \epsilon \frac{\Delta x}{\Delta u} + \eta \frac{\Delta y}{\Delta u} \qquad (b)$$

As $\Delta u$ approaches zero as its limit, by definition of a derivative,

Five variables: $\qquad \dfrac{\partial z}{\partial u}\bigg)_w = \dfrac{\partial z}{\partial x}\bigg)_y \dfrac{\partial x}{\partial u}\bigg)_w + \dfrac{\partial z}{\partial y}\bigg)_x \dfrac{\partial y}{\partial u}\bigg)_w \qquad (A\text{-}6a)$

since both $\epsilon$ and $\eta$ become zero when $\Delta u$ becomes zero. The partial derivatives $\partial z/\partial x)_y$ and $\partial z/\partial y)_x$ are constants in this limiting process since they are the values of the derivatives at the point where the limit is reached; they are not related to $\Delta x$, $\Delta y$, or $\Delta u$.

For the case where $x = x(y,w)$, the foregoing steps could be modified by dividing by $\Delta y$ at constant $w$ to yield, finally,

Four variables:
$$\left.\frac{\partial z}{\partial y}\right)_w = \left.\frac{\partial z}{\partial x}\right)_y \left.\frac{\partial x}{\partial y}\right)_w + \left.\frac{\partial z}{\partial y}\right)_x \tag{A-6b}$$

Equations (A-6) are readily memorized by noting that they are equivalent to the basic equation (A-5b) with *apparent* division by $\partial u)_w$ or $\partial y)_w$, etc.

Formulas can be derived to facilitate calculation of partial derivatives from implicit functions. Consider, first,

$$u = F(x,y,z) \tag{c}$$

Then, † by Eq. (A-5b),

$$du = \frac{\partial F}{\partial x} dx + \frac{\partial F}{\partial y} dy + \frac{\partial F}{\partial z} dz \tag{d}$$

where the differentials $dx$, $dy$, and $dz$ are simply increments since $x$, $y$, and $z$ are all independent variables. For the case where $u$ is a constant, $du = 0$, and either $x$, $y$, or $z$ can become the dependent variable. Upon selecting $z$, by Eq. (A-5a),

$$dz = \frac{\partial z}{\partial x} \Delta x + \frac{\partial z}{\partial y} \Delta y \tag{e}$$

When (e) is substituted into (d) with $du = 0$ (and $x$ and $y$ as the independent variables),

$$\frac{\partial F}{\partial x} \Delta x + \frac{\partial F}{\partial y} \Delta y + \frac{\partial F}{\partial z} \left( \frac{\partial z}{\partial x} \Delta x + \frac{\partial z}{\partial y} \Delta y \right) = 0$$

and
$$\left[ \frac{\partial F}{\partial x} + \frac{\partial F}{\partial z} \frac{\partial z}{\partial x} \right] \Delta x + \left[ \frac{\partial F}{\partial y} + \frac{\partial F}{\partial z} \frac{\partial z}{\partial y} \right] \Delta y = 0 \tag{f}$$

Since $\Delta x$ and $\Delta y$ are independent variables, each bracketed quantity must be zero to satisfy the equation; therefore,

$$\left.\frac{\partial z}{\partial x}\right)_y = - \frac{\left.\dfrac{\partial F}{\partial x}\right)_{y,z}}{\left.\dfrac{\partial F}{\partial z}\right)_{y,x}} \tag{A-7}$$

and similarly for $\partial z/\partial y)_x$. Equation (A-7) is the tool that enables a particular derivative to be easily obtained from an implicit function of the variables.

When the foregoing analysis is repeated, but with $x$ selected as the dependent variable, the reciprocal of Eq. (A-7) is obtained. This shows (what seems self-evident) that the partial derivatives can always be inverted if the same variable is held constant,

$$\left.\frac{\partial z}{\partial x}\right)_y = \frac{1}{\left.\dfrac{\partial x}{\partial z}\right)_y} \qquad \left[\left.\frac{\partial x}{\partial z}\right)_y \neq 0\right] \tag{A-8}$$

Each of the three derivatives $\partial y/\partial z)_x$, $\partial z/\partial x)_y$, $\partial x/\partial y)_z$ can be found from equations similar to Eq. (A-7). When the three equations are multiplied together,

Three variables:
$$\left.\frac{\partial y}{\partial z}\right)_x \left.\frac{\partial z}{\partial x}\right)_y \left.\frac{\partial x}{\partial y}\right)_z = -1 \tag{A-9}$$

† Here the symbol $F$ is introduced to stand for the function $F(x,y,z)$ in place of the symbol $u$.

It is well to note that partial derivatives and total derivatives differ in one important respect: the total derivative, which is the limit of a ratio, is also equal to the ratio of two differentials; the partial derivative, however, although it, too, is the limit of a ratio, cannot be picked apart as a ratio of differentials. Consider the product $\dfrac{\partial z}{\partial x}\Big)_y \dfrac{\partial x}{\partial y}\Big)_z$. Here, if $\partial x$ is construed to be the same quantity in both terms, cancellation would yield $+\dfrac{\partial z}{\partial y}\Big)_x$ and this answer has the wrong sign [Eq. (A-9)].

Consider that $x = x(y,w)$; by Eq. (A-5b),

$$dx = \frac{\partial x}{\partial y}\Big)_w dy + \frac{\partial x}{\partial w}\Big)_y dw \qquad (g)$$

And if $y = y(z,w)$

$$dy = \frac{\partial y}{\partial z}\Big)_w \Delta z + \frac{\partial y}{\partial w}\Big)_z \Delta w \qquad (h)$$

By substituting $(h)$ into $(g)$ (and considering that $z$ and $w$ are independent),

$$dx = \frac{\partial x}{\partial y}\Big)_w \frac{\partial y}{\partial z}\Big)_w \Delta z + \left[\frac{\partial x}{\partial y}\Big)_w \frac{\partial y}{\partial w}\Big)_z + \frac{\partial x}{\partial w}\Big)_y\right] \Delta w \qquad (i)$$

But $x$ is also a function of $z$ and $w$,

$$dx = \frac{\partial x}{\partial z}\Big)_w \Delta z + \frac{\partial x}{\partial w}\Big)_z \Delta w \qquad (j)$$

Since $z$ and $w$ are independent variables, the coefficients of $\Delta z$ and $\Delta w$ in $(i)$ and $(j)$ must be equal. [The $\Delta w$ terms will yield Eq. (A-6b).] For the $\Delta z$ terms,

$$\frac{\partial x}{\partial z}\Big)_w = \frac{\partial x}{\partial y}\Big)_w \frac{\partial y}{\partial z}\Big)_w$$

or

Four variables: $\qquad\qquad \dfrac{\partial x}{\partial y}\Big)_w \dfrac{\partial y}{\partial z}\Big)_w \dfrac{\partial z}{\partial x}\Big)_w = 1 \qquad$ (A-10)

Equations (A-9) and (A-10) are easily memorized by noting the apparent cancellation of the parts of the derivatives: if the variables held constant all differ (and there are three variables), then the product is $-1$; if only one variable is held constant (and there are four variables), the product is $+1$.

The partial derivatives can be differentiated a second time to obtain the second partial derivatives,

$$\frac{\partial}{\partial x}\frac{\partial z}{\partial x} = \frac{\partial^2 z}{\partial x^2} \qquad \frac{\partial}{\partial y}\frac{\partial z}{\partial y} = \frac{\partial^2 z}{\partial y^2}$$

and the second cross partials,

$$\frac{\partial}{\partial x}\frac{\partial z}{\partial y} = \frac{\partial^2 z}{\partial x\,\partial y} \qquad \frac{\partial}{\partial y}\frac{\partial z}{\partial x} = \frac{\partial^2 z}{\partial y\,\partial x}$$

If all derivatives are continuous in the interval considered, the cross partials are equal (*test of exactness*) and the order of differentiation is immaterial.

## PROBLEMS

**1.** Find an expression for the period of a pendulum by dimensional analysis.

**2.** Determine by dimensional analysis the form of the equation in the engineering $FML\tau$ system for the velocity of sound in a gas if the variables are velocity, density, and pressure of the gas.

**3.** Repeat Prob. 2, but use the $FL\tau$ system.

**4.** Show that the pressure gradient for liquid flowing in a pipe is given by

$$\frac{\Delta p}{\Delta L} = \frac{\rho V^2}{Dg_c} f\left(\frac{\mu_m}{DV\rho}\right)$$

where $D$ is the diameter and $L$ is the length of the pipe.

**5.** Air is flowing through a pipe of 1.820 in. diameter at a rate of 1.173 $\text{lb}_m$/sec. The temperature of the air is 60°F. Determine the value for **Re**. *Ans.* 815,000.

**6.** Show that **Re** is a ratio of two forces, and discuss.

**7.** Write out, in words, exactly what is meant by the lower and upper critical **Re**.

**8.** Deduce why the acoustic velocity at high frequencies is expressed better by an isothermal limit in Eq. (A-2b).

**9.** If $y = f(x)$ and $x = a$ is an inflection point, then $d^2y/dx^2 = 0$ at $a$. Show (using $y = x^4$) that the converse statement—if $d^2y/dx^2(a) = 0$, then $a$ is an inflection point—is false. Hence, $d^2y/dx^2(a) = 0$ is a necessary but not a sufficient condition for an inflection point at $a$. What conditions, in addition to $d^2y/dx^2(a) = 0$, are sufficient conditions for an inflection point at $a$?

## SELECTED REFERENCES

1. Osgood, W. F.: "Advanced Calculus," The Macmillan Company, New York, 1937 (in particular, p. 453).
2. Menger, K.: "Calculus," Ginn & Company, Boston, 1955.
3. Courant, R.: "Differential and Integral Calculus," Interscience Publishers, Inc., New York, 1937.
4. Ludovici, B.: New System of Physical Units and Standards, *Am. J. Phys.*, **24**: 400 (May, 1956).
5. Kayser, R.: Analogue among Heat, Mass, and Momentum Transfer, *Ind. Eng. Chem.*, **45**: 2634 (December, 1953).
6. Herzfeld, W. F., and F. O. Rice: Dispersion and Absorption of High-frequency Sound Waves, *Phys. Rev.*, **31**: 691 (April, 1928).
7. Obert, E. F.: "Concepts of Thermodynamics," McGraw-Hill Book Company, Inc., New York, 1960.

# APPENDIX B

# TABLES AND CHARTS

TABLE B-1. DEFINITIONS AND CONVERSION FACTORS[†]

*Gravitational Acceleration*

$g_0$ = standard acceleration of gravity = 32.1739 ft/sec$^2$ = 980.665 cm/sec$^2$

$g$ = local acceleration of gravity

*Force*

1 poundal will accelerate 1 pound mass ($lb_m$) at rate of 1 ft/sec$^2$.

1 pound force ($lb_f$) will accelerate 1 $lb_m$ at rate of $g_0$ ft/sec$^2$.

1 $lb_f$ will accelerate 1 slug at rate of 1 ft/sec$^2$.

1 gravitational $lb_f$ will accelerate 1 $lb_m$ at rate of $g$ ft/sec$^2$.

1 dyne will accelerate 1 $g_m$ at rate of 1 cm/sec$^2$.

1 kilogram force ($kg_f$) will accelerate 1 kilogram mass ($kg_m$) at rate of $g_0$ cm/sec$^2$.

| | |
|---|---|
| 32.1739 poundals/$lb_f$ | 1 $lb_f$ = 32.1739 poundals |
| 980,665 dynes/$kg_f$ | 1 $kg_f$ = 980,665 dynes |
| 13,825 dynes/poundal | 1 poundal = 13,825 dynes |
| 444,805 dynes/$lb_f$ | 1 $lb_f$ = 444,805 dynes |
| 980.665 dynes/$g_f$ | 1 $g_f$ = 980.665 dynes |

*Length*

1 cm = 0.01 m = 0.3937 in. (int.) = 10$^4$ microns ($\mu$) = 10$^8$ angstroms

1 in. = 2.54000 cm

1 ft = 12 in. = 30.4801 cm

1 yd = 3 ft = 0.9144 m

1 mile = 5,280 ft = 1.609 km

1 nautical mile = 6,080.27 ft

1 m = 39.37 in.

1 mile/hr (mph) = 88 ft/min = 44.70 cm/sec

1 knot = 1 nautical mile/hr

*Temperature*

$$°F = 1.8°C + 32 \qquad\qquad °C = \frac{°F - 32}{1.8}$$

Degree Rankine (°R) = °F + 459.69

Degree Kelvin (°K) = °C + 273.16

1.8°R/°K

$$1 \frac{\text{IT cal}}{g_m \, °K} = 1 \frac{\text{Btu}}{lb_m \, °R}$$

$$1 \frac{\text{cal}}{g_m \, °K} = 0.999346 \frac{\text{Btu}}{lb_m \, °R}$$

[†] American Petroleum Institute Research Project 44, National Bureau of Standards, Washington, 1952.

TABLE B-1. DEFINITIONS AND CONVERSION FACTORS (*Continued*)

*Mass Equivalents*

1 dram (dr) (avoirdupois) = 27.34 $g_m$
1 ounce (oz) (avoirdupois) = 16 dr
1 $lb_m$ (int.) = 453.59237 $g_m$ = 16 oz = 7,000 grains
1 ton (short) = 2,000 $lb_m$
1 $g_m$ = 15.432 grains
1 $kg_m$ = 2.200462 $lb_m$
1 slug = 32.1739 $lb_m$

*Pressure Equivalents*

1 $lb_f$/in.$^2$ = 2.03601 in. Hg at 32°F = 2.307 ft $H_2O$ at 4°C = 0.0703067 $kg_f$/cm$^2$
1 in. Hg = 33,864 dynes/cm$^2$ = 0.0334211 atm = 0.491157 $lb_f$/in.$^2$
1 in. $H_2O$ (4°C) = 0.07354 in. Hg (32°F) = 0.03612 $lb_f$/in.$^2$ = 5.201 $lb_f$/ft$^2$
1 $lb_f$/ft$^2$ = 4.882 $kg_f$/mm$^2$
1 $kg_f$/cm$^2$ = 14.2234 $lb_f$/in.$^2$
1 atm = 14.6960 $lb_f$/in.$^2$ = 760 mm Hg (32°F) = 29.9212 in. Hg (32°F)
    = 1.03323 $kg_f$/cm$^2$ = 1,013,250 dynes/cm$^2$ = 33.934 ft $H_2O$ (60°F)
1 bar = 10$^6$ dynes/cm$^2$ = 0.9869 atm
Absolute pressure = (barometric pressure) + (gauge reading)
        = (barometric pressure) − (vacuum reading)

*Work—Energy—Power*

1 foot-poundal (ft-poundal) is work done by 1 poundal exerted through a distance of 1 ft.
1 foot-pound force (ft-$lb_f$) is work done by 1 $lb_f$ exerted through a distance of 1 ft.
1 erg is work done by 1 dyne exerted through a distance of 1 cm.

| | |
|---|---|
| 10$^7$ ergs/joule | 1 joule/(sec)(watt) |
| 1.000165 joules/int. joule | 44.261 ft-$lb_f$/(min)(watt) |
| 3,600(10$^3$) joules/kwhr | 860.421 cal/watthr |
| 3,600(10$^{10}$) ergs/kwhr | 859.858 IT cal/watthr |
| 2,615,218 ft-$lb_f$/kwhr | 745.701 watts/hp |
| 1.000165 watts/int. watt | 2544.48 Btu/hp-hr |
| 1,055.040 joules/Btu | 1.341 hp-hr/kwhr |
| 778.16 ft-$lb_f$/Btu | 3412.19 Btu/kwhr |

251.996 IT cal/Btu
252.161 cal/Btu

1.000654 cal/IT cal
42.408 Btu/(min)(hp)

33,000 ft-$lb_f$/(min)(hp)
550 ft-$lb_f$/(sec)(hp)

$$\frac{1 \text{ Btu}}{lb_m} = \frac{0.555919 \text{ cal}}{g_m}$$

$$\frac{1 \text{ cal}}{g_m} = \frac{1.798823 \text{ Btu}}{lb_m}$$

$$\frac{1 \text{ IT cal}}{g_m} = \frac{1.8 \text{ Btu}}{lb_m}$$

*Notes:*

Btu without prefix is understood to be the international (IT) Btu (see page 19).
    Joule (or watt or kilowatthour) without prefix is understood to be the absolute joule (or watt or kilowatthour).
    Calorie without prefix is understood to be the defined, or thermochemical, calorie.
    See Art. A-1 for discussion of electrical units.

TABLE B-1. DEFINITIONS AND CONVERSION FACTORS (*Continued*)

*Area*

1 cm² = 0.155 in.²
1 in.² = 6.45163 cm² = 6.94444(10⁻³) ft²
1 ft² = 929.034 cm² = 144 in.²

*Density*

1 $lb_m/ft^3$ = 0.0160184 $g_m/cm^3$ = 5.78704(10⁻⁴) $lb_m/in.^3$
1 $g_m/cm^3$ = 1.000028 $g_m/ml$ = 0.0361275 $lb_m/in.^3$ = 62.428 $lb_m/ft^3$
         = 8.3455 $lb_m/gal$

*Volume*

1 liter = 0.0353154 ft³ = 1,000.028 cm³ = 1,000 ml = 0.264178 gal
         = 61.0251 in.³
1 ft³ =    28.3162 liters = 7.48052 gal
1 gal = 231 in.³ = 0.133681 ft³ = 3,785.43 cm³
1 in.³ = 16.3872 cm³

*Miscellaneous*

$\ln_e x$ = 2.3025851 $\log_{10} x$      1 radian = 57.296°

CONVERSION FACTORS FOR VISCOSITY[†]

| To convert viscosity in centipoises to viscosity in: | poises = $\dfrac{1 \text{ dyne sec}}{cm^2}$ = $\dfrac{1 \text{ g}}{\text{sec cm}}$ | $\dfrac{lb_m}{ft\ sec}$ | $\dfrac{lb_m}{ft\ hr}$ | $\dfrac{lb_f\ sec}{ft^2}$ = $\dfrac{slug}{ft\ sec}$ | $\dfrac{kg_m}{m\ sec}$ | $\dfrac{kg_f\ sec}{m^2}$ |
|---|---|---|---|---|---|---|
| Multiply by | $\dfrac{1}{100}$ | 0.000672 | 2.42 | 0.0000209 | $\dfrac{1}{1,000}$ | 0.000102 |

[†] Reproduced from J. H. Perry (ed.), "Chemical Engineers' Handbook," 3d ed., McGraw-Hill Book Company, Inc., New York, 1950.

To convert Saybolt Universal viscosity readings into absolute viscosity readings,

$$\frac{\mu}{\rho} = 2.20 \times 10^{-3}\tau - \frac{1.80}{\tau}$$

where $\mu$ = absolute viscosity, poises
      $\rho$ = density, g/cm³
      $\tau$ = time of efflux (Saybolt Universal sec)
The above quantities are all to be measured at room temperature.

TABLE B-2. HEAT-CAPACITY EQUATIONS FOR THE IDEAL-GAS STATE

| Gas or vapor[†] | Equation $c_p$, Btu/(mole)(°R) | Range, °R | Maximum % error |
|---|---|---|---|
| $O_2$..... | $c_p = 11.515 - \dfrac{172}{\sqrt{T}} + \dfrac{1{,}530}{T}$ | 540–5000 | 1.1 |
| $N_2$..... | $c_p = 9.47 - \dfrac{3.47(10^3)}{T} + \dfrac{1.16(10^6)}{T^2}$ | 540–9000 | 1.7 |
| $CO$.... | $c_p = 9.46 - \dfrac{3.29(10^3)}{T} + \dfrac{1.07(10^6)}{T^2}$ | 540–9000 | 1.1 |
| $H_2O$... | $c_p = 19.86 - \dfrac{597}{\sqrt{T}} + \dfrac{7{,}500}{T}$ | 540–5400 | 1.8 |
| $CO_2$.... | $c_p = 16.2 - \dfrac{6.53(10^3)}{T} + \dfrac{1.41(10^6)}{T^2}$ | 540–6300 | 0.8 |

[†] R. L. Sweigert and M. W. Beardsley, Empirical Specific Heat Equations Based upon Spectroscopic Data, *Georgia School Technol. Bull.*, vol. 1, no. 3, June, 1938.

$$c_p = a + b(10^{-3})T + c(10^{-6})T^2 + d(10^{-9})T^3 \quad (T, °K)$$
$$= \text{cal/(g mole)(°K) or closely Btu/(lb}_m \text{ mole)(°R)}$$

| Gas or vapor[†] | $a$ | $b$ | $c$ | $d$ | Range, °K | Maximum % error |
|---|---|---|---|---|---|---|
| Air......... | 6.713 | 0.4697 | 1.147 | −0.4696 | 273–1800 | 0.72 |
|  | 6.557 | 1.477 | −0.2148 | 0 | 273–3800 | 1.64 |
| CO........ | 6.726 | 0.4001 | 1.283 | −0.5307 | 273–1800 | 0.89 |
|  | 6.480 | 1.566 | −0.2387 | 0 | 273–3800 | 1.86 |
| $CO_2$....... | 5.316 | 14.285 | −8.362 | 1.784 | 273–1800 | 0.67 |
| $c_p = 18.036 - 0.00004474T - 158.08(T)^{-\frac{1}{2}}$ | | | | | 273–3800 | 2.65 |
| $H_2$......... | 6.952 | −0.4576 | 0.9563 | −0.2079 | 273–1800 | 1.01 |
|  | 6.424 | 1.039 | −0.07804 | 0 | 273–3800 | 2.14 |
| $H_2O$....... | 7.700 | 0.4594 | 2.521 | −0.8587 | 273–1800 | 0.53 |
|  | 6.970 | 3.464 | −0.4833 | 0 | 273–3800 | 2.03 |
| $O_2$......... | 6.085 | 3.631 | −1.709 | 0.3133 | 273–1800 | 1.19 |
|  | 6.732 | 1.505 | −0.1791 | 0 | 273–3800 | 3.24 |
| $N_2$......... | 6.903 | −0.3753 | 1.930 | −6.861 | 273–1800 | 0.59 |
|  | 6.529 | 1.488 | −0.2271 | 0 | 273–3800 | 2.05 |
| $NH_3$....... | 6.5846 | 6.1251 | 2.3663 | −1.5981 | 273–1500 | 0.91 |
| $CH_4$....... | 4.750 | 12.00 | 3.030 | −2.630 | 273–1500 | 1.33 |
| $C_3H_8$....... | −0.966 | 72.79 | −37.55 | 7.580 | 273–1500 | 0.40 |
| $C_4H_{10}$...... | 0.945 | 88.73 | −43.80 | 8.360 | 273–1500 | 0.54 |
| $C_6H_6$....... | −8.650 | 115.78 | −75.40 | 18.54 | 273–1500 | 0.34 |
| $C_2H_2$....... | 5.21 | 22.008 | −15.59 | 4.349 | 273–1500 | 1.46 |
| $CH_3OH$..... | 4.55 | 21.86 | −2.91 | −1.92 | 273–1000 | 0.18 |

[†] K. A. Kobe, Thermochemistry for the Petrochemical Industry, *Petrol. Refiner*, January, 1949–November, 1954.

TABLE B-3. MISCELLANEOUS PHYSICAL CONSTANTS

HEAT CAPACITY AND DENSITY
(At about 70°F)

|  | $c$ <br> Btu/$(\text{lb}_m)$(°F) | $\rho$ <br> $\text{lb}_m/\text{ft}^3$ |
|---|---|---|
| Aluminum............... | 0.224 | 169 |
| Copper.................. | 0.0918 | 558 |
| Iron.................... | 0.122 | 492 |
| Lead................... | 0.0302 | 708 |
| Nickel................. | 0.111 | 556 |
| Water (liquid)........... | 1.00 | 62 |
| Water (ice at 32°F)....... | 0.487 | 56 |

HEAT OF FUSION
(All values in Btu/$\text{lb}_m$)

Lead..............    9.8
Glycerin..........    76.5
Ice...............    144.0

TUBE-THICKNESS DATA

| Size No., BWG† | Thickness, in. |
|---|---|
| 20 | 0.035 |
| 18 | 0.049 |
| 16 | 0.065 |
| 14 | 0.083 |
| 12 | 0.109 |
| 10 | 0.134 |
| 8 | 0.165 |

† Birmingham Wire Gauge.

STANDARD PIPE

| Nominal diameter, in. | Actual external diameter, in. | Approximate internal diameter, in. | Nominal weight per foot, $\text{lb}_m$ |
|---|---|---|---|
| ⅛ | 0.405 | 0.27 | 0.24 |
| ¼ | 0.540 | 0.36 | 0.42 |
| ½ | 0.840 | 0.62 | 0.85 |
| 1 | 1.315 | 1.05 | 1.68 |
| 1½ | 1.900 | 1.61 | 2.72 |
| 2 | 2.375 | 2.07 | 3.65 |
| 4 | 4.500 | 4.03 | 10.79 |
| 8 | 8.625 | 8.07 | 24.69 |
| 10 | 10.75 | 10.19 | 31.20 |
| 12 | 12.75 | 12.09 | 43.77 |

## TABLE B-4. STEAM TABLES†
### Steam Table 1.   Saturation: Temperatures

| Temp, °F | Abs press. Psia | Specific volume ft³/lb_m | | | Enthalpy Btu/lb_m | | | Entropy Btu/(lb_m)(°F) | | | Temp, °F |
|---|---|---|---|---|---|---|---|---|---|---|---|
| | | Sat. liquid | Evap. | Sat. vapor | Sat. liquid | Evap. | Sat. vapor | Sat. liquid | Evap. | Sat. vapor | |
| $t$ | $p$ | $v_f$ | $v_{fg}$ | $v_g$ | $h_f$ | $h_{fg}$ | $h_g$ | $s_f$ | $s_{fg}$ | $s_g$ | $t$ |
| 32° | 0.08854 | 0.01602 | 3306 | 3306 | 0.00 | 1075.8 | 1075.8 | 0.0000 | 2.1877 | 2.1877 | 32° |
| 35 | 0.09995 | 0.01602 | 2947 | 2947 | 3.02 | 1074.1 | 1077.1 | 0.0061 | 2.1709 | 2.1770 | 35 |
| 40 | 0.12170 | 0.01602 | 2444 | 2444 | 8.05 | 1071.3 | 1079.3 | 0.0162 | 2.1435 | 2.1597 | 40 |
| 50 | 0.17811 | 0.01603 | 1703.2 | 1703.2 | 18.07 | 1065.6 | 1083.7 | 0.0361 | 2.0903 | 2.1264 | 50 |
| 60 | 0.2563 | 0.01604 | 1206.6 | 1206.7 | 28.06 | 1059.9 | 1088.0 | 0.0555 | 2.0393 | 2.0948 | 60 |
| 70° | 0.3631 | 0.01606 | 867.8 | 867.9 | 38.04 | 1054.3 | 1092.3 | 0.0745 | 1.9902 | 2.0647 | 70° |
| 80 | 0.5069 | 0.01608 | 633.1 | 633.1 | 48.02 | 1048.6 | 1096.6 | 0.0932 | 1.9428 | 2.0360 | 80 |
| 90 | 0.6982 | 0.01610 | 468.0 | 468.0 | 57.99 | 1042.9 | 1100.9 | 0.1115 | 1.8972 | 2.0087 | 90 |
| 100 | 0.9492 | 0.01613 | 350.3 | 350.4 | 67.97 | 1037.2 | 1105.2 | 0.1295 | 1.8531 | 1.9826 | 100 |
| 110 | 1.2748 | 0.01617 | 265.3 | 265.4 | 77.94 | 1031.6 | 1109.5 | 0.1471 | 1.8106 | 1.9577 | 110 |
| 120° | 1.6924 | 0.01620 | 203.25 | 203.27 | 87.92 | 1025.8 | 1113.7 | 0.1645 | 1.7694 | 1.9339 | 120° |
| 130 | 2.2225 | 0.01625 | 157.32 | 157.34 | 97.90 | 1020.0 | 1117.9 | 0.1816 | 1.7296 | 1.9112 | 130 |
| 150 | 3.718 | 0.01634 | 97.06 | 97.07 | 117.89 | 1008.2 | 1126.1 | 0.2149 | 1.6537 | 1.8685 | 150 |
| 160 | 4.741 | 0.01639 | 77.27 | 77.29 | 127.89 | 1002.3 | 1130.2 | 0.2311 | 1.6174 | 1.8485 | 160 |
| 170 | 5.992 | 0.01645 | 62.04 | 62.06 | 137.90 | 996.3 | 1134.2 | 0.2473 | 1.5822 | 1.8295 | 170 |
| 180° | 7.510 | 0.01651 | 50.21 | 50.23 | 147.92 | 990.2 | 1138.1 | 0.2630 | 1.5480 | 1.8109 | 180° |
| 200 | 11.526 | 0.01663 | 33.62 | 33.64 | 167.99 | 977.9 | 1145.9 | 0.2938 | 1.4824 | 1.7762 | 200 |
| 210 | 14.123 | 0.01670 | 27.80 | 27.82 | 178.05 | 971.6 | 1149.7 | 0.3090 | 1.4508 | 1.7598 | 210 |
| 212 | 14.696 | 0.01672 | 26.78 | 26.80 | 180.07 | 970.3 | 1150.4 | 0.3120 | 1.4446 | 1.7566 | 212 |
| 220 | 17.186 | 0.01677 | 23.13 | 23.15 | 188.13 | 965.2 | 1153.4 | 0.3239 | 1.4201 | 1.7440 | 220 |
| 240° | 24.969 | 0.01692 | 16.306 | 16.323 | 208.34 | 952.2 | 1160.5 | 0.3531 | 1.3609 | 1.7140 | 240° |
| 250 | 29.825 | 0.01700 | 13.804 | 13.821 | 218.48 | 945.5 | 1164.0 | 0.3675 | 1.3323 | 1.6998 | 250 |
| 260 | 35.429 | 0.01709 | 11.746 | 11.763 | 228.64 | 938.7 | 1167.3 | 0.3817 | 1.3043 | 1.6860 | 260 |
| 270 | 41.858 | 0.01717 | 10.044 | 10.061 | 238.84 | 931.8 | 1170.6 | 0.3958 | 1.2769 | 1.6727 | 270 |
| 290 | 57.556 | 0.01735 | 7.444 | 7.461 | 259.31 | 917.5 | 1176.8 | 0.4234 | 1.2238 | 1.6472 | 290 |
| 300° | 67.013 | 0.01745 | 6.449 | 6.466 | 269.59 | 910.1 | 1179.7 | 0.4369 | 1.1980 | 1.6350 | 300° |
| 310 | 77.68 | 0.01755 | 5.609 | 5.626 | 279.92 | 902.6 | 1182.5 | 0.4504 | 1.1727 | 1.6231 | 310 |
| 320 | 89.66 | 0.01765 | 4.896 | 4.914 | 290.28 | 894.9 | 1185.2 | 0.4637 | 1.1478 | 1.6115 | 320 |
| 340 | 118.01 | 0.01787 | 3.770 | 3.788 | 311.13 | 879.0 | 1190.1 | 0.4900 | 1.0992 | 1.5891 | 340 |
| 350 | 134.63 | 0.01799 | 3.324 | 3.342 | 321.63 | 870.7 | 1192.3 | 0.5029 | 1.0754 | 1.5783 | 350 |
| 370° | 173.37 | 0.01823 | 2.606 | 2.625 | 342.79 | 853.5 | 1196.3 | 0.5286 | 1.0287 | 1.5573 | 370° |
| 390 | 220.37 | 0.01850 | 2.0651 | 2.0836 | 364.17 | 835.4 | 1199.6 | 0.5539 | 0.9832 | 1.5371 | 390 |
| 400 | 247.31 | 0.01864 | 1.8447 | 1.8633 | 374.97 | 826.0 | 1201.0 | 0.5664 | 0.9608 | 1.5272 | 400 |
| 420 | 308.83 | 0.01894 | 1.4811 | 1.5000 | 396.77 | 806.3 | 1203.1 | 0.5912 | 0.9166 | 1.5078 | 420 |
| 440 | 381.59 | 0.01926 | 1.1979 | 1.2171 | 418.90 | 785.4 | 1204.3 | 0.6158 | 0.8730 | 1.4887 | 440 |
| 450° | 422.6 | 0.0194 | 1.0799 | 1.0993 | 430.1 | 774.5 | 1204.6 | 0.6280 | 0.8513 | 1.4793 | 450° |
| 470 | 514.7 | 0.0198 | 0.8811 | 0.9009 | 452.8 | 751.5 | 1204.3 | 0.6523 | 0.8083 | 1.4606 | 470 |
| 490 | 621.4 | 0.0202 | 0.7221 | 0.7423 | 476.0 | 726.8 | 1202.8 | 0.6766 | 0.7653 | 1.4419 | 490 |
| 500 | 680.8 | 0.0204 | 0.6545 | 0.6749 | 487.8 | 713.9 | 1201.7 | 0.6887 | 0.7438 | 1.4325 | 500 |
| 540 | 962.5 | 0.0215 | 0.4434 | 0.4649 | 536.6 | 656.6 | 1193.2 | 0.7374 | 0.6568 | 1.3942 | 540 |
| 580° | 1325.8 | 0.0228 | 0.2989 | 0.3217 | 588.9 | 588.4 | 1177.3 | 0.7872 | 0.5659 | 1.3532 | 580° |
| 600 | 1542.9 | 0.0236 | 0.2432 | 0.2668 | 617.0 | 548.5 | 1165.5 | 0.8131 | 0.5176 | 1.3307 | 600 |
| 640 | 2059.7 | 0.0260 | 0.1538 | 0.1798 | 678.6 | 452.0 | 1130.5 | 0.8679 | 0.4110 | 1.2789 | 640 |
| 680 | 2708.1 | 0.0305 | 0.0810 | 0.1115 | 757.3 | 309.9 | 1067.2 | 0.9351 | 0.2719 | 1.2071 | 680 |
| 700 | 3093.7 | 0.0369 | 0.0392 | 0.0761 | 823.3 | 172.1 | 995.4 | 0.9905 | 0.1484 | 1.1389 | 700 |
| 705.4 | 3206.2 | 0.0503 | 0 | 0.0503 | 902.7 | 0 | 902.7 | 1.0580 | 0 | 1.0580 | 705.4 |

† Abstracted by permission from "Thermodynamic Properties of Steam" by Joseph H. Keenan and Frederick G. Keyes; published by John Wiley & Sons, Inc., New York.

TABLE B-4. STEAM TABLES (*Continued*)
Steam Table 2.  Saturation: Pressures

| Abs press., psia | Temp, °F | Specific volume | | Enthalpy | | | Entropy | | | Internal energy | | Abs press., psia |
|---|---|---|---|---|---|---|---|---|---|---|---|---|
| | | Sat. liquid | Sat. vapor | Sat. liquid | Evap. | Sat. vapor | Sat. liquid | Evap. | Sat. vapor | Sat. liquid | Sat. vapor | |
| $p$ | $t$ | $v_f$ | $v_g$ | $h_f$ | $h_{fg}$ | $h_g$ | $s_f$ | $s_{fg}$ | $s_g$ | $u_f$ | $u_g$ | $p$ |
| 1.0 | 101.74 | 0.01614 | 333.6 | 69.70 | 1036.3 | 1106.0 | 0.1326 | 1.8456 | 1.9782 | 69.70 | 1044.3 | 1.0 |
| 2.0 | 126.08 | 0.01623 | 173.73 | 93.99 | 1022.2 | 1116.2 | 0.1749 | 1.7451 | 1.9200 | 93.98 | 1051.9 | 2.0 |
| 3.0 | 141.48 | 0.01630 | 118.71 | 109.37 | 1013.2 | 1122.6 | 0.2008 | 1.6855 | 1.8863 | 109.36 | 1056.7 | 3.0 |
| 5.0 | 162.24 | 0.01640 | 73.52 | 130.13 | 1001.0 | 1131.1 | 0.2347 | 1.6094 | 1.8441 | 130.12 | 1063.1 | 5.0 |
| 6.0 | 170.06 | 0.01645 | 61.98 | 137.96 | 996.2 | 1134.2 | 0.2472 | 1.5820 | 1.8292 | 137.94 | 1065.4 | 6.0 |
| 7.0 | 176.85 | 0.01649 | 53.64 | 144.76 | 992.1 | 1136.9 | 0.2581 | 1.5586 | 1.8167 | 144.74 | 1067.4 | 7.0 |
| 8.0 | 182.86 | 0.01653 | 47.34 | 150.79 | 988.5 | 1139.3 | 0.2674 | 1.5383 | 1.8057 | 150.77 | 1069.2 | 8.0 |
| 10 | 193.21 | 0.01659 | 38.42 | 161.17 | 982.1 | 1143.3 | 0.2835 | 1.5041 | 1.7876 | 161.14 | 1072.2 | 10 |
| 14.696 | 212.00 | 0.01672 | 26.80 | 180.07 | 970.3 | 1150.4 | 0.3120 | 1.4446 | 1.7566 | 180.02 | 1077.5 | 14.696 |
| 15 | 213.03 | 0.01672 | 26.29 | 181.11 | 969.7 | 1150.8 | 0.3135 | 1.4415 | 1.7549 | 181.06 | 1077.8 | 15 |
| 20 | 227.96 | 0.01683 | 20.089 | 196.16 | 960.1 | 1156.3 | 0.3356 | 1.3962 | 1.7319 | 196.10 | 1081.9 | 20 |
| 30 | 250.33 | 0.01701 | 13.746 | 218.82 | 945.3 | 1164.1 | 0.3680 | 1.3313 | 1.6993 | 218.73 | 1087.8 | 30 |
| 35 | 259.28 | 0.01708 | 11.898 | 227.91 | 939.2 | 1167.1 | 0.3807 | 1.3063 | 1.6870 | 227.80 | 1090.1 | 35 |
| 40 | 267.25 | 0.01715 | 10.498 | 236.03 | 933.7 | 1169.7 | 0.3919 | 1.2844 | 1.6763 | 235.90 | 1092.0 | 40 |
| 50 | 281.01 | 0.01727 | 8.515 | 250.09 | 924.0 | 1174.1 | 0.4110 | 1.2474 | 1.6585 | 249.93 | 1095.3 | 50 |
| 55 | 287.07 | 0.01732 | 7.787 | 256.30 | 919.6 | 1175.9 | 0.4193 | 1.2316 | 1.6509 | 256.12 | 1096.7 | 55 |
| 60 | 292.71 | 0.01738 | 7.175 | 262.09 | 915.5 | 1177.6 | 0.4270 | 1.2168 | 1.6438 | 261.90 | 1097.9 | 60 |
| 65 | 297.97 | 0.01743 | 6.655 | 267.50 | 911.6 | 1179.1 | 0.4342 | 1.2032 | 1.6374 | 267.29 | 1099.1 | 65 |
| 70 | 302.92 | 0.01748 | 6.206 | 272.61 | 907.9 | 1180.6 | 0.4409 | 1.1906 | 1.6315 | 272.38 | 1100.2 | 70 |
| 80 | 312.03 | 0.01757 | 5.472 | 282.02 | 901.1 | 1183.1 | 0.4531 | 1.1676 | 1.6207 | 281.76 | 1102.1 | 80 |
| 85 | 316.25 | 0.01761 | 5.168 | 286.39 | 897.8 | 1184.2 | 0.4587 | 1.1571 | 1.6158 | 286.11 | 1102.9 | 85 |
| 90 | 320.27 | 0.01766 | 4.896 | 290.56 | 894.7 | 1185.3 | 0.4641 | 1.1471 | 1.6112 | 290.27 | 1103.7 | 90 |
| 95 | 324.12 | 0.01770 | 4.652 | 294.56 | 891.7 | 1186.2 | 0.4692 | 1.1376 | 1.6068 | 294.25 | 1104.5 | 95 |
| 100 | 327.81 | 0.01774 | 4.432 | 298.40 | 888.8 | 1187.2 | 0.4740 | 1.1286 | 1.6026 | 298.08 | 1105.2 | 100 |
| 120 | 341.25 | 0.01789 | 3.728 | 312.44 | 877.9 | 1190.4 | 0.4916 | 1.0962 | 1.5878 | 312.05 | 1107.6 | 120 |
| 140 | 353.02 | 0.01802 | 3.220 | 324.82 | 868.2 | 1193.0 | 0.5069 | 1.0682 | 1.5751 | 324.35 | 1109.6 | 140 |
| 144 | 355.21 | 0.01805 | 3.134 | 327.13 | 866.3 | 1193.4 | 0.5097 | 1.0631 | 1.5728 | 326.65 | 1109.9 | 144 |
| 150 | 358.42 | 0.01809 | 3.015 | 330.51 | 863.6 | 1194.1 | 0.5138 | 1.0556 | 1.5694 | 330.01 | 1110.5 | 150 |
| 160 | 363.53 | 0.01815 | 2.834 | 335.93 | 859.2 | 1195.1 | 0.5204 | 1.0436 | 1.5640 | 335.39 | 1111.2 | 160 |
| 170 | 368.41 | 0.01822 | 2.675 | 341.09 | 854.9 | 1196.0 | 0.5266 | 1.0324 | 1.5590 | 340.52 | 1111.9 | 170 |
| 180 | 373.06 | 0.01827 | 2.532 | 346.03 | 850.8 | 1196.9 | 0.5325 | 1.0217 | 1.5542 | 345.42 | 1112.5 | 180 |
| 200 | 381.79 | 0.01839 | 2.288 | 355.36 | 843.0 | 1198.4 | 0.5435 | 1.0018 | 1.5453 | 354.68 | 1113.7 | 200 |
| 250 | 400.95 | 0.01865 | 1.8438 | 376.00 | 825.1 | 1201.1 | 0.5675 | 0.9588 | 1.5263 | 375.14 | 1115.8 | 250 |
| 300 | 417.33 | 0.01890 | 1.5433 | 393.84 | 809.0 | 1202.8 | 0.5879 | 0.9225 | 1.5104 | 392.79 | 1117.1 | 300 |
| 400 | 444.59 | 0.0193 | 1.1613 | 424.0 | 780.5 | 1204.5 | 0.6214 | 0.8360 | 1.4844 | 422.6 | 1118.5 | 400 |
| 500 | 467.01 | 0.0197 | 0.9278 | 449.4 | 755.0 | 1204.4 | 0.6487 | 0.8147 | 1.4634 | 447.6 | 1118.6 | 500 |
| 600 | 486.21 | 0.0201 | 0.7698 | 471.6 | 731.6 | 1203.2 | 0.6720 | 0.7734 | 1.4454 | 469.4 | 1117.7 | 600 |
| 700 | 503.10 | 0.0205 | 0.6554 | 491.5 | 709.7 | 1201.2 | 0.6925 | 0.7371 | 1.4296 | 488.8 | 1116.3 | 700 |
| 800 | 518.23 | 0.0209 | 0.5687 | 509.7 | 688.9 | 1198.6 | 0.7108 | 0.7045 | 1.4153 | 506.6 | 1114.4 | 800 |
| 900 | 531.98 | 0.0212 | 0.5006 | 526.6 | 668.8 | 1195.4 | 0.7275 | 0.6744 | 1.4020 | 523.1 | 1112.1 | 900 |
| 1000 | 544.61 | 0.0216 | 0.4456 | 542.4 | 649.4 | 1191.8 | 0.7430 | 0.6467 | 1.3897 | 538.4 | 1109.4 | 1000 |
| 1200 | 567.22 | 0.0223 | 0.3619 | 571.7 | 611.7 | 1183.4 | 0.7711 | 0.5956 | 1.3667 | 566.7 | 1103.0 | 1200 |
| 1400 | 587.10 | 0.0231 | 0.3012 | 598.7 | 574.7 | 1173.4 | 0.7963 | 0.5491 | 1.3454 | 592.7 | 1095.4 | 1400 |
| 1500 | 596.23 | 0.0235 | 0.2765 | 611.6 | 556.3 | 1167.9 | 0.8082 | 0.5269 | 1.3351 | 605.1 | 1091.2 | 1500 |
| 2000 | 635.82 | 0.0257 | 0.1878 | 671.7 | 463.4 | 1135.1 | 0.8619 | 0.4230 | 1.2849 | 662.2 | 1065.6 | 2000 |
| 3000 | 695.36 | 0.0346 | 0.0858 | 802.5 | 217.8 | 1020.3 | 0.9731 | 0.1885 | 1.1615 | 783.4 | 972.7 | 3000 |
| 3206.2 | 705.40 | 0.0503 | 0.0503 | 902.7 | 0 | 902.7 | 1.0580 | 0 | 1.0580 | 872.9 | 872.9 | 3206.2 |
| 1.00 (in. Hg) | 79.03 | 0.01608 | 652.3 | 47.05 | 1049.2 | 1096.3 | 0.0914 | 1.9473 | 2.0387 | 47.05 | 1037.0 | 1.00 (in. Hg) |

## TABLE B-4. STEAM TABLES (*Continued*)
### Steam Table 3.  Superheated Vapor

| Abs press., psia (sat. temp) | | Temperature, °F | | | | | | | | | | |
|---|---|---|---|---|---|---|---|---|---|---|---|---|
| | | 200° | 300° | 400° | 500° | 600° | 700° | 800° | 900° | 1000° | 1200° | 1600° |
| 1 (101.74) | v | 392.6 | 452.3 | 512.0 | 571.6 | 631.2 | 690.8 | 750.4 | 809.9 | 869.5 | 988.7 | 1227.0 |
| | h | 1150.4 | 1195.8 | 1241.7 | 1288.3 | 1335.7 | 1383.8 | 1432.8 | 1482.7 | 1533.5 | 1637.7 | 1857.5 |
| | s | 2.0512 | 2.1153 | 2.1720 | 2.2233 | 2.2702 | 2.3137 | 2.3542 | 2.3923 | 2.4283 | 2.4952 | 2.6137 |
| 5 (162.24) | v | 78.16 | 90.25 | 102.26 | 114.22 | 126.16 | 138.10 | 150.03 | 161.95 | 173.87 | 197.71 | 245.4 |
| | h | 1148.8 | 1195.0 | 1241.2 | 1288.0 | 1335.4 | 1383.6 | 1432.7 | 1482.6 | 1533.4 | 1637.7 | 1857.4 |
| | s | 1.8718 | 1.9370 | 1.9942 | 2.0456 | 2.0927 | 1.1361 | 2.1767 | 2.2148 | 2.2509 | 2.3178 | 2.4363 |
| 10 (193.21) | v | 38.85 | 45.00 | 51.04 | 57.05 | 63.03 | 69.01 | 74.98 | 80.95 | 86.92 | 98.84 | 122.69 |
| | h | 1146.6 | 1193.9 | 1240.6 | 1287.5 | 1335.1 | 1383.4 | 1432.5 | 1482.4 | 1533.2 | 1637.6 | 1857.3 |
| | s | 1.7927 | 1.8595 | 1.9172 | 1.9689 | 2.0160 | 2.0596 | 2.1002 | 2.1383 | 2.1744 | 2.2413 | 2.3598 |
| 14.696 (212.00) | v | ...... | 30.53 | 34.68 | 38.78 | 42.86 | 46.94 | 51.00 | 55.07 | 59.13 | 67.25 | 83.48 |
| | h | ...... | 1192.8 | 1239.9 | 1287.1 | 1334.8 | 1383.2 | 1432.3 | 1482.3 | 1533.1 | 1637.5 | 1857.3 |
| | s | ...... | 1.8160 | 1.8743 | 1.9261 | 1.9734 | 2.0170 | 2.0576 | 2.0958 | 2.1319 | 2.1989 | 2.3174 |
| 20 (227.96) | v | ...... | 22.36 | 25.43 | 28.46 | 31.47 | 34.47 | 37.46 | 40.45 | 43.44 | 49.41 | 61.34 |
| | h | ...... | 1191.6 | 1239.2 | 1286.6 | 1334.4 | 1382.9 | 1432.1 | 1482.1 | 1533.0 | 1637.4 | 1857.2 |
| | s | ...... | 1.7808 | 1.8396 | 1.8918 | 1.9392 | 1.9829 | 2.0235 | 2.0618 | 2.0978 | 2.1648 | 2.2834 |
| 40 (267.25) | v | ...... | 11.040 | 12.628 | 14.168 | 15.688 | 17.198 | 18.702 | 20.20 | 21.70 | 24.69 | 30.66 |
| | h | ...... | 1186.8 | 1236.5 | 1284.8 | 1333.1 | 1381.9 | 1431.3 | 1481.4 | 1532.4 | 1637.0 | 1857.0 |
| | s | ...... | 1.6994 | 1.7608 | 1.8140 | 1.8619 | 1.9058 | 1.9467 | 1.9850 | 2.0212 | 2.0883 | 2.2069 |
| 60 (292.71) | v | ...... | 7.259 | 8.357 | 9.403 | 10.427 | 11.441 | 12.449 | 13.452 | 14.454 | 16.451 | 20.44 |
| | h | ...... | 1181.6 | 1233.6 | 1283.0 | 1331.8 | 1380.9 | 1430.5 | 1480.8 | 1531.9 | 1636.6 | 1856.7 |
| | s | ...... | 1.6492 | 1.7135 | 1.7678 | 1.8162 | 1.8605 | 1.9015 | 1.9400 | 1.9762 | 2.0434 | 2.1621 |
| 80 (312.03) | v | ...... | ...... | 6.220 | 7.020 | 7.797 | 8.562 | 9.322 | 10.077 | 10.830 | 12.332 | 15.325 |
| | h | ...... | ...... | 1230.7 | 1281.1 | 1330.5 | 1379.9 | 1429.7 | 1480.1 | 1531.3 | 1636.2 | 1856.5 |
| | s | ...... | ...... | 1.6791 | 1.7346 | 1.7836 | 1.8281 | 1.8694 | 1.9079 | 1.9442 | 2.0115 | 2.1303 |
| 100 (327.81) | v | ...... | ...... | 4.937 | 5.589 | 6.218 | 6.835 | 7.446 | 8.052 | 8.656 | 9.860 | 12.258 |
| | h | ...... | ...... | 1227.6 | 1279.1 | 1329.1 | 1378.9 | 1428.9 | 1479.5 | 1530.8 | 1635.7 | 1856.2 |
| | s | ...... | ...... | 1.6518 | 1.7085 | 1.7581 | 1.8029 | 1.8443 | 1.8829 | 1.9193 | 1.9867 | 2.1056 |
| 120 (341.25) | v | ...... | ...... | 4.081 | 4.636 | 5.165 | 5.683 | 6.195 | 6.702 | 7.207 | 8.212 | 10.213 |
| | h | ...... | ...... | 1224.4 | 1277.2 | 1327.7 | 1377.8 | 1428.1 | 1478.8 | 1530.2 | 1635.3 | 1856.0 |
| | s | ...... | ...... | 1.6287 | 1.6869 | 1.7370 | 1.7822 | 1.8237 | 1.8625 | 1.8990 | 1.9664 | 2.0854 |
| 140 (353.02) | v | ...... | ...... | 3.468 | 3.954 | 4.413 | 4.861 | 5.301 | 5.738 | 6.172 | 7.035 | 8.752 |
| | h | ...... | ...... | 1221.1 | 1275.2 | 1326.4 | 1376.8 | 1427.3 | 1478.2 | 1529.7 | 1634.9 | 1855.7 |
| | s | ...... | ...... | 1.6087 | 1.6683 | 1.7190 | 1.7645 | 1.8063 | 1.8451 | 1.8817 | 1.9493 | 2.0683 |
| 144 (355.21) | v | ...... | ...... | 3.366 | 3.840 | 4.288 | 4.724 | 5.152 | 5.577 | 6.000 | 6.839 | 8.508 |
| | h | ...... | ...... | 1220.4 | 1274.8 | 1326.1 | 1376.6 | 1427.1 | 1478.0 | 1529.6 | 1634.8 | 1855.7 |
| | s | ...... | ...... | 1.6050 | 1.6649 | 1.7157 | 1.7613 | 1.8031 | 1.8420 | 1.8785 | 1.9461 | 2.0652 |
| 160 (363.53) | v | ...... | ...... | 3.008 | 3.443 | 3.849 | 4.244 | 4.631 | 5.015 | 5.396 | 6.152 | 7.656 |
| | h | ...... | ...... | 1217.6 | 1273.1 | 1325.0 | 1375.7 | 1426.4 | 1477.5 | 1529.1 | 1634.5 | 1855.5 |
| | s | ...... | ...... | 1.5908 | 1.6519 | 1.7033 | 1.7491 | 1.7911 | 1.8301 | 1.8667 | 1.9344 | 2.0535 |
| 180 (373.06) | v | ...... | ...... | 2.649 | 3.044 | 3.411 | 3.764 | 4.110 | 4.452 | 4.792 | 5.466 | 6.804 |
| | h | ...... | ...... | 1214.0 | 1271.0 | 1323.5 | 1374.7 | 1425.6 | 1476.8 | 1528.6 | 1634.1 | 1855.2 |
| | s | ...... | ...... | 1.5745 | 1.6373 | 1.6894 | 1.7355 | 1.7776 | 1.8167 | 1.8534 | 1.9212 | 2.0404 |
| 200 (381.79) | v | ...... | ...... | 2.361 | 2.726 | 3.060 | 3.380 | 3.693 | 4.002 | 4.309 | 4.917 | 6.123 |
| | h | ...... | ...... | 1210.3 | 1268.9 | 1322.1 | 1373.6 | 1424.8 | 1476.2 | 1528.0 | 1633.7 | 1855.0 |
| | s | ...... | ...... | 1.5594 | 1.6240 | 1.6767 | 1.7232 | 1.7655 | 1.8048 | 1.8415 | 1.9094 | 2.0287 |
| 220 (389.86) | v | ...... | ...... | 2.125 | 2.465 | 2.772 | 3.066 | 3.352 | 3.634 | 3.913 | 4.467 | 5.565 |
| | h | ...... | ...... | 1206.5 | 1266.7 | 1320.7 | 1372.6 | 1424.0 | 1475.5 | 1527.5 | 1633.3 | 1854.7 |
| | s | ...... | ...... | 1.5453 | 1.6117 | 1.6652 | 1.7120 | 1.7545 | 1.7939 | 1.8308 | 1.8987 | 2.0181 |
| 240 (397.37) | v | ...... | ...... | 1.9276 | 2.247 | 2.533 | 2.804 | 3.068 | 3.327 | 3.584 | 4.093 | 5.100 |
| | h | ...... | ...... | 1202.5 | 1264.5 | 1319.2 | 1371.5 | 1423.2 | 1474.8 | 1526.9 | 1632.9 | 1854.5 |
| | s | ...... | ...... | 1.5319 | 1.6003 | 1.6546 | 1.7017 | 1.7444 | 1.7839 | 1.8209 | 1.8889 | 2.0084 |
| 260 (404.42) | v | ...... | ...... | ...... | 2.063 | 2.330 | 2.582 | 2.827 | 3.067 | 3.305 | 3.776 | 4.707 |
| | h | ...... | ...... | ...... | 1262.3 | 1317.7 | 1370.4 | 1422.3 | 1474.2 | 1526.3 | 1632.5 | 1854.2 |
| | s | ...... | ...... | ...... | 1.5897 | 1.6447 | 1.6922 | 1.7352 | 1.7748 | 1.8118 | 1.8799 | 1.9995 |
| 280 (411.05) | v | ...... | ...... | ...... | 1.9047 | 2.156 | 2.392 | 2.621 | 2.845 | 3.066 | 3.504 | 4.370 |
| | h | ...... | ...... | ...... | 1260.0 | 1316.2 | 1369.4 | 1421.5 | 1473.5 | 1525.8 | 1632.1 | 1854.0 |
| | s | ...... | ...... | ...... | 1.5796 | 1.6354 | 1.6834 | 1.7265 | 1.7662 | 1.8033 | 1.8716 | 1.9912 |
| 300 (417.33) | v | ...... | ...... | ...... | 1.7675 | 2.005 | 2.227 | 2.442 | 2.652 | 2.859 | 3.269 | 4.078 |
| | h | ...... | ...... | ...... | 1257.6 | 1314.7 | 1368.3 | 1420.6 | 1472.8 | 1525.2 | 1631.7 | 1853.7 |
| | s | ...... | ...... | ...... | 1.5701 | 1.6268 | 1.6751 | 1.7184 | 1.7582 | 1.7954 | 1.8638 | 1.9835 |
| 350 (431.72) | v | ...... | ...... | ...... | 1.4923 | 1.7036 | 1.8980 | 2.084 | 2.266 | 2.445 | 2.798 | 3.493 |
| | h | ...... | ...... | ...... | 1251.5 | 1310.9 | 1365.5 | 1418.5 | 1471.1 | 1523.8 | 1630.7 | 1853.1 |
| | s | ...... | ...... | ...... | 1.5481 | 1.6070 | 1.6563 | 1.7002 | 1.7403 | 1.7777 | 1.8463 | 1.9663 |

TABLE B-4. STEAM TABLES (Continued)
Steam Table 3.   Superheated Vapor (Continued)

| Abs press., psia (sat. temp) | | Temperature, °F | | | | | | | | | | |
|---|---|---|---|---|---|---|---|---|---|---|---|---|
| | | 500° | 550° | 600° | 620° | 640° | 660° | 680° | 700° | 800° | 1000° | 1600° |
| 400 (444.59) | $v$.... | 1.2851 | 1.3843 | 1.4770 | 1.5128 | 1.5480 | 1.5827 | 1.6169 | 1.6508 | 1.8161 | 2.134 | 3.055 |
| | $h$.... | 1245.1 | 1277.2 | 1306.9 | 1318.3 | 1329.6 | 1340.8 | 1351.8 | 1362.7 | 1416.4 | 1522.4 | 1852.5 |
| | $s$.... | 1.5281 | 1.5606 | 1.5894 | 1.6001 | 1.6105 | 1.6205 | 1.6303 | 1.6398 | 1.6842 | 1.7623 | 1.9513 |
| 500 (467.01) | $v$.... | 0.9927 | 1.0800 | 1.1591 | 1.1893 | 1.2188 | 1.2478 | 1.2763 | 1.3044 | 1.4405 | 1.6996 | 2.442 |
| | $h$.... | 1231.3 | 1266.8 | 1298.6 | 1310.7 | 1322.6 | 1334.2 | 1345.7 | 1357.0 | 1412.1 | 1519.6 | 1851.3 |
| | $s$.... | 1.4919 | 1.5280 | 1.5588 | 1.5701 | 1.5810 | 1.5915 | 1.6016 | 1.6115 | 1.6571 | 1.7363 | 1.9262 |
| 600 (486.21) | $v$.... | 0.7947 | 0.8753 | 0.9463 | 0.9729 | 0.9988 | 1.0241 | 1.0489 | 1.0732 | 1.1899 | 1.4096 | 2.033 |
| | $h$.... | 1215.7 | 1255.5 | 1289.9 | 1302.7 | 1315.2 | 1327.4 | 1339.3 | 1351.1 | 1407.7 | 1516.7 | 1850.0 |
| | $s$.... | 1.4586 | 1.4990 | 1.5323 | 1.5443 | 1.5558 | 1.5667 | 1.5773 | 1.5875 | 1.6343 | 1.7147 | 1.9056 |
| 800 (518.23) | $v$.... | ...... | 0.6154 | 0.6779 | 0.7006 | 0.7223 | 0.7433 | 0.7635 | 0.7833 | 0.8763 | 1.0470 | 1.5214 |
| | $h$.... | ...... | 1229.8 | 1270.7 | 1285.4 | 1299.4 | 1312.9 | 1325.9 | 1338.6 | 1398.6 | 1511.0 | 1847.5 |
| | $s$.... | ...... | 1.4467 | 1.4863 | 1.5000 | 1.5129 | 1.5250 | 1.5366 | 1.5476 | 1.5972 | 1.6801 | 1.8729 |
| 1000 (544.61) | $v$.... | ...... | 0.4533 | 0.5140 | 0.5350 | 0.5546 | 0.5733 | 0.5912 | 0.6084 | 0.6878 | 0.8294 | 1.2146 |
| | $h$.... | ...... | 1198.3 | 1248.8 | 1265.9 | 1281.9 | 1297.0 | 1311.4 | 1325.3 | 1389.2 | 1505.1 | 1845.0 |
| | $s$.... | ...... | 1.3961 | 1.4450 | 1.4610 | 1.4757 | 1.4893 | 1.5021 | 1.5141 | 1.5670 | 1.6525 | 1.8474 |
| 1200 (567.22) | $v$.... | ...... | ...... | 0.4016 | 0.4222 | 0.4410 | 0.4586 | 0.4752 | 0.4909 | 0.5617 | 0.6843 | 1.0101 |
| | $h$.... | ...... | ...... | 1223.5 | 1243.9 | 1262.4 | 1279.6 | 1295.7 | 1311.0 | 1379.3 | 1499.2 | 1842.5 |
| | $s$.... | ...... | ...... | 1.4052 | 1.4243 | 1.4413 | 1.4568 | 1.4710 | 1.4843 | 1.5409 | 1.6293 | 1.8263 |
| 1600 (604.90) | $v$.... | ...... | ...... | ...... | 0.2733 | 0.2936 | 0.3112 | 0.3271 | 0.3417 | 0.4034 | 0.5027 | 0.7545 |
| | $h$.... | ...... | ...... | ...... | 1187.8 | 1215.2 | 1238.7 | 1259.6 | 1278.7 | 1358.4 | 1487.0 | 1837.5 |
| | $s$.... | ...... | ...... | ...... | 1.3489 | 1.3741 | 1.3952 | 1.4137 | 1.4303 | 1.4964 | 1.5914 | 1.7926 |
| 2000 (635.82) | $v$.... | ...... | ...... | ...... | ...... | 0.1936 | 0.2161 | 0.2337 | 0.2489 | 0.3074 | 0.3935 | 0.6011 |
| | $h$.... | ...... | ...... | ...... | ...... | 1145.6 | 1184.9 | 1214.8 | 1240.0 | 1335.5 | 1474.5 | 1832.5 |
| | $s$.... | ...... | ...... | ...... | ...... | 1.2945 | 1.3300 | 1.3564 | 1.3783 | 1.4576 | 1.5603 | 1.7660 |
| 3000 (695.36) | $v$.... | ...... | ...... | ...... | ...... | ...... | ...... | ...... | 0.0984 | 0.1760 | 0.2476 | 0.3966 |
| | $h$.... | ...... | ...... | ...... | ...... | ...... | ...... | ...... | 1060.7 | 1267.2 | 1441.8 | 1819.9 |
| | $s$.... | ...... | ...... | ...... | ...... | ...... | ...... | ...... | 1.1966 | 1.3690 | 1.4984 | 1.7163 |
| 3206.2 (705.40) | $v$.... | ...... | ...... | ...... | ...... | ...... | ...... | ...... | ...... | 0.1583 | 0.2288 | 0.3703 |
| | $h$.... | ...... | ...... | ...... | ...... | ...... | ...... | ...... | ...... | 1250.5 | 1434.7 | 1817.2 |
| | $s$.... | ...... | ...... | ...... | ...... | ...... | ...... | ...... | ...... | 1.3508 | 1.4874 | 1.7080 |
| 4000 | $v$.... | ...... | ...... | ...... | ...... | ...... | ...... | ...... | 0.0287 | 0.1052 | 0.1743 | 0.2943 |
| | $h$.... | ...... | ...... | ...... | ...... | ...... | ...... | ...... | 763.8 | 1174.8 | 1406.8 | 1807.2 |
| | $s$.... | ...... | ...... | ...... | ...... | ...... | ...... | ...... | 0.9347 | 1.2757 | 1.4482 | 1.6795 |
| 5000 | $v$.... | ...... | ...... | ...... | ...... | ...... | ...... | ...... | 0.0268 | 0.0593 | 0.1303 | 0.2329 |
| | $h$.... | ...... | ...... | ...... | ...... | ...... | ...... | ...... | 746.4 | 1047.1 | 1369.5 | 1794.5 |
| | $s$.... | ...... | ...... | ...... | ...... | ...... | ...... | ...... | 0.9152 | 1.1622 | 1.4034 | 1.6499 |

Steam Table 4.   Compressed Liquid

| Abs press., psi (sat. temp) | | | Temperature, °F | | | |
|---|---|---|---|---|---|---|
| | | | 32° | 100° | 200° | 300° |
| | Saturated liquid | $p$ | 0.08854 | 0.9492 | 11.526 | 67.013 |
| | | $v_f$ | 0.016022 | 0.016132 | 0.016634 | 0.017449 |
| | | $h_f$ | 0 | 67.97 | 167.99 | 269.59 |
| | | $s_f$ | 0 | 0.12948 | 0.29382 | 0.43694 |
| 200 (381.79) | $(v - v_f)(10^5)$ | | −1.1 | −1.1 | −1.1 | −1.1 |
| | $(h - h_f)$ | | +0.61 | +0.54 | +0.41 | +0.23 |
| | $(s - s_f)(10^3)$ | | +0.03 | −0.05 | −0.21 | −0.21 |
| 400 (444.59) | $(v - v_f)(10^5)$ | | −2.3 | −2.1 | −2.2 | −2.8 |
| | $(h - h_f)$ | | +1.21 | +1.09 | +0.88 | +0.61 |
| | $(s - s_f)(10^3)$ | | +0.04 | −0.16 | −0.47 | −0.56 |
| 600 (486.21) | $(v - v_f)(10^5)$ | | −3.5 | −3.2 | −3.4 | −4.3 |
| | $(h - h_f)$ | | +1.80 | +1.67 | +1.31 | +0.97 |
| | $(s - s_f)(10^3)$ | | +0.07 | −0.27 | −0.74 | −0.94 |
| 800 (518.23) | $(v - v_f)(10^5)$ | | −4.6 | −4.0 | −4.4 | −5.6 |
| | $(h - h_f)$ | | +2.39 | +2.17 | +1.78 | +1.35 |
| | $(s - s_f)(10^3)$ | | +0.10 | −0.40 | −0.97 | −1.27 |
| 1000 (544.61) | $(v - v_f)(10^5)$ | | −5.7 | −5.1 | −5.4 | −6.9 |
| | $(h - h_f)$ | | +2.99 | +2.70 | +2.21 | +1.75 |
| | $(s - s_f)(10^3)$ | | +0.15 | −0.53 | −1.20 | −1.64 |

TABLE B-5. PROPERTIES OF MERCURY VAPOR†

($h$ and $s$ are measured from 32°F)

| Pressure $p$ psia | Temp $t$ °F | Specific volume $v_g$ ft³/lb$_m$ | Enthalpy, Btu/lb$_m$ | | | Entropy, Btu/(lb$_m$)(°R) | | |
|---|---|---|---|---|---|---|---|---|
| | | | Saturated liquid $h_f$ | Vaporization $h_{fg}$ | Saturated vapor $h_g$ | Saturated liquid $s_f$ | Vaporization $s_{fg}$ | Saturated vapor $s_g$ |
| 0.4 | 402.3 | 114.5 | 13.81 | 128.1 | 141.9 | 0.02094 | 0.1486 | 0.1696 |
| 0.6 | 426.1 | 78.23 | 14.70 | 127.6 | 142.3 | 0.02195 | 0.1441 | 0.1660 |
| 0.8 | 443.8 | 59.71 | 15.36 | 127.2 | 142.6 | 0.02269 | 0.1408 | 0.1635 |
| 1.0 | 458.1 | 48.45 | 15.89 | 126.9 | 142.8 | 0.02328 | 0.1382 | 0.1615 |
| 1.5 | 485.1 | 33.14 | 16.90 | 126.3 | 143.2 | 0.02436 | 0.1337 | 0.1580 |
| 2 | 505.2 | 25.31 | 17.65 | 125.8 | 143.5 | 0.02514 | 0.1304 | 0.1556 |
| 3 | 535.4 | 17.34 | 18.78 | 125.2 | 144.0 | 0.02629 | 0.1258 | 0.1521 |
| 4 | 558.0 | 13.26 | 19.62 | 124.7 | 144.3 | 0.02714 | 0.1225 | 0.1497 |
| 5 | 576.2 | 10.77 | 20.30 | 124.3 | 144.6 | 0.02780 | 0.1200 | 0.1478 |
| 6 | 591.4 | 9.096 | 20.87 | 123.9 | 144.8 | 0.02834 | 0.1179 | 0.1462 |
| 7 | 605.0 | 7.882 | 21.37 | 123.6 | 145.0 | 0.02882 | 0.1161 | 0.1450 |
| 8 | 616.8 | 6.963 | 21.81 | 123.4 | 145.2 | 0.02923 | 0.1146 | 0.1439 |
| 9 | 627.5 | 6.244 | 22.21 | 123.2 | 145.4 | 0.02960 | 0.1133 | 0.1429 |
| 10 | 637.3 | 5.661 | 22.58 | 122.9 | 145.5 | 0.02993 | 0.1121 | 0.1420 |
| 15 | 676.5 | 3.892 | 24.04 | 122.1 | 146.1 | 0.03124 | 0.1074 | 0.1387 |
| 20 | 706.2 | 2.983 | 25.15 | 121.4 | 146.6 | 0.03220 | 0.1041 | 0.1363 |
| 25 | 730.4 | 2.429 | 26.05 | 120.9 | 146.9 | 0.03297 | 0.1016 | 0.1345 |
| 30 | 750.9 | 2.053 | 26.81 | 120.4 | 147.2 | 0.03360 | 0.09953 | 0.1331 |
| 35 | 769.0 | 1.781 | 27.49 | 120.0 | 147.5 | 0.03416 | 0.09774 | 0.1319 |
| 40 | 784.8 | 1.576 | 28.08 | 119.7 | 147.8 | 0.03464 | 0.09621 | 0.1308 |
| 45 | 799.3 | 1.414 | 28.62 | 119.4 | 148.0 | 0.03507 | 0.09486 | 0.1299 |
| 50 | 812.5 | 1.284 | 29.11 | 119.1 | 148.2 | 0.03546 | 0.09364 | 0.1291 |
| 60 | 836.1 | 1.086 | 29.99 | 118.6 | 148.6 | 0.03614 | 0.09154 | 0.1276 |
| 70 | 856.6 | 0.9436 | 30.75 | 118.1 | 148.9 | 0.03672 | 0.08976 | 0.1264 |
| 80 | 874.8 | 0.8349 | 31.43 | 117.7 | 149.1 | 0.03725 | 0.08824 | 0.1254 |
| 90 | 891.6 | 0.7497 | 32.06 | 117.3 | 149.4 | 0.03771 | 0.08687 | 0.1245 |
| 100 | 906.9 | 0.6811 | 32.63 | 117.0 | 149.6 | 0.03813 | 0.08565 | 0.1237 |
| 120 | 934.4 | 0.5767 | 33.60 | 116.4 | 150.1 | 0.03887 | 0.08353 | 0.1224 |
| 140 | 958.3 | 0.5012 | 34.55 | 115.9 | 150.4 | 0.03951 | 0.08175 | 0.1212 |
| 160 | 979.9 | 0.4438 | 35.35 | 115.4 | 150.8 | 0.04007 | 0.08019 | 0.1202 |
| 180 | 999.6 | 0.3990 | 36.09 | 115.0 | 151.1 | 0.04058 | 0.07881 | 0.1193 |

† Reproduced from L. S. Marks (ed.), "Mechanical Engineers' Handbook," 6th ed., McGraw-Hill Book Company, Inc., New York, 1958.

TABLE B-6. GAS-CONSTANT VALUES†

| Substance | Symbol | $M$ | $R$ ft-lb$_f$ / lb$_m$ °R | $c_p$ Btu / lb$_m$ °R at 77°F | $c_v$ Btu / lb$_m$ °R at 77°F | $k$ $\dfrac{c_p}{c_v}$ |
|---|---|---|---|---|---|---|
| Acetylene............ | $C_2H_2$ | 26.038 | 59.39 | 0.4030 | 0.3267 | 1.234 |
| Air................... | ....... | 28.967 | 53.36 | 0.2404 | 0.1718 | 1.399 |
| Ammonia............. | $NH_3$ | 17.032 | 90.77 | 0.5006 | 0.3840 | 1.304 |
| Argon............... | A | 39.944 | 38.73 | 0.1244 | 0.0746 | 1.668 |
| Benzene............. | $C_6H_6$ | 78.114 | 19.78 | 0.2497 | 0.2243 | 1.113 |
| n-Butane............ | $C_4H_{10}$ | 58.124 | 26.61 | 0.4004 | 0.3662 | 1.093 |
| Isobutane........... | $C_4H_{10}$ | 58.124 | 26.59 | 0.3979 | 0.3637 | 1.094 |
| 1-Butene............ | $C_4H_8$ | 56.108 | 27.545 | 0.3646 | 0.3282 | 1.111 |
| Carbon dioxide........ | $CO_2$ | 44.011 | 35.12 | 0.2015 | 0.1564 | 1.288 |
| Carbon monoxide....... | CO | 28.011 | 55.19 | 0.2485 | 0.1776 | 1.399 |
| Carbon tetrachloride... | $CCl_4$ | 153.839 | | | | |
| n-Deuterium......... | $D_2$ | 4.029 | | | | |
| Dodecane............ | $C_{12}H_{26}$ | 170.340 | 9.074 | 0.3931 | 0.3814 | 1.031 |
| Ethane.............. | $C_2H_6$ | 30.070 | 51.43 | 0.4183 | 0.3522 | 1.188 |
| Ethyl ether.......... | $C_4H_{10}O$ | 74.124 | | | | |
| Ethylene............ | $C_2H_4$ | 28.054 | 55.13 | 0.3708 | 0.3000 | 1.236 |
| Freon, F-12.......... | $CCl_2F_2$ | 120.925 | 12.78 | 0.1369 | 0.1204 | 1.136 |
| Helium.............. | He | 4.003 | 386.33 | 1.241 | 0.7446 | 1.667 |
| n-Heptane........... | $C_7H_{16}$ | 100.205 | 15.42 | 0.3956 | 0.3758 | 1.053 |
| n-Hexane............ | $C_6H_{14}$ | 86.178 | 17.93 | 0.3966 | 0.3736 | 1.062 |
| Hydrogen............ | $H_2$ | 2.016 | 766.53 | 3.416 | 2.431 | 1.405 |
| Hydrogen sulfide....... | $H_2S$ | 34.082 | | | | |
| Mercury............. | Hg | 200.610 | | | | |
| Methane............. | $CH_4$ | 16.043 | 96.40 | 0.5318 | 0.4079 | 1.304 |
| Methyl fluoride........ | $CH_3F$ | 34.035 | | | | |
| Neon................ | Ne | 20.183 | 76.58 | 0.2460 | 0.1476 | 1.667 |
| Nitric oxide.......... | NO | 30.008 | 51.49 | 0.2377 | 0.1715 | 1.386 |
| Nitrogen............ | $N_2$ | 28.016 | 55.15 | 0.2483 | 0.1774 | 1.400 |
| Octane.............. | $C_8H_{18}$ | 114.232 | 13.54 | 0.3949 | 0.3775 | 1.046 |
| Oxygen............. | $O_2$ | 32.000 | 48.29 | 0.2191 | 0.1570 | 1.396 |
| n-Pentane........... | $C_5H_{12}$ | 72.151 | 21.42 | 0.3980 | 0.3705 | 1.074 |
| Isopentane........... | $C_5H_{12}$ | 72.151 | 21.42 | 0.3972 | 0.3697 | 1.074 |
| Propane............. | $C_3H_8$ | 44.097 | 35.07 | 0.3982 | 0.3531 | 1.128 |
| Propylene............ | $C_3H_6$ | 42.081 | 36.72 | 0.3627 | 0.3055 | 1.187 |
| Sulfur dioxide........ | $SO_2$ | 64.066 | 24.12 | 0.1483 | 0.1173 | 1.264 |
| Water vapor.......... | $H_2O$ | 18.016 | 85.80 | 0.4452 | 0.3349 | 1.329 |
| Xenon............... | Xe | 131.300 | 11.78 | 0.03781 | 0.02269 | 1.667 |

† Data selected from J. F. Masi, *Trans. ASME*, **76**: 1067 (October, 1954); *Natl. Bur. Standards (U.S.) Circ.* 500, February, 1952; API Research Project 44, National Bureau of Standards, Washington, December, 1952.

TABLE B-7. CRITICAL CONSTANTS†

| Substance | Ref. date | Symbol | $M$ | $T_c$ °K | $p_c$ atm | $v_c$ cm³/g mole | $v_c$ ft³/ mole | $z_c$ |
|---|---|---|---|---|---|---|---|---|
| Acetylene.......... | 1928 | $C_2H_2$ | 26.038 | 309.5 | 61.6 | 113 | ...... | 0.274 |
| Air............... | 1917 | ...... | 28.967 | 132.41 | 37.25 | 93.25 | | |
| Ammonia........... | 1920 | $NH_3$ | 17.032 | 405.4 | 111.3 | 72.5 | 1.16 | 0.243 |
| Argon............. | 1958 | A | 39.944 | 150.87 | 48.34 | 74.56 | 1.19 | 0.291 |
| Benzene........... | 1948 | $C_6H_6$ | 78.114 | 562.6 | 48.6 | 260 | 4.17 | 0.274 |
| n-Butane.......... | 1939 | $C_4H_{10}$ | 58.124 | 425.17 | 37.47 | 255 | 4.08 | 0.274 |
| Isobutane.......... | 1910 | $C_4H_{10}$ | 58.124 | 408.14 | 36.00 | 263 | 4.21 | 0.283 |
| 1-Butene.......... | 1950 | $C_4H_8$ | 56.108 | 419.6 | 39.7 | 240 | 3.84 | 0.277 |
| Carbon dioxide....... | 1950 | $CO_2$ | 44.011 | 304.20 | 72.90 | 94 | 1.51 | 0.275 |
| Carbon monoxide..... | 1936 | CO | 28.011 | 132.91 | 34.529 | 93 | 1.49 | 0.294 |
| Carbon tetrachloride.. | 1931 | $CCl_4$ | 153.839 | 556.4 | 45.0 | 276 | ...... | 0.272 |
| n-Deuterium......... | 1951 | $D_2$ | 4.029 | 38.43 | 16.421 | | | |
| Dodecane........... | 1953 | $C_{12}H_{26}$ | 170.340 | 659 | 17.9 | ...... | 11.5 | 0.237 |
| Ethane............. | 1939 | $C_2H_6$ | 30.070 | 305.43 | 48.20 | 148 | 2.37 | 0.285 |
| Ethyl ether......... | 1929 | $C_4H_{10}O$ | 74.124 | 467.8 | 35.6 | 282.9 | | |
| Ethylene........... | 1939 | $C_2H_4$ | 28.054 | 283.06 | 50.50 | 124 | 1.99 | 0.270 |
| Freon, F-12......... | 1957 | $CCl_2F_2$ | 120.925 | 385.16 | 40.6 | 217 | 3.47 | 0.279 |
| Helium............. | 1936 | He | 4.003 | 5.19 | 2.26 | 58 | 0.929 | 0.308 |
| n-Heptane.......... | 1937 | $C_7H_{16}$ | 100.205 | 540.17 | 27.00 | 426 | 6.82 | 0.260 |
| n-Hexane.......... | 1946 | $C_6H_{14}$ | 86.178 | 507.9 | 29.94 | 368 | 5.89 | 0.264 |
| Hydrogen.......... | 1951 | $H_2$ | 2.016 | 33.24 | 12.797 | 65 | 1.04 | 0.304 |
| Hydrogen sulfide..... | 1948 | $H_2S$ | 34.082 | 373.7 | 88.8 | 98 | 1.57 | 0.284 |
| Mercury............ | 1953 | Hg | 200.610 | | | | | |
| Methane........... | 1953 | $CH_4$ | 16.043 | 190.7 | 45.8 | 99 | 1.59 | 0.290 |
| Methyl fluoride...... | 1932 | $CH_3F$ | 34.035 | 317.71 | 58.0 | | | |
| Neon.............. | 1936 | Ne | 20.183 | 44.39 | 26.86 | 41.7 | 0.668 | 0.308 |
| Nitric oxide.......... | 1951 | NO | 30.008 | 179.2 | 65.0 | 58 | 0.929 | 0.256 |
| Nitrogen........... | 1951 | $N_2$ | 28.016 | 126.2 | 33.54 | 90 | 1.44 | 0.291 |
| Octane............ | 1931 | $C_8H_{18}$ | 114.232 | 569.4 | 24.64 | 486 | 7.77 | 0.256 |
| Oxygen............ | 1948 | $O_2$ | 32.000 | 154.78 | 50.14 | 74 | 1.19 | 0.292 |
| n-Pentane.......... | 1899 | $C_5H_{12}$ | 72.151 | 469.78 | 33.31 | 311 | 4.98 | 0.269 |
| Isopentane.......... | 1910 | $C_5H_{12}$ | 72.151 | 461.0 | 32.92 | 308 | 4.93 | 0.268 |
| Propane........... | 1940 | $C_3H_8$ | 44.097 | 370.01 | 42.1 | 200 | 3.20 | 0.277 |
| Propylene.......... | 1953 | $C_3H_6$ | 42.081 | 365.1 | 45.40 | 181 | 2.90 | 0.274 |
| Sulfur dioxide........ | 1945 | $SO_2$ | 64.066 | 430.7 | 77.8 | 122 | ...... | 0.269 |
| Water............. | 1934 | $H_2O$ | 18.016 | 647.27 | 218.167 | 56 | 0.897 | 0.230 |
| Xenon............. | 1951 | Xe | 131.300 | 289.81 | 58.0 | 118.8 | 1.90 | 0.290 |

† Bibliographies of certain of the data sources are in Ref. 9 of Chap. 10.

TABLE B-8. CONSTANTS FOR EQUATIONS OF STATE

| Gas | van der Waals[†] | | Beattie-Bridgeman[‡] | | | | | |
|---|---|---|---|---|---|---|---|---|
| | $a$ $\dfrac{\text{atm ft}^6}{\text{mole}^2}$ | $b$ $\dfrac{\text{ft}^3}{\text{mole}}$ | $A_0$ $\dfrac{\text{atm ft}^6}{\text{mole}^2}$ | $a$ $\dfrac{\text{ft}^3}{\text{mole}}$ | $B_0$ $\dfrac{\text{ft}^3}{\text{mole}}$ | $b$ $\dfrac{\text{ft}^3}{\text{mole}}$ | $10^{-6}c$ $\dfrac{\text{ft}^3\ ^\circ\text{R}^3}{\text{mole}}$ | Range $^\circ$R |
| Air....... | 343.8 | 0.585 | 334.1 | 0.309 | 0.739 | −0.716 | 4.05 | 230–850 |
| CO....... | 374.7 | 0.630 | 344.9 | 0.419 | 0.808 | −0.111 | 3.92 | |
| CO$_2$...... | 924.2 | 0.685 | 1,284.9 | 1.143 | 1.678 | 1.159 | 61.65 | 490–670 |
| CH$_4$...... | 578.9 | 0.684 | 584.6 | 0.297 | 0.895 | −0.254 | 11.98 | 490–850 |
| C$_3$H$_8$..... | 2,374 | 1.446 | 305.8 | 1.173 | 2.90 | 0.688 | 112.12 | |
| C$_4$H$_{10}$..... | 3,675 | 1.944 | 456.5 | 1.948 | 3.944 | 1.51 | 327.02 | |
| H$_2$........ | 63.02 | 0.427 | 50.57 | −0.0811 | 0.336 | −0.698 | 0.0471 | 50–850 |
| N$_2$........ | 346.0 | 0.618 | 344.92 | 0.419 | 0.808 | −0.111 | 3.92 | |
| NH$_3$...... | 1,076 | 0.598 | 613.91 | 2.729 | 0.547 | 3.062 | 445.6 | |
| O$_2$........ | 349.5 | 0.510 | 382.53 | 0.410 | 0.741 | 0.0674 | 4.48 | |

Benedict-Webb-Rubin Equation[§]
[Units: atm liters/(g mole)($^\circ$K)]

| Gas | $A_0$ | $B_0(10^2)$ | $C_0(10^{-6})$ | $a$ | $b(10^2)$ | $c(10^{-6})$ | $\alpha(10^3)$ | $\gamma(10^2)$ |
|---|---|---|---|---|---|---|---|---|
| N$_2$........ | 1.19250 | 4.580 | 0.005889 | 0.0149 | 0.1982 | 0.000548 | 0.2915 | 0.7500 |
| CH$_4$....... | 1.855 | 4.260 | 0.0226 | 0.0494 | 0.338 | 0.00254 | 0.124 | 0.600 |
| C$_2$H$_6$....... | 4.156 | 6.277 | 0.1796 | 0.345 | 1.112 | 0.0328 | 0.243 | 1.180 |
| C$_3$H$_8$....... | 6.872 | 9.731 | 0.5083 | 0.948 | 2.25 | 0.129 | 0.607 | 2.200 |
| C$_4$H$_{10}$...... | 10.08 | 12.44 | 0.9928 | 1.882 | 4.00 | 0.3164 | 1.101 | 3.400 |

VALUES FOR THE GAS CONSTANT $R_0$

$$R_0 = 1{,}545.33\ \frac{\text{ft-lb}_f}{\text{mole } ^\circ\text{R}} \qquad R_0 = 1.98588\ \frac{\text{IT cal}}{\text{g mole } ^\circ\text{K}}$$

$$R_0 = 10.7315\ \frac{\text{psia ft}^3}{\text{mole } ^\circ\text{R}} \qquad R_0 = 0.0820544\ \frac{\text{atm liter}}{\text{g mole } ^\circ\text{K}}$$

$$R_0 = 0.73023\ \frac{\text{atm ft}^3}{\text{mole } ^\circ\text{R}} \qquad R_0 = 82.0544\ \frac{\text{atm ml}}{\text{g mole } ^\circ\text{K}}$$

$$R_0 = 1.98588\ \frac{\text{IT Btu}}{\text{mole } ^\circ\text{R}} \qquad R_0 = 82.0567\ \frac{\text{atm cm}^3}{\text{g mole } ^\circ\text{K}}$$

† Evaluated from critical data in the manner of Art. 10-2.
‡ J. A. Beattie and O. C. Bridgeman, *Proc. Am. Acad. Arts Sci.*, **63**: 229–308 (1928);
*J. ACS*, **50**: 3133 (1928).
§ Values mainly from Chap. 10, Refs. 4 and 7.

$$1\ \frac{\text{atm ft}^6}{\text{mole}^2} = 0.003898\ \frac{\text{atm liters}^2}{\text{g mole}^2} \qquad 1\ \frac{\text{ft}^3}{\text{mole}} = 0.06242\ \frac{\text{liter}}{\text{g mole}}$$

## TABLE B-9. DRY AIR AT LOW PRESSURES†
(Btu and lb$_m$ units)

| $T$ | $t$ | $h$ | $p_r$ | $u$ | $v_r$ | $\phi$ |
|---|---|---|---|---|---|---|
| 100 | −359.7 | 23.74 | 0.003841 | 16.88 | 9,643 | 0.19174 |
| 200 | −259.7 | 47.67 | 0.04320 | 33.96 | 1,714.9 | 0.36303 |
| 300 | −159.7 | 71.61 | 0.17795 | 51.04 | 624.5 | 0.46007 |
| 330 | −129.7 | 78.78 | 0.24819 | 56.16 | 492.6 | 0.48287 |
| 400 | −59.7 | 95.53 | 0.4858 | 68.11 | 305.0 | 0.52890 |
| 468 | 8.3 | 111.82 | 0.8405 | 79.73 | 206.26 | 0.55648 |
| 500 | 40.3 | 119.48 | 1.0590 | 85.20 | 174.90 | 0.58233 |
| 520 | 60.3 | 124.27 | 1.2147 | 88.62 | 158.58 | 0.59173 |
| 528 | 68.3 | 126.18 | 1.2813 | 89.99 | 152.65 | 0.59539 |
| 537 | 77.3 | 128.34 | 1.3593 | 91.53 | 146.34 | 0.59945 |
| 560 | 100.3 | 133.86 | 1.5742 | 95.47 | 131.78 | 0.60950 |
| 600 | 140.3 | 143.47 | 2.005 | 102.34 | 110.88 | 0.62607 |
| 700 | 240.3 | 167.56 | 3.446 | 119.58 | 75.25 | 0.66321 |
| 800 | 340.3 | 191.81 | 5.526 | 136.97 | 53.63 | 0.69558 |
| 900 | 440.3 | 216.26 | 8.411 | 154.57 | 39.64 | 0.72438 |
| 996 | 536.3 | 239.99 | 12.121 | 171.71 | 30.44 | 0.74942 |
| 1000 | 540.3 | 240.98 | 12.298 | 172.43 | 30.12 | 0.75042 |
| 1050 | 590.3 | 253.45 | 14.686 | 181.47 | 26.48 | 0.76259 |
| 1100 | 640.3 | 265.99 | 17.413 | 190.58 | 23.40 | 0.77426 |
| 1200 | 740.3 | 291.30 | 24.01 | 209.05 | 18.514 | 0.79628 |
| 1300 | 840.3 | 316.94 | 32.39 | 227.83 | 14.868 | 0.81680 |
| 1400 | 940.3 | 342.90 | 42.88 | 246.93 | 12.095 | 0.83604 |
| 1500 | 1040.3 | 369.17 | 55.86 | 266.34 | 9.948 | 0.85416 |
| 1510 | 1050.3 | 371.82 | 57.30 | 268.30 | 9.761 | 0.85592 |
| 1600 | 1140.3 | 395.74 | 71.73 | 286.06 | 8.263 | 0.87130 |
| 1700 | 1240.3 | 422.59 | 90.95 | 306.06 | 6.924 | 0.88758 |
| 1800 | 1340.3 | 449.71 | 114.03 | 326.32 | 5.847 | 0.90308 |
| 1900 | 1440.3 | 477.09 | 141.51 | 346.85 | 4.974 | 0.91788 |
| 2000 | 1540.3 | 504.71 | 174.00 | 367.61 | 4.258 | 0.93205 |
| 2100 | 1640.3 | 532.55 | 212.1 | 388.60 | 3.667 | 0.94564 |
| 2200 | 1740.3 | 560.59 | 256.6 | 409.78 | 3.176 | 0.95868 |
| 2300 | 1840.3 | 588.82 | 308.1 | 431.16 | 2.765 | 0.97123 |
| 2400 | 1940.3 | 617.22 | 367.6 | 452.70 | 2.419 | 0.98331 |
| 2500 | 2040.3 | 645.78 | 435.7 | 474.40 | 2.125 | 0.99497 |
| 2600 | 2140.3 | 674.49 | 513.5 | 496.26 | 1.8756 | 1.00623 |
| 2700 | 2240.3 | 703.35 | 601.9 | 518.26 | 1.6617 | 1.01712 |
| 2800 | 2340.3 | 732.33 | 702.0 | 540.40 | 1.4775 | 1.02767 |
| 2900 | 2440.3 | 761.45 | 814.8 | 562.66 | 1.3184 | 1.03788 |
| 3000 | 2540.3 | 790.68 | 941.4 | 585.04 | 1.1803 | 1.04779 |
| 3250 | 2790.3 | 864.24 | 1,327.5 | 641.46 | 0.9069 | 1.07134 |
| 3450 | 2990.3 | 923.52 | 1,718.7 | 687.04 | 0.7436 | 1.08904 |
| 3500 | 3040.3 | 938.40 | 1,829.3 | 698.48 | 0.7087 | 1.09332 |
| 3750 | 3290.3 | 1,030.09 | 2,471.1 | 756.04 | 0.5621 | 1.11393 |
| 4000 | 3540.3 | 1,088.26 | 3,280 | 814.06 | 0.4518 | 1.13334 |
| 4250 | 3790.3 | 1,163.87 | 4,285 | 872.53 | 0.3674 | 1.15168 |
| 4500 | 4040.3 | 1,239.86 | 5,521 | 931.39 | 0.3019 | 1.16905 |
| 4750 | 4290.3 | 1,316.21 | 7,026 | 990.60 | 0.2505 | 1.18556 |
| 5000 | 4540.3 | 1,392.87 | 8,837 | 1,050.12 | 0.20959 | 1.20129 |
| 5250 | 4790.3 | 1,469.83 | 11,002 | 1,109.95 | 0.17677 | 1.21631 |
| 5500 | 5040.3 | 1,547.07 | 13,568 | 1,170.04 | 0.15016 | 1.23068 |
| 5750 | 5290.3 | 1,624.57 | 16,588 | 1,230.41 | 0.12840 | 1.24445 |
| 6000 | 5540.3 | 1,702.29 | 20,120 | 1,291.00 | 0.11047 | 1.25769 |
| 6250 | 5790.3 | 1,780.27 | 24,228 | 1,351.83 | 0.09556 | 1.27042 |
| 6500 | 6040.3 | 1,858.44 | 28,974 | 1,412.87 | 0.08310 | 1.28268 |

† Abstracted, by permission, from J. Keenan and J. Kaye, "Thermodynamic Properties of Air," John Wiley & Sons, Inc., New York, 1948.

## TABLE B-10. INTERNAL ENERGY OF IDEAL GASES†
(Btu/mole; datum, 520°R)

| °R | O₂ | N₂ | Air | CO₂ | H₂O | H₂ | CO | C₈H₁₈ | C₁₂H₂₆ | pv/778.16 |
|---|---|---|---|---|---|---|---|---|---|---|
| 520 | 0 | 0 | 0 | 0 | 0 | 0 | 0 | 0 | 0 | 1,033 |
| 536.7 | 83 | 81 | 81 | 115 | 101 | 80 | 81 | 640 | 980 | 1,066 |
| 540 | 100 | 97 | 97 | 139 | 122 | 96 | 97 | 756 | 1,181 | 1,072 |
| 560 | 200 | 196 | 196 | 280 | 244 | 193 | 196 | 1,536 | 2,491 | 1,112 |
| 580 | 301 | 295 | 295 | 424 | 357 | 291 | 295 | 2,340 | 3,931 | 1,152 |
| 600 | 402 | 395 | 395 | 570 | 490 | 390 | 396 | 3,167 | 5,481 | 1,192 |
| 700 | 920 | 896 | 897 | 1,320 | 1,110 | 887 | 896 | 7,668 | 13,223 | 1,390 |
| 800 | 1,449 | 1,399 | 1,403 | 2,120 | 1,734 | 1,386 | 1,402 | 12,768 | 22,044 | 1,589 |
| 900 | 1,989 | 1,905 | 1,915 | 2,965 | 2,366 | 1,886 | 1,913 | 18,471 | 31,771 | 1,787 |
| 1000 | 2,539 | 2,416 | 2,431 | 3,852 | 3,009 | 2,387 | 2,430 | 24,773 | 42,277 | 1,986 |
| 1100 | 3,101 | 2,934 | 2,957 | 4,778 | 3,666 | 2,889 | 2,954 | 31,677 | 53,468 | 2,185 |
| 1200 | 3,675 | 3,461 | 3,492 | 5,736 | 4,339 | 3,393 | 3,485 | 39,182 | 65,290 | 2,383 |
| 1300 | 4,262 | 3,996 | 4,036 | 6,721 | 5,030 | 3,899 | 4,026 | 47,288 | 77,706 | 2,582 |
| 1400 | 4,861 | 4,539 | 4,587 | 7,731 | 5,740 | 4,406 | 4,580 | 55,995 | 90,688 | 2,780 |
| 1500 | 5,472 | 5,091 | 5,149 | 8,764 | 6,468 | 4,916 | 5,145 | 65,303 | 104,209 | 2,979 |
| 1600 | 6,092 | 5,652 | 5,720 | 9,819 | 7,212 | 5,429 | 5,720 | 74,825 | 118,240 | 3,178 |
| 1700 | 6,718 | 6,224 | 6,301 | 10,896 | 7,970 | 5,945 | 6,305 | 84,901 | 132,757 | 3,376 |
| 1800 | 7,349 | 6,805 | 6,889 | 11,993 | 8,741 | 6,464 | 6,899 | 95,503 | 147,735 | 3,575 |
| 1900 | 7,985 | 7,393 | 7,485 | 13,105 | 9,526 | 6,988 | 7,501 | ...... | ....... | 3,773 |
| 2000 | 8,629 | 7,989 | 8,087 | 14,230 | 10,327 | 7,517 | 8,109 | ...... | ....... | 3,972 |
| 2100 | 9,279 | 8,592 | 8,698 | 15,368 | 11,146 | 8,053 | 8,722 | ...... | ....... | 4,171 |
| 2200 | 9,934 | 9,203 | 9,314 | 16,518 | 11,983 | 8,597 | 9,339 | ...... | ....... | 4,369 |
| 2300 | 10,592 | 9,817 | 9,934 | 17,680 | 12,835 | 9,147 | 9,961 | ...... | ....... | 4,568 |
| 2400 | 11,252 | 10,435 | 10,558 | 18,852 | 13,700 | 9,703 | 10,588 | ...... | ....... | 4,766 |
| 2500 | 11,916 | 11,056 | 11,185 | 20,033 | 14,578 | 10,263 | 11,220 | ...... | ....... | 4,965 |
| 2600 | 12,584 | 11,682 | 11,817 | 21,222 | 15,469 | 10,827 | 11,857 | ...... | ....... | 5,164 |
| 2700 | 13,257 | 12,313 | 12,453 | 22,419 | 16,372 | 11,396 | 12,499 | ...... | ....... | 5,362 |
| 2800 | 13,937 | 12,949 | 13,095 | 23,624 | 17,288 | 11,970 | 13,144 | ...... | ....... | 5,561 |
| 2900 | 14,622 | 13,590 | 13,742 | 24,836 | 18,217 | 12,549 | 13,792 | ...... | ....... | 5,759 |
| 3000 | 15,309 | 14,236 | 14,394 | 26,055 | 19,160 | 13,133 | 14,443 | ...... | ....... | 5,958 |
| 3100 | 16,001 | 14,888 | 15,051 | 27,281 | 20,117 | 13,723 | 15,097 | ...... | ....... | 6,157 |
| 3200 | 16,693 | 15,543 | 15,710 | 28,513 | 21,086 | 14,319 | 15,754 | ...... | ....... | 6,355 |
| 3300 | 17,386 | 16,199 | 16,369 | 29,750 | 22,066 | 14,921 | 16,414 | ...... | ....... | 6,554 |
| 3400 | 18,080 | 16,855 | 17,030 | 30,991 | 23,057 | 15,529 | 17,078 | ...... | ....... | 6,752 |
| 3500 | 18,776 | 17,512 | 17,692 | 32,237 | 24,057 | 16,143 | 17,744 | ...... | ....... | 6,951 |
| 3600 | 19,475 | 18,171 | 18,356 | 33,487 | 25,067 | 16,762 | 18,412 | ...... | ....... | 7,150 |
| 3700 | 20,179 | 18,833 | 19,022 | 34,741 | 26,085 | 17,385 | 19,082 | ...... | ....... | 7,348 |
| 3800 | 20,887 | 19,496 | 19,691 | 35,998 | 27,110 | 18,011 | 19,755 | ...... | ....... | 7,547 |
| 3900 | 21,598 | 20,162 | 20,363 | 37,258 | 28,141 | 18,641 | 20,430 | ...... | ....... | 7,745 |
| 4000 | 22,314 | 20,830 | 21,037 | 38,522 | 29,178 | 19,274 | 21,107 | ...... | ....... | 7,944 |
| 4100 | 23,034 | 21,500 | 21,714 | 39,791 | 30,221 | 19,911 | 21,784 | ...... | ....... | 8,143 |
| 4200 | 23,757 | 22,172 | 22,393 | 41,064 | 31,270 | 20,552 | 22,462 | ...... | ....... | 8,341 |
| 4300 | 24,482 | 22,845 | 23,073 | 42,341 | 32,326 | 21,197 | 23,143 | ...... | ....... | 8,540 |
| 4400 | 25,209 | 23,519 | 23,755 | 43,622 | 33,389 | 21,845 | 23,823 | ...... | ....... | 8,738 |
| 4500 | 25,938 | 24,194 | 24,437 | 44,906 | 34,459 | 22,497 | 24,503 | ...... | ....... | 8,937 |
| 4600 | 26,668 | 24,869 | 25,120 | 46,193 | 35,535 | 23,154 | 25,186 | ...... | ....... | 9,136 |
| 4700 | 27,401 | 25,546 | 25,805 | 47,483 | 36,616 | 23,816 | 25,868 | ...... | ....... | 9,334 |
| 4800 | 28,136 | 26,224 | 26,491 | 48,775 | 37,701 | 24,480 | 26,533 | ...... | ....... | 9,533 |
| 4900 | 28,874 | 26,905 | 27,180 | 50,069 | 38,791 | 25,148 | 27,219 | ...... | ....... | 9,731 |
| 5000 | 29,616 | 27,589 | 27,872 | 51,365 | 39,885 | 25,819 | 27,907 | ...... | ....... | 9,930 |
| 5100 | 30,361 | 28,275 | 28,566 | 52,663 | 40,983 | 26,492 | 28,597 | ...... | ....... | 10,129 |
| 5200 | 31,108 | 28,961 | 29,262 | 53,963 | 42,084 | 27,166 | 29,288 | ...... | ....... | 10,327 |
| 5300 | 31,857 | 29,648 | 29,958 | 55,265 | 43,187 | 27,842 | 29,980 | ...... | ....... | 10,526 |
| 5400 | 32,607 | 30,337 | 30,655 | 56,569 | 44,293 | 28,519 | 30,674 | ...... | ....... | 10,724 |
| 5500 | 33,386 | 31,026 | 31,353 | 57,875 | 45,402 | 29,298 | 31,369 | ...... | ....... | 10,923 |
| 5600 | 34,161 | 31,726 | 32,051 | 59,183 | 46,513 | 29,978 | 32,065 | ...... | ....... | 11,121 |
| 5700 | 34,900 | 32,428 | 32,750 | 60,491 | 47,627 | 30,659 | 32,762 | ...... | ....... | 11,320 |
| 5800 | 35,673 | 33,130 | 33,449 | 61,891 | 48,744 | 31,342 | 33,461 | ...... | ....... | 11,519 |
| 5900 | 36,412 | 33,833 | 34,150 | 63,293 | 49,863 | 32,026 | 34,161 | ...... | ....... | 11,717 |
| 6000 | 37,149 | 34,537 | 34,852 | 64,297 | 50,985 | 32,712 | 34,863 | ...... | ....... | 11,916 |
| 6500 | ...... | ...... | 38,364 | ....... | ....... | ...... | ...... | ...... | ....... | 12,908 |
| 7000 | ...... | ...... | 41,893 | ....... | ....... | ...... | ...... | ...... | ....... | 13,901 |

† From L. C. Lichty, "Internal Combustion Engines," McGraw-Hill Book Company, Inc., New York, 1939, and based upon data of A. Hershey, J. Eberhardt, and H. Hottel, *Trans. SAE*, **39**: 409 (October, 1936).

TABLE B-11. HEAT OF COMBUSTION[†]
$(-\Delta H^\circ$ at 77°F)

| Substance | Symbol | $h$ ($h_{fg}$) of vaporization, Btu/lb$_m$ | H$_2$O($l$) and CO$_2$($g$) | | H$_2$O($g$) and CO$_2$($g$) | |
|---|---|---|---|---|---|---|
| | | | kcal g mole | Btu lb$_m$ | kcal g mole | Btu lb$_m$ |
| Acetylene......... | C$_2$H$_2$($g$) | ... | 310.62 | 21,460 | 300.10 | 20,734 |
| Benzene........... | C$_6$H$_6$($g$) | 186 | 789.08 | 18,172 | 757.52 | 17,446 |
| $n$-Butane.......... | C$_4$H$_{10}$($g$) | 156 | 687.65 | 21,283 | 635.05 | 19,655 |
| Isobutane........ | C$_4$H$_{10}$($g$) | 141 | 685.65 | 21,221 | 633.05 | 19,593 |
| 1-Butene.......... | C$_4$H$_{10}$($g$) | 156 | 649.45 | 20,824 | 607.37 | 19,475 |
| Carbon............ | C(graphite) | ... | 94.0518 | 14,086 | | |
| Carbon monoxide.. | CO($g$) | ... | 67.6361 | 4,343.6 | | |
| $n$-Decane.......... | C$_{10}$H$_{22}$($g$) | 155 | 1632.34 | 20,638 | 1516.63 | 19,175 |
| $n$-Dodecane ....... | C$_{12}$H$_{26}$($g$) | 155 | 1947.23 | 20,564 | 1810.48 | 19,120 |
| Ethane............ | C$_2$H$_6$($g$) | ... | 372.82 | 22,304 | 341.26 | 20,416 |
| Ethylene.......... | C$_2$H$_4$($g$) | ... | 337.23 | 21,625 | 316.20 | 20,276 |
| $n$-Heptane........ | C$_7$H$_{16}$($g$) | 157 | 1160.01 | 20,825 | 1075.85 | 19,314 |
| $n$-Hexane......... | C$_6$H$_{14}$($g$) | 157 | 1002.57 | 20,928 | 928.93 | 19,391 |
| Hydrogen......... | H$_2$($g$) | ... | 68.3174 | 60,957 | 57.7979 | 51,571 |
| Methane.......... | CH$_4$($g$) | ... | 212.80 | 23,861 | 191.76 | 21,502 |
| $n$-Nonane........ | C$_9$H$_{20}$($g$) | 156 | 1474.90 | 20,687 | 1369.70 | 19,211 |
| $n$-Octane......... | C$_8$H$_{18}$($g$) | 156 | 1317.45 | 20,747 | 1222.77 | 19,256 |
| $n$-Pentane........ | C$_5$H$_{12}$($g$) | 157 | 845.16 | 21,072 | 782.04 | 19,499 |
| Isopentane........ | C$_5$H$_{12}$($g$) | 147 | 843.24 | 21,025 | 780.12 | 19,451 |
| Propane.......... | C$_3$H$_8$($g$) | 147 | 530.6 | 21,646 | 488.53 | 19,929 |
| Propylene......... | C$_3$H$_6$($g$) | ... | 491.99 | 21,032 | 460.43 | 19,683 |

† Data from API Research Project 44, National Bureau of Standards, Washington, December, 1952.

TABLE B-12. THERMAL CONDUCTIVITY OF VARIOUS SUBSTANCES†

| Substance | $t$ °F | $k$ $\dfrac{\text{Btu}}{\text{hr ft °F}}$ |
|---|---|---|
| Pure metals and alloys:‡ | | |
| Silver............................................. | −451 | 9840 |
| | −370 | 445 |
| | −260 | 289 |
| | −148 | 242 |
| | 32 | 241 |
| | 392 | 238 |
| Copper........................................... | −451 | 1445 |
| | −370 | 724 |
| | −260 | 283 |
| | −148 | 235 |
| | 32 | 224 |
| | 572 | 212 |
| | 1112 | 204 |
| Aluminum ....................................... | −451 | 2110 |
| | −370 | 578 |
| | −260 | 156 |
| | 32 | 132 |
| | 572 | 132 |
| | 752 | 131 |
| Zinc............................................. | 32 | 65 |
| | 392 | 62 |
| | 752 | 54 |
| Brass (70% copper, 30% zinc)...................... | 32 | 56 |
| | 392 | 63 |
| | 752 | 67 |
| Nickel (pure, 99.9%)............................. | −148 | 60 |
| | 32 | 54 |
| | 392 | 42 |
| | 752 | 34 |
| Nickel (impure, 99.2%)........................... | 32 | 40 |
| | 392 | 34 |
| | 752 | 30 |
| Cast iron........................................ | 32 | 32 |
| | 392 | 28 |
| | 752 | 25 |
| Steel (mild, 1% carbon).......................... | 212 | 26 |
| | 572 | 25 |
| | 1112 | 21 |
| 18-8 stainless steel.............................. | −451 | 0.2 |
| | −370 | 3.6 |
| | −260 | 6.0 |
| | 32 | 9.4 |
| | 752 | 11 |
| | 1472 | 15 |
| | 1832 | 18 |

TABLE B-12. THERMAL CONDUCTIVITY OF VARIOUS SUBSTANCES† (*Continued*)

| Substance | $t$ °F | $k$ Btu / hr ft °F |
|---|---|---|
| Uranium (cast).................................. | 200 | 15 |
| | 600 | 18 |
| Constantan (60% copper, 40% nickel)............... | −451 | 0.7 |
| | −370 | 9.8 |
| | −260 | 12 |
| | −148 | 12 |
| | 68 | 13.1 |
| | 392 | 15 |
| Insulating materials: | | |
| Corrugated asbestos paper (four plies per inch)........ | 100 | 0.050 |
| | 300 | 0.069 |
| 85% magnesia, 15% asbestos (apparent density 13 $lb_m$/ft³) | 100 | 0.034 |
| | 200 | 0.036 |
| | 400 | 0.040 |
| Glass wool (apparent density 4 $lb_m$/ft³)............... | 20 | 0.0179 |
| | 100 | 0.0239 |
| | 200 | 0.0317 |
| Diatomaceous earth, powdered (apparent density 10 $lb_m$/ft³) | 100 | 0.024 |
| | 300 | 0.034 |
| | 600 | 0.048 |
| Same but apparent density 18 $lb_m$/ft³............... | 100 | 0.039 |
| | 300 | 0.044 |
| | 600 | 0.054 |
| | 1000 | 0.068 |
| Miscellaneous solids: | | |
| Water (ice)......................................... | 32 | 1.3 |
| Window glass....................................... | ..... | 0.3−0.61 |
| Building brick..................................... | 68 | 0.4 |
| Wallboard (density, 14.8 $lb_m$/ft³)..................... | 70 | 0.028 |
| Wallboard (density, 43 $lb_m$/ft³)....................... | ..... | 0.04 |
| Wood (density, 40 $lb_m$/ft³)............................ | ..... | 0.1−0.2 |
| Concrete (1:4 sand, dry)............................ | ..... | 0.44 |
| Cork, ground....................................... | 86 | 0.025 |
| Paper.............................................. | ..... | 0.075 |
| Marble............................................. | 68 | 1.6 |
| Magnesite brick (apparent density 158 $lb_m$/ft³)......... | 399 | 2.2 |
| | 1202 | 1.6 |
| | 2192 | 1.1 |
| Chrome brick (apparent density 200 $lb_m$/ft³).......... | 392 | 0.67 |
| | 1202 | 0.85 |
| | 2399 | 1.0 |
| Liquids: | | |
| Ethylene glycol.................................... | 32 | 0.140 |
| | 140 | 0.150 |
| | 212 | 0.152 |

TABLE B-12. THERMAL CONDUCTIVITY OF VARIOUS SUBSTANCES† (*Continued*)

| Substance | $t$ °F | $k$ $\dfrac{\text{Btu}}{\text{hr ft °F}}$ |
|---|---|---|
| Freon-12 | −58 | 0.039 |
| | −4 | 0.041 |
| | 86 | 0.041 |
| Water (saturated liquid) | 32 | 0.343 |
| | 300 | 0.395 |
| | 620 | 0.275 |
| Sodium (melting point, 208°F) | 200 | 49.8 |
| | 700 | 41.8 |
| | 1300 | 34.5 |
| Mercury (melting point, −38°F) | 50 | 4.7 |
| | 300 | 6.7 |
| | 600 | 8.1 |
| Gases: | | |
| Oxygen (atmospheric pressure) | −298 | 0.0047 |
| | 80 | 0.0155 |
| | 296 | 0.0209 |
| | 620 | 0.0280 |
| Nitrogen (atmospheric pressure) | −280 | 0.0055 |
| | 80 | 0.0151 |
| | 296 | 0.0201 |
| | 800 | 0.0296 |
| Water (saturated vapor) | 212 | 0.014 |
| | 572 | 0.0246 |
| | 932 | 0.0434 |
| Hydrogen (atmospheric pressure) | −370 | 0.0209 |
| | 80 | 0.1050 |
| | 620 | 0.1819 |

† Data for temperatures above −200°F from various sources.   Data below −200°F from R. L. Powell and W. A. Blanpied, Thermal Conductivity of Metals and Alloys at Low Temperatures, *Natl. Bur. Standards Circ.* 556.

‡ At temperatures of a few degrees above absolute zero, very large values of $k$ are found for pure metals (but not for alloys—note stainless steel and constantan).   Pure metals exhibit maximum values of thermal conductivity in a range of temperatures from 5 to 40°R, and then $k$ seems to decrease toward zero as the temperature approaches absolute zero.   For pure metals at low temperatures, it is found that the value of $k$ is quite dependent upon the past history of the metal.   For example, the following $k$ values are cited by Powell and Blanpied for silver at −451°F: unannealed, $k = 31.8$; annealed at 650°F, $k = 9,840$; cold drawn, $k = 231$; cold drawn and then annealed, $k = 5,040$.

TABLE B-13. HEAT OF FORMATION, ABSOLUTE ENTROPY, AND FREE
ENERGY OF FORMATION AT 25°C (77°F) AND 1 ATM†

| Substance | Symbol | State | $h_f°$ kcal / g mole | $s°$ cal / g mole °K | $g_f°$ kcal / g mole |
|---|---|---|---|---|---|
| Acetylene............... | $C_2H_2$ | Gas | 54.194 | 47.997 | 50.000 |
| Ammonia............... | $NH_3$ | Gas | −11.04 | 46.01 | −3.976 |
| Argon.................. | A | Gas | 0 | 36.983 | 0 |
| Benzene............... | $C_6H_6$ | Gas | 19.820 | 64.34 | 30.989 |
| n-Butane............... | $C_4H_{10}$ | Gas | −30.15 | 74.12 | −4.10 |
| 1-Butene............... | $C_4H_8$ | Gas | −0.03 | 73.04 | 17.09 |
| Carbon................. | C | Graphite | 0 | 1.3609 | 0 |
|  |  | Gas | 171.698 | 37.7611 | 160.845 |
| Carbon dioxide........... | $CO_2$ | Gas | −94.0518 | 51.061 | −94.2598 |
| Carbon monoxide......... | CO | Gas | −26.4157 | 47.300 | −32.8077 |
| Carbon tetrachloride...... | $CCl_4$ | Gas | −25.5 | 73.95 | −15.3 |
| n-Dodecane............. | $C_{12}H_{26}$ | Gas | −69.52 | 148.79 | 11.98 |
| Ethane................. | $C_2H_6$ | Gas | −20.236 | 54.85 | −7.860 |
| Ethylene............... | $C_2H_4$ | Gas | 12.496 | 52.45 | 16.282 |
| Helium................. | He | Gas | 0 | 30.126 | 0 |
| n-Heptane.............. | $C_7H_{16}$ | Gas | −44.89 | 102.24 | 1.94 |
| n-Hexane............... | $C_6H_{14}$ | Gas | −39.96 | 92.83 | −0.07 |
| Hydrogen............... | $H_2$ | Gas | 0 | 31.211 | 0 |
| Hydrogen sulfide.......... | $H_2S$ | Gas | −4.815 | 49.15 | −7.892 |
| Krypton................ | Kr | Gas | 0 | 39.19 | 0 |
| Mercury................ | Hg | Gas | 14.54 | 41.80 | 7.59 |
| Methane................ | $CH_4$ | Gas | −17.889 | 44.50 | −12.140 |
| Neon................... | Ne | Gas | 0 | 34.948 | 0 |
| Nitric oxide............. | NO | Gas | 21.600 | 50.339 | 20.719 |
| Nitrogen................ | $N_2$ | Gas | 0 | 45.767 | 0 |
| n-Octane............... | $C_8H_{18}$ | Gas | −49.82 | 111.55 | 3.95 |
|  |  | Liquid | −59.74 | 86.23 | 1.58 |
| Oxygen................. | $O_2$ | Gas | 0 | 49.003 | 0 |
| n-Pentane.............. | $C_5H_{12}$ | Gas | −35.00 | 83.40 | −2.00 |
| Propane................ | $C_3H_8$ | Gas | −24.820 | 64.51 | −5.614 |
| Sulfur dioxide........... | $SO_2$ | Gas | −70.96 | 59.40 | −71.79 |
| Water.................. | $H_2O$ | Gas | −57.7979 | 45.106 | −54.6351 |
|  |  | Liquid | −68.3174 | 16.716 | −56.6899 |
| Xenon.................. | Xe | Gas | 0 | 40.53 | 0 |

† Data from *Natl. Bur. Standards* (*U.S.*) *Circ.* 500, February, 1952, and from API
Research Project 44, National Bureau of Standards, Washington, December, 1952.

ELEMENTS OF THERMODYNAMICS

## TABLE B-14. PROPERTIES OF DICHLORODIFLUOROMETHANE, F-12†

| Sat. temp, °F | Abs press., psia | Volume, ft³/lb$_m$ | | Specific enthalpy and entropy taken from −40°F | | | | | | | |
|---|---|---|---|---|---|---|---|---|---|---|---|
| | | | | Specific enthalpy, h | | Entropy, s | | 25°F superheat | | 50°F superheat | |
| | | Liquid | Vapor | Liquid | Vapor | Liquid | Vapor | h | s | h | s |
| −40 | 9.32 | 0.0106 | 3.911 | 0.0 | 73.5 | 0.0 | 0.17517 | 76.85 | 0.18277 | 80.25 | 0.19020 |
| −30 | 12.02 | 0.0107 | 3.088 | 2.03 | 74.7 | 0.00471 | 0.17387 | 78.05 | 0.18147 | 81.45 | 0.18890 |
| −20 | 15.28 | 0.0108 | 2.474 | 4.07 | 75.87 | 0.00940 | 0.17275 | 79.22 | 0.18035 | 82.62 | 0.18778 |
| −10 | 19.2 | 0.0109 | 2.003 | 6.14 | 77.05 | 0.01403 | 0.17175 | 80.48 | 0.17918 | 84.01 | 0.18639 |
| 0 | 23.87 | 0.0110 | 1.637 | 8.25 | 78.21 | 0.01869 | 0.17091 | 81.71 | 0.17829 | 85.26 | 0.18547 |
| 2 | 24.89 | 0.0110 | 1.574 | 8.67 | 78.44 | 0.01961 | 0.17075 | 81.94 | 0.17812 | 85.51 | 0.18529 |
| 4 | 25.96 | 0.0111 | 1.514 | 9.10 | 78.67 | 0.02052 | 0.17060 | 82.17 | 0.17795 | 85.76 | 0.18511 |
| 5 | 26.51 | 0.0111 | 1.485 | 9.32 | 78.79 | 0.02097 | 0.17052 | 82.29 | 0.17786 | 85.89 | 0.18502 |
| 6 | 27.05 | 0.0111 | 1.457 | 9.53 | 78.90 | 0.02143 | 0.17045 | 82.41 | 0.17778 | 86.01 | 0.18494 |
| 8 | 28.18 | 0.0111 | 1.403 | 9.96 | 79.13 | 0.02235 | 0.17030 | 82.66 | 0.17763 | 86.26 | 0.18477 |
| 10 | 29.35 | 0.0112 | 1.351 | 10.39 | 79.36 | 0.02328 | 0.17015 | 82.90 | 0.17747 | 86.51 | 0.18460 |
| 12 | 30.56 | 0.0112 | 1.301 | 10.82 | 79.59 | 0.02419 | 0.17001 | 83.14 | 0.17733 | 86.76 | 0.18444 |
| 14 | 31.80 | 0.0112 | 1.253 | 11.26 | 79.82 | 0.02510 | 0.16987 | 83.38 | 0.17720 | 87.01 | 0.18429 |
| 16 | 33.08 | 0.0112 | 1.207 | 11.70 | 80.05 | 0.02601 | 0.16974 | 83.61 | 0.17706 | 87.26 | 0.18413 |
| 18 | 34.40 | 0.0113 | 1.163 | 12.12 | 80.27 | 0.02692 | 0.16961 | 83.85 | 0.17693 | 87.51 | 0.18397 |
| 20 | 35.75 | 0.0113 | 1.121 | 12.55 | 80.49 | 0.02783 | 0.16949 | 84.09 | 0.17679 | 87.76 | 0.18382 |
| 22 | 37.15 | 0.0113 | 1.081 | 13.00 | 80.72 | 0.02873 | 0.16938 | 84.32 | 0.17666 | 88.00 | 0.18369 |
| 24 | 38.58 | 0.0113 | 1.043 | 13.44 | 80.95 | 0.02963 | 0.16926 | 84.55 | 0.17652 | 88.24 | 0.18355 |
| 26 | 40.07 | 0.0114 | 1.007 | 13.88 | 81.17 | 0.03053 | 0.16913 | 84.79 | 0.17639 | 88.49 | 0.18342 |
| 28 | 41.59 | 0.0114 | 0.973 | 14.32 | 81.39 | 0.03143 | 0.16900 | 85.02 | 0.17625 | 88.73 | 0.18328 |
| 30 | 43.16 | 0.0115 | 0.939 | 14.76 | 81.61 | 0.03233 | 0.16887 | 85.25 | 0.17612 | 88.97 | 0.18315 |
| 32 | 44.77 | 0.0115 | 0.908 | 15.21 | 81.83 | 0.03323 | 0.16876 | 85.48 | 0.17600 | 89.21 | 0.18303 |
| 34 | 46.42 | 0.0115 | 0.877 | 15.65 | 82.05 | 0.03413 | 0.16865 | 85.71 | 0.17589 | 89.45 | 0.18291 |
| 36 | 48.13 | 0.0116 | 0.848 | 16.10 | 82.27 | 0.03502 | 0.16854 | 85.95 | 0.17577 | 89.68 | 0.18280 |
| 38 | 49.88 | 0.0116 | 0.819 | 16.55 | 82.49 | 0.03591 | 0.16843 | 86.18 | 0.17566 | 89.92 | 0.18268 |
| 39 | 50.78 | 0.0116 | 0.806 | 16.77 | 82.60 | 0.03635 | 0.16838 | 86.29 | 0.17560 | 90.04 | 0.18262 |
| 40 | 51.68 | 0.0116 | 0.792 | 17.00 | 82.71 | 0.03680 | 0.16833 | 86.41 | 0.17554 | 90.16 | 0.18256 |
| 41 | 52.70 | 0.0116 | 0.779 | 17.23 | 82.82 | 0.03725 | 0.16828 | 86.52 | 0.17549 | 90.28 | 0.18251 |
| 42 | 53.51 | 0.0116 | 0.767 | 17.46 | 82.93 | 0.03770 | 0.16823 | 86.64 | 0.17544 | 90.40 | 0.18245 |
| 44 | 55.40 | 0.0117 | 0.742 | 17.91 | 83.15 | 0.03859 | 0.16813 | 86.86 | 0.17534 | 90.65 | 0.18235 |
| 46 | 57.35 | 0.0117 | 0.718 | 18.36 | 83.36 | 0.03948 | 0.16803 | 87.09 | 0.17525 | 90.89 | 0.18224 |
| 48 | 59.35 | 0.0117 | 0.695 | 18.82 | 83.57 | 0.04037 | 0.16794 | 87.31 | 0.17515 | 91.14 | 0.18214 |
| 50 | 61.39 | 0.0118 | 0.673 | 19.27 | 83.78 | 0.04126 | 0.16785 | 87.54 | 0.17505 | 91.38 | 0.18203 |
| 52 | 63.49 | 0.0118 | 0.652 | 19.72 | 83.99 | 0.04215 | 0.16776 | 87.76 | 0.17496 | 91.61 | 0.18193 |
| 54 | 65.63 | 0.0118 | 0.632 | 20.18 | 84.20 | 0.04304 | 0.16767 | 87.98 | 0.17486 | 91.83 | 0.18184 |
| 56 | 67.84 | 0.0119 | 0.612 | 20.64 | 84.41 | 0.04392 | 0.16758 | 88.20 | 0.17477 | 92.06 | 0.18174 |
| 58 | 70.10 | 0.0119 | 0.593 | 21.11 | 84.62 | 0.04480 | 0.16749 | 88.42 | 0.17467 | 92.28 | 0.18165 |
| 60 | 72.41 | 0.0119 | 0.575 | 21.57 | 84.82 | 0.04568 | 0.16741 | 88.64 | 0.17458 | 92.51 | 0.18155 |
| 62 | 74.77 | 0.0120 | 0.557 | 22.03 | 85.02 | 0.04657 | 0.16733 | 88.86 | 0.17450 | 92.74 | 0.18147 |

TABLE B-14. PROPERTIES OF DICHLORODIFLUOROMETHANE, F-12† (*Continued*)

| Sat. temp °F | Abs press., psia | Volume, ft³/lb_m | | Specific enthalpy and entropy taken from −40°F | | | | | | | |
|---|---|---|---|---|---|---|---|---|---|---|---|
| | | | | Specific enthalpy, h | | Entropy, s | | 25°F superheat | | 50°F superheat | |
| | | Liquid | Vapor | Liquid | Vapor | Liquid | Vapor | h | s | h | s |
| 64 | 77.20 | 0.0120 | 0.540 | 22.49 | 85.22 | 0.0474 | 0.16725 | 89.07 | 0.17442 | 92.97 | 0.18139 |
| 66 | 79.67 | 0.0120 | 0.524 | 22.95 | 85.42 | 0.04833 | 0.16717 | 89.29 | 0.17433 | 93.20 | 0.18130 |
| 68 | 82.24 | 0.0121 | 0.508 | 23.42 | 85.62 | 0.04921 | 0.16709 | 89.50 | 0.17425 | 93.43 | 0.18122 |
| 70 | 84.82 | 0.0121 | 0.493 | 23.90 | 85.82 | 0.05009 | 0.16701 | 89.72 | 0.17417 | 93.66 | 0.18114 |
| 72 | 87.50 | 0.0121 | 0.479 | 24.37 | 86.02 | 0.05097 | 0.16693 | 89.93 | 0.17409 | 93.99 | 0.18106 |
| 74 | 90.20 | 0.0122 | 0.464 | 24.84 | 86.22 | 0.05185 | 0.16685 | 90.14 | 0.17402 | 94.12 | 0.18098 |
| 76 | 93.00 | 0.0122 | 0.451 | 25.32 | 86.42 | 0.05272 | 0.16677 | 90.36 | 0.17394 | 94.34 | 0.18091 |
| 78 | 95.85 | 0.0123 | 0.438 | 25.80 | 86.61 | 0.05359 | 0.16669 | 90.57 | 0.17387 | 94.57 | 0.18083 |
| 80 | 98.76 | 0.0123 | 0.425 | 26.28 | 86.80 | 0.05446 | 0.16662 | 90.78 | 0.17379 | 94.80 | 0.18075 |
| 82 | 101.70 | 0.0123 | 0.413 | 26.76 | 86.99 | 0.05534 | 0.16655 | 90.98 | 0.17372 | 95.01 | 0.18068 |
| 84 | 104.8 | 0.0124 | 0.401 | 27.24 | 87.18 | 0.05621 | 0.16648 | 91.18 | 0.17365 | 95.22 | 0.18061 |
| 86 | 107.9 | 0.0124 | 0.389 | 27.72 | 87.37 | 0.05708 | 0.16640 | 91.37 | 0.17358 | 95.44 | 0.18054 |
| 88 | 111.1 | 0.0124 | 0.378 | 28.21 | 87.56 | 0.05795 | 0.16632 | 91.57 | 0.17351 | 95.65 | 0.18047 |
| 90 | 114.3 | 0.0125 | 0.368 | 28.70 | 87.74 | 0.05882 | 0.16624 | 91.77 | 0.17344 | 95.86 | 0.18040 |
| 92 | 117.7 | 0.0125 | 0.357 | 29.19 | 87.92 | 0.05969 | 0.16616 | 91.97 | 0.17337 | 96.07 | 0.18033 |
| 94 | 121.0 | 0.0126 | 0.347 | 29.68 | 88.10 | 0.06056 | 0.16608 | 92.16 | 0.17330 | 96.28 | 0.18026 |
| 96 | 124.5 | 0.0126 | 0.338 | 30.18 | 88.28 | 0.06143 | 0.16600 | 92.36 | 0.17322 | 96.50 | 0.18018 |
| 98 | 128.0 | 0.0126 | 0.328 | 30.67 | 88.45 | 0.06230 | 0.16592 | 92.55 | 0.17315 | 96.71 | 0.18011 |
| 100 | 131.6 | 0.0127 | 0.319 | 31.16 | 88.62 | 0.06316 | 0.16584 | 92.75 | 0.17308 | 96.92 | 0.18004 |
| 102 | 135.3 | 0.0127 | 0.310 | 31.65 | 88.79 | 0.06403 | 0.16576 | 92.93 | 0.17301 | 97.12 | 0.17998 |
| 104 | 139.0 | 0.0128 | 0.302 | 32.15 | 88.95 | 0.06490 | 0.16568 | 93.11 | 0.17294 | 97.32 | 0.17993 |
| 106 | 142.8 | 0.0128 | 0.293 | 32.65 | 89.11 | 0.06577 | 0.16560 | 93.30 | 0.17288 | 97.53 | 0.17987 |
| 108 | 146.8 | 0.0129 | 0.285 | 33.15 | 89.27 | 0.06663 | 0.16551 | 93.48 | 0.17281 | 97.73 | 0.17982 |
| 110 | 150.7 | 0.0129 | 0.277 | 33.65 | 89.43 | 0.06749 | 0.16542 | 93.66 | 0.17274 | 97.93 | 0.17976 |
| 112 | 154.8 | 0.0130 | 0.269 | 34.15 | 89.58 | 0.06836 | 0.16533 | 93.82 | 0.17266 | 98.11 | 0.17969 |
| 114 | 158.9 | 0.0130 | 0.262 | 34.65 | 89.73 | 0.06922 | 0.16524 | 93.98 | 0.17258 | 98.29 | 0.17961 |
| 116 | 163.1 | 0.0131 | 0.254 | 35.15 | 89.87 | 0.07008 | 0.16515 | 94.15 | 0.17249 | 98.48 | 0.17954 |
| 118 | 167.4 | 0.0131 | 0.247 | 35.65 | 90.01 | 0.07094 | 0.16505 | 94.31 | 0.17241 | 98.66 | 0.17946 |
| 120 | 171.8 | 0.0132 | 0.240 | 36.16 | 90.15 | 0.07180 | 0.16495 | 94.47 | 0.17233 | 98.84 | 0.17939 |
| 122 | 176.2 | 0.0132 | 0.233 | 36.66 | 90.28 | 0.07266 | 0.16484 | 94.63 | 0.17224 | 99.01 | 0.17931 |
| 124 | 180.8 | 0.0133 | 0.227 | 37.16 | 90.40 | 0.07352 | 0.16473 | 94.78 | 0.17215 | 99.18 | 0.17922 |
| 126 | 185.4 | 0.0133 | 0.220 | 37.67 | 90.52 | 0.07437 | 0.16462 | 94.94 | 0.17206 | 99.35 | 0.17914 |
| 128 | 190.1 | 0.0134 | 0.214 | 38.18 | 90.64 | 0.07522 | 0.16450 | 95.09 | 0.17196 | 99.53 | 0.17906 |
| 130 | 194.9 | 0.0134 | 0.208 | 38.69 | 90.76 | 0.07607 | 0.16438 | 95.25 | 0.17186 | 99.70 | 0.17897 |
| 132 | 199.8 | 0.0135 | 0.202 | 39.19 | 90.86 | 0.07691 | 0.16425 | 95.41 | 0.17176 | 99.87 | 0.17889 |
| 134 | 204.8 | 0.0135 | 0.196 | 39.70 | 90.96 | 0.07775 | 0.16411 | 95.56 | 0.17166 | 100.04 | 0.17881 |
| 136 | 209.9 | 0.0136 | 0.191 | 40.21 | 91.06 | 0.07858 | 0.16396 | 95.72 | 0.17156 | 100.22 | 0.17873 |
| 138 | 215.0 | 0.0137 | 0.185 | 40.72 | 91.15 | 0.07941 | 0.16380 | 95.87 | 0.17145 | 100.39 | 0.17864 |
| 140 | 220 2 | 0.0138 | 0.180 | 41.24 | 91.24 | 0.08024 | 0.16363 | 96.03 | 0.17134 | 100.56 | 0.17856 |

† International (IT) Btu and pound mass units; data are mainly from "Heating, Ventilating, Air Conditioning Guide," American Society of Heating and Ventilating Engineers, New York, 1959.

TABLE B-15. PROPERTIES OF AIR AT ATMOSPHERIC PRESSURE†

| $t$ | $\rho$ | $c_p$ | $\mu_m$ | $k$ |
|---|---|---|---|---|
| °F | $\dfrac{\text{lb}_m}{\text{ft}^3}$ | $\dfrac{\text{Btu}}{\text{lb}_m \ °\text{F}}$ | $\dfrac{\text{lb}_m}{\text{hr ft}}$ | $\dfrac{\text{Btu}}{\text{hr ft } °\text{F}}$ |
| −64 | 0.1003 | 0.2402 | 0.0348 | 0.0114 |
| 8 | 0.0848 | 0.2401 | 0.0399 | 0.0133 |
| 80 | 0.0735 | 0.2404 | 0.0447 | 0.0152 |
| 116 | 0.0690 | 0.2406 | 0.0469 | 0.0161 |
| 152 | 0.0648 | 0.2409 | 0.0491 | 0.0169 |
| 188 | 0.0612 | 0.2412 | 0.0512 | 0.0178 |
| 224 | 0.0580 | 0.2417 | 0.0533 | 0.0186 |
| 260 | 0.0551 | 0.2422 | 0.0553 | 0.0194 |
| 296 | 0.0525 | 0.2428 | 0.0572 | 0.0203 |
| 332 | 0.0501 | 0.2435 | 0.0592 | 0.0210 |
| 368 | 0.0479 | 0.2443 | 0.0610 | 0.0218 |
| 404 | 0.0459 | 0.2451 | 0.0628 | 0.0226 |
| 440 | 0.0441 | 0.2460 | 0.0646 | 0.0234 |
| 512 | 0.0408 | 0.2479 | 0.0681 | 0.0248 |
| 584 | 0.0380 | 0.2500 | 0.0714 | 0.0262 |
| 656 | 0.0355 | 0.2522 | 0.0746 | 0.0276 |
| 728 | 0.0334 | 0.2545 | 0.0776 | 0.0290 |
| 800 | 0.0315 | 0.2568 | 0.0806 | 0.0303 |
| 872 | 0.0298 | 0.2591 | 0.0835 | 0.0315 |
| 944 | 0.0282 | 0.2613 | 0.0863 | 0.0328 |

† From Joseph Hilsenrath and others, Tables of Thermal Properties of Gases, *Natl. Bur. Standards Circ.* 564.

## TABLE B-16. PROPERTIES OF AMMONIA†

| Sat. temp., °F | Abs. press., psia | Volume, ft³/lbm | | Specific enthalpy and entropy taken from −40°F | | | | | | | |
| | | | | Specific enthalpy h, Btu/lbm | | Entropy s, Btu/(lbm)(°F) | | 100°F superheat | | 200°F superheat | |
| | | Liquid | Vapor | Liquid | Vapor | Liquid | Vapor | h | s | h | s |
|---|---|---|---|---|---|---|---|---|---|---|---|
| −40 | 10.41 | 0.02322 | 24.86 | 0.0 | 597.6 | 0.0 | 1.4242 | 649.3 | 1.5353 | 700.9 | 1.6261 |
| −30 | 13.9 | 0.02345 | 18.97 | 10.7 | 601.4 | 0.0250 | 1.4001 | 654.0 | 1.5101 | 706.0 | 1.6000 |
| −20 | 18.3 | 0.02369 | 14.68 | 21.4 | 605.0 | 0.0497 | 1.3774 | 658.3 | 1.4868 | 710.7 | 1.5760 |
| −10 | 23.74 | 0.02393 | 11.5 | 32.1 | 608.5 | 0.0738 | 1.3558 | 662.7 | 1.4647 | 715.6 | 1.5531 |
| 0 | 30.42 | 0.02419 | 9.116 | 42.9 | 611.8 | 0.0975 | 1.3352 | 666.8 | 1.4439 | 720.3 | 1.5317 |
| 2 | 31.92 | 0.02424 | 8.714 | 45.1 | 612.4 | 0.1022 | 1.3312 | 667.6 | 1.4400 | 721.2 | 1.5277 |
| 4 | 33.47 | 0.02430 | 8.333 | 47.2 | 613.0 | 0.1069 | 1.3273 | 668.4 | 1.4360 | 722.2 | 1.5236 |
| 5 | 34.27 | 0.02432 | 8.150 | 48.3 | 613.3 | 0.1092 | 1.3253 | 668.8 | 1.4340 | 722.6 | 1.5216 |
| 6 | 35.09 | 0.02435 | 7.971 | 49.4 | 613.6 | 0.1115 | 1.3234 | 669.3 | 1.4321 | 723.1 | 1.5196 |
| 8 | 36.77 | 0.02440 | 7.629 | 51.6 | 614.3 | 0.1162 | 1.3195 | 670.1 | 1.4281 | 724.1 | 1.5155 |
| 10 | 38.51 | 0.02446 | 7.304 | 53.8 | 614.9 | 0.1208 | 1.3157 | 670.9 | 1.4242 | 725.0 | 1.5115 |
| 12 | 40.31 | 0.02451 | 6.996 | 56.0 | 615.5 | 0.1254 | 1.3118 | 671.7 | 1.4205 | 725.9 | 1.5077 |
| 14 | 42.18 | 0.02457 | 6.703 | 58.2 | 616.1 | 0.1300 | 1.3081 | 672.5 | 1.4168 | 726.8 | 1.5039 |
| 16 | 44.12 | 0.02462 | 6.425 | 60.3 | 616.6 | 0.1346 | 1.3043 | 673.4 | 1.4130 | 727.8 | 1.5001 |
| 18 | 46.13 | 0.02468 | 6.161 | 62.5 | 617.2 | 0.1392 | 1.3006 | 674.2 | 1.4093 | 728.7 | 1.4963 |
| 20 | 48.21 | 0.02474 | 5.910 | 64.7 | 617.8 | 0.1437 | 1.2969 | 675.0 | 1.4056 | 729.6 | 1.4925 |
| 22 | 50.36 | 0.02479 | 5.671 | 66.9 | 618.3 | 0.1483 | 1.2933 | 675.8 | 1.4021 | 730.5 | 1.4889 |
| 24 | 52.59 | 0.02485 | 5.443 | 69.1 | 618.9 | 0.1528 | 1.2897 | 676.6 | 1.3985 | 731.4 | 1.4853 |
| 26 | 54.90 | 0.02491 | 5.227 | 71.3 | 619.4 | 0.1573 | 1.2861 | 677.3 | 1.3950 | 732.4 | 1.4816 |
| 28 | 57.28 | 0.02497 | 5.021 | 73.5 | 619.9 | 0.1618 | 1.2825 | 678.1 | 1.3914 | 733.3 | 1.4780 |
| 30 | 59.74 | 0.02503 | 4.825 | 75.7 | 620.5 | 0.1663 | 1.2790 | 678.9 | 1.3879 | 734.2 | 1.4744 |
| 32 | 62.29 | 0.02508 | 4.637 | 77.9 | 621.0 | 0.1708 | 1.2755 | 679.7 | 1.3846 | 735.1 | 1.4710 |
| 34 | 64.91 | 0.02514 | 4.459 | 80.1 | 621.5 | 0.1753 | 1.2721 | 680.4 | 1.3812 | 736.0 | 1.4676 |
| 36 | 67.63 | 0.02521 | 4.289 | 82.3 | 622.0 | 0.1797 | 1.2686 | 681.2 | 1.3779 | 736.8 | 1.4643 |
| 38 | 70.43 | 0.02527 | 4.126 | 84.6 | 622.5 | 0.1841 | 1.2652 | 681.9 | 1.3745 | 737.7 | 1.4609 |
| 39 | 71.87 | 0.02530 | 4.048 | 85.7 | 622.7 | 0.1863 | 1.2635 | 682.3 | 1.3729 | 738.2 | 1.4592 |
| 40 | 73.32 | 0.02533 | 3.971 | 86.8 | 623.0 | 0.1885 | 1.2618 | 682.7 | 1.3712 | 738.6 | 1.4575 |
| 41 | 74.80 | 0.02536 | 3.897 | 87.9 | 623.2 | 0.1908 | 1.2602 | 683.1 | 1.3696 | 739.0 | 1.4559 |
| 42 | 76.31 | 0.02539 | 3.823 | 89.0 | 623.4 | 0.1930 | 1.2585 | 683.4 | 1.3680 | 739.5 | 1.4542 |
| 44 | 79.38 | 0.02545 | 3.682 | 91.2 | 623.9 | 0.1974 | 1.2552 | 684.2 | 1.3648 | 740.4 | 1.4510 |
| 46 | 82.55 | 0.02551 | 3.547 | 93.5 | 624.4 | 0.2018 | 1.2519 | 684.9 | 1.3616 | 741.3 | 1.4477 |
| 48 | 85.82 | 0.02558 | 3.418 | 95.7 | 624.8 | 0.2062 | 1.2486 | 685.6 | 1.3584 | 742.2 | 1.4445 |
| 50 | 89.19 | 0.02564 | 3.294 | 97.9 | 625.2 | 0.2105 | 1.2453 | 686.4 | 1.3552 | 743.1 | 1.4412 |
| 52 | 92.66 | 0.02571 | 3.176 | 100.2 | 625.7 | 0.2149 | 1.2421 | 687.1 | 1.3521 | 744.0 | 1.4382 |
| 54 | 96.23 | 0.02577 | 3.063 | 102.4 | 626.1 | 0.2192 | 1.2389 | 687.8 | 1.3491 | 744.8 | 1.4351 |
| 56 | 99.91 | 0.02584 | 2.954 | 104.7 | 626.5 | 0.2236 | 1.2357 | 688.5 | 1.3460 | 745.7 | 1.4321 |
| 58 | 103.7 | 0.02590 | 2.851 | 106.9 | 626.9 | 0.2279 | 1.2325 | 689.2 | 1.3430 | 746.5 | 1.4290 |
| 60 | 107.6 | 0.02597 | 2.751 | 109.2 | 627.3 | 0.2322 | 1.2294 | 689.9 | 1.3399 | 747.4 | 1.4260 |
| 62 | 111.6 | 0.02604 | 2.656 | 111.5 | 627.7 | 0.2365 | 1.2262 | 690.6 | 1.3370 | 748.2 | 1.4231 |

TABLE B-16. PROPERTIES OF AMMONIA† (Continued)

| Sat. temp., °F | Abs. press., psia | Volume, ft³/lbm Liquid | Vapor | Specific enthalpy h, Btu/lbm Liquid | Vapor | Entropy s, Btu/(lbm)(°F) Liquid | Vapor | 100°F superheat h | s | 200°F superheat h | s |
|---|---|---|---|---|---|---|---|---|---|---|---|
| | | | | | | Specific enthalpy and entropy taken from −40°F | | | | | |
| 64 | 115.7 | 0.02611 | 2.565 | 113.7 | 628.0 | 0.2408 | 1.2231 | 691.3 | 1.3341 | 749.1 | 1.4202 |
| 66 | 120.0 | 0.02618 | 2.477 | 116.0 | 628.4 | 0.2451 | 1.2201 | 691.9 | 1.3312 | 749.9 | 1.4172 |
| 68 | 124.3 | 0.02625 | 2.393 | 118.3 | 628.8 | 0.2494 | 1.2170 | 692.6 | 1.3283 | 750.8 | 1.4143 |
| 70 | 128.8 | 0.02632 | 2.312 | 120.5 | 629.1 | 0.2537 | 1.2140 | 693.3 | 1.3254 | 751.6 | 1.4114 |
| 72 | 133.4 | 0.02639 | 2.235 | 122.8 | 629.4 | 0.2579 | 1.2110 | 694.0 | 1.3226 | 752.4 | 1.4086 |
| 74 | 138.1 | 0.02646 | 2.161 | 125.1 | 629.8 | 0.2622 | 1.2080 | 694.6 | 1.3199 | 753.3 | 1.4059 |
| 76 | 143.0 | 0.02653 | 2.089 | 127.4 | 630.1 | 0.2664 | 1.2050 | 695.3 | 1.3171 | 754.1 | 1.4031 |
| 78 | 147.9 | 0.02661 | 2.021 | 129.7 | 630.4 | 0.2706 | 1.2020 | 695.9 | 1.3144 | 755.0 | 1.4004 |
| 80 | 153.0 | 0.02668 | 1.955 | 132.0 | 630.7 | 0.2749 | 1.1991 | 696.6 | 1.3116 | 755.8 | 1.3976 |
| 82 | 158.3 | 0.02675 | 1.892 | 134.3 | 631.0 | 0.2791 | 1.1962 | 697.2 | 1.3089 | 756.6 | 1.3949 |
| 84 | 163.7 | 0.02684 | 1.831 | 136.6 | 631.3 | 0.2833 | 1.1933 | 697.8 | 1.3063 | 757.4 | 1.3923 |
| 86 | 169.2 | 0.02691 | 1.772 | 138.9 | 631.5 | 0.2875 | 1.1904 | 698.5 | 1.3040 | 758.3 | 1.3896 |
| 88 | 174.8 | 0.02699 | 1.716 | 141.2 | 631.8 | 0.2917 | 1.1875 | 699.1 | 1.3010 | 759.1 | 1.3870 |
| 90 | 180.6 | 0.02707 | 1.661 | 143.5 | 632.0 | 0.2958 | 1.1846 | 699.7 | 1.2983 | 759.9 | 1.3843 |
| 92 | 186.6 | 0.02715 | 1.609 | 145.8 | 632.2 | 0.3000 | 1.1818 | 700.3 | 1.2957 | 760.7 | 1.3818 |
| 94 | 192.7 | 0.02723 | 1.559 | 148.2 | 632.5 | 0.3041 | 1.1789 | 700.9 | 1.2932 | 761.5 | 1.3793 |
| 96 | 198.9 | 0.02731 | 1.510 | 150.5 | 632.6 | 0.3083 | 1.1761 | 701.5 | 1.2906 | 762.2 | 1.3768 |
| 98 | 205.3 | 0.02739 | 1.464 | 152.9 | 632.9 | 0.3125 | 1.1733 | 702.1 | 1.2881 | 763.0 | 1.3743 |
| 100 | 211.9 | 0.02747 | 1.419 | 155.2 | 633.0 | 0.3166 | 1.1705 | 702.7 | 1.2855 | 763.8 | 1.3718 |
| 102 | 218.6 | 0.02756 | 1.375 | 157.6 | 633.2 | 0.3207 | 1.1677 | 703.3 | 1.2830 | 764.6 | 1.3693 |
| 104 | 225.4 | 0.02764 | 1.334 | 159.9 | 633.4 | 0.3248 | 1.1649 | 703.8 | 1.2805 | 765.3 | 1.3668 |
| 106 | 232.5 | 0.02773 | 1.293 | 162.3 | 633.5 | 0.3289 | 1.1621 | 704.3 | 1.2780 | 766.1 | 1.3643 |
| 108 | 239.7 | 0.02782 | 1.254 | 164.6 | 633.6 | 0.3330 | 1.1593 | 705.0 | 1.2755 | 766.9 | 1.3619 |
| 110 | 247.0 | 0.02790 | 1.217 | 167.0 | 633.7 | 0.3372 | 1.1566 | 705.5 | 1.2731 | 767.6 | 1.3596 |
| 112 | 254.5 | 0.02799 | 1.180 | 169.4 | 633.8 | 0.3413 | 1.1538 | 706.1 | 1.2708 | 768.3 | 1.3573 |
| 114 | 262.2 | 0.02808 | 1.145 | 171.8 | 633.9 | 0.3453 | 1.1510 | 706.6 | 1.2684 | 769.1 | 1.3550 |
| 116 | 270.1 | 0.02817 | 1.112 | 174.2 | 634.0 | 0.3495 | 1.1483 | 707.2 | 1.2661 | 769.8 | 1.3527 |
| 118 | 278.2 | 0.02827 | 1.079 | 176.6 | 634.0 | 0.3535 | 1.1455 | 707.7 | 1.2636 | 770.5 | 1.3503 |
| 120 | 286.4 | 0.02836 | 1.047 | 179.0 | 634.0 | 0.3576 | 1.1427 | 708.2 | 1.2612 | 771.3 | 1.3479 |
| 122 | 294.8 | 0.02846 | 1.017 | 181.4 | 634.0 | 0.3618 | 1.1400 | 708.6 | 1.2587 | 772.0 | 1.3455 |
| 124 | 303.4 | 0.02855 | 0.987 | 183.9 | 634.0 | 0.3659 | 1.1372 | 709.1 | 1.2563 | 772.8 | 1.3431 |
| 126 | 312.2 | 0.02865 | 0.958 | 186.3 | 633.9 | 0.3700 | 1.1344 | 709.6 | 1.2538 | 773.5 | 1.3407 |
| 128 | 321.2 | 0.02875 | 0.931 | 188.8 | 633.9 | 0.3741 | 1.1316 | 710.0 | 1.2513 | 774.2 | 1.3383 |

† Data are mainly from "Heating, Ventilating, Air Conditioning Guide," American Society of Heating and Ventilating Engineers, New York, 1947.

## TABLE B-17. ISENTROPIC FLOW
(Ideal gas, $k = 1.4$)

| M | $\dfrac{V}{V^*}$ | $\dfrac{A}{A^*}$ | $\dfrac{p}{p_0}$ | $\dfrac{\rho}{\rho_0}$ | $\dfrac{T}{T_0}$ | $\dfrac{F}{F^*}$ |
|---|---|---|---|---|---|---|
| 0.00 | 0.000 | $\infty$ | 1.000 | 1.000 | 1.000 | $\infty$ |
| 0.05 | 0.0548 | 11.6 | 0.998 | 0.999 | 0.999 | 9.16 |
| 0.10 | 0.109 | 5.82 | 0.993 | 0.995 | 0.998 | 4.62 |
| 0.15 | 0.164 | 3.91 | 0.984 | 0.989 | 0.996 | 3.13 |
| 0.20 | 0.218 | 2.96 | 0.972 | 0.980 | 0.992 | 2.40 |
| 0.25 | 0.272 | 2.40 | 0.957 | 0.969 | 0.988 | 1.97 |
| 0.30 | 0.326 | 2.04 | 0.939 | 0.956 | 0.982 | 1.70 |
| 0.40 | 0.431 | 1.59 | 0.896 | 0.924 | 0.969 | 1.37 |
| 0.50 | 0.535 | 1.34 | 0.843 | 0.885 | 0.952 | 1.20 |
| 0.60 | 0.635 | 1.19 | 0.784 | 0.840 | 0.933 | 1.10 |
| 0.70 | 0.732 | 1.09 | 0.721 | 0.792 | 0.911 | 1.05 |
| 0.80 | 0.825 | 1.04 | 0.656 | 0.740 | 0.887 | 1.02 |
| 0.90 | 0.915 | 1.01 | 0.591 | 0.687 | 0.861 | 1.00+ |
| 1.00 | 1.00 | 1.00 | 0.528 | 0.634 | 0.833 | 1.00 |
| 1.10 | 1.08 | 1.01 | 0.468 | 0.582 | 0.805 | 1.00+ |
| 1.20 | 1.16 | 1.03 | 0.412 | 0.531 | 0.776 | 1.01 |
| 1.30 | 1.23 | 1.07 | 0.361 | 0.483 | 0.747 | 1.02 |
| 1.40 | 1.30 | 1.11 | 0.314 | 0.437 | 0.718 | 1.03 |
| 1.50 | 1.36 | 1.18 | 0.272 | 0.395 | 0.690 | 1.05 |
| 1.60 | 1.43 | 1.25 | 0.235 | 0.356 | 0.661 | 1.06 |
| 1.70 | 1.48 | 1.34 | 0.203 | 0.320 | 0.634 | 1.08 |
| 1.80 | 1.54 | 1.44 | 0.174 | 0.287 | 0.607 | 1.09 |
| 1.90 | 1.59 | 1.56 | 0.149 | 0.257 | 0.581 | 1.11 |
| 2.00 | 1.63 | 1.69 | 0.128 | 0.230 | 0.556 | 1.12 |
| 2.10 | 1.68 | 1.84 | 0.109 | 0.206 | 0.531 | 1.14 |
| 2.20 | 1.72 | 2.00 | 0.0935 | 0.184 | 0.508 | 1.15 |
| 2.30 | 1.76 | 2.19 | 0.0800 | 0.165 | 0.486 | 1.16 |
| 2.40 | 1.79 | 2.40 | 0.0684 | 0.147 | 0.465 | 1.18 |
| 2.50 | 1.83 | 2.64 | 0.0585 | 0.132 | 0.444 | 1.19 |
| 2.60 | 1.86 | 2.90 | 0.0501 | 0.118 | 0.425 | 1.20 |
| 2.70 | 1.89 | 3.18 | 0.0429 | 0.106 | 0.407 | 1.21 |
| 2.80 | 1.91 | 3.50 | 0.0368 | 0.0946 | 0.389 | 1.22 |
| 2.90 | 1.94 | 3.85 | 0.0316 | 0.0849 | 0.373 | 1.23 |
| 3.00 | 1.96 | 4.23 | 0.0272 | 0.0762 | 0.357 | 1.24 |
| 3.50 | 2.06 | 6.79 | 0.0131 | 0.0452 | 0.290 | 1.27 |
| 4.00 | 2.14 | 10.7 | 0.00658 | 0.0277 | 0.238 | 1.30 |
| 4.50 | 2.19 | 16.6 | 0.00346 | 0.0174 | 0.198 | 1.32 |
| 5.00 | 2.24 | 25.0 | 0.00189 | 0.0113 | 0.167 | 1.34 |
| 6.00 | 2.30 | 53.2 | $633(10^{-6})$ | 0.00519 | 0.122 | 1.37 |
| 7.00 | 2.33 | 104 | $242(10^{-6})$ | 0.00261 | 0.0926 | 1.38 |
| 8.00 | 2.36 | 190 | $102(10^{-6})$ | 0.00141 | 0.0725 | 1.39 |
| 9.00 | 2.38 | 327 | $474(10^{-7})$ | 0.000815 | 0.0581 | 1.40 |
| 10.00 | 2.39 | 536 | $236(10^{-7})$ | 0.000495 | 0.0476 | 1.40 |
| $\infty$ | 2.45 | $\infty$ | 0 | 0 | 0 | 1.43 |

TABLE B-18. FANNO FLOW

(Ideal gas, $k = 1.4$)

| M | $\dfrac{T}{T^*}$ | $\dfrac{p}{p^*}$ | $\dfrac{p_0}{p_0{}^*}$ | $\dfrac{V}{V^*} = \dfrac{\rho^*}{\rho}$ | $\dfrac{F}{F^*}$ | $4\dfrac{fL_{max}}{D}$ |
|---|---|---|---|---|---|---|
| 0.00 | 1.20 | ∞ | ∞ | 0 | ∞ | ∞ |
| 0.05 | 1.20 | 21.9 | 11.6 | 0.0548 | 9.16 | 280 |
| 0.10 | 1.20 | 10.9 | 5.82 | 0.109 | 4.62 | 66.9 |
| 0.15 | 1.19 | 7.29 | 3.91 | 0.164 | 3.13 | 27.9 |
| 0.20 | 1.19 | 5.46 | 2.96 | 0.218 | 2.40 | 14.5 |
| 0.25 | 1.19 | 4.35 | 2.40 | 0.272 | 1.97 | 8.48 |
| 0.30 | 1.18 | 3.62 | 2.04 | 0.326 | 1.70 | 5.30 |
| 0.40 | 1.16 | 2.70 | 1.59 | 0.431 | 1.37 | 2.31 |
| 0.50 | 1.14 | 2.14 | 1.34 | 0.535 | 1.20 | 1.07 |
| 0.60 | 1.12 | 1.76 | 1.19 | 0.635 | 1.10 | 0.491 |
| 0.70 | 1.09 | 1.49 | 1.09 | 0.732 | 1.05 | 0.208 |
| 0.80 | 1.06 | 1.29 | 1.04 | 0.825 | 1.02 | 0.0723 |
| 0.90 | 1.03 | 1.13 | 1.01 | 0.915 | 1.00+ | 0.0145 |
| 1.00 | 1.00 | 1.00 | 1.00 | 1.00 | 1.00 | 0.000 |
| 1.10 | 0.966 | 0.894 | 1.01 | 1.08 | 1.00+ | 0.00993 |
| 1.20 | 0.932 | 0.804 | 1.03 | 1.16 | 1.01 | 0.0336 |
| 1.30 | 0.897 | 0.728 | 1.07 | 1.23 | 1.02 | 0.0648 |
| 1.40 | 0.862 | 0.663 | 1.11 | 1.30 | 1.03 | 0.0997 |
| 1.50 | 0.828 | 0.606 | 1.18 | 1.36 | 1.05 | 0.136 |
| 1.60 | 0.794 | 0.557 | 1.25 | 1.42 | 1.06 | 0.172 |
| 1.70 | 0.760 | 0.513 | 1.34 | 1.48 | 1.08 | 0.208 |
| 1.80 | 0.728 | 0.474 | 1.44 | 1.54 | 1.09 | 0.242 |
| 1.90 | 0.697 | 0.439 | 1.56 | 1.59 | 1.11 | 0.274 |
| 2.00 | 0.667 | 0.408 | 1.69 | 1.63 | 1.12 | 0.305 |
| 2.10 | 0.638 | 0.380 | 1.84 | 1.68 | 1.14 | 0.334 |
| 2.20 | 0.610 | 0.355 | 2.00 | 1.72 | 1.15 | 0.361 |
| 2.30 | 0.583 | 0.332 | 2.19 | 1.76 | 1.16 | 0.386 |
| 2.40 | 0.558 | 0.311 | 2.40 | 1.79 | 1.18 | 0.410 |
| 2.50 | 0.533 | 0.292 | 2.64 | 1.83 | 1.19 | 0.432 |
| 2.60 | 0.510 | 0.275 | 2.90 | 1.86 | 1.20 | 0.453 |
| 2.70 | 0.488 | 0.259 | 3.18 | 1.89 | 1.21 | 0.472 |
| 2.80 | 0.467 | 0.244 | 3.50 | 1.91 | 1.22 | 0.490 |
| 2.90 | 0.447 | 0.231 | 3.85 | 1.94 | 1.23 | 0.507 |
| 3.00 | 0.429 | 0.218 | 4.23 | 1.96 | 1.24 | 0.522 |
| 3.50 | 0.348 | 0.168 | 6.79 | 2.06 | 1.27 | 0.586 |
| 4.00 | 0.286 | 0.134 | 10.7 | 2.14 | 1.30 | 0.633 |
| 4.50 | 0.238 | 0.108 | 16.6 | 2.19 | 1.32 | 0.668 |
| 5.00 | 0.200 | 0.0894 | 25.0 | 2.24 | 1.34 | 0.694 |
| 6.00 | 0.146 | 0.0638 | 53.2 | 2.30 | 1.37 | 0.730 |
| 7.00 | 0.111 | 0.0476 | 104 | 2.33 | 1.38 | 0.753 |
| 8.00 | 0.0870 | 0.0369 | 190 | 2.36 | 1.39 | 0.768 |
| 9.00 | 0.0698 | 0.0293 | 327 | 2.38 | 1.40 | 0.779 |
| 10.00 | 0.0571 | 0.0239 | 536 | 2.39 | 1.40 | 0.787 |
| ∞ | 0.00 | 0.00 | ∞ | 2.45 | 1.43 | 0.822 |

TABLE B-19. NORMAL SHOCK RELATIONSHIPS
(Ideal gas, $k = 1.4$)

| $M_x$ | $M_y$ | $\dfrac{p_y}{p_x}$ | $\dfrac{\rho_y}{\rho_x}$ | $\dfrac{T_y}{T_x}$ | $\dfrac{p_{0y}}{p_{0x}}$ | $\dfrac{p_{0y}}{p_x}$ |
|---|---|---|---|---|---|---|
| 1.00 | 1.000 | 1.00 | 1.00 | 1.00 | 1.00 | 1.89 |
| 1.05 | 0.953 | 1.12 | 1.08 | 1.03 | 1.00 − | 2.01 |
| 1.10 | 0.912 | 1.24 | 1.17 | 1.06 | 0.999 | 2.13 |
| 1.15 | 0.875 | 1.38 | 1.25 | 1.10 | 0.997 | 2.27 |
| 1.20 | 0.842 | 1.51 | 1.34 | 1.13 | 0.993 | 2.41 |
| 1.25 | 0.813 | 1.66 | 1.43 | 1.16 | 0.987 | 2.56 |
| 1.30 | 0.786 | 1.80 | 1.52 | 1.19 | 0.979 | 2.71 |
| 1.40 | 0.740 | 2.12 | 1.69 | 1.25 | 0.958 | 3.05 |
| 1.50 | 0.701 | 2.46 | 1.86 | 1.32 | 0.930 | 3.41 |
| 1.60 | 0.668 | 2.82 | 2.03 | 1.39 | 0.895 | 3.80 |
| 1.70 | 0.641 | 3.20 | 2.20 | 1.46 | 0.856 | 4.22 |
| 1.80 | 0.616 | 3.61 | 2.36 | 1.53 | 0.813 | 4.67 |
| 1.90 | 0.596 | 4.04 | 2.52 | 1.61 | 0.767 | 5.14 |
| 2.00 | 0.577 | 4.50 | 2.67 | 1.69 | 0.721 | 5.64 |
| 2.10 | 0.561 | 4.98 | 2.81 | 1.77 | 0.674 | 6.17 |
| 2.20 | 0.547 | 5.48 | 2.95 | 1.86 | 0.628 | 6.72 |
| 2.30 | 0.534 | 6.00 | 3.08 | 1.95 | 0.583 | 7.29 |
| 2.40 | 0.523 | 6.55 | 3.21 | 2.04 | 0.540 | 7.90 |
| 2.50 | 0.513 | 7.12 | 3.33 | 2.14 | 0.499 | 8.53 |
| 2.60 | 0.504 | 7.72 | 3.45 | 2.24 | 0.460 | 9.18 |
| 2.70 | 0.496 | 8.34 | 3.56 | 2.34 | 0.424 | 9.86 |
| 2.80 | 0.488 | 8.98 | 3.66 | 2.45 | 0.389 | 10.6 |
| 2.90 | 0.481 | 9.64 | 3.76 | 2.56 | 0.358 | 11.3 |
| 3.00 | 0.475 | 10.3 | 3.86 | 2.68 | 0.328 | 12.1 |
| 3.50 | 0.451 | 14.1 | 4.26 | 3.31 | 0.213 | 16.2 |
| 4.00 | 0.435 | 18.5 | 4.57 | 4.05 | 0.139 | 21.1 |
| 4.50 | 0.424 | 23.5 | 4.81 | 4.88 | 0.0917 | 26.5 |
| 5.00 | 0.415 | 29.0 | 5.00 | 5.80 | 0.0617 | 32.7 |
| 6.00 | 0.404 | 41.8 | 5.27 | 7.94 | 0.0296 | 46.8 |
| 7.00 | 0.397 | 57.0 | 5.44 | 10.5 | 0.0153 | 63.6 |
| 8.00 | 0.393 | 74.5 | 5.57 | 13.4 | 0.00849 | 82.9 |
| 9.00 | 0.390 | 94.3 | 5.65 | 16.7 | 0.00496 | 105 |
| 10.00 | 0.388 | 116 | 5.71 | 20.4 | 0.00304 | 129 |
| ∞ | 0.378 | ∞ | 6.00 | ∞ | 0 | ∞ |

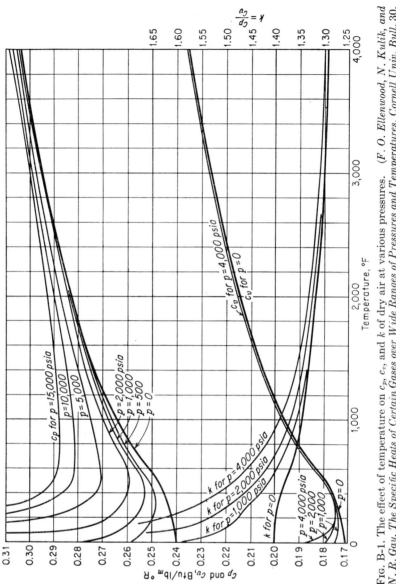

Fig. B-1. The effect of temperature on $c_p$, $c_v$, and $k$ of dry air at various pressures. (F. O. Ellenwood, N. Kulik, and N. R. Gay, The Specific Heats of Certain Gases over Wide Ranges of Pressures and Temperatures, Cornell Univ. Bull. 30, October, 1942.)

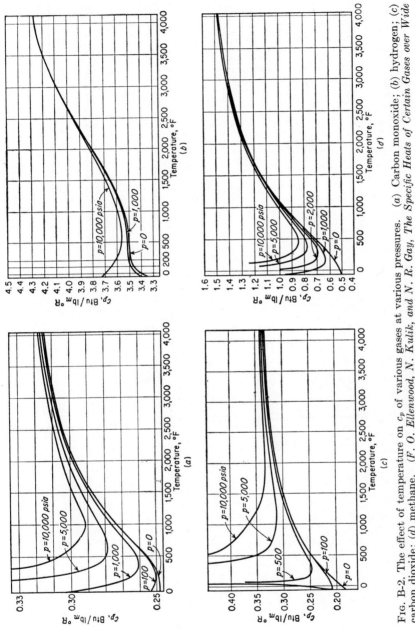

FIG. B-2. The effect of temperature on $c_p$ of various gases at various pressures. (a) Carbon monoxide; (b) hydrogen; (c) carbon dioxide; (d) methane. (F. O. Ellenwood, N. Kulik, and N. R. Gay, *The Specific Heats of Certain Gases over Wide Ranges of Pressures and Temperatures, Cornell Univ. Bull. 30, October, 1942.*)

523

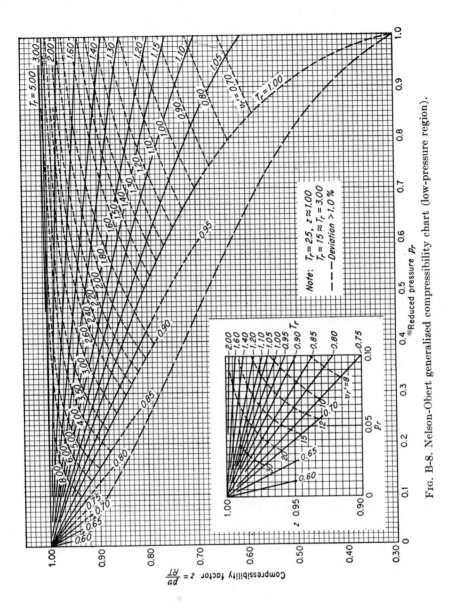

FIG. B-8. Nelson-Obert generalized compressibility chart (low-pressure region).

Note: $T_r = 2.5$, $z \approx 1.00$
$T_r = 15 \approx T_r = 3.00$
— — — Deviation > 1.0 %

525

# INDEX

527